Structural Steel Design: ASD Method

101 Hudson Street, Jersey City, N.J.
(Courtesy of Owen Steel Company, Inc.)

Structural Steel Design: ASD Method

FOURTH EDITION

Jack C. McCormac
Clemson University

HarperCollins*Publishers*

To my wife, Mary, and daughters, Mary Christine and Ann Rebecca

Sponsoring Editor: Don Childress
Project Editor: Janet Tilden/Kristin Syverson
Assistant Art Director: Julie Anderson
Text Design: Julie Anderson
Cover Design: Lucy Lesiak Design
Cover Photo: John Moore
Photo Researcher: Nina Page
Production Administrator: Kathy Donnelly
Compositor: Interactive Composition Corporation
Printer and Binder: R. R. Donnelley & Sons Company
Cover Printer: The Lehigh Press, Inc.

Structural Steel Design, Fourth Edition (ASD Method)

Library of Congress Cataloging-in-Publication Data

McCormac, Jack C.
　　Structural steel design : ASD method / Jack C. McCormac.—4th
　ed.
　　　p.　　cm.
　　Includes bibliographical references and index.
　　ISBN 0-06-500060-9
　　1. Building, Iron and steel.　2. Steel, Structural.　I. Title.
TA684.M383　1992
624.1'821—dc20　　　　　　　　　　　　　　　　　　91-24826
　　　　　　　　　　　　　　　　　　　　　　　　　　CIP

　92　　93　94　9　8　7　6　5　4　3　2

Contents

Appendix

Preface

The primary objective of this edition is to update the text to reflect the latest revisions of the allowable-stress design (ASD) specification of the American Institute of Steel Construction dated June 1, 1989, and contained in the ninth edition of the *Manual of Steel Construction Allowable Stress Design*. In addition, the last part of the text presents introductory information concerning plastic analysis and design, and load and resistance factor design (LRFD).

The author has added or appreciably expanded the topics of block shear, stiffness reduction factors for columns, flexural-torsional buckling, eccentrically loaded connections, and so on. The homework problems have been substantially changed and their numbers increased.

Today there is a great deal of interest in the LRFD method. As a result, we devote four chapters (20–23) to an introduction to this subject. The LRFD design of tension members, columns, beams, and connections is presented. If additional material on LRFD is required, please consult the author's 1989 edition of *Structural Steel Design Using the LRFD Method* (also published by Harper-Collins).

The author wishes to thank the following persons who reviewed this edition:

Fouad H. Fouad
University of Alabama at Birmingham

Achintya Haldar
University of Arizona

Joseph A. Recktenwald
Ohio University

John J. Schemmel
University of Arkansas

He also thanks the many reviewers and users of the previous editions. He is always very grateful to anyone who writes to him concerning any part of the book.

Jack C. McCormac
Clemson University

Structural Steel Design: ASD Method

1

Introduction to Structural Steel Design

1-1 ADVANTAGES OF STEEL AS A STRUCTURAL MATERIAL

A person traveling in the United States might quite understandably decide that steel was the perfect structural material. He or she would see an endless number of steel bridges, buildings, towers, and other structures—comprising, in fact, a list too lengthy to enumerate. After seeing these numerous steel structures this traveler might be quite surprised to learn that steel was not economically made in the United States until late in the nineteenth century and the first wide-flange beams were not rolled until 1908.

The assumption of the perfection of this metal, perhaps the most versatile of structural materials, would appear to be even more reasonable when its great strength, light weight, ease of fabrication, and many other desirable properties are considered. These and other advantages of structural steel are discussed in detail in the following paragraphs.

High Strength

The high strength of steel per unit of weight means that structure weights will be small. This fact is of great importance for long-span bridges, tall buildings, and structures having poor foundation conditions.

Uniformity

The properties of steel do not change appreciably with time as do those of a reinforced-concrete structure.

Elasticity

Steel behaves closer to design assumptions than most materials because it follows Hooke's law up to fairly high stresses. The moments of inertia of a steel structure can be definitely calculated, while the values obtained for a reinforced-concrete structure are rather indefinite.

Permanence

Steel frames that are properly maintained will last indefinitely. Research on some of the newer steels indicates that under certain conditions no painting maintenance whatsoever will be required.

Ductility

The property of a material by which it can withstand extensive deformation without failure under high tensile stresses is said to be its *ductility*. When a *mild* or *low-carbon* steel member is being tested in tension, a considerable reduction in cross section and a large amount of elongation will occur at the point of failure before the actual fracture occurs. A material that does not have this property is probably hard and brittle and might break if subjected to a sudden shock.

In structural members under normal loads, high stress concentrations develop at various points. The ductile nature of the usual structural steels enables them to yield locally at those points, thus preventing premature failures. A further advantage of ductile structures is that, when they are overloaded, their large deflections give visible evidence of impending failure (sometimes jokingly referred to as "running time").

Fracture Toughness

Structural steels are tough; that is, they have both strength and ductility. A steel member loaded until it has large deformations will still be able to withstand large forces. This is a very important characteristic because it means that steel members can be subjected to large deformations during fabrication and erection without fracture—thus allowing them to be bent, hammered, sheared, and punched with holes without visible damage. The ability of a material to absorb energy in large amounts is called *toughness*.

Additions to Existing Structures

Steel structures are quite well suited to having additions made to them. New bays or even entire new wings can be added to existing steel-frame buildings, and steel bridges may often be widened.

Miscellaneous

Several other important advantages of structural steel are (a) ability to be fastened together by several simple connection devices including welds, bolts, and rivets, (b) adaptation to prefabrication, (c) speed of erection, (d) ability to be rolled into a wide variety of sizes and shapes as described in Section 1-4, (e) fatigue strength, (f) possible reuse after a structure is disassembled, and (g) scrap value even though not reusable in its existing form.

1-2 DISADVANTAGES OF STEEL AS A STRUCTURAL MATERIAL

In general, steel has the following disadvantages.

Maintenance Costs

Most steels are susceptible to corrosion when freely exposed to air and water and must therefore be periodically painted. The use of weathering steels, in suitable design applications, tends to eliminate this cost.

Fireproofing Costs

Although structural members are incombustible, their strength is tremendously reduced at temperatures commonly reached in fires when the other materials in a building burn. Many disastrous fires have occurred in empty buildings where the only fuel for the fire were the buildings themselves. Furthermore steel is an excellent heat conductor: nonfireproofed steel members may transmit enough heat from a burning section or compartment of a building to ignite materials with which they are in contact in adjoining sections of the building. As a result of these facts the steel frame of a building may have to be protected by materials with certain insulating characteristics, or the building may have to include a sprinkler system if it is to meet the building code requirements of the locality in question.

Susceptibility to Buckling

The longer and slenderer the compression members, the greater the danger of buckling. As previously indicated, steel has a high strength per unit of weight but when used for columns is sometimes not very economical because considerable material has to be used merely to stiffen the columns against buckling.

Fatigue

Another undesirable property of steel is that its strength may be reduced if the steel is subjected to a large number of stress reversals or even to a large number of variations of tensile stress. (We have fatigue problems only when tension is involved.) The present practice is to reduce the estimated strengths of such members if it is anticipated that they will have more than a prescribed number of cycles of stress variation.

Brittle Fracture

Under certain conditions steel may lose its ductility, and brittle fracture may occur at places of stress concentration. Fatigue-type loadings and very low temperatures aggravate the situation.

1-3 EARLY USES OF IRON AND STEEL

Although the first metal used by human beings was probably some type of copper alloy such as bronze (made with copper and tin and perhaps some other additives), the most important metal developments throughout history have occurred in the manufacture and use of iron and its famous alloy, steel. Today, iron and steel compose almost 95 percent of all the tonnage of metal produced in the world.[1]

Despite diligent efforts for many decades, archaeologists have been unable to discover when iron was first used. They did find in the Great Pyramid in Egypt an iron dagger and an iron bracelet which they claim had been there undisturbed for at least 5000 years. The use of iron has had a great influence on the course of civilization since the earliest times and may very well continue to do so in the centuries ahead. Since the beginning of the Iron Age in about 1000 B.C. the progress of civilization in peace and war has been heavily dependent on what people have been able to make with iron. On many occasions its use has decidedly affected the outcome of military engagements. For instance in 490 B.C. in Greece at the Battle of Marathon the greatly outnumbered Athenians killed 6400 Persians and lost only 192 of their own men. Each of the victors wore 57 pounds of iron armor in the battle. (This was the battle where the runner Pheidippides ran the approximately 25 miles to Athens and died while shouting news of the victory.) This battle supposedly saved Greek civilization for many years.

---

[1] American Iron and Steel Institute, *The Making of Steel* (Washington, D.C., not dated), p. 6.

According to the classic theory concerning the first production of iron in the world, there was once a great forest fire on Mount Ida in ancient Troy (now Turkey) near the Aegean Sea. The land surface supposedly had a rich content of iron, and the heat of the fire is said to have produced a rather crude form of iron which could be hammered into various shapes. Many historians believe, however, that human beings first learned to use iron that fell to the earth in the form of meteorites. Frequently the iron in meteorites is combined with nickel, producing a harder metal. Perhaps early human beings were able to hammer and chip this material into the shape of crude tools and weapons.

Steel is defined as a combination of iron with a small amount of carbon, usually less than 1 percent. It also contains small percentages of some other elements. Although some steel has been made for at least 2000 or 3000 years, there was really no economical production method available until the middle of the nineteenth century.

The first steel was surely obtained when the other elements necessary for producing it were accidentally present when iron was heated. As the years went by, steel was probably made by heating iron in contact with charcoal. The surface of the iron absorbed some carbon from the charcoal which was then hammered into the hot iron. Repeating this process several times resulted in a case-hardened exterior of steel. In this way the famous swords of Toledo and Damascus were produced.

The first large-volume process for producing steel was named after Sir Henry Bessemer of England. He received an English patent for his process in 1855, but his efforts to obtain a U.S. patent for the process in 1856 were unsuccessful, as it was proved that William Kelly of Eddyville, Kentucky, had made steel by the same process seven years before Bessemer applied for his English patent. Although Kelly was given the patent, the name Bessemer was used for the process.[2]

Kelly and Bessemer learned that a blast of air through molten iron burned out most of the impurities in the metal. Unfortunately at the same time the blow eliminated some desirable elements, such as carbon and manganese. It was later learned that these needed elements could be restored by adding spiegeleisen, which is an alloy of iron, carbon, and manganese. It was further learned that the addition of limestone in the converter resulted in the removal of the phosphorus and most of the sulfur.

The Bessemer converter was commonly used in the United States until after the turn of the century, but since that time it has been replaced with better methods, such as the open-hearth process and the basic oxygen process.

As a result of the Bessemer process structural carbon steel could be produced in quantity by 1870, and by 1890 steel had become the principal structural metal used in the United States.

The first use of metal for a sizable structure occurred in England in Shropshire (about 140 miles northwest of London) in 1779 when cast iron was used

---

[2] American Iron and Steel Institute, *Steel '76* (Washington, D.C., 1976), pp. 5–11.

for the construction of the 100-ft Coalbrookdale Arch Bridge over the River Severn. It is said that this bridge (which still stands) was a turning point in engineering history because it changed the course of the Industrial Revolution by introducing iron as a structural material. This iron was supposedly 4 times as strong as stone and 30 times as strong as wood.[3]

A number of other cast-iron bridges were constructed in the following decades, but soon after 1840 the more malleable wrought iron began to replace cast iron. The development of the Bessemer process and subsequent advances such as the open-hearth process permitted the manufacture of steel at competitive prices, which encouraged the beginning of the almost unbelievable developments of the last 100 years with structural steel.

1-4 STEEL SECTIONS

The first structural shapes made in the United States were angle irons rolled in 1819. Steel I beams were first rolled in the United States in 1884, and the first skeleton frame structure (the Home Insurance Company Building of Chicago) was erected that same year. Credit for inventing the skyscraper is usually given to an engineer, William LeBaron Jenney, who planned this building apparently during a bricklayers' strike. Up until this time tall buildings in the United States were constructuted with several-foot-thick load-bearing brick walls.

For the exterior walls of this ten-story building Jenney used cast-iron columns encased in brick. The beams for the lower six floors were made from wrought iron, while structural steel beams were used for the upper floors. The first building completely framed with steel was the second Rand-McNally building completed in Chicago in 1890.

An important feature of the 985-ft wrought-iron Eiffel tower constructed in 1889 was the use of mechanically operated passenger elevators. The availability of these machines along with Jenney's skeleton frame idea led to the construction of thousands of high-rise buildings throughout the world in the next 100 years.

During these early years the various mills rolled their own shapes and published catalogs which provided the dimensions, weight, and other properties of these shapes. In 1896 the Association of American Steel Manufacturers (now the American Iron and Steel Institute, AISI) made the first efforts to standardize shapes. Today nearly all structural shapes are standardized, though their exact dimensions may vary just a little from mill to mill.[4]

Structural steel can be economically rolled into a wide variety of shapes and sizes without appreciably changing its physical properties. Usually the

---

[3] M. H. Sawyer, "World's First Iron Bridge," *Civil Engineering* (New York: ASCE, December 1979), pp. 46–49.

[4] W. McGuire, *Steel Structures* (Englewood Cliffs, N.J.: Prentice Hall, 1968), pp. 19–21.

Erection of tower for the Holy Name Church, Edens-
burg, Pa. (Courtesy of Lincoln Electric Company.)

most desirable members have large moments of inertia in proportion to their areas. The **I, T,** and [shapes, so commonly used, fall into this class.

Steel sections are usually designated by the shapes of their cross sections. As examples, there are angles, tees, zees, and plates. It is necessary, however, to make a definite distinction between American standard beams (called S beams) and wide-flange beams (called W beams), as they are both I-shaped. The inner surface of the flange of a W section is either parallel to the outer surface or nearly so with a maximum slope of 1 to 20 on the inner surface, depending on the manufacturer.

The S beams, which were the first beam sections rolled in America, have a slope on their inside flange surfaces of 1 to 6. It might be noted that the constant, or nearly constant, thickness of W beam flanges as compared to the tapered S beam flanges may facilitate connections. Wide-flange beams compose nearly 50 percent of the tonnage of structural steel shapes rolled today. The W and S sections are shown in Fig. 1-1 together with several other familiar steel sections. The uses of these various shapes will be discussed in detail in the chapters to follow.

Constant reference is made throughout this book to the *Manual of Steel Construction Allowable Stress Design* published by the American Institute of Steel Construction (AISC). This manual, which provides detailed information for structural steel shapes, is referred to hereafter as the ASD Manual or just the Manual or the Handbook. Reference is made here to the 9th edition of the Handbook, which is based on the "Specification for Structural Steel Buildings Allowable Stress Design and Plastic Design" dated June 1, 1989. Structural shapes are abbreviated by a certain system described in the Handbook for use in drawings, specifications, and designs. This system is standardized so that all steel mills can use the same identification for purposes of ordering, billing, and so on. In addition so much work is handled today with computers and other automated equipment that it is necessary to have a letter and number system which can be printed out with a standard keyboard (as opposed to the old system where certain symbols were used for angles, channels, etc.). Examples of this abbreviation system are as follows:

1. A W27×114 is a W section approximately 27 in. deep weighing 114 lb/ft.
2. An S12×35 is an S section 12 in. deep weighing 35 lb/ft.
3. An HP12×74 is a bearing pile section approximately 12 in. deep weighing 74 lb/ft. Bearing piles are made with the regular W rolls but with thicker webs to provide better resistance to the impact of pile

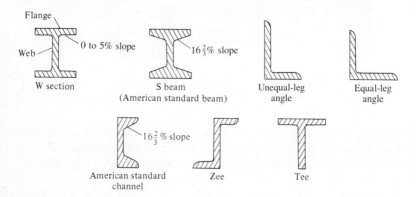

Figure 1-1 Rolled-steel shapes.

driving. The width and depth of these sections are approximately equal, and the flanges and webs have equal or almost equal thicknesses.

4. An M8×6.5 is a miscellaneous section 8 in. deep weighing 6.5 lb/ft. It is one of a group of doubly symmetrical H-shaped members which cannot be classified as a W, S, or HP section by dimensions.
5. A C10×30 is a channel 10 in. deep weighing 30 lb/ft.
6. An MC18×58 is a miscellaneous channel 18 in. deep weighing 58 lb/ft which cannot be classified as a C shape by dimensions.
7. An L6×6×$\frac{1}{2}$ is an equal-leg angle, each leg being 6 in. long and $\frac{1}{2}$ in. thick.
8. A WT18×140 is a tee obtained by splitting a W36×280. This type of section is known as a structural tee or a split tee.

The student should refer to the ASD Manual for information concerning other rolled shapes, such as the distinction between bars and plates, pipe and structural tubing, and other items. Additional sections will be mentioned here as it becomes necessary.

In Part 1 of the ASD Manual the dimensions and properties of W, S, C, and other shapes are given. The dimensions are given in decimals (for the use of designers) and in fractions to the nearest sixteenth of an inch (for the use of craftworkers and steel detailers or drafters). Also provided for the use of designers are such items as moments of inertia, section moduli, radii of gyration, and other cross-sectional properties discussed later in this text.

There are variations present in any manufacturing process, and the steel industry is certainly no exception. As a result the cross-sectional dimensions of steel members may vary somewhat from the values specified in the ASD

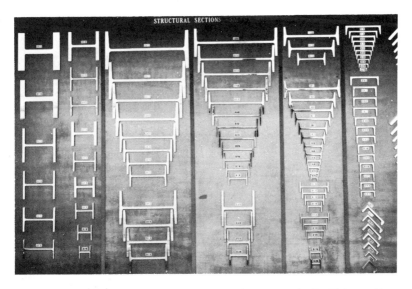

Structural shapes rolled at Bethlehem Steel Company's Bethlehem, Pa., plant. (Courtesy of Bethlehem Steel Corporation.)

Manual. Maximum tolerances for the rolling of steel shapes are set up by the American Society for Testing and Materials (ASTM) A6 Specification and are quoted in Part 1 of the Manual. As a result calculations can be made on the basis of the properties given in the Manual regardless of the manufacturer.

Many of the heavier W sections are shown shaded in the Manual to indicate that they are not available from domestic producers. Nevertheless, there is no difficulty at the present time in obtaining any of these sections.

Through the years there have been changes in the sizes of steel sections. For instance, there may be insufficient demand to continue rolling a certain shape, an existing shape may be dropped because a similar-size but more efficient shape has been developed, and so forth. Occasionally the designer may need to know the properties of one of the discontinued shapes which are no longer listed in the latest edition of the Manual or in other tables normally available to him or her. For example, it may be necessary to add another floor to an existing building which was constructed with shapes no longer rolled. In 1953 the AISC published a book entitled *Iron and Steel Beams 1873 to 1952* which gives a complete listing of iron and steel beams and their properties rolled in the United States during that period. Since that book was published, there have been many additional shape changes. As a result the wise structural designer will carefully preserve old editions of the Manual so as to have them available when needed.

1-5 COLD-FORMED LIGHT-GAGE STEEL SHAPES

In addition to the hot-rolled steel shapes discussed in the last section there are some cold-formed steel shapes. These are made by bending thin sheets of carbon or low-alloy steels into almost any desired cross section, such as the ones shown in Fig. 1-2.[5] These shapes, which may be used for light members in roofs, floors, and walls, vary in thickness from about 0.01 in. up to about

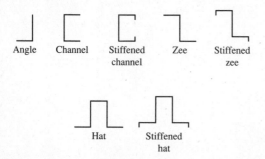

Angle   Channel   Stiffened channel   Zee   Stiffened zee

Hat   Stiffened hat

Figure 1-2 Cold-formed shapes.

---

[5] *Specification for the Design of Cold-formed Steel Structural Members* (Washington, D.C.: American Iron and Steel Institute, 1986).

0.25 in. The thinner shapes are most often used for some types of panels. Though cold-working does somewhat reduce ductility, it causes some strength increases. Under certain conditions, specifications will permit the use of these higher strengths.

1-6 STRESS-STRAIN RELATIONSHIPS IN STRUCTURAL STEEL

To understand the behavior of steel structures it is absolutely essential for the designer to be familiar with the properties of steel. Stress-strain diagrams present a valuable part of the information necessary to understand how steel will behave in a given situation. Satisfactory steel design methods cannot be developed unless complete information is available concerning the stress-strain relationships of the material being used.

If a piece of mild structural steel is subjected to a tensile force, it will begin to elongate. If the tensile force is increased at a constant rate, the amount of elongation will increase constantly within certain limits. In other words, elongation will double when the stress goes from 6,000 to 12,000 pounds per

Steel erection for the Chase Manhattan Bank Building, New York City. (Courtesy of Bethlehem Steel Corporation.)

square inch (psi). When the tensile stress reaches a value roughly equal to one-half of the ultimate strength of the steel, the elongation will begin to increase at a greater rate without a corresponding increase in the stress.

The largest stress for which Hooke's law applies or the highest point on the straight-line portion of the stress-strain diagram (Fig. 1-3) is the *proportional limit*. The largest stress which a material can withstand without being permanently deformed is called the *elastic limit*. This value is seldom actually measured, and for most engineering materials, including structural steel, is synonymous with the proportional limit. For this reason the term *proportional elastic limit* is sometimes used.

The stress at which there is a decided increase in the elongation or strain without a corresponding increase in stress is said to be the *yield point*. It is the first point on the stress-strain diagram where a tangent to the curve is horizontal. The yield point is probably the most important property of steel to the designer, as so many design procedures are based on its value. Beyond the yield point there is a range in which a considerable increase in strain occurs without increase in stress. The strain that occurs before the yield point is referred to as the *elastic strain;* the strain that occurs after the yield point, with no increase in stress, is referred to as the *plastic strain*. Plastic strains are usually from 10 to 15 times the elastic strains.

Yielding of steel without stress increase may be thought to be a severe disadvantage when in actuality it is a very useful characteristic. It has often performed the wonderful service of preventing failure due to omissions or mistakes on the designer's part. Should the stress at one point in a ductile steel structure reach the yield point, that part of the structure will yield locally without stress increase, thus preventing premature failure. This ductility allows the

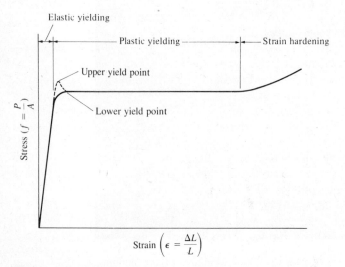

Figure 1-3 Typical stress-strain diagram for a mild or low-carbon structural steel.

Newport Bridge between Jamestown and Newport, R.I. (Courtesy of Bethlehem Steel Corporation.)

stresses in a steel structure to readjust. Another way of describing this phenomenon is to say that very high stresses caused by fabrication, erection, or loading will tend to equalize themselves. It might also be said that a steel structure has a reserve of plastic strain that enables it to resist overloads and sudden shocks. If it did not have this ability, it might suddenly fracture, like glass or other vitreous substances.

Following the plastic strain there is a range where additional stress is necessary to produce additional strain, and this is called *strain hardening*. This portion of the diagram is not too important to today's designer because the strains are so large. A familiar stress-strain diagram for mild or low-carbon structural steel is shown in Fig. 1-3. Only the initial part of the curve is shown here because of the great deformation which occurs before failure. At failure in the mild steels the total strains are from 150 to 200 times the elastic strains. The curve will actually continue up to its maximum stress value and then "tail off" before failure. A sharp reduction in the cross section of the member takes place (called necking), followed by failure.

The stress-strain curve of Fig. 1-3 is typical of the usual ductile structural steel and is assumed to be the same for members in tension or compression. (The compression members must be stocky because slender compression members subjected to compression loads tend to bend laterally, and their properties are greatly affected by the bending moments so produced.) The shape of the diagram varies with the speed of loading, the type of steel, and the temperature. One such variation is shown in the figure by the dotted line which is marked *upper yield*. This shape is the result when a mild steel has the load applied rapidly, while the *lower yield* is the result for slow loading.

A very important property of a structure which has not been stressed beyond its yield point is that it will return to its original length when the loads are removed. Should it be stressed beyond this point, it will return only part of the way to its original position. This knowledge leads to the possibility of testing an existing structure by loading and unloading. If after the loads are removed the structure will not resume its original dimensions, it has been stressed beyond its yield point.

Steel is an alloy consisting almost entirely of iron (usually over 98 percent). It also contains small quantities of carbon, silicon, manganese, sulfur, phosphorus, and other elements. Carbon is the material that has the greatest effect on the properties of steel. The hardness and strength increase as the carbon percentage is increased, but unfortunately the resulting steel is more brittle and its weldability is adversely affected. A smaller amount of carbon will make the steel softer and more ductile but also weaker. The addition of such elements as chromium, silicon, and nickel produces steels with considerably higher strengths. These steels, however, are appreciably more expensive and are often not as easy to fabricate.

A typical stress-strain diagram for a brittle steel is shown in Fig. 1-4. Unfortunately, low ductility or brittleness is a property usually associated with high strengths in steels (although not entirely confined to high-strength steels). As it is desirable to have both high strength and ductility, the designer may have to decide between the two extremes or compromise between them. A brittle steel may fail suddenly without warning when overstressed, and during erection could fail due to the shock of erection procedures.

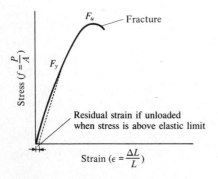

Figure 1-4 Typical stress-strain diagram for a brittle steel.

F_y = minimum yield stress

F_u = minimum ultimate tensile strength

Brittle steels have a considerable range where stress is proportional to strain but do not have clearly defined yield stresses. Yet to apply many of the formulas given in structural steel design specifications, it is necessary to have definite yield stress values regardless of whether the steels are ductile or brittle.

If a mild steel member is strained beyond its elastic limit and then unloaded, it will not return to a condition of zero strain. As it is unloaded, its stress-strain diagram will follow a new path (shown by the dotted line in Fig. 1-4) parallel to the initial straight line. The result is a permanent or residual strain.

The yield stress for a brittle steel is usually defined as the stress at the point of unloading which corresponds to some arbitrarily defined residual strain (0.002 being the common value). In other words, we increase the strain by a designated amount and draw a line from that point, parallel to the straight-line portion of the stress-strain diagram, until the new line intersects the old. This intersection is the yield stress at that particular strain. If 0.002 is used, the intersection is usually referred to as the yield stress at 0.2 percent offset strain.

1-7 MODERN STRUCTURAL STEELS

The properties of steel can be greatly changed by varying the quantities of carbon present and by adding other elements such as silicon, nickel, manganese, and copper. A steel that has a significant amount of the latter elements is referred to as an alloy steel. Although these elements do have a great effect on the properties of steel, the actual quantities of carbon or other alloying elements are quite small. For instance, the carbon content of steel is almost always less than 0.5 percent by weight and is normally between 0.2 and 0.3 percent.

The chemistry of steel is extremely important in its effect on the properties of the steel such as weldability, corrosion resistance, resistance to brittle fracture, and so on. The ASTM specifies the exact maximum percentages of carbon, manganese, silicon, and so on, that are permissible for a number of structural steels. Although the physical and mechanical properties of steel sections are primarily determined by their chemical composition, they are also influenced to a certain degree by the rolling process and by their stress history and heat treatment.

Perhaps 50 percent of the steel used in engineering structures in the United States is a structural carbon steel designated as A36 by the ASTM, but many other steels are available, and demand for them is rising rapidly. A higher-strength steel designated as A572 by the ASTM and described later in this section is today being used almost as much as A36.

In recent decades the engineering and architecture professions have been continually requesting stronger steels, steels with more corrosion resistance, better welding properties, and various other improvements. Research by the

TABLE 1-1 PROPERTIES OF STRUCTURAL STEELS

ASTM DESIGNATION	TYPE OF STEEL	FORMS	RECOMMENDED USES	MINIMUM YIELD STRESS F_y, KSI[a]	SPECIFIED MINIMUM TENSILE STRENGTH F_u, KSI[b]
A36	Carbon	Shapes, bars, and plates	Bolted, welded, or riveted buildings and bridges and other structural uses	36, but 32 if thickness $>$ 8 in.	58–80
A529	Carbon	Shapes and plates to $\frac{1}{2}$ in.	Similar to A36	42	60–85
A441	High-strength low-alloy	Shapes, plates, and bars to 8 in.	Similar to A36	40–50	60–70
A572	High-strength low-alloy	Shapes, plates, and bars to 6 in.	Bolted, welded, and riveted construction. Not for welded bridges for F_y grades 55 and above	42–65	60–80
A242	Atmospheric corrosion-resistant high-strength low-alloy	Shapes, plates, and bars to 4 in.	Bolted, welded, or riveted construction; welding technique very important	42–50	63–70

ASTM DESIGNATION	TYPE OF STEEL	FORMS	RECOMMENDED USES	MINIMUM YIELD STRESS F_y, KSI[a]	SPECIFIED MINIMUM TENSILE STRENGTH F_u, KSI[b]
A588	Atmospheric corrosion-resistant high-strength low-alloy	Plates and bars	Bolted and riveted construction	42–50	63–70
A852	Quenched and tempered low-alloy	Plates only to 4 in.	Welded, riveted, or bolted construction, primarily for welded bridges and buildings. Welding technique of fundamental importance.	70	90–110
A514	Quenched and tempered alloy	Plates only to 4 in.	Welded structures with great attention to technique, discouraged if ductility important	90–100	100–130

[a] F_y values vary with thickness and group (see Tables 1 and 2, Part 1, ASD Manual.)

[b] F_u values vary by grade and type.

steel industry during this period has supplied several groups of new steels which satisfy many of the demands, and today there are quite a few structural steels designated by the ASTM and included in the ASD Specification.

Structural steels are generally grouped into several major ASTM classifications: the *all-purpose carbon steel* (A36), the *structural carbon steel* (A529), the *high-strength low-alloy structural steels* (A441 and A572), the *atmospheric corrosion-resistant high-strength low-alloy structural steels* (A242 and A588) and *quenched and tempered alloy steel plate* (A514 and A852).

In the paragraphs that follow, a few general remarks are made concerning these steel classifications, and the eight ASTM steels mentioned are listed in Table 1-1 with a few special notes concerning their uses and properties. (As you look at this table, notice the phenomenon that the thinner steel is rolled the stronger it becomes. Thicker members tend to be more brittle, and their slower cooling rate causes a coarser microstructure of the steel.)

One half of a 170-ft clear-span roof truss for the Athletic and Convention Center, Lehigh University, Bethlehem, Pa. (Courtesy of Bethlehem Steel Corporation.)

Carbon Steels

These steels have as their principal strengthening agents carefully controlled quantities of carbon and manganese. Carbon steels have their contents limited to the following maximum percentages: 1.7 percent carbon, 1.65 percent manganese, 0.60 percent silicon, and 0.60 percent copper. These steels are divided into four categories depending on carbon percentages as follows:

1. Low-carbon steel $<$ 0.15 percent.
2. Mild steel 0.15 to 0.29 percent. (The structural carbon steels fall into this category.)
3. Medium-carbon steel 0.30 to 0.59 percent.
4. High-carbon steel 0.60 to 1.70 percent.

The very common steel A36, which has a yield stress of 36 kips per square inch (ksi), is suitable for bolted, welded, or riveted bridges and buildings. It is used for the majority of the design problems in this text.

High-Strength Low-Alloy Steels

There are a large number of high-strength low-alloy steels, and they are included under several ASTM numbers. In addition to containing carbon and manganese, these steels obtain their higher strengths and other properties by the addition of one or more alloying agents such as columbium, vanadium, chromium, silicon, copper, and nickel. Included are steels with yield stresses as low as 40 ksi and as high as 70 ksi. These steels generally have much greater atmospheric corrosion resistance than the carbon steels.

The term *low-alloy* is used arbitrarily to describe steels for which the total of all the alloying elements does not exceed 5 percent of the total composition of the steel.

Atmospheric Corrosion-Resistance High-Strength Low-Alloy Structural Steels

When steels are alloyed with small percentages of copper, they become more corrosion-resistant. When exposed to the atmosphere, the surfaces of these steels oxidize and form a very tightly adherent film (sometimes referred to as a "tightly bound patina") which prevents further oxidation and thus eliminates the need for painting. After this process takes place within 18 months to 3 years (depending on the type of exposure—rural, industrial, direct or indirect sunlight, etc.), the steel reaches a deep reddish, brown or black color.

Supposedly the first steel of this type was developed in 1933 by the U.S. Steel Corporation (now the USX Corporation) to provide resistance to the severe corrosive conditions of railroad coal cars.

You can see the many uses which can be made of such a steel, particularly for structures with exposed members that are difficult to paint—bridges, electrical transmission towers, and others. This steel is not considered to be satisfactory for use where it is frequently subject to saltwater sprays or fogs, or continually submerged in water (fresh or salt) or the ground, or where there are severe corrosive industrial fumes. It is also not satisfactory in very dry areas, as in some western parts of the United States. For the patina to form, the steel must be subjected to a wetting and drying cycle. Otherwise it will continue to look like unpainted steel.

Quenched and Tempered Alloy Steels

These steels have alloying agents in excess of those used in the carbon steels, and they are heat-treated by quenching and tempering to obtain strong and tough steels with yield strengths in the 70 to 110 ksi range. Quenching consists of rapid cooling of the steel with water or oil from temperatures of at least 1650°F down to about 300 or 400°F. In tempering, the steel is reheated to at least 1150°F and allowed to cool.

The quenched and tempered steels do not exhibit well-defined yield points as do the carbon steels and the high-strength low-alloy steels. As a result their yield strengths are usually defined as the stress at 0.2 percent offset strain. The quenched and tempered steel plates and bars in Table 1-1 are A852 with a yield stress of 70 ksi and A514 with yield stresses of 90 or 100 ksi depending

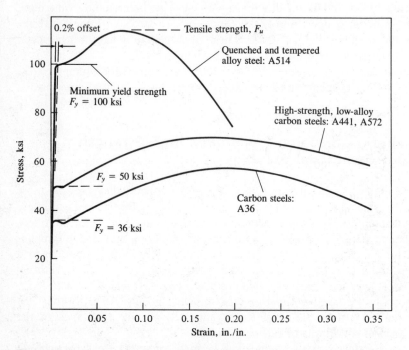

Figure 1-5 Stress-strain curves. (From C. G. Salmon and J. E. Johnson, *Steel Structures,* 3d ed., Copyright © 1990, Harper & Row, Publishers, Inc., p. 46.)

on thickness. One quenched and tempered steel in Table 1-1 is A514, which has yield stresses of 90 or 100 ksi, depending on thickness.

In Section A3.1 of Part 5 of the ASD Manual eight other ASTM grades of steel are listed (A53, A500, A501, A570, A606, A607, A618, and A709). These grades cover pipe, cold- and hot-formed tubing, sheet, strip, and structural steel for bridges.

A set of actual stress-strain curves for the three major types of steels described here (carbon, high-strength low-alloy, and quenched and tempered alloy) are shown in Fig. 1-5. The reader will be able to clearly see that the first two of these steels have well-defined yield points while the quenched and tempered ones do not.

1-8 USES OF HIGH-STRENGTH STEELS

There are indeed other groups of high-strength steels such as the ultra-high-strength steels which have yield strengths from 160 to 300 ksi. These steels have not been included in the ASD Manual because they have not been assigned ASTM numbers.

It is said that there are today more than 200 steels on the market that provide yield stresses in excess of 36 ksi. The steel industry is now experimenting with steels with yield stresses varying from 200 to 300 ksi, and this may be only the beginning. Many people in the steel industry feel that steels with 500 ksi yield strengths will be made available within a few years. The theoretical binding force between iron atoms has been estimated to be in excess of 4000 ksi.[6]

Although the prices of steels increase with increasing yield points, the percentage of price increase does not keep up with the percentage of yield-point increase. The result is that the use of the stronger steels will quite frequently be economical for tension members, beams, and columns. Perhaps the greatest economy can be realized with tension members (particularly those without bolt or rivet holes). They may provide a great deal of saving for beams if deflections are not important or if deflections can be controlled by methods described in later chapters. In addition considerable economy can frequently be achieved with high-strength steels for short and medium-length stocky columns. Another application which can provide considerable saving is hybrid construction. In this type of construction two or more steels of different strengths are used, the weaker steels being used where stresses are smaller and the stronger steels where stresses are higher.

Among the other factors that might lead to the use of high-strength steels are the following:

1. Superior corrosion resistance.
2. Possible savings in shipping, erection, and foundation costs caused by weight saving.

---

[6] L. S. Beedle et al., *Structural Steel Design* (New York: Ronald Press, 1964), p. 44.

3. Use of shallower beams permitting smaller floor depths.

4. Possible savings in fireproofing because smaller members can be used.

The first thought of most engineers in choosing a type of steel is the direct cost of the members. Such a comparison can be made quite easily, but determining which strength grade is most economical requires consideration of weights, sizes, deflections, maintenance, and fabrication. To make an accurate general comparison of the steels is probably impossible—rather, it is necessary to consider the specific job.

1-9 MEASUREMENT OF TOUGHNESS

The fracture toughness of steel is used as a general measure of its impact resistance or its ability to absorb sudden increases in stress at a notch. The more ductile steel is, the greater will be its toughness. On the other hand, the lower its temperature, the higher will be its brittleness.

There are several procedures available for estimating notch toughness, but the Charpy V-notch test is the most commonly used. Although this test (which is described in ASTM Specification A6) is somewhat inaccurate, it does help identify brittle steels. The energy required to fracture a small bar of rectangular cross section with a specified notch (see Fig. 1-6) is measured.

The bar is fractured with a pendulum swung from a certain height. The amount of energy needed to fracture the bar is determined from the height to which the pendulum rises after the blow. The test may be repeated for different temperatures and plotted as a graph as shown in Fig. 1-7. Such a graph clearly shows the relationship between temperature, ductility, and brittleness. The temperature at the point of steepest slope is the *transition temperature*.

Different structural steels have different specifications for absorbed energy levels (say 20 ft-lb at 20°F) depending on the temperature, stress, and loading conditions under which they are to be used. The topic of brittleness is continued in the next section.

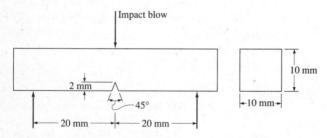

Figure 1-6 Specimen for Charpy V-notch test.

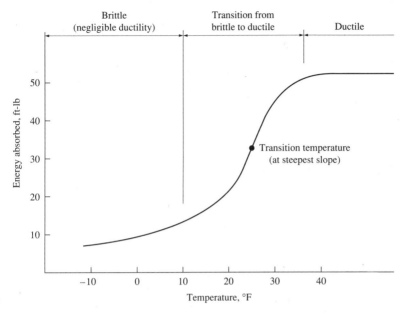

Figure 1-7 Results of Charpy V-notch test.

1-10 JUMBO SECTIONS

On page 1–8 of the ASD Manual, steel shapes are grouped from one to five depending on their flange and web thicknesses. The heavy W sections in groups 4 and 5 (and structural tees cut from them) are often referred to as *jumbo sections*.

Jumbo sections were originally developed for use as compression members and as such are quite satisfactory. However, designers have frequently used them for tension or flexural members. In these uses, flange and web areas have had some serious cracking problems where welding or thermal cutting has been used. These cracks may result in smaller load carrying capacities and problems related to fatigue.[7]

Thick pieces of steel tend to be more brittle than thin ones. Some of the reasons for this are that the core areas of thicker shapes (shown in Fig. 1-8) are subject to less rolling, have higher carbon contents (needed to produce required yield stresses), and have higher tensile stresses from cooling. These topics are discussed in later chapters.

Jumbo sections spliced with welds can be satisfactorily used for axial tension or flexural situations if the procedures listed in ASD Specification A3.1c are carefully followed. Included among the requirements are the following:

---

[7] R. Bjorhovde, "Solutions for the Use of Jumbo Shapes," *Proceedings 1988 National Steel Construction Conference,* AISC, Chicago, June 8–11, 1988, pp. 2-1 to 2-20.

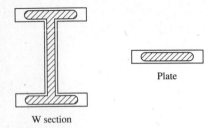

Plate

W section

Figure 1-8 Core areas where brittle failure may be a problem.

1. The steel used must have certain absorbed energy levels as determined by the Charpy V-notch test (20 ft-lb at 70°F). It is absolutely necessary that the tests be made on specimens taken from the core areas (shown in Fig. 1-8), where brittle fracture has proved to be a problem.

2. Temperature must be controlled during welding, and the work must follow a certain sequence.

3. Special splice details are required.

1-11 LAMELLAR TEARING

The steel specimens used for testing and developing stress-strain curves usually have their longitudinal axes in the direction that the steel was rolled. Should specimens be taken with their longitudinal axes transverse to the rolling direction "through the thickness" of the steel, the results will be lower ductility and toughness. Fortunately this is a matter of little significance for almost all situations. It can, however, be quite important when thick plates and heavy structural shapes are used in highly restrained welded joints. (It can be a problem for thin members, too, but it is much more likely to give trouble in thick ones.)

If a joint is highly restrained, the shrinkage of the welds in the through-the-thickness direction cannot be adequately redistributed, and the result can be a tearing of the steel called *lamellar tearing*. (*Lamellar* means "consisting of thin layers.") The situation is aggravated by the application of external tension. Lamellar tearing may show up as fatigue cracking after a number of cycles of load applications.

The lamellar tearing problem can be eliminated or greatly minimized with appropriate weld details and weld procedures. For example, the welds should be detailed so that shrinkage occurs as much as possible in the direction the steel was rolled. Several steel companies produce steels with enhanced through-the-thickness properties which provide much greater resistance to lamellar tearing. Even if such steels are used for heavy restrained joints the special joint details mentioned above are still necessary.[8]

---

[8] "Commentary on Highly Restrained Welded Connections," *Engineering Journal,* AISC, 10, no. 3 (3d quarter, 1973), pp. 61–73.

1-12 FURNISHING OF STRUCTURAL STEEL

The furnishing of structural steel consists of the rolling of the steel shapes, the fabrication of the shapes for the particular job (including cutting to the proper dimensions and punching the holes necessary for field connections), and their erection. Very rarely will a company perform all three of these functions, and the average company performs only one or two of them. For instance, many companies fabricate structural steel and erect it, while others may only be steel fabricators or steel erectors. There are approximately 400 to 500 companies in the United States which make up the fabricating industry for structural steel. Most of them do fabrication and erection.

Steel fabricators normally carry very little steel in stock because of the high interest and storage charges. When they get a job, they may order the shapes to certain lengths directly from the rolling mill, or they may obtain them from service centers. Service centers, which are an increasingly important factor in the supply of structural steel, buy and stock large quantities of structural steel which they buy at the best prices they can find anywhere in the world.

Structural steel is usually designed by an engineer in collaboration with an architectural firm. The designer makes design drawings that show member sizes, controlling dimensions, and any unusual connections. The company that is to fabricate the steel makes the detailed drawings subject to the engineer's approval. These drawings provide all the information necessary to fabricate the members correctly. They show the dimensions for each member, the locations and sizes of holes, the positions and sizes of connections, and the like.

The erection of steel buildings is more a matter of assembly than nearly any other line of construction work. Each of the members is marked in the shop with letters and numbers to distinguish it from the other members to be used. The erection is performed in accordance with a set of erection plans. These plans are not detailed drawings but are simple line diagrams showing the position of the various members in the building. A piece mark is placed on the left end of each member as it appears on the detail drawing. Directions (north, south, east, or west) are usually painted on column faces. These marks enable steelworkers to quickly find the position of each member in the field. (People performing steel erection are often called *ironworkers,* which is a name held over from the days before structural steel.)

1-13 THE STRUCTURAL DESIGNER

Structural designers can take great pride in their part in the development of our country. Cities, farmlands, and industrial areas of the United States are filled with the amazing structures designed by members of that profession. But even this remarkable array of structures is merely child's play compared to the structural endeavors of the next few generations. These structures of the future should provide many opportunities in the structural field for young engineers.

The structural designer arranges and proportions structures and their parts so that they will satisfactorily support the loads to which they may feasily be subjected. It might be said that he or she is involved with the general layout of structures; studies of the possible structural forms that can be used; consideration of loading conditions; analysis of stresses, deflections, and so on; design of parts; and the preparation of design drawings. More precisely, the word *design* pertains to the proportioning of the various parts of a structure after the forces have been calculated, and it is this process which will be emphasized throughout the text, using structural steel as the material.

1-14 OBJECTIVES OF THE STRUCTURAL DESIGNER

The structural designer must learn to arrange and proportion the parts of structures so that they can be practically erected and will have sufficient strength and reasonable economy. These items are discussed briefly below.

Safety

Not only must the frame of a structure safely support the loads to which it is subjected, but it must support them in such a manner that deflections and vibrations are not so great as to frighten the occupants or to cause unsightly cracks.

Cost

The designer needs to keep in mind the factors that can lower cost without sacrifice of strength. These items, which are discussed in more detail throughout the text, include the use of standard-size members, simple connections and details, and members and materials that will not require an unreasonable amount of maintenance through the years.

Practicality

Another objective is the design of structures that can be fabricated and erected without great problems arising. Designers need to understand fabrication methods and should try to fit their work to the fabrication facilities available.

Designers should learn everything possible about the detailing, fabrication, and field erection of steel. The more the designer knows about the problems, tolerances, and clearances in shop and field the more probable it is that reasonable, practical, and economical designs will be produced. This knowledge should include information concerning the transportation of the materials to the job site (such as the largest pieces that can be transported practically by rail or truck), labor conditions, and the equipment available for erection. Perhaps the designer should ask, "Could I get this thing together if I were sent out to do it?"

Steel erection for Transamerica Pyramid, San Francisco, Calif. (Courtesy of Kaiser Steel Corporation.)

Finally, he or she needs to proportion the parts of the structure so that they will not unduly interfere with the mechanical features of the structure (pipes, ducts, etc.) or the architectural effects.

1-15 ECONOMICAL DESIGN OF STEEL MEMBERS

The design of a steel member involves much more than a calculation of the properties required to support the loads and the selection of the lightest section providing these properties. Although at first glance this procedure would seem to give the most economical designs, many other factors need to be considered. Among these are:

1. The designer needs to select steel sections of sizes which are usually rolled. Steel beams and bars and plates of unusual sizes will be difficult to obtain during boom periods and will be expensive during any period. A little study on the designer's part will enable him or her to avoid these expensive shapes. Steel fabricators are constantly supplied with information from the steel companies and the steel warehousers as to the sizes and lengths of sections available. (Most structural shapes can be obtained in lengths from 60 to 75 ft depending on the producer, while it is possible under certain conditions to obtain some shapes up to 120 ft in length.)

2. A blind assumption that the lightest section is the cheapest one may be in considerable error. A building frame designed by the "lightest-section" procedure will consist of a large number of different shapes and sizes of members. Trying to connect these many-sized members and fit them in the building will be quite complicated, and the pound price of the steel will in all probability be rather high. A more reasonable approach would be to smooth out the sizes by selecting many members of the same sizes although some of them may be slightly overdesigned.

3. The beams usually selected for the floors in buildings will normally be the deeper sections, because these sections for the same weights have the largest moments of inertia and the greatest resisting moments. As building heights increase, however, it may be economical to modify this practice. As an illustration, consider the erection of a 20-story building, for which each floor has a minimum clearance. It is assumed that the depths of the floor beams may be reduced by 6 in. without an unreasonable increase in beam weights. The beams will cost more, but the building height will be reduced by 20 × 6 in. = 120 in., or 10 ft, with resulting savings in walls, elevator shafts, column heights, plumbing, wiring, and footings.

4. For larger sections, particularly the built-up ones, the designer needs to have information pertaining to transportation problems. The desired information includes the greatest lengths and depths that can be shipped by truck or rail (see next section), clearances available under bridges and power lines leading to the project, and allowable loads on bridges. It may be possible to fabricate a steel roof truss in one piece, but is it possible to transport it to the job site and erect it in one piece?

5. Sections should be selected which are reasonably easy to erect and which have no conditions that will make them difficult to maintain. As an example, it is necessary to have access to all exposed surfaces of steel bridge members so that they may be periodically painted (unless one of the special corrosion-resistant steels is being used).

6. Buildings are often filled with an amazing conglomeration of pipes, ducts, conduits, and other items. Every effort should be made to select steel members that will fit in with the requirements made by these items.

7. The members of a steel structure are often exposed to the public, particularly in the case of steel bridges and auditoriums. Appearance may often be the major factor in selecting the type of structure, as where a bridge is desired that will fit in and actually contribute to the appearance of an area. Exposed members may be surprisingly graceful when a simple arrangement and perhaps curved members are used, but other arrangements may create a terrible eye-

Steel erection for Transamerica Pyramid, San Francisco, Calif. (Courtesy of Kaiser Steel Corporation.)

sore. The student has certainly seen illustrations of each case. It is very interesting to know that beautiful structures in steel are usually quite reasonable in cost.

The question is often asked, *How do we achieve economy in structural steel design? The answer is simple: it lies in what the steel fabricator does not have to do.* (In other words economy can be realized when fabrication is minimized.)

1-16 HANDLING AND SHIPPING STRUCTURAL STEEL

The following general rules apply to the sizes and weights of structural steel pieces which can be fabricated in the shop, shipped to the job and erected.

1. The maximum weights and lengths that can be handled in the shop and at a construction site are roughly 90 tons and 120 ft respectively.

2. Pieces as large as 8 ft high, 8 ft wide and 60 ft long can be shipped on trucks with no difficulty (provided the axle or gross weights do not exceed the permissible values given by public agencies along the designated routes).

3. There are few problems in railroad shipment if pieces are no larger than 10 ft high, 8 ft wide, 60 ft long and weigh no more than 20 tons.

4. Routes should be carefully studied and carriers consulted for weights and sizes exceeding the values given in 2 and 3 above.

1-17 CALCULATION ACCURACY

A most important point which many students with their superb pocket calculators and microcomputers have difficulty in understanding is that structural design is not an exact science for which answers can confidently be calculated to eight places. Among the reasons for this fact are: the methods of analysis are based on partly true assumptions, the strengths of materials used vary appreciably, and maximum loadings can only be approximated. With respect to this last sentence, how many of the users of this book could estimate within 10 percent the maximum load in pounds per square foot that will ever occur on the building floor which they are now occupying? Calculations to more than two or three significant figures are obviously of little value and may actually be harmful in that they mislead the student by giving him or her a fictitious sense of precision.

2

Specifications, Loads, and Methods of Design

2-1 SPECIFICATIONS AND BUILDING CODES

The design of most structures is controlled by specifications. Even if the specifications do not control, the designer will probably refer to them as a guide. No matter how many structures a person has designed, it is impossible for him or her to have encountered every situation, and by referring to specifications he or she is making use of the best available material on the subject. Engineering specifications which are developed by various organizations present the best opinion of those organizations as to what represents good practice.

Municipal and state governments concerned with the safety of the public have established building codes with which they control the construction of various structures within their jurisdiction. These codes, which are actually laws or ordinances, specify design loads, design stresses, construction types, material quality, and other factors. They vary considerably from city to city, a fact that causes some confusion among architects and engineers.

Several organizations publish recommended practices for regional or national use. Their specifications are not legally enforceable unless they are embodied in the local building code or made a part of a particular contract. Among these organizations are the AISC and American Association of State

South Fork Feather River Bridge in northern California, being erected by use of a 1626-ft-long cableway strung from 210-ft-high masts anchored on each side of the canyon. (Courtesy of Bethlehem Steel Corporation.)

Highway and Transportation Officials (AASHTO). Nearly all municipal and state building codes have adopted an AISC Specification, and nearly all state highway and transportation departments have adopted the AASHTO Specifications.

Some people feel that specifications prevent engineers from thinking for themselves—and there may be some basis for the criticism. They say that the ancient engineers who built the great pyramids, the Parthenon, and the great Roman bridges were controlled by few specifications, which is certainly true. On the other hand, it should be said that only a few score of these great projects were built over many centuries, and they were apparently built without regard to cost of material, labor, or human life. They were probably built by intuition and by certain rules of thumb developed by observing the minimum size or strength of members which would just fail under given conditions. Their probably numerous failures are not recorded in history; only their successes endured.

Today, however, there are hundreds of projects being constructed at any one time in the United States that rival in importance and magnitude the famous structures of the past. It appears that, if all engineers in our country were allowed to design projects such as these without restrictions, there would be many disastrous failures. *The important thing to remember about specifications, therefore, is that they are not written for the purpose of restricting engineers but for the purpose of protecting the public.*

No matter how many specifications are written, it is impossible for them to cover every possible situation. As a result, no matter which code or specification is or is not being used, the ultimate responsibility for the design of a safe structure lies with the structural designer.

2-2 LOADS

Perhaps the most important and most difficult task faced by the structural designer is the accurate estimation of the loads that may be applied to a structure during its life. No loads that may reasonably be expected to occur may be overlooked. After loads are estimated, the next problem is to decide the worst possible combinations of these loads which might occur at one time. For instance, would a highway bridge completely covered with ice and snow be simultaneously subjected to fast-moving lines of heavily loaded trailer trucks in every lane and to a 90-mile lateral wind, or is some lesser combination of these loads more feasible?

The next two sections of this chapter provide a brief introduction to the types of loads with which the structural designer needs to be familiar. The purpose of these sections is not to discuss loads in great detail but rather to give the reader a feel for the subject. As will be seen, loads are classed as being dead loads or live loads.

2-3 DEAD LOADS

Dead loads are loads of constant magnitude that remain in one position. They consist of the structural frame's own weight and other loads that are permanently attached to the frame. For a steel-frame building some dead loads are the frame, walls, floors, roof, plumbing, and fixtures.

To design a structure it is necessary for the weights or dead loads of the various parts to be estimated for use in the analysis. The exact sizes and weights of the parts are not known until the structural analysis is made and the members of the structure selected. The weights, as determined from the actual design, must be compared with the estimated weights. If large discrepancies are present, it will be necessary to repeat the analysis and design using better estimated weights.

Reasonable estimates of structure weights may be obtained by referring to similar types of structures or to various formulas and tables available in several locations. The weights of many materials are given in Part 6 of the ASD Manual. Even more detailed information on dead loads is provided in Tables C1 and C2 of the American Society of Civil Engineers (ASCE) Standard.[1] An experienced designer can estimate very closely the weights of most materials and will spend little time repeating designs because of poor estimates.

2-4 LIVE LOADS

Live loads are loads that may change in position and magnitude. Simply stated, all loads that are not dead loads are live loads. Live loads that move under their own power are moving loads, such as trucks, people, and cranes, whereas those loads that may be moved are movable loads, such as furniture, warehouse materials, and snow. Other live loads include those caused by construction operations, wind, rain, earthquakes, blasts, soils, and temperature changes. A brief description of some of these loads follows:

1. *Floor loads*. The minimum gravity live loads to be used for building floors are usually clearly specified by the applicable building code. Unfortunately, however, the values given in these various codes vary from city to city, and the designer must be sure that his or her designs meet the requirements of that locality. A few of the typical values for floor loadings are listed in Table 2-1. These values were adopted from the ASCE Standard.[2] In the absence of a governing code this seems to be an excellent one to follow.

2. *Snow and ice*. In the colder states snow and ice loads are often quite important. One inch of snow is equivalent to approximately 0.5 pounds per square foot (psf), but it may be higher at lower elevations where snow is denser. For roof designs, snow loads of from 10 to 40 psf are used, the magnitude depending primarily on the slope of the roof and to a lesser degree on the character of the roof surface. The larger values are used for flat roofs and the smaller ones for sloped roofs. Snow tends to slide off sloped roofs, particularly those with metal or slate surfaces. A load of approximately 10 psf might be used for 45° slopes and a 40-psf load for flat roofs. Studies of snowfall records in areas with severe winters may indicate the occurrence of snow loads much greater than 40 psf, with values as high as 100 psf in northern Maine.

Snow is a variable load which may cover an entire roof or only part of it. There may be drifts against walls or buildup in valleys or between parapets. Snow may slide off one roof onto a lower one. The wind may blow it off one side of a sloping roof, or the snow may crust over and remain in position even during very heavy winds.

Bridges are generally not designed for snow loads, since the weight of the snow is usually not significant in comparison with truck and train loads. In any

---

[1] *American Society of Civil Engineers Minimum Design Loads for Buildings and Other Structures,* ASCE 7-88 (formerly ANSI A58.1) (New York: ASCE), pp. 49–50.

[2] *American Society of Civil Engineers Minimum Design Loads for Buildings and Other Structures,* ASCE 7-88 (New York: ASCE), p. 4.

TABLE 2-1 TYPICAL MINIMUM UNIFORM LIVE LOADS
FOR DESIGN OF BUILDINGS

TYPE OF BUILDING	LIVE LOAD (PSF)
Apartment houses	
Apartments	40
Public rooms	100
Dining rooms and restaurants	100
Garages (passenger cars only)	50
Gymnasiums, main floors, and balconies	100
Office buildings	
Lobbies	100
Offices	50
Schools	
Classrooms	40
Corridors first floor	100
Corridors above first floor	80
Storage warehouses	
Light	125
Heavy	250
Stores (retail)	
First floor	100
Other floors	75

case it is doubtful that a full load of snow and maximum traffic would be present at the same time. Bridges and towers are sometimes covered with layers of ice from 1 to 2 in. thick. The weight of the ice runs up to about 10 psf. Another factor to be considered is the increased surface area of the ice-coated members as it pertains to wind loads.

3. *Rain.* Though snow loads are a more severe problem than rain loads for the usual roof, the situation may be reversed for flat roofs, particularly those in warmer climates. If water on a flat roof accumulates faster than it runs off, the result is called *ponding* because the increased load causes the roof to deflect into a dish shape that can hold more water, which causes greater deflections, and so on. This process continues until equilibrium is reached or until collapse occurs. Ponding is a serious matter, as illustrated by the large number of flat-roof failures that occur during rainstorms every year in the United States.

Ponding will occur on almost any flat roof to a certain degree even though roof drains are present. Roof drains may very well be used but they may be inadequate during severe storms or they may become partially or completely clogged. The best method of preventing ponding is to have an appreciable slope of the roof.

4. *Traffic loads for bridges.* Bridges are subjected to series of concentrated loads of varying magnitude caused by groups of truck or train wheels.

5. *Impact loads.* Impact loads are caused by the vibration of moving or movable loads. It is obvious that a crate dropped on the floor of a warehouse or

Cold-storage warehouse, Grand Junction, Colo. (Courtesy of American Institute of Steel Construction, Inc.)

a truck bouncing on the uneven pavement of a bridge causes greater forces than would occur if the loads were applied gently and gradually. Impact loads are equal to the difference between the magnitude of the loads actually caused and the magnitude of the loads had they been dead loads. They are significant in only a very few buildings (usually some types of industrial buildings) but are routinely considered in bridge design.

The ASD Specification (A4.2) states that structures that are to support live loads tending to cause impact must have their assumed live loads for design increased by no less than the following percentages if not otherwise specified:

For supports of elevators	100%
For cab-operated traveling-crane support girders and their connections	25%
For pendant-operated traveling-crane support girders and their connections	10%
For supports of light machinery, shaft or motor-driven	20%
For supports of reciprocating machinery or power-driven units	50%
For hangers supporting floors and balconies	33%

6. *Lateral loads.* Lateral loads are of two main types, wind and earthquake. A survey of engineering literature for the past 150 years reveals many references to structural failures caused by wind. Perhaps the most infamous of these have been bridge failures such as those of the Tay Bridge in Scotland in 1879 (which caused the deaths of 75 persons) and the Tacoma Narrows Bridge (Tacoma, Wash.) in 1940. But there have also been some disastrous building failures due to wind during the same period, such as that of the Union Carbide Building in Toronto in 1958. It is important to realize that a large percentage of building failures due to wind have occurred during their erection.[3]

(a) *Wind loads.* A great deal of research has been conducted in recent years on the subject of wind loads. Nevertheless a great deal more work needs to be done, as the estimation of these forces can by no means be classified as an exact science. The magnitudes of wind loads vary with geographical locations, heights above ground, types of terrain surrounding the buildings including other nearby structures, and other factors. Wind is present most of the time, but wind loads of the magnitudes considered in design occur quite rarely and as a result are not considered to be fatigue-type loads.

Wind pressures are usually assumed to be uniformly applied to the windward surfaces of buildings and are assumed to be capable of coming from any direction. These assumptions are not very accurate because wind pressures are not very even over large areas, the pressures near the corners of buildings being probably greater than elsewhere due to wind rushing around the corners, and so on. From a practical standpoint, therefore, all of the possible variations cannot be considered in design, although today's specifications are becoming more and more detailed in their requirements.

When the designer working with large stationary buildings makes poor wind estimates, the results are not too serious, but this is not usually the case when tall slender buildings (or long flexible bridges) are being considered. For many years the average designer ignored wind forces for buildings whose heights were not at least twice their least lateral dimensions. For such cases as these it was felt that the floors and walls provided sufficient lateral stiffnesses to eliminate the need for definite wind-bracing systems. A better practice for designers to follow, however, is to consider all the possible loading conditions that a particular structure may have to resist. If one or more of these conditions (as perhaps wind loading) seem of little significance, they may then be neglected. Should buildings have their walls and floors constructed with modern lightweight materials and/or should they be subjected to unusually high wind loads (as in coastal or mountainous areas), they will probably have to be designed for wind loads even if the height/least lateral dimension ratio is less than 2.

---

[3] "Wind Forces on Structures," Task Committee on Wind Forces, Committee on Loads and Stresses, Structural Division, ASCE, Final Report, *Transactions ASCE* 126, Part II (1961), pp. 1124–1125.

Access bridge, Renton, Wash. (Courtesy of Bethlehem Steel Corporation.)

Building codes do not usually provide for the estimated forces during tornadoes. The average designer considers the forces created directly in the paths of tornadoes to be so violent that it is not economically feasible to design buildings to resist them. This feeling is undergoing some change, however, as it has been found that the wind resistance of structures (even for small buildings, including houses) can be greatly increased at very reasonable costs by using better practices of tying the parts of structures together from the roofs through the walls to the footings, and by making appropriate connections of window frames to walls and perhaps other parts of the structure.[4,5]

Wind forces act as pressures on vertical windward surfaces, pressures or suction on sloping windward surfaces (depending on the slope), and suction on flat surfaces and on leeward vertical and sloping surfaces (due to the creation of negative pressures or vacuums). The student may have noticed this definite suction effect where shingles or other roof coverings have been lifted from the leeward roof surfaces of buildings. Suction or uplift can easily be demonstrated by holding a piece of paper at two of its corners and blowing above it. For some common structures uplift may be as large as 20 to 30 psf or even more.

---

[4] P. R. Sparks, "Wind Induced Instability in Low-Rise Buildings," *Proceedings of the 5th U.S. National Conference on Wind Engineering,* Lubbock, Tex., November 6–18, 1985.

[5] P. R. Sparks, "The Risk of Progressive Collapse of Single-Story Buildings in Severe Storms," *Proceedings of the ASCE Structures Congress,* Orlando, Fla., August 17–20, 1987.

During the passing of a tornado or hurricane, a sharp reduction in atmospheric pressure occurs. This decrease in pressure is not reflected inside airtight buildings, and the inside pressures, being greater than the external pressures, cause outward forces against the roofs and walls. Nearly everyone has heard stories of the walls of a building "exploding" outward during a storm.

The average building code in the United States makes no reference to wind velocities in the area or to shapes of the building or to other factors. It may require the use of some specified wind pressure in design, such as 20 psf on projected area in elevation up to 300 ft with an increase of 2.5 psf for each additional 100-ft increase in height. The values given in these codes are thought to be rather inaccurate for modern structural design.

For a period of several years the ASCE Task Committee on Wind Forces made a detailed study of existing information concerning wind forces. A splendid report[6] of this information was presented by the committee in 1961. As stated in the report, its purpose was to provide a compact source of information which could be used practically by the design profession.

The report presents much information concerning wind-pressure coefficients for various types of structures, information concerning maximum wind velocities for particular geographical areas, and much additional data which can be of great value in realistically estimating wind forces.

The wind pressure on a building can be estimated with the expression to follow in which p is the pressure in pounds per square foot acting on vertical surfaces, C_s is a shape coefficient, and V is the basic wind velocity in miles per hour estimated from weather bureau records for that area of the country.

$$p = 0.002558 \, C_s V^2$$

The coefficient C_s depends on the shape of the structure, primarily the roof. For box-type structures C_s is 1.3, of which 0.8 is for the pressure on the windward side and 0.5 is for suction on the leeward side. For such a building the total pressure on the two surfaces equals 20 psf for a wind velocity of 77.8 mph.

As described in the foregoing paragraphs, the accurate determination of the most critical wind loads on a building or bridge is an extremely involved problem; however, sufficient information is available today to permit satisfactory estimates on a reasonably simple basis.

(b) *Earthquake loads.* Many areas of the world fall in earthquake territory, and in these areas it is necessary to consider seismic forces in design for all types of structures. Through the centuries there have been catastrophic failures of buildings and bridges during earthquakes. It has been estimated that as many as 50,000 people lost their lives in the 1988 earthquake in Armenia. In

---

[6] "Wind Forces on Structures," Task Committee on Wind Forces, Committee on Loads and Stresses, Structural Division, ASCE, Final Report, *Transactions ASCE* 126, Part II (1961), pp. 1124–1198.

1989 the Loma Priata earthquake in California caused billions of dollars of property damage as well as some loss of life.[7]

Most buildings can be designed with little extra expense to withstand the forces caused during an earthquake of fairly severe intensity. On the other hand, earthquakes during recent years have clearly shown that the average building or bridge that is not designed for earthquake forces can be destroyed by earthquakes that are not particularly severe. Furthermore, the cost of providing seismic resistance to existing structures (often called *retrofitting*) can be extremely high.

During an earthquake there is an acceleration of the ground surface. This acceleration can be broken down into vertical and horizontal components. Usually the vertical component of the acceleration is assumed to be negligible, but the horizontal component can be severe. Various formulas are used to change earthquake accelerations into static horizontal forces that are dependent on the mass of the structure. The forces are expressed as a percentage of the gravity load of the structure and its contents and depend on the location of the structure on an earthquake probability map (see Fig. 2-1), the type of framing system, and other items. This kind of approach is usually adequate for

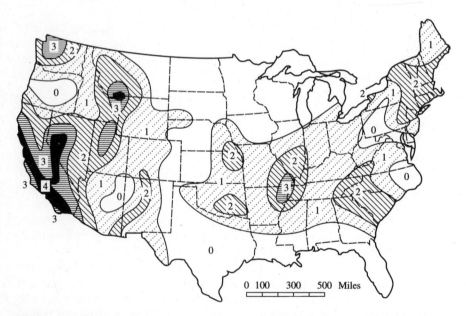

Figure 2-1 Map of seismic zones of the United States. The severity of risk ranges from a low in zone 0 to a high in zone 4. (From ASCE Standard 7-88, reprinted with the permission of the American Society of Civil Engineers.)

---

[7] V. Fairweather, "The Next Earthquake," *Civil Engineering* (New York: ASCE, March 1990), pp. 54–57.

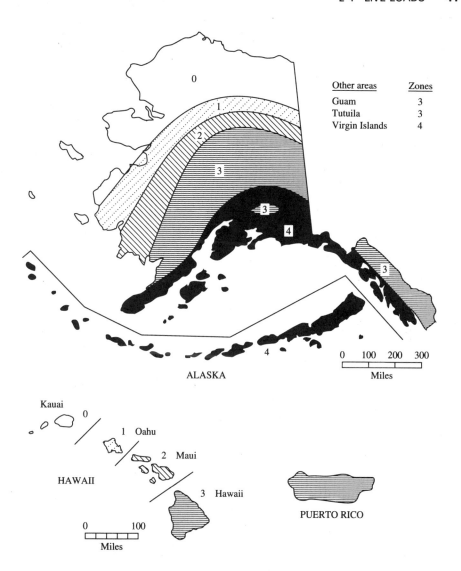

Other areas	Zones
Guam	3
Tutuila	3
Virgin Islands	4

ALASKA

HAWAII

PUERTO RICO

low-rise buildings, but for high-rise buildings complete dynamic analyses are needed.

An excellent reference on the subject of seismic forces is a publication by the Structural Engineers Association of California entitled *Recommended Lateral Force Requirements and Commentary.* [8]

---

[8] *Recommended Lateral Force Requirements and Commentary* (San Francisco, Calif.: Structural Engineers Association of California, 1987).

It should be noted that many persons look upon the seismic loads to be used in designs as being merely percentage increases of the wind loads. This thought is not correct, however, as seismic loads are different in their action and are not proportional to the exposed area but to the building weight above the level in question.

The effect of the horizontal acceleration increases with the distance above the ground because of the "whipping effect" of the earthquake, and design loads should be increased accordingly. Obviously, towers, water tanks, and penthouses on building roofs occupy precarious positions during an earthquake.

Fig. 2-1 shows the seismic risk zones in various locations in the United States. These zones were established on the basis of past earthquakes.[9] This map indicates that the most severe risks are in zones numbered 4 and the least in those marked 0.

Another factor to be considered in seismic design is the soil condition. Almost all of the structural damage and loss of life in the Loma Priata earthquake occurred in areas that have soft clay soils. Apparently these soils amplified the motions of the underlying rock.[10]

7. *Longitudinal loads.* Longitudinal loads also must be considered in designing some structures. Stopping a train on a railroad bridge or a truck on a highway bridge causes longitudinal forces to be applied. It is not difficult to imagine the tremendous longitudinal force developed when the driver of a 40-ton trailer truck traveling 60 mph suddenly has to apply the brakes while crossing a bridge. There are other longitudinal load situations, such as ships running into docks and the movement of traveling cranes that are supported by building frames.

8. *Other live loads.* Among the other types of live loads with which the structural designer will have to contend are *soil pressures* (such as the exertion of lateral earth pressures on walls or upward pressures on foundations); *hydrostatic pressures* (such as water pressure on dams, inertia forces of large bodies of water during earthquakes, and uplift pressures on tanks and basement structures); *blast loads* (caused by explosions, sonic booms, and military weapons); *thermal forces* (due to changes in temperature resulting in structural deformations and resulting structural forces); and *centrifugal forces* (such as those caused on curved bridges by trucks and trains or similar effects on roller coasters).

2-5 SELECTION OF DESIGN LOADS

To assist the designer in estimating the magnitudes of live loads with which he or she should proportion structures, various records have been assembled through the years in the form of building codes and specifications. These pub-

---

[9] *American Society of Civil Engineers Minimum Design Loads for Buildings and Other Structures,* ASCE 7-88 (New York: ASCE), pp. 33–34.

[10] V. Fairweather, "The Next Earthquake," *Civil Engineering* (New York; ASCE, March 1990), pp. 54–57.

lications provide conservative estimates of live load magnitudes for various situations. One of the most widely used design load specifications for buildings is the previously mentioned ASCE Standard. Some other commonly used specifications are

1. For railroad bridges, American Railway Engineering Association (AREA).[11]
2. For highway bridges, AASHTO.[12]
3. For buildings, the Uniform Building Code (UBC).[13]

These specifications will on many occasions clearly prescribe the loads for which structures are to be designed. Despite the availability of this information the designer's ingenuity and knowledge of the situation are often needed to predict what loads a particular structure will have to support in years to come. Over the past several decades insufficient estimates of future traffic loads by bridge designers have resulted in a great amount of replacement with wider and stronger structures.

2-6 ELASTIC AND PLASTIC DESIGN METHODS DEFINED

Almost all existing steel structures were designed by the *elastic design* methods. The designer estimates the *working* or *service* loads, that is, the loads that the structure may feasibly have to support, and proportions the members on the basis of certain allowable stresses. These allowable stresses are usually some fraction of the specified minimum yield point of the steel. Although the term "elastic design" is very commonly used to describe this method, the terms *allowable-stress design* or *working-stress design* are definitely more appropriate. Many of the provisions of the specifications for this method are actually based on plastic or ultimate-strength behavior and not on elastic behavior.

The ductility of steel has been shown to give it a reserve strength, and the realization of this fact is the theory behind *plastic design*. In this method the working loads are estimated and multiplied by certain load or overcapacity or safety factors and the members designed on the basis of collapse strengths. Another name for this method is *collapse design*.

The design profession has long been aware that the major portion of the stress-strain curve lies beyond the steel's elastic limit. Furthermore, tests through the years have made it clear that steels can resist stresses appreciably beyond their yield points and that, in cases of overload, statically indeterminate structures have the happy facility of spreading out the load, due to the steel's ductility. On the basis of this information many plastic-design proposals

---

[11] *Specifications for Steel Railway Bridges* (Chicago: AREA, 1985).

[12] *Standard Specifications for Highway Bridges*, 14th ed. (Washington, D.C.: AASHTO, 1989).

[13] *Uniform Building Code*, 1988 ed. (Whittier, CA: UBC, 1988).

Erection of steel joists. (Courtesy of Vulcraft.)

have been made in recent decades. It is undoubtedly true that for certain types of structures, plastic design results in a more economical use of steel than does elastic design. Nevertheless, the plastic-design method has never "caught on" with the steel design profession, and it is seldom used in practice. Despite this, a general knowledge of the plastic method is extremely helpful to the designer in his or her understanding of steel-structure behavior.

The specifications for the two design methods are presented in Part 5 of the ASD Manual. In Chapters A–M the Manual includes the ASD Specification, while the specification for plastic design is in Chapter N.

2-7 LOAD AND RESISTANCE FACTOR DESIGN

The AISC publishes another handbook, *Manual of Steel Construction Load and Resistance Factor Design*. This method of design, abbreviated LRFD, combines the calculation of ultimate or limit states of strength and serviceability with a probability-based approach to safety.

LRFD is similar to plastic design in that the failure or ultimate-strength condition is considered. The loads are multiplied by load or overcapacity factors (larger than 1.0), and the members are designed to provide sufficient strength to resist these factored loads. In addition, however, the theoretical or nominal capacity of each member is multiplied by an undercapacity or resistance factor less than 1.0 (to account for variations in material properties and member dimensions). The LRFD criterion may be expressed as

Design or usable strength $\geq \Sigma$ Factored load effects

In Canada a probability-based LRFD method has been used since 1974, and is the only method used in that country since 1978 while much progress has been made in Europe toward the goal of formulating LRFD procedures for various national codes. In 1986 the AISC adopted the LRFD Specification and its use in the United States has gradually increased since that date. The ASD method, however, is by far the predominant method used by American designers. Efforts are being made to prepare similar procedures for timber and reinforced-concrete structures in both Europe and America.[14] As a result great increases in the use of LRFD will undoubtedly be made during the next decade.

The average person looking at this material will ask, will LRFD save money compared with allowable stress design (ASD)? The answer is that it often will, particularly if the live loads are small compared with the dead loads.

It can be shown that for the lower range of live-load to dead-load values, that is, less than 3, LRFD can produce steel weight savings perhaps as high as one-sixth for tension members and columns and one-tenth for beams. On the other hand, if we have very high live-load to dead-load ratios, there is almost no increase in steel weight with LRFD as compared with ASD.[15]

The last several chapters of this book present an introduction to the LRFD method.

2-8 FACTOR OF SAFETY

The factor of safety of a structural member is defined as the ratio of strength of the member to its maximum anticipated stress. The strength of a member used in determining the factor of safety may be thought of as being the ultimate strength of the member, but often some lesser value is used. For instance, failure may be assumed to occur when the members become excessively deformed. If this is the case, the safety factor might be determined by dividing the yield-point stress by the maximum anticipated stress. For ductile materials the safety factor is usually based on yield-point stresses while for brittle materials it is usually based on ultimate strengths.

Students may feel that it is quite foolish to build a structure with a strength of several times that which is theoretically required. As the years go by, however, they will learn that safety factors are subject to so many uncertainties that they may spend sleepless nights wondering if those factors used are sufficient (and they may join other designers in calling them "factors of igno-

[14] M. K. Ravindra and T. V. Galambos, "Load and Resistance Factor Design for Steel," *Journal of Structural Division,* ASCE, *104,* ST9 (September 1978), pp. 1337–1353.

[15] J. A. Edinger, "Introduction to the Proposed AISC Load and Resistance Factor Design Specification," *Engineering Journal,* AISC, 21, no. 1 (1st quarter, 1984), pp. 62–65.

rance" rather than factors of safety). Some of the uncertainties affecting safety factors are

1. Material strengths may initially vary appreciably from their assumed values, and they will vary more with time, due to creep, corrosion, and fatigue.

2. The methods of analysis are often subject to appreciable errors.

3. The so-called beggaries of nature or acts of God (hurricanes, earthquakes, etc.) cause conditions difficult to predict.

4. The stresses produced during fabrication and erection are often severe. Workers in shop and field seem to treat steel shapes with reckless abandon. They drop them. They ram them. They force the members into position to line up the bolt holes. In fact, the stresses during fabrication and erection may exceed those which occur after the structure is completed. The floors for the rooms of apartment houses and office buildings are probably designed for live loads varying from 40 to 80 psf. During the erection of such buildings the contractor may have 10 ft of bricks or concrete blocks or other construction materials or equipment piled up on some of the floors, causing loads of several hundred psf. This discussion is not intended to criticize the practice (not that it is a good one), but rather to make the student aware of the things that happen during construction. (It is probable that the majority of steel structures are overloaded somewhere during construction, but hardly any of them fail.)

5. Technological changes affect the magnitude of live loads. The constantly increasing traffic loads applied to bridges through the years is an illustration.

6. Although the dead loads of a structure can usually be estimated quite closely, the estimate of the live loads is more inaccurate. This is particularly true in estimating the worst possible combination of live loads occurring at any one time. For instance, in estimating the load supported by a column in the bottom level of a 30-story building—would 100 percent live load be assumed to exist on every one of the 30 floors at the same time or is some lesser percentage more realistic?

7. Other uncertainties are the presence of residual stresses and stress concentrations, variations in dimensions of member cross sections, and so on.

The magnitude of the safety factors used will be affected by the preceding uncertainties and also by the answers to the following questions:

1. Is the structure to be permanent or temporary?

2. Is it a public or private structure?

3. What is the penalty for failure? Will there be loss of life or extensive property damage or great inconvenience while the structure is out of use?

4. Is a particular member a main one or a secondary one? (It may be reasonable to use high safety factors for the design of main members and low values for secondary members.)

2-9 FAILURE OF STRUCTURES

Many people who are superstitious do not discuss flat tires or make their wills because they fear they would be tempting fate. These same people would probably not care to discuss the subject of engineering failures. Despite the prevalence of this superstition the author feels that an awareness of the items which have most frequently caused failures in the past is invaluable to experienced and inexperienced designers alike. Perhaps a study of past failures is more important than a study of past successes. Benjamin Franklin supposedly made the observation that "a wise man learns more from failures than from success."

The designer of little experience particularly needs to know where the most attention should be given and where outside advice is needed. The vast majority of designers, experienced and inexperienced, select members of sufficient size and strength. The collapse of structures is usually due to insufficient attention to the details of connections, deflections, erection problems, and foundation settlements. Rarely if ever do steel structures fail due to faults in the material, but rather due to its improper use.

A frequent fault displayed by designers is that after carefully designing the members of a structure they carelessly select connections which may or may not be of sufficient size. They may even turn the job of selecting the connections over to drafters, who may not have sufficient understanding of the difficulties that can arise in connection design. Perhaps the most common mistake made in connection design is to neglect some of the forces acting on the connection, such as twisting moments. In a truss for which the members have been designed for axial forces only, the connections may be eccentrically loaded, resulting in moments that cause increasing stresses. These secondary stresses are occasionally so large that they need to be considered in design.

Another source of failure occurs where beams supported on walls have insufficient bearing or anchorage. Imagine a beam of this type supporting a flat roof on a rainy night when the roof drains are not functioning properly. As the water begins to form puddles on the roof, it tends to cause the beam to sag in the middle, causing a pocket to catch more rain, which will cause more beam sag, and so on. As the beam deflects, it pushes out against the walls, possibly causing collapse of walls or slippage of beam ends off the wall. Picture a 60-ft steel beam supported on a wall with only an inch or two of bearing when the temperature drops 50 or 60 degrees overnight. A collapse due to a combination of beam contraction, outward deflection of walls, and vertical deflection caused by precipitation loads is not difficult to visualize; furthermore, actual cases in engineering literature are not difficult to find.

Foundation settlements cause a large number of structural failures, probably more than any other factor. Most foundation settlements do not result in collapse, but they very often cause unsightly cracks and depreciation of the structure. If all parts of the foundation of a structure settle equally, the stresses in the structure theoretically will not change. The designer, usually not able to prevent settlement, has the goal of designing foundations in such a manner that equal settlements occur. Equal settlements may be an impossible goal, and consideration should be given to the stresses that would be produced if settlement variations occurred. The student's background in structural analysis will tell him or her that uneven settlements in statically indeterminate structures may cause extreme stress variations. Where foundation conditions are poor, it is desirable, if feasible, to use statically determinate structures whose stresses are not appreciably changed by support settlements. (The student will learn in subsequent discussions that the ultimate strength of steel structures is usually affected only slightly by uneven support settlements.)

Some structural failures occur because inadequate attention is given to deflections, fatigue of members, bracing against swaying, vibrations, and the possibility of buckling of compression members or the compression flanges of beams. The usual structure when completed is sufficiently braced with floors, walls, connections, and special bracing, but there are times during construction when many of these items are not present. As previously indicated, the worst conditions may well occur during erection, and special temporary bracing may be required.

3

Analysis of Tension Members

3-1 INTRODUCTION

Tension members are found in bridge and roof trusses, towers, bracing systems, and situations where they are used as tie-rods. The selection of a section to be used as a tension member is one of the simplest problems encountered in design. As there is no danger of buckling, the designer needs only to compute the estimated maximum force to be carried by the member and divide that force by an allowable stress to determine the effective cross-sectional area required (A_e reqd $= T/F_t$). Then it is necessary to select a steel section that provides the required area. Though these introductory calculations for tension members are quite simple, they do serve the important tasks of getting students started with design ideas and getting their feet wet with the massive ASD Manual.

One of the simplest forms of tension members is the round rod, but there is some difficulty in connecting it to many structures. The rod has been used frequently in the past but has only occasional uses today in bracing systems, light trusses, and timber construction. One important reason why the rod is not popular with designers is that it has been used improperly so often in the past that it has a bad name; but if rods are designed and installed correctly, they are satisfactory for many situations.

49

The average-size rod has very little stiffness and may quite easily sag under its own weight, injuring the appearance of the structure. The threaded rods formerly used in bridges often worked loose and rattled. Another disadvantage of rods is the difficulty of fabricating them to the exact length required and the consequent difficulties of installation.

When rods are used in wind bracing, it is a good practice to produce initial tension in them, as this will tighten up the structure and reduce rattling and swaying. To obtain initial tension the members may be detailed shorter than their required lengths, a method that gives the steel fabricator very little trouble. A common rule of thumb is to detail the rods about $\frac{1}{16}$ in. short for each 20 ft. of length. (Approximate stress $f = \epsilon E = [\frac{1}{16}/(12)(20)] \ (29 \times 10^6) = 7550$ psi.) Another very satisfactory method involves tightening the rods with some sort of sleeve net or turnbuckle. Part 4 of the ASD Manual provides detailed information for these devices.

The preceding discussion on rods should illustrate why rolled shapes such as angles have supplanted rods for most applications. In the early days of steel structures, tension members consisted of rods, bars, and perhaps cables. Today, although the use of cables is increasing for suspended-roof structures, tension members usually consist of single angles, double angles, tees, channels, W sections, or sections built up from plates or rolled shapes. These members look better than the old ones and are stiffer and easier to connect. Another type of tension section often used is the welded tension plate or flat bar, which is very satisfactory for use in transmission towers, signs, footbridges, and similar structures.

The tension members of steel roof trusses may consist of single angles as small as $2\frac{1}{2} \times 2 \times \frac{1}{4}$ in. for minor members. A more satisfactory member is made from two angles placed back-to-back with sufficient space between them to permit the insertion of plates for connection purposes. Where steel sections are used back-to-back in this manner, they should be connected every 4 or 5 ft to prevent rattling, particularly in bridge trusses. Single angles and double angles are probably the most common types of tension members in use. Structural tees make very satisfactory chord members for welded trusses because web members can conveniently be connected to them.

For bridges and large roof trusses tension members may consist of channels, W or S shapes, or even sections built up from some combination of angles, channels, and plates. Single channels are frequently used, as they have little eccentricity and are conveniently connected. Although, for the same weight, W sections are stiffer than S sections, they may have a connection disadvantage in their varying depths. For instance, the W12×79, W12×72, and W12×65 all have slightly different depths. (12.38 in., 12.25 in., and 12.12 in., respectively), while the S sections of a certain nominal size all have the same depths. For instance, the S12×50, the S12×40.8, and the S12×35 all have 12.00-in. depths.

Although single structural shapes are a little more economical than built-up sections, the latter are occasionally used when the designer is unable to obtain sufficient area or rigidity from single shapes. Where built-up sections are used, it is important to remember that field connections will have to be made

Erection of Federal Reserve Bank Building, Minneapolis, Minn. (Courtesy of IR Construction Products Co.)

and paint applied; therefore, sufficient space must be available to accomplish these tasks.

Members consisting of more than one section need to be tied together. Tie plates (also called tie bars) located at various intervals or perforated cover plates serve to hold the various pieces in their correct positions. These plates correct any unequal distribution of loads between the various parts. They also keep the slenderness ratios (to be discussed) of the individual parts within desired limitations, and they may permit easier handling of the built-up members. Long individual members such as angles may be inconvenient to handle due to flexibility, but when four angles are laced together into one member as shown in Fig. 3-1, the member has considerable stiffness. None of the intermittent tie plates may be considered to increase the effective areas of the sections. As they theoretically do not carry portions of the force in the main sections, their sizes are usually governed by specifications and perhaps by some judgment on the designer's part. Perforated cover plates (see Fig. 6-6) are an exception to this rule, as part of their areas can be considered as being effective in resisting axial load.

A few of the various types of tension members in general use are illustrated in Fig. 3-1. In this figure the dotted lines represent the intermittent tie plates or bars used to connect the shapes.

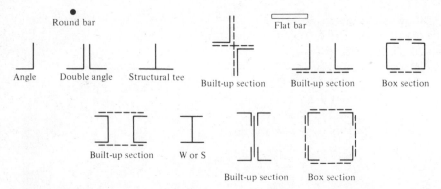

Figure 3-1 Types of tension members.

Federal Reserve Bank Building, Minneapolis, Minn. (Courtesy of IR Construction Products Co.)

Steel cables are made with special steel alloy wire ropes which are cold-drawn to the desired diameter. The resulting wires with strengths of about 200,000 to 250,000 psi can be economically used for suspension bridges, cable-supported roofs, ski lifts, and other similar applications.

Normally, to select a cable tension member the designer uses a manufacturer's catalog. From the catalog the yield point of the steel and the cable size required for the design force are determined. It is also possible to select clevises or other devices to use for connectors at the cable ends.

3-2 ALLOWABLE TENSILE STRESSES AND LOADS

A ductile steel member without holes and subject to a tensile load can resist without fracture a load larger than its gross cross-sectional area times its yield stress, because of strain hardening. However, a tension member loaded until strain hardening is reached will lengthen a great deal before fracture, a fact that will in all probability take away the member's usefulness and may even cause failure of the structural system of which the member is a part.

If we have a tension member with bolt holes, on the other hand, it can possibly fail by fracture at the net section through the holes. This failure load may very well be smaller than the load required to yield the gross section away from the holes. But note that the portion of the member where we have a reduced cross-sectional area due to the presence of holes is normally very short compared with the total length of the member. Though the strain-hardening situation is quickly reached at the net section portion of the member, yielding there may not really be a matter of significance, because the overall change in length of the member due to yielding in this small part of the member length may be negligible.

As a result of the preceding, the ASD Specification (D-1) gives an allowable tensile stress of $0.60F_y$ on the gross cross-sectional areas of members at sections where there are no holes. For sections where there are holes for bolts or rivets, the allowable tensile *stress* is $0.50F_u$, applicable to the effective net areas. Thus a safety factor of 1.67 is provided against yielding of the gross section ($F_y/0.60F_y = 1.67$), and a value of 2.0 is provided against the fracture of the member through its smallest effective net area ($F_u/0.50F_u = 2.0$).

The actual effective net area A_e that can be considered to resist tension at the section through the holes may be somewhat smaller than the actual net area A_n, because of stress concentrations and other factors which are discussed in Section 3-5 of this chapter. Thus the allowable capacity of a tensile member with bolt or rivet holes permitted by the ASD Specification is equal to the smaller of the following two values:

$$T = 0.60F_y A_g \quad \text{or} \quad T = 0.50F_u A_e$$

On some very rare occasions tension members may have slots or holes that extend over a significant length of the member. The ASD Commentary (D1) suggests that a "significant length" is a distance equal to or greater than the depth of the member. For such situations the designer must judge whether yielding over that length is a serious matter. If a decision is made that such yielding would be serious, an allowable stress of $0.60F_y$ will probably be used for the effective net area.

The preceding values are not applicable to threaded steel rods or to members with pin holes (as in eyebars). For members with pin holes an allowable tensile stress of $0.60F_y$ is permitted on the gross area of the member, while a value of $0.45F_y$ is permitted on the net area through the pin hole. Tension members made with threaded steel rods and their allowable stresses are discussed in Section 4-3.

The bridge specifications allow somewhat smaller allowable tensile stresses than does the AISC. For instance, the 1989 AASHTO Specifications permit an allowable tensile stress of $0.55F_y$ for members without holes and the smaller of $0.55F_y$ or $0.50 F_u$ for members with bolt or rivet holes ($0.46F_u$ on high-yield-strength quenched and tempered alloy steels).

It is not likely that stress fluctuations will be a problem in the average building frame because the changes in load in such structures usually occur only occasionally and produce relatively minor stress variations. Full design wind or earthquake loads occur so infrequently that they are not considered in fatigue design. Should there be frequent variations or even reversals in stress, however, the matter of fatigue must be considered. This subject is presented in Section 4-5.

3-3 NET AREAS

The presence of a hole obviously increases the unit stress in a tension member even if the hole is occupied by a rivet or bolt. There is less area of steel to which the load can be distributed, and there will be some concentration of stress along the edges of the hole.

Tension is assumed to be uniformly distributed over the net section of a tension member, although photoelastic studies show there is a decided increase in stress intensity around the edges of holes, sometimes equaling several times the stresses if the holes were not present. For ductile materials, however, an assumption of uniform stress distribution is reasonable when the material is loaded beyond its yield point. Should the fibers around the holes be stressed to their yield point, they will yield without further stress increase, with the result that there is a redistribution or balancing of stresses. At ultimate load it is reasonable to assume a uniform stress distribution. The subject of plastic theory will be discussed at length in later chapters. The importance of ductility on the strength of riveted or bolted tension members has been clearly demonstrated in tests. Tension members (with rivet or bolt holes) made from ductile steels have proved to be as much as one-fifth to one-sixth stronger than similar members made from brittle steels with the same ultimate strengths.

It is possible by a laborious mathematical process to consider stress concentrations occurring around holes and at other sudden changes in the dimensions of the cross section. The calculated values are quite approximate, and it is doubtful if they are appreciably more valuable than those obtained by dividing the total load by the cross-sectional area after any holes are subtracted.

This discussion is applicable only for tension members subjected to relatively static loading. For tension members designed for structures subjected to fatigue-type loadings, considerable effort should be made to minimize the items causing stress concentrations such as points of sudden change of cross section and sharp corners. In addition, as described in Section 4-5 the members may have to be enlarged.

The term *net cross-sectional area* or simply *net area* refers to the gross cross-sectional area of a member minus any holes, notches, or other indentations. In considering the area of such items as these it is important to realize that it is usually necessary to subtract an area a little larger than the actual hole. For instance, in fabricating structural steel which is to be connected with bolts, the holes are usually punched $\frac{1}{16}$ in. larger than the diameter of the bolt. Furthermore, the punching of the hole is assumed to damage or even destroy $\frac{1}{16}$ in. (1.6 mm) more of the surrounding metal; therefore, the area of the holes subtracted is $\frac{1}{8}$ in. (3 mm) larger than the diameter of the bolt. The area of the holes subtracted is rectangular and equals the diameter of the hole times the thickness of the metal.

For steel much thicker than bolt diameters it is difficult to punch out the holes to the full sizes required without excessive deformation of the surrounding material. These holes may be subpunched (with diameters $\frac{3}{16}$ in. undersized) and then reamed out to full size after the pieces are assembled. Very little material is damaged by this quite expensive process, as the holes are even and smooth, and it is considered unnecessary to subtract the $\frac{1}{16}$ in. for damage to the sides. Sometimes when very thick pieces are being connected, the holes may be drilled to the diameter of the bolts or rivets plus $\frac{1}{32}$ in. This is a very expensive process and should be avoided if possible.

It may be necessary to have an even greater latitude in meeting dimensional tolerances during erection, and for high-strength bolts larger than $\frac{5}{8}$ in. in diameter, holes larger than the standard ones may be used without reducing the performance of the connections. These oversized holes can be short-slotted or long-slotted as described in Chapter 11. For calculating net areas when slotted holes are used, $\frac{1}{16}$ in. is added to the width of each slot.

Example 3-1 illustrates the calculations necessary for determining the net area of a plate type of tension member.

■ Example 3-1

Determine the net area of the $\frac{3}{8} \times 8$-in plate shown in Fig. 3-2. The plate is connected at its end with two lines of $\frac{3}{4}$-in. bolts.

Solution.   Net area $= A_n \left(\frac{3}{8}\right)(8) - (2)(\frac{3}{4} + \frac{1}{8})(\frac{3}{8}) = 2.34$ in.^{2}(1510 mm^2)   ■

The connections of tension members should be arranged so that no eccentricity is present. (An exception to this rule is permitted by the ASD

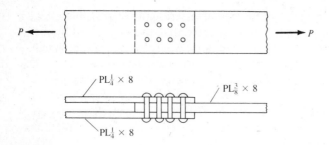

Figure 3-2

Specification for certain bolted and welded connections as described in Chapter 13.) If this arrangement is possible, the stress is assumed to be spread uniformly across the net section of a member. Should the connections have eccentricities, moments will be produced that will cause additional stresses in the vicinity of the connection. Unfortunately it is often quite difficult to arrange connections without eccentricity. Although specifications cover some situations, the designer may have to give consideration to eccentricities in some cases by making special estimates.

The lines of action of truss members meeting at a joint are assumed to coincide. Should they not coincide, eccentricity is present and secondary stresses are the result. The centers of gravity of truss members are assumed to coincide with the lines of action of their respective forces. No problem is present in a symmetrical member, as its center of gravity is at its center line, but for unsymmetrical members the problem is a little more difficult. When the center line is not the center of gravity, the usual practice is to arrange the members at a joint so their gage lines coincide. If a member has more than one gage line, the one closest to the actual center of gravity of the member is used in detailing. Fig. 3-3 shows a truss joint in which the gage lines of all the members pass through the same point.

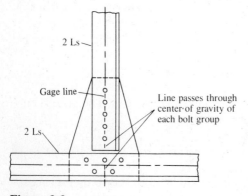

Figure 3-3

Two International Place, Boston, Mass. (Courtesy of Owen Steel Company, Inc.)

3-4 EFFECT OF STAGGERED HOLES

When the bolts or rivets for the end connection of a tension member are placed in a single line, we obtain the maximum net area. If space is not available for a single line of connectors, it may be necessary to use more than one line. If this is the case, it is often desirable to stagger them in order to provide as large a net area as possible at any one section to resist the load. In the preceding paragraphs tensile members have been assumed to fail transversely as along line *AB* in Fig. 3-4(a) and (b). Fig. 3-4(c) shows a member in which a nontransverse failure is possible. The holes are staggered, and a failure along section *ABCD* is possible unless the holes are a large distance apart.

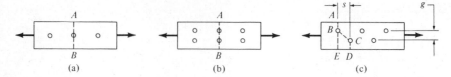

Figure 3-4

To determine the critical net area in Fig. 3-4(c), it might seem logical to compute the area of a section transverse to the member (as AE) less the area of one hole and then the area along section $ABCD$ less two holes. The smallest value obtained along these sections would be the critical value, but this method is at fault. Along the diagonal line from B to C there is a combination of direct stress and shear, and a somewhat smaller area should be used. The strength of the member along section $ABCD$ is obviously somewhere between the strength obtained by using a net area computed by subtracting one hole from the transverse cross-sectional area and the value obtained by subtracting two holes from section $ABCD$.

Tests on joints show that little is gained by using complicated theoretical formulas to consider the staggered-hole situation, and the problem is usually handled with an empirical equation. The ASD Specification (B2) and other steel specifications use a very simple method for computing the net width of a tension member along a zigzag section.[1] The method is to take the gross width of the member regardless of the line along which failure might occur, subtract the diameter of the holes along the zigzag section being considered, and add for each inclined line the quantity given by the expression $s^2/4g$.

In this expression s is the longitudinal spacing (or pitch) of any two holes, and g is the transverse spacing (or gage) of the same holes. The values of s and g are shown in Fig. 3-4(c). There may be several paths, any one of which may be critical at a particular joint. Each possibility should be considered, and the one giving the least value should be used. The smallest net width obtained is multiplied by the plate thickness to give the net area A_n. Example 3-2 illustrates the method of computing the critical net area of a section which has three lines of bolts. (For angles, the gage for holes in opposite legs is considered to be the sum of the gages from the back of the angle minus the thickness of the angle.)

Holes for bolts and rivets are normally punched in steel angles at certain standard locations. These locations or gages are dependent on the angle-leg widths and on the number of lines of holes. Table 3-1, which is taken from Part 1 of the ASD Manual, shows these gages. It is unwise for the designer to require different gages from those given in the table unless unusual situations are present, because of the appreciably higher fabrication costs that will result.

---

[1] V. H. Cochrane, "Rules for Riveted Hole Deductions in Tension Members," *Engineering News-Record* (November 16, 1922), pp. 847–848.

TABLE 3-1 USUAL GAGES FOR ANGLES, IN INCHES

Leg	8	7	6	5	4	$3\frac{1}{2}$	3	$2\frac{1}{2}$	2	$1\frac{3}{4}$	$1\frac{1}{2}$	$1\frac{3}{8}$	$1\frac{1}{4}$	1
g	$4\frac{1}{2}$	4	$3\frac{1}{2}$	3	$2\frac{1}{2}$	2	$1\frac{3}{4}$	$1\frac{3}{8}$	$1\frac{1}{8}$	1	$\frac{7}{8}$	$\frac{7}{8}$	$\frac{3}{4}$	$\frac{5}{8}$
g_1	3	$2\frac{1}{2}$	$2\frac{1}{4}$	2										
g_2	3	3	$2\frac{1}{2}$	$1\frac{3}{4}$										

■ **Example 3-2**

Determine the critical net area of the $\frac{1}{2}$-in.-thick plate shown in Fig. 3-5, using the ASD Specification (B2). The holes are punched for $\frac{3}{4}$-in. bolts.

Solution.   The critical section could be *ABCD*, *ABCEF*, or *ABEF*. Hole diameters to be subtracted are $\frac{3}{4} + \frac{1}{8} = \frac{7}{8}$ in. The net widths for each case are

$$ABCD = 11 - (2)\left(\frac{7}{8}\right) = 9.25 \text{ in.}$$

$$ABCEF = 11 - (3)\left(\frac{7}{8}\right) + \frac{(3)^2}{(4)(3)} = 9.125 \text{ in.} \qquad \text{(controls)}$$

$$ABEF = 11 - (2)\left(\frac{7}{8}\right) + \frac{(3)^2}{(4)(6)} = 9.625 \text{ in.}$$

The reader should note that it is a waste of time to check route *ABEF* for this plate. Two holes need to be subtracted for routes *ABCD* and *ABEF*. As *ABCD* is a shorter route, it obviously controls over *ABEF*

$$A_n = (9.125)(\tfrac{1}{2}) = \underline{4.56 \text{ in.}^2} \qquad ■$$

The problem of determining the minimum pitch of staggered bolts such that no more than a certain number of holes need be subtracted to determine the net section is handled in Example 3-3. The ASD Handbook (page 4–99) has a chart entitled "Net Section of Tension Members," which can be used to

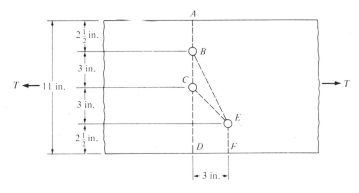

Figure 3-5

determine the values of $s^2/4g$. This chart can also be used to handle the type of problem solved in Example 3-3.

■ Example 3-3

For the two lines of bolt holes shown in Fig. 3-6 determine the pitch that will give a net area $DEFG$ equal to the one along ABC. The problem may also be stated as follows: determine the pitch that will give a net area equal to the gross area less one bolt hole. The holes are punched for $\frac{3}{4}$-in. bolts, thus $\frac{7}{8}$ in. is subtracted for each hole.

Solution.

$$ABC = 6 - (1)\left(\frac{7}{8}\right) = 5.125 \text{ in.}$$

$$DEFG = 6 - (2)\left(\frac{7}{8}\right) + \frac{s^2}{(4)(2)} = 4.25 + \frac{s^2}{8}$$

$$ABC = DEFG$$

$$5.125 = 4.25 + \frac{s^2}{8}$$

$$\underline{\underline{s = 2.65 \text{ in.}}}$$ ■

The $s^2/4g$ rule is merely an approximation or simplification of the complex stress variations which occur in members with staggered arrangements of bolts or rivets. Steel specifications can only provide minimum standards, and designers will have to apply such information logically to complicated situations which the specifications could not cover in their attempts at brevity and simplicity. The next few paragraphs present a discussion and numerical examples of the $s^2/4g$ rule applied to situations not specifically addressed in the ASD Specification.

The ASD Specification does not include a method to be used for determining the net widths of sections other than plates and angles. For channels, W sections, S sections, and others the web and flange thicknesses are not the same. As a result it is necessary to work with net areas rather than net widths. If the holes are placed in straight lines across such a member, the net area can

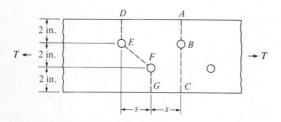

Figure 3-6

simply be obtained by subtracting the cross-sectional areas of the holes from the gross area of the member. If the holes are staggered, it is necessary to multiply the $s^2/4g$ values by the applicable thickness to change it to an area. Such a procedure is illustrated for a W section in Example 3-4 where bolts pass through the web only.

■ Example 3-4

Determine the net area of the W12×16 (A_g = 4.71 in.²) shown in Fig. 3-7, assuming the holes are for 1-in. bolts.

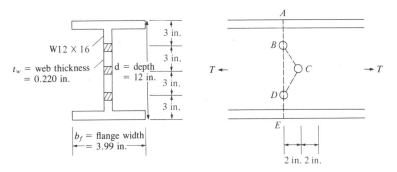

Figure 3-7

Solution.   Net areas:

$$ABDE = 4.71 - (2)(1\tfrac{1}{8})(0.220) = 4.21 \text{ in.}^2$$

$$ABCDE = 4.71 - (3)(1\tfrac{1}{8})(0.220) + (2)\frac{(2)^2}{(4)(3)}(0.220) = \underline{\underline{4.11 \text{ in.}^2}} \quad ■$$

If the zigzag line goes from a web hole to a flange hole, the thickness changes at the junction of the flange and web. In Example 3-5 the author has computed the net area of a channel that has bolt holes staggered in its flanges and web. The channel is assumed to be flattened out into a single plate as shown in Fig. 3-8(b) and (c). The net area along route *ABCDEF* is determined by taking the area of the channel minus the area of the holes along the route in the flanges and web plus the $s^2/4g$ values for each zigzag line times the appropriate thickness. For line *CD*, $s^2/4g$ has been multiplied by the thickness of the web. For lines *BC* and *DE* (which run from holes in the web to holes in the flange) an approximate procedure has been used in which the $s^2/4g$ values have been multiplied by the average of the web and flange thicknesses.

■ Example 3-5

Determine the net area along route *ABCDEF* for the C15×33.9 (A_g = 9.96 in.²) shown in Fig. 3-8. Holes are for $\tfrac{3}{4}$-in. bolts.

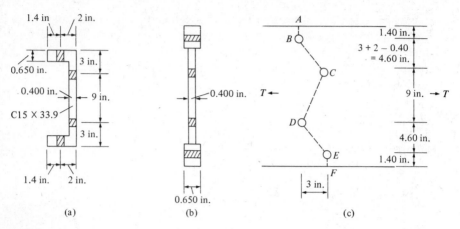

Figure 3-8

Solution.

Approximate net A along *ABCDEF*

$$= 9.96 - (2)\left(\frac{7}{8}\right)(0.650) - (2)\left(\frac{7}{8}\right)(0.400) + \frac{(3)^2}{(4)(9)}(0.400)$$

$$+ (2)\frac{(3)^2}{(4)(4.60)}\left(\frac{0.650 + 0.400}{2}\right)$$

$$= \underline{8.736 \text{ in.}^2} \qquad\qquad \blacksquare$$

3-5 EFFECTIVE NET AREAS

When a member other than a flat plate or bar is loaded in axial tension until failure occurs across its net section, its actual tensile failure stress will probably be less than the tensile strength of the steel *unless all of the various elements which make up the section are connected so stress is transferred uniformly across the section.* The reason for the reduced strength of the member is the concentration of shear stress, called *shear lag,* in the vicinity of the connection. In such a situation the flow of tensile stress between the full member cross section and the smaller connected cross section is not 100 percent effective. As a result the ASD Specification (B3) states that the effective net area A_e of such a member is to be determined by multiplying its net area (if bolted or riveted) or its gross area (if welded) by a reduction factor U. The use of a factor such as U accounts for the nonuniform stress distribution in a simple manner. An explanation of the way in which U factors are determined follows.

The angle shown in Fig. 3-9(a) is connected at its ends to only one leg. You can easily see that its area effective in resisting tension can be appreciably increased by shortening the width of the unconnected leg and lengthening the width of the connected leg as shown in Fig. 3-9(b).

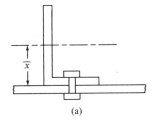
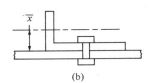

(a)   (b)

Figure 3-9

Investigators have found that a convenient measure of the effectiveness of a member such as an angle connected by one leg is the distance $\bar{x}$ measured from the plane of the connection to the centroid of the area of the whole section.[2,3] The smaller the value of $\bar{x}$, the larger is the effective area of the member.

The values of U in ASD Specification B3 were computed from the empirical expression $1 - \bar{x}/L$, where L is the length of the connection. The specification in effect reduces the length L of a connection with shear lag to a shorter effective length L'. The value of U then equals L'/L or $1 - \bar{x}/L$. Several values of $\bar{x}$ are shown in Fig. 3-10. The U values given in the specification are the approximate lower bounds of the results obtained when the $1 - \bar{x}/L$ expression is computed for different shear lag situations. Using this expression to determine U will often result in a more conservative value than is given in the specification. Should this happen the designer is permitted to use the computed value or the specification value.

To calculate U for a W section connected by its flanges only, we will assume first that the section is split into two structural tees. Then the value of $\bar{x}$ used in the $1 - \bar{x}/L$ expression will be the distance from the outside edge of the flange to the center of gravity of the structural tee as shown in Fig. 3-10(d).

Detailed ASD requirements for tension members for which all of the parts are not connected follow.

General

If the force is transmitted directly to each of the cross-sectional elements of a member by connectors, the effective net area A_e is equal to its net area A_n.

[2] E. H. Gaylord, Jr., and C. N. Gaylord, *Design of Steel Structures,* 2d ed. (New York: McGraw-Hill, 1972), pp. 119–123.

[3] W. H. Munse and E. Chesson, Jr., "Riveted and Bolted Joints: Net Section Design," *Journal of Structural Division,* ASCE, 89, ST2 (February 1963), pp. 107–126.

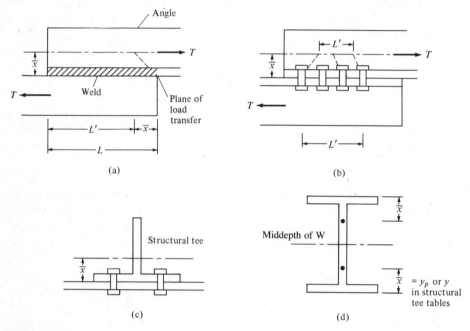

Figure 3-10

Bolted or Riveted Members

If the load is transmitted by bolts or rivets through some but not all of the elements of the member, the value of A_e is to be determined with the expression

$$A_e = UA_n \qquad \text{(ASD Equation B3-1)}$$

The following values of U (from ASD Manual, page 5-34) are to be used unless larger values can be justified on the basis of tests or recognized theory. It will be noted from the values shown that, as the number of fasteners in a line is increased, the shear lag decreases.

    a. W, M, or S shapes with flange widths not less than two-thirds the depth, and structural tees cut from these shapes, provided the connection is to the flanges. Bolted or riveted connections shall have no fewer than three fasteners per line in the direciton of stress $U = 0.90$
    b. W, M, or S shapes not meeting the conditions of subparagraph a, structural tees cut from these shapes and all other shapes, including built-up cross sections. Bolted or riveted connections shall have no fewer than three fasteners per line in the direction of stress $U = 0.85$
    c. All members with bolted or riveted connections having only two fasteners per line in the direction of stress $U = 0.75$

Welded Members

1. If the load is transmitted by welds through some but not all of the elements of a tension member, the effective net area is to be determined by multiplying the reduction coefficient U times the *gross area* of the member.

$$A_e = UA_g \qquad \text{(ASD Equation B3-2)}$$

The values of U to be used are the same as the ones for bolted or riveted members except that the provision about members having only two fasteners per line has no significance. In addition there are some special requirements for cases where loads are transmitted by transverse or longitudinal welds. These situations are described in paragraphs 2, 3 and 4.

2. If a tensile load is transmitted by transverse welds to some but not all of the elements of W, M, or S shapes and structural tees cut from those shapes, the effective net area A_e is to be set equal to the area of the directly connected parts.

3. Tests have shown that when flat plates or bars connected by longitudinal fillet welds (a term described in Chapter 13) are used as tension members they may fail prematurely by shear lag at the corners if the welds are too far apart.[4] Therefore, the ASD Specification states that when such situations are encountered the length of the welds may not be less than the width of the plates or bars and the effective net area is to equal UA_g. For such situations the following values of U are to be used (from ASD Manual, page 5-34):

When $L > 2w$, $U = 1.0$
When $2w > L > 1.5w$, $U = 0.87$
When $1.5w > L > w$, $U = 0.75$

where

L = weld length, in.

w = plate width (distance between welds), in.

4. When longitudinal welds are used along the edges of some but not all of the elements of steel shapes (other than plates) U is to be handled as it was for bolted or riveted members.

Example 3-6 illustrates the calculations necessary for determining the effective net area of a bolted W section connected only to its flanges. In addition the allowable tensile strength of the member is computed.

■ Example 3-6

Determine the allowable tensile load for a W10×45 with two lines of $\frac{3}{4}$-in.-diameter bolts in each flange, using A36 steel and the ASD Specification.

---

[4] G. L. Kulak, J. W. Fisher, and J. H. A. Struik, *Guide to Design Criteria for Bolted and Riveted Joints,* 2d ed. (New York: John Wiley & Sons, 1987).

There are at least three bolts in each line, and the bolts are not staggered with respect to each other.

Solution.   Using a W10×45 (A_g = 13.3 in.^{2}, d = 10.10 in., b_f = 8.020 in.)

(a) $T = 0.60F_y A_g = (0.60)(36)(13.3) = 287.3$ k $\leftarrow$

(b) $A_n = 13.3 - (4)(\frac{7}{8})(0.620) = 11.13$ in.^{2}

$U = 0.90$ since $b_f > \frac{2}{3}d$

$A_e = UA_n = (0.90)(11.13) = 10.02$ in.^{2}

$T = 0.50\,F_u A_e = (0.50)(58)(10.02) = 290.6$ k

Allowable $T = 287.3$ k (1278 kN)   ■

3-6 CONNECTING ELEMENTS FOR TENSION MEMBERS

When relatively short fittings such as splice or gusset plates are used as statically loaded tensile connecting elements, the A_n to be used to compute the allowable tensile load on the fitting may not exceed 85 percent of A_g. Tests have shown for decades that bolted or riveted tension connection elements rarely have an efficiency greater than 85 percent even if the holes represent a very small percentage of the gross area of the elements. In Example 3-7 the strength of a pair of tensile connecting plates is computed.

■ Example 3-7

The A36 tension member of Example 3-6 is assumed to be connected at its ends with two $\frac{3}{8}$×12-in. plates as shown in Fig. 3-11. If two lines of $\frac{3}{4}$-in. bolts are used in each plate, determine the maximum allowable tensile load that the plates can transfer.

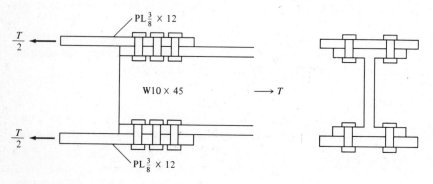

Figure 3-11

Solution.

$$T = 0.60F_y A_g = (0.60)(36)(2 \times \tfrac{3}{8} \times 12) = 194.4 \text{ k} \leftarrow$$
$$A_n \text{ of 2 plates} = (\tfrac{3}{8} \times 12 - \tfrac{7}{8} \times 2 \times \tfrac{3}{8}) 2 = 7.69 \text{ in.}^2$$
$$0.85 \, A_g = (0.85)(2 \times \tfrac{3}{8} \times 12) = 7.65 \text{ in.}^2 = A_n$$
$$T = 0.50F_u A_m = (0.50)(58)(7.65) = 221.8 \text{ k}$$
$$\underline{\underline{T = 194.4 \text{ k}}}$$ ∎

3-7 BLOCK SHEAR

The allowable tensile load of a member is not always controlled by $0.60F_y A_g$ or $0.50F_u A_e$ or by the allowable load on the bolts or welds with which the member is connected. It may instead be controlled by its allowable *block shear* strength as described in this section.

The failure of a member in block shear may occur along a path involving tension on one plane and shear on a perpendicular plane; Fig. 3-12 shows several possible block shear failures in tension members.

The ASD Specification (J4) states that the allowable block shear strength of a particular member is determined by computing the allowable shear stress $0.30F_u$ times the net shear area A_v plus the allowable tensile stress $0.50F_u$ times the net tension area A_t.

$$T_{bs} = 0.30F_u A_v + 0.50F_u A_t$$

The ASD Specification states that a similar calculation is made for the periphery of welded connections.

Examples 3-8 and 3-9 illustrate the determination of the allowable block shear strengths for two tension members. For consistency throughout this book, the bolt hole diameters are considered to be equal to the bolt diameter plus $\tfrac{1}{8}$ in. (even though the ASD Manual [page 4-8] uses fastener diameters plus $\tfrac{1}{16}$ in. for block shear calculations).

■ Example 3-8

The A36 tension member shown in Fig. 3-13 is connected with three $\tfrac{3}{4}$-in. bolts. Determine its allowable strength considering both block shear and tensile capacity.

Solution.   Block shear strength:

$$T = (0.30)(58)\left(10.00 - 2\tfrac{1}{2} \times \tfrac{7}{8}\right)\left(\tfrac{1}{2}\right) + (0.50)(58)\left(2\tfrac{1}{2} - \tfrac{1}{2} \times \tfrac{7}{8}\right)\left(\tfrac{1}{2}\right)$$

$$T = 97.9 \text{ k}$$

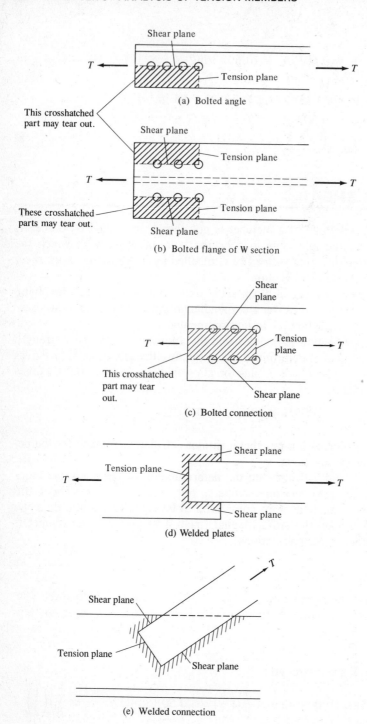

Figure 3-12 Block shear examples.

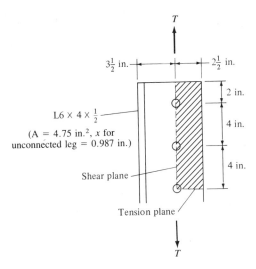

Figure 3-13

Tensile strength of angle:

$$T = 0.60F_y A_g = (0.60)(36)(4.75) = 102.6 \text{ k}$$
$$A_n = 4.75 - (1)(\tfrac{7}{8})(\tfrac{1}{2}) = 4.31 \text{ in.}^2$$
$$U = 1 - \frac{0.987}{8} = 0.88 \quad (0.85 \text{ given in ASD})$$
$$T = 0.50F_u A_e = (0.50)(58)(0.85 \times 4.31) = 106.2 \text{ k}$$
$$\underline{\underline{T = 97.9 \text{ k}}}$$ ■

■ Example 3-9

Determine the allowable design strength of the A36 welded member shown in Fig. 3-14. Include block shear.

Solution. Allowable block shear strength:

$$T = (0.30)(58)(2)(4 \times \tfrac{1}{2}) + (0.50)(58)(8 \times \tfrac{1}{2}) = 185.6 \text{ k}$$

Allowable tensile strength of plate:

$$T = (0.60)(36)(8 \times \tfrac{1}{2}) = 86.4 \text{ k} \leftarrow$$
$$\underline{\underline{T = 86.4 \text{ k}}}$$ ■

Sometimes cases are encountered where it is not altogether clear what sections should be considered for block shear calculations. For such situations designers will have to use their own judgment. One such case is shown in Fig.

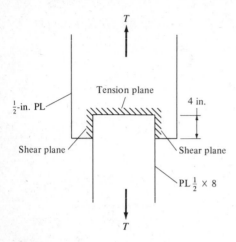

Figure 3-14

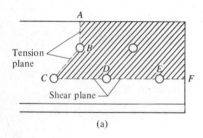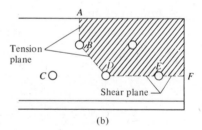

Figure 3-15

3-15. In part (a) of the figure it is assumed that web tear-out will occur along the lines *ABCDEF*. An alternate tear-out possibility for the same member along lines *ABDEF* is shown in part (b) of the figure. For this connection it is assumed that the load to be resisted is distributed equally among the five bolts. Thus, when web tear-out is considered for case (b), we will assume only $\frac{4}{5} T$ is carried by the section in question because one of the bolts is outside of the tear-out area. To compute the width of the tension planes *ABC* and *ABD* for these two cases, it seems logical to make use of the $s^2/4g$ expression presented in Section 3-4.

Block shear very seldom controls the design of tension members, but it may very well control the design of connections at the ends of beams with thin webs. This topic is continued in Chapters 13 and 14.

PROBLEMS (Use standard-size bolt holes for all problems)

3-1 to 3-7   Compute the net area of each of the given members.

3-1 An L8×4×$\frac{9}{16}$ with one line of $\frac{3}{4}$-in. bolts in each leg. (*Ans.* 5.45 in.^{2})

3-2 A pair of 9×4×$\frac{1}{2}$ Ls with two rows of $\frac{7}{8}$-in. bolts in the long legs and one row in the short legs.

3-3 A W21×101 with two holes in each flange and two in the web all for $\frac{3}{4}$-in. bolts. (*Ans.* 26.12 in.^{2})

3-4 The $\frac{7}{8}$×12 plate shown in the illustration. The holes are for $\frac{7}{8}$-in. bolts.

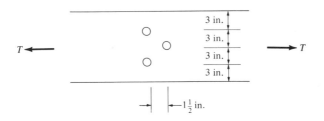

3-5 The $\frac{1}{2}$×9 plate shown in the illustration. The holes are for $\frac{3}{4}$-in. bolts. (*Ans.* 3.79 in.^{2})

3-6 The $\frac{3}{4}$×12 plate shown in the illustration. The holes are for $\frac{7}{8}$-in. bolts.

3-7 The 8×6×$\frac{1}{2}$ angle shown has one line of $\frac{3}{4}$-in. bolts in each leg. The bolts are 3 in. on center in each line and are staggered $1\frac{1}{2}$ in. with respect to each other. (*Ans.* 5.91 in.)

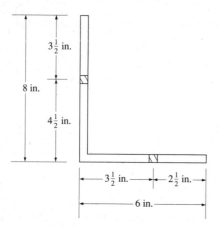

3-8 For the plate shown find the minimum pitch s for which only two bolts need be subtracted at any one section in calculating the net area. Bolts are $\frac{3}{4}$ in.

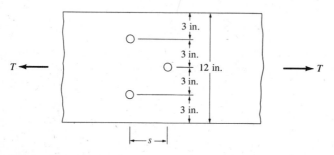

3-9 Find the minimum pitch s in the plate of Prob. 3-8 so that only two and one-half bolts need be subtracted at any one section. (*Ans.* 1.62 in.)

3-10 An L8×8×$\frac{3}{4}$ is used as a tension member with one gage line of 1-in. bolts in each leg at the usual gage location. What is the minimum amount of stagger necessary so that only one bolt need be subtracted from the gross area of the angle? Compute the net area of this member if the holes are staggered at 2 in.

3-11 An L7×4×$\frac{5}{8}$ is shown. Two rows of $\frac{7}{8}$-in. bolts are used in the long leg and one in the short leg. Determine the minimum stagger (or pitch s in the figure) necessary so that only two holes need be subtracted in determining the net area. (*Ans.* 2.67 in.)

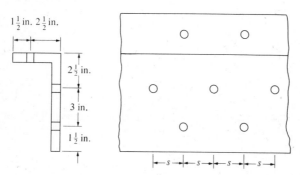

3-12 A $8 \times 6 \times \frac{9}{16}$ angle has one line of holes in each leg for $\frac{3}{4}$-in. bolts. Determine the minimum pitch so that only one and one-half holes need to be deducted to obtain the net area. (Use the usual gage for angles as given in the ASD Manual and Table 3-1 of this chapter.)

3-13 Determine the effective net area of the MC12×45 shown. Assume the holes are for 1-in. bolts. (*Ans.* 9.52 in.²)

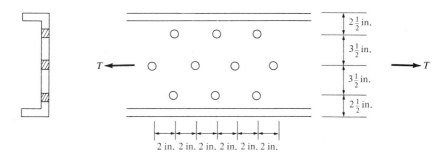

2 in. 2 in. 2 in. 2 in. 2 in. 2 in.

3-14 Determine the effective net cross-sectional area of the C15×50 shown. Holes are for $\frac{3}{4}$-in bolts.

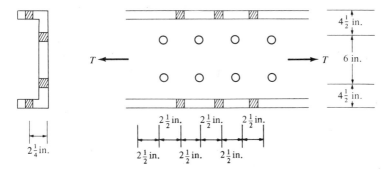

3-15 Compute the effective net area of the built-up section shown if the holes are punched for $\frac{7}{8}$-in. bolts. Assume at least three bolts in each line. (*Ans.* 19.30 in.²)

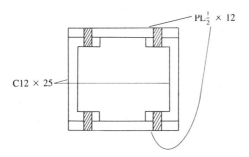

3-16 A C12×20.7 is connected through its web with three gage lines of $\frac{3}{4}$-in. bolts. The gage lines are 3 in. on centers, and the bolts are spaced 4 in. on centers along the gage line. If the center row of bolts is staggered with respect to the outer row, determine the effective net cross-sectional area of the channel. Assume there are at least three bolts in each line.

3-17 Using A36 steel determine the allowable tensile strength T of a W12×45 with two lines of $\frac{3}{4}$-in. bolts in each flange (three bolts in each line). Neglect block shear strength. (*Ans.* 285.1 k)

3-18 Determine the allowable tensile strength T of a W18×119 (A36 steel) if it has two lines of 1-in. bolts in each flange (at least three bolts in each line). Neglect block shear.

3-19 A single-angle tension member (7×4×$\frac{3}{4}$) has two gage lines in its long leg and one in the short leg for $\frac{3}{4}$-in. bolts arranged as shown. Determine the allowable tensile strength T of this member if A36 steel is used and if block shear is neglected. (*Ans.* 166.1 k)

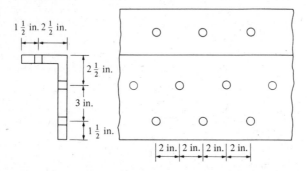

3-20 Determine the allowable tensile strength T of the pair of 6×6×$\frac{3}{4}$ angles shown, made from A242 steel. Standard gages are to be used as determined from Table 3-1 or the ASD Manual for the $\frac{3}{4}$-in. bolts. Neglect block shear. There are at least three bolts in each line.

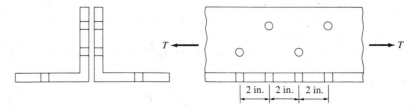

3-21 The 7×4×$\frac{5}{8}$ angle shown is connected with three 1-in. bolts. If the angle consists

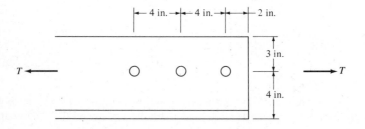

of A36 steel, determine its allowable block shear strength. Compare the results with the allowable design tensile strength of the member. (*Ans.* 122.3 k)

3-22 A W12×53 is connected at its ends with the plates shown. Determine the allowable block shear strength of the member if A36 steel is used and if it is connected with six $\frac{7}{8}$-in. bolts in each flange as shown. Compare the result with the allowable tensile design strength of the member. Do not check plate strength.

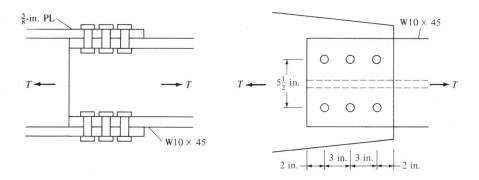

3-23 Repeat Prob. 3-17 if A242 steel is used and block shear is included. (*Ans.* 348.0 k)

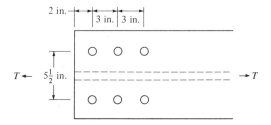

3-24 Compute the allowable tensile strength of the 6×6×$\frac{1}{2}$ angle shown if it consists of A36 steel. Consider block shear as well as the tensile strength of the angle.

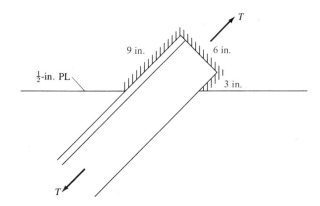

3-25 Compute the allowable tensile strength of the bolted connection shown. Include the allowable block shear strength as well as the allowable tensile strength of the angle. The angle is made of A36 steel and the bolts are $\frac{3}{4}$ in. (*Ans.* 72.0 k)

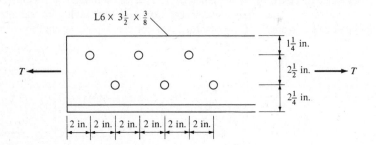

4

Design of Tension Members

4-1 SELECTION OF SECTIONS

The determination of the allowable design strengths of various tension members was presented in Chapter 3. In this chapter the selection of members to support given tension loads is described. Although the designer has considerable freedom in the selection, the resulting members should have the following properties: (a) compactness, (b) dimensions that fit into the structure with reasonable relation to the dimensions of the other members, and (c) connections to as many parts of the sections as possible to minimize shear lag.

The choice of member type is often affected by the type of connections used for the structure. Some steel sections are not very convenient to bolt together with the required gusset plates, while the same sections may be welded together with little difficulty. Tension members consisting of angles, channels, or W or S sections will probably be used when the connections are made with bolts. The same shapes may be used when the connections are welded, but plates, structural tees, and other shapes are also commonly used.

Various types of sections are selected for tension members in the examples to follow, and in each case where bolts are used some allowance is made for holes. Should the connections be made *entirely* by welding, no holes have to be added to the net areas to give the required gross area. *The student should realize, however, that very often welded members may have holes punched in them for temporary bolting during field erection before the permanent field welds are made. These holes need to be considered in tension member design.* It is also to be remembered that in the allowable tensile strength expression $(T = 0.50F_u A_e)$ the value of A_e may be less than A_g, depending on the arrangement of welds and on whether all of the parts of the members are connected.

The slenderness ratio of a member is the ratio of its unsupported length (L) to its least radius of gyration (r). Steel specifications give preferable maximum values of slenderness ratios for both tension and compression members. The purpose of such limitations for tension members is to ensure the use of sections with sufficient stiffness to prevent undesirable lateral deflections or vibrations. Although tension members are not subject to buckling under normal loads, stress reversal may occur during shipping and erection and perhaps due to wind or other loads. The specifications recommend that slenderness ratios be kept below certain maximum values in order that some minimum compressive strengths be provided in the members. For tension members other than rods the ASD Specification (B7) recommends maximum slenderness ratios of 300.

This same specification states that members that have been designed to support tensile loads but that experience some compression loading do not have to satisfy the preferable maximum effective slenderness ratio of 200 given by the ASD for compression members.

It should be noted that out-of-straightness does not affect the strength of tension members very much because the tension loads tend to straighten the members. (The same statement cannot be made for compression members.) For this reason the ASD Specification is a little more liberal in its consideration of tension members, including those subject to some compressive forces due to transient loads such as wind or earthquake.

The *recommended* maximum slenderness ratio of 300 is not applicable to tension rods. Maximum L/r values for rods are left to the designer's judgment. If a maximum value of 300 was specified for them, they would seldom be used because of their extremely small radii of gyration.

The 1989 AASHTO Specifications provide a maximum slenderness ratio of 120 for main compression members and 200 for main tension members, and these requirements are *mandatory*.

Example 4-1 illustrates the design of a bolted tension member with a W section, while Example 4-2 illustrates the selection of a bolted single-angle tension member. In both cases the ASD Specification is used. The allowable design strength T is the least of (a) $0.60F_y A_g$ or (b) $0.50F_u A_e$.

To satisfy the first of these expressions, the minimum gross area must be at least

$$\min A_g = \frac{T}{0.60F_y} \tag{1}$$

To satisfy the second expression the minimum value of A_e must be at least

$$\min A_e = \frac{T}{0.50F_u}$$

And since $A_e = UA_n$ the minimum value of A_n is

$$\min A_n = \frac{\min A_e}{U} = \frac{T}{0.50F_u U}$$

Then the minimum A_g for the second expression must at least equal the minimum value of A_n plus the estimated hole areas.

$$\min A_g = \frac{T}{0.50F_u U} + \text{estimated hole areas} \tag{2}$$

The designer can substitute into Equations (1) and (2), taking the larger value of A_g so obtained for an initial size estimate. It is, however, well to notice that the maximum preferable slenderness ratio L/r is 300. From this value it is easy to compute the preferable minimum value of r for a particular design, that is, the value of r for which the slenderness ratio will be exactly 300. It is undesirable to consider a section whose least r is less than this value, because its slenderness ratio would exceed the preferable maximum value of 300.

Transfer truss, 150 Federal Street, Boston, Mass. (Courtesy of Owen Steel Company, Inc.)

$$\min r = \frac{L}{300} \tag{3}$$

In Example 4-1 a W section is selected for a given set of tensile loads. For this first application of the tension design formulas, the author has narrowed the problem down to one series of W shapes so the reader can concentrate on the application of the formulas and not become lost in considering W8s, W10s,

Bridge over Allegheny River at Kittaning, Pa. (Courtesy of American Bridge Company.)

W14s, and so on. Exactly the same procedure can be used for trying these other series as is used here for the W12.

■ Example 4-1

Select a 30-ft-long W12 section of A36 steel to support a total tensile load of 220 k. As shown in Fig. 4-1, the member is to have two lines of bolts in each flange for $\frac{7}{8}$-in. bolts (at least three in a line).

Solution.   Computing the minimum A_g required:

$$\text{Min } A_g = \frac{T}{0.60F_y} = \frac{220}{(0.60)(36)} = 10.19 \text{ in.}^2$$

$$\text{Min } A_g = \frac{T}{0.50F_u U} + \text{estimated hole areas}$$

Assume $U = 0.90$ and assume flange thickness is about 0.515 in. after looking at W12 sections in the ASD Manual, which have areas of about 10.19 in.^{2}

$$\text{Min } A_g = \frac{220}{(0.50)(58)(0.90)} + (4)(\tfrac{7}{8} + \tfrac{1}{8})(0.515) = 10.49 \text{ in.}^2$$

$$\text{Preferable min } r = \frac{L}{300} = \frac{(12)(30)}{300} = 1.2 \text{ in.}$$

$$\underline{\text{Try W12} \times 40\ (A_g = 11.8 \text{ in.}^2,\ d = 11.94 \text{ in.},}$$
$$\underline{b_f = 8.005 \text{ in.},\ t_f = 0.515 \text{ in.},\ r_y = 1.93 \text{ in.})}$$

Checking:

$$T = 0.60F_y A_g = (0.60)(36)(11.8) = 254.9 \text{ k} > 220 \text{ k} \qquad \text{OK}$$

$$T = 0.50\ F_u U A_n \text{ with } U = 0.90 \text{ since } \frac{b_f}{d} > \frac{2}{3},$$

$$\text{and } A_n = 11.8 - (4)(1.00)(0.515) = 9.74 \text{ in.}^2$$

$$T = (0.50)(58)(0.90)(9.74) = 254.2 \text{ k} > 220 \text{ k} \qquad \text{OK}$$

$$\frac{L}{r} = \frac{(12)(30)}{1.93} = 187 < 300 \qquad \text{OK}$$

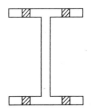

Figure 4-1

A subsequent check shows a W12 × 35 will not do.

Use W12 × 40   ∎

In Example 4-2 a broader example is presented: the lightest satisfactory angle in the ASD Manual is selected for a given set of tensile loads.

∎ Example 4-2

Design a 9-ft single-angle tension member to support a total tensile load of 70 k. The member is to be connected to one leg only with $\frac{7}{8}$-in. bolts (at least three in a line). Assume that only one bolt is to be located at any one cross section. Use A36 steel and the ASD Specification.

Solution.   There are many different angles listed in the ASD Manual which will support the load for the conditions described. As a result it may seem rather difficult to arrive at the absolutely lightest satisfactory section. It is possible, however, to set up a table with which the various possible sections may be methodically considered for various angle thicknesses. In the table presented with this solution the author has listed various possible thicknesses of angles which will provide sufficient areas. He has then found the lightest angle for each thickness. Finally a glance at the completed table reveals the lightest angle or the one with the smallest cross-sectional area. A similar process could be followed for designing tension members from other steel sections.

The limiting values are calculated as follows.

$$\text{Min } A_g \text{ required} = \frac{T}{0.60F_y} = \frac{70}{(0.60)(36)} = 3.24 \text{ in.}^2$$

From Section B3 of Manual, $U = 0.85$

$$\text{Min } A_n \text{ required} = \frac{T}{0.50F_u U} = \frac{70}{(0.50)(58)(0.85)} = 2.84 \text{ in.}^2$$

$$\text{Min } r = \frac{L}{300} = \frac{(12)(9)}{300} = 0.36 \text{ in.}$$

ANGLE t (IN.)	AREA OF ONE 1-IN. HOLE (IN.²)	GROSS AREA REQUIRED[a] (IN.²)	LIGHTEST ANGLES AVAILABLE, THEIR AREAS (IN.²) AND LEAST RADII OF GYRATION (IN.)
$\frac{5}{16}$	0.312	3.24	$6 \times 6 \times \frac{5}{16}$ ($A = 3.65$, $r = 1.20$)
$\frac{3}{8}$	0.375	3.24	$6 \times 3\frac{1}{2} \times \frac{3}{8}$ ($A = 3.42$, $r = 0.767$)
$\frac{7}{16}$	0.438	3.28	$5 \times 3 \times \frac{7}{16}$ ($A = 3.31$, $r = 0.651$) $4 \times 4 \times \frac{7}{16}$ ($A = 3.31$, $r = 0.785$)
$\frac{1}{2}$	0.500	3.34	$4 \times 3\frac{1}{2} \times \frac{1}{2}$ ($A = 3.50$, $r = 0.722$)

[a] Larger of $T/0.60F_y$ and $T/0.50 F_u$, plus hole area.

Use L4 × 4 × $\frac{7}{16}$ or L5 × 3 × $\frac{7}{16}$ ∎

4-2 BUILT-UP TENSION MEMBERS

Section D2 of the ASD Specification provides a definite set of rules describing how the different parts of built-up tension members are to be connected together.

1. When a tension member is built up from elements in continuous contact with each other such as a plate and a shape, or two plates, the longitudinal spacing of connectors between those elements must not exceed 24 times the thickness of the thinner plate or 12 in. if the member is to be painted or if it is not to be painted and not to be subjected to corrosive conditions.

2. Should the member consist of unpainted weathering steel elements in continuous contact, the maximum permissible connector spacings are 14 times the thickness of the thinner plate or 7 in.

3. The longitudinal spacing of fasteners and intermittent welds connecting two or more steel shapes may not be greater than 24 in.

4. Should a tension member be built up from two or more shapes separated by intermittent fillers, the shapes must be connected to each other at intervals such that the slenderness ratio of the individual shapes between the fasteners does not exceed 300.

Example 4-3 reviews a tension member that is built up from two channels that are separated from each other. Included in the problem is the design of tie plates or tie bars to hold the channels together as shown in Fig. 4-2(b). Section D2 of the ASD Specification states that tie plates (or perforated cover plates) must be used on the open sides of built-up tension members. In this specification empirical rules for the design of ties are presented. These rules are based on many decades of experience with built-up tension members.

In Fig. 4-2 the location of the bolts, that is, the usual gage for these channels, is shown as being $1\frac{3}{4}$ in. from the back of the channels. The ASD Manual does not provide the usual gages except for angles. For other rolled steel shapes such as Cs, Ws, and Ss it is necessary to refer to a manufacturer's catalog or to old AISC Steel Manuals. The gages necessary to solve problems in this text are furnished by the author. (Though rather standard values of these gages are used by the steel industry for the various rolled shapes, they are not specified in the ASD Manual, in order to give steel fabricators more freedom in placing the holes.)

In Fig. 4-2 the distance between the lines of bolts connecting the tie plates to the channels equals 8.50 in. The ASD Specification (D2) states that the length of tie plates (lengths are always measured parallel to the long direction of the members in this text) may not be less than two-thirds the distance between the lines of connectors. Furthermore their thickness may not be less than one-fiftieth of this distance.

The minimum permissible width of tie plates (not mentioned in the specification) is the width between the lines of connectors plus the necessary edge distance on each side to keep the bolts from splitting the plate. For this example this minimum edge distance is taken as $1\frac{1}{2}$ in. from Table J3.5 of the ASD Specification. (Detailed information concerning edge distances for bolts

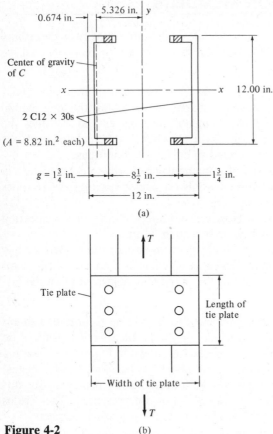

Center of gravity of C

2 C12 × 30s

(A = 8.82 in.² each)

$g = 1\frac{3}{4}$ in.

0.674 in.

5.326 in.

x — x 12.00 in.

$8\frac{1}{2}$ in. $1\frac{3}{4}$ in.

12 in.

(a)

Tie plate

Length of tie plate

Width of tie plate

T

(b)

Figure 4-2

and rivets is provided in Chapter 11.) The plate dimensions are rounded off to agree with the plate sizes available from the steel mills as given in the Bars and Plates section of Part 1 of the ASD Manual. It is usually cheaper to select standard thicknesses and widths rather than picking odd ones that will require cutting or other operations.

The ASD Specification (D2) provides a desirable maximum spacing between tie plates by stating that the L/r of each individual component of a built-up member running along by itself between tie plates preferably should not exceed 300. If the designer substitutes into this expression ($L/r = 300$) the least r of an individual component of the built-up member, the value of L may be computed. This will be the maximum spacing of the tie plates permitted by the ASD Specification for this member.

■ Example 4-3

Two C12 × 30s (see Fig. 4-2) have been selected to support a total tensile load of 360 k. The member is 30 ft long, consists of A36 steel, and has one line of at least three $\frac{7}{8}$-in. bolts in each channel flange. Using the ASD

Specification, determine if the member is satisfactory and design the necessary tie plates. Assume centers of bolt holes are $1\frac{3}{4}$ in. from the backs of the channels.

Solution.   Using C12 × 30s ($A_g = 8.82$ in.^{2} each, $t_f = 0.501$ in., $I_x = 162$ in.^{4} each, $I_y = 5.14$ in.^{4} each, y axis 0.674 in. from back of C, $r_y = 0.763$ in.)

Allowable tensile load:

$$T = 0.60F_y A_g = (0.60)(36)\ (2 \times 8.82) = 381.0\text{ k} > 360\text{ k} \qquad \text{OK}$$
$$A_n = [8.82 - (2)(1.0)(0.501)]\ 2 = 15.64\text{ in.}^2$$
$$T = 0.50F_u A_n U = (0.50)(58)(15.64)(0.85) = 385.5\text{ k} > 360\text{ k} \qquad \text{OK}$$

Slenderness ratio:

$$I_x = (2)(162) = 324\text{ in.}^4$$
$$I_y = (2)(5.14) + (2)(8.82)(5.33)^2 = 511\text{ in.}^4$$
$$r_x = \sqrt{\frac{324}{(2)(8.82)}} = 4.29\text{ in.}$$
$$\frac{L}{r} = \frac{(12 \times 30)}{4.29} = 83.9 < 300$$

Design of tie plates (ASD Specification D2):

Distance between lines of bolts $= 12.00 - (2)(1\frac{3}{4}) = 8.50$ in.

Minimum length of tie plates $= (\frac{2}{3})(8.50) = 5.67$ in. (say 6 in.)

Minimum thickness of tie plates $= (\frac{1}{50})(8.50) = 0.17$ in. (say $\frac{3}{16}$ in.)

Minimum width of tie plates $= 8.50 + (2)(1\frac{1}{2}) = 11.5$ in. (say 12 in.)

(ASD Specification Table J3.5)

Maximum preferable spacing of tie plates

least r of one C = 0.763 in.

Maximum preferable $\dfrac{L}{r} = 300$

$$\frac{(12)(L)}{0.763} = 300$$

$$L = 19.08\text{ ft}$$

Use $\frac{3}{16}$ × 6 × 1 ft 0 in. tie plates 15 ft 0 in. on center   ∎

4-3 RODS AND BARS

When rods and bars are used as tension members, they may be simply welded at their ends, or they may be threaded and held in place with nuts. The ASD nominal tensile design stress for threaded rods is given in their Table J3.2. It equals $0.33\ F_u$ and is to be applied to the gross area of the rod A_D computed with the major thread diameter, that is, the diameter to the outer extremity of the thread. The area required for a particular tensile load can then be calculated from the following expression:

$$A_D = \frac{T}{0.33F_u}$$

In the table "Threaded Fasteners Screw Threads" the ASD Manual (Part 4) presents properties of standard threaded rods. Example 4-4 illustrates the selection of a rod using this table. Note that the ASD Specification (J1.6) states that the tensile load T used for connection design may not be less than 6 k except for lacing, sag rods, or girts.

■ Example 4-4

Using A36 steel and the ASD Specification, select a threaded rod of A36 steel to support a total tensile load of 30 k.

Solution.

$$A_D = \frac{T}{0.33F_u} = \frac{30}{(0.33)(58)} = 1.57 \text{ in.}^2$$

Use $1\frac{1}{2}$-in.-diameter rod with 6 threads per in. ($A_D = 1.767 \text{ in.}^2$)   ■

Sometimes upset rods are used, where the ends are made larger than the regular rod and the threads are placed in the enlarged section so that the area at the root of the thread is larger than that of the regular bar. As a footnote to Table J3.2 the ASD Specification states that the allowable tensile capacity of the upset end is equal to $0.33F_u$ times the cross-sectional area (A_b) at its major thread diameter. This value must be larger than the nominal body area of the rod before upsetting times $0.60F_y$. Upsetting permits the designer to use the entire area of the cross section; however, the use of upset bars is probably not economical and should be avoided unless a large order is being made.

A common example of the use of tension rods occurs in steel-frame industrial buildings which have purlins running between their roof trusses to support the roof surface. These types of buildings will also frequently have girts running between the columns along the vertical walls. Sag rods may be required to provide support for the purlins parallel to the roof surface and vertical support for the girts along the walls. For roofs with steeper slopes than 1 vertically to 4 horizontally, sag rods are often considered necessary to provide lateral support for the purlins, particularly where the purlins consist of steel channels. Steel channels are commonly used as purlins, but they have very little resistance to lateral bending. Although the resisting moment needed parallel to the roof surface is small, an extremely large channel is required to provide such a moment. The use of sag rods for providing lateral support to purlins made from channels may be economical because of the bending weakness of channels about their y axes. For light roofs (as where trusses support corru-

New Albany Bridge crossing the Ohio River between Louisville, Ky., and New Albany, Ind. (Courtesy of Lincoln Electric Company.)

gated steel roofs) sag rods will probably be needed at the one-third points if the trusses are more than 20 ft on centers. Sag rods at the midpoints are sufficient if the trusses are less than 20 ft on centers. For heavier roofs such as those made of slate, cement tile, or clay tile, sag rods will probably be needed at closer intervals. The one-third points will probably be necessary if the trusses are spaced at greater intervals than 14 ft and the midpoints will be satisfactory if truss spacings are less than 14 ft. Some designers assume that the load components parallel to the roof surface can be taken by the roof, particularly if it consists of corrugated steel sheets, and that tie-rods are unnecessary. This assumption, however, is open to some doubt and definitely should not be followed if the roof is very steep.

Designers have to use their own judgment in limiting the slenderness values for rods, as they will usually be several times the limiting values mentioned for other types of tension members. A common practice of many designers is to use rod diameters no less than $\frac{1}{500}$ of their lengths to obtain some rigidity even though design calculations may permit much smaller sizes.

It is usually desirable to limit the minimum size of sag rods to $\frac{5}{8}$ in. because smaller rods than these are often injured during construction. The threads on smaller rods are quite easily injured by overtightening, which seems to be a frequent habit of workers. Sag rods are designed for the purlins of a roof truss in Example 4-5. The rods are assumed to support the simple beam reactions for the components of the gravity loads (roofing, purlins, snow and ice) parallel to the roof surface. Wind forces are assumed to act perpendicular to the roof surfaces and theoretically will not affect the sag rod forces. The maximum force in a sag rod will occur in the top sag rod because it must support the sum of the forces in the lower sag rods. It is theoretically possible

to use smaller rods for the lower sag rods, but this reduction in size will probably be impractical.

In Fig. 4-3 notice how the sag rods are offset about 6 in. from each other
in the plan view to accommodate their installation.

■ Example 4-5

Design the sag rods for the purlins of the truss shown in Fig. 4-3. Purlins are
to be supported at their one-third points between the trusses, which are spaced
21 ft on centers. Use A36 steel and assume a minimum size rod of $\frac{5}{8}$ in. is permitted. A clay tile roof weighing 16 psf (0.77 kN/m^2) of roof surface is used
and supports a snow load of 20 psf (0.96 kN/m^2) of horizontal projection of
roof surface. Details of the purlins and the sag rods and their connections are
shown in Figs. 4-3 and 4-4. In Fig. 4-3 the dotted lines represent a common

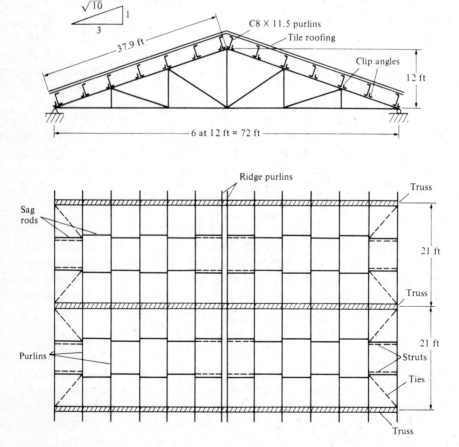

Figure 4-3 Plan of two bays of roof.

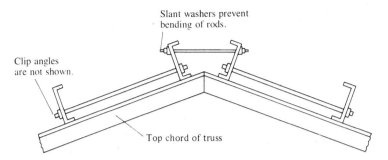

Figure 4-4 Details of sag rod connections.

practice of using ties and struts in the end panels in the plane of the roof to give greater resistance to loads located on one side of the roof (a loading situation which might occur when snow is blown off one side of the roof during a severe windstorm).

Solution.   Gravity loads in psf of roof surface:

$$\text{Purlins} = \frac{7 \times 11.5}{37.9} = 2.1 \text{ psf}$$

$$\text{Snow} = 20\frac{3}{\sqrt{10}} = 19.0 \text{ psf}$$

$$\text{Tile roofing} = \frac{16.0 \text{ psf}}{37.1 \text{ psf}}$$

$$\text{Component of loads} \perp \text{to roof surface} = \left(\frac{1}{\sqrt{10}}\right)(37.1) = 11.7 \text{ psf}$$

$$\text{Load on sag rod} = (37.9)(7)(11.7) = 3110 \text{ lb}$$

$$A_D = \frac{3.110}{(0.33)(58)} = 0.162 \text{ in.}^2$$

$$\underline{\text{Use } \tfrac{5}{8}\text{-in. rod (minimum practical size),}}$$
$$\underline{\text{11 threads per in. } (A_D = 0.307 \text{ in.}^2)}$$

Force in tie-rod between ridge purlins:

$$T = \frac{\sqrt{10}}{3}(3110) = 3270 \text{ lb}$$

$$A_D = \frac{3.270}{(0.33)(58)} = 0.171 \text{ in.}^2$$

$$\underline{\text{Use } \tfrac{5}{8}\text{-in. rod}}$$

∎

4-4 PIN-CONNECTED MEMBERS

Until the early years of the twentieth century nearly all bridges in the United States were pin-connected, but today pin-connected bridges are used infrequently because of the advantages of bolted and welded connections. One trouble with the old pin-connected trusses was the wearing of the pins in the holes, which caused looseness of the joints.

Pin-connected eyebars are used occasionally today as tension members for long-span bridges and as hangers for some types of bridges and other structures where they are normally subjected to very large dead loads. As a result the eyebars are usually prevented from rattling and wearing under live loads.

Eyebars are generally made not by forging but by thermally cutting them from plates. As stated in the ASD Commentary (D3), extensive testing has shown that thermally cut members result in more balanced designs. The heads of eyebars are specially shaped so as to provide optimum stress flow around the holes. These proportions are based on long experience and testing with forged eyebars and the results are rather conservative for today's thermally cut members.

The ASD Specification (D3) provides detailed requirements for pin-connected members as to allowable stresses and proportions of the pins and plates. An allowable stress of $0.60F_y$ is permitted on the gross areas of members, while $0.45F_y$ is permitted on the net areas through the pin holes.

It is also necessary to check these members for the bearing produced by the contact of the pins. An average bearing stress is determined by dividing the total tensile load in a member by its projected rectangular pin area (equal to the pin diameter times the plate thickness at the hole). The ASD Specification (J8) provides for this situation an allowable bearing stress of $0.90\ F_y$. If the parts in contact have different yield stresses, F_y will be the smaller value.

In addition to the other requirements mentioned, the ASD specifies certain proportions between the pins and the eyebars. These values are based on long experience in the steel industry, and experimental work by B. G. Johnston.[1]

It has been found that when eyebars and pin-connected members are made from steels with yield stresses greater than 70 kips per square inch (ksi) there may be a possibility of "dishing" (a complicated inelastic stability failure where the head of the eyebar tends to curl laterally into a dish-like shape). For this reason the ASD Specification requires stockier member proportions for such situations (hole diameter not to exceed five times the plate thickness and the width of the eyebar reduced accordingly).

---

[1] B. G. Johnston, "Pin-Connected Plate Links," *Transactions ASCE* 104 (1939).

4-5 DESIGN FOR FATIGUE LOADS

It is not likely that fatigue stresses will be a problem in the average building frame because the changes in load in such structures usually occur only occasionally and produce relatively minor stress variations. Should there be frequent variations or even reversals of stress, however, the matter of fatigue must be considered. Fatigue can be a problem in buildings when crane runway girders or heavy vibrating or moving machinery or equipment is supported.

If steel members are subjected to loads that are applied and then removed or changed significantly many thousands of times, cracks may occur and may spread so much as to cause fatigue failures. These cracks tend to occur where there are stress concentrations, such as holes or rough or damaged edges or poor welds. *Furthermore, they are more likely to occur in tension members.*

The ASD Specification Appendix K provides a simple design method for considering repeated loads. With this procedure the number of stress cycles, the expected range of stress (that is, the difference between the maximum and minimum estimated stresses), and the type and location of the member are considered. Based on this information a maximum allowable-stress range is given *for service or working loads.*

In detail the maximum calculated stress in a member by the ASD Specification may not exceed the basic allowable stress in that type of member, nor may its maximum range of stress exceed the permissible stress range provided in Appendix K of the Specification.

If it is anticipated that there will be less than 20,000 cycles of loading during the estimated life of the structure (normally 25 years), no consideration needs to be given to fatigue. If the number of loading cycles exceeds 20,000, a permissible stress range is determined as follows:

1. The loading condition is determined from Table A-K4.1 of Appendix K of the ASD Specification. For instance, if it is anticipated that there will be no less than 100,000 cycles of loading (that is, roughly ten applications a day for 25 years) and no more than 500,000 cycles, loading condition 2 is to be used as determined from the table.

2. The type and location of material is selected from Fig. A-K4 of the Appendix to the ASD Specification. If a tension member consists of two angles fillet-welded to a plate, it should be classified as illustrative example 17 in the figure. (The term *fillet weld* will be defined in Chapter 13. In such a weld one member is lapped over another, and they are welded together.)

3. From Table A-K4.2 the stress category A, B, C, D, E, or F is selected. For a fillet-welded tension connection classified as illustrative example 17, the stress category is E.

4. Finally from Table A-K4.3 of the Appendix the allowable-stress range for the service loads for stress category E and loading condition 2 is $F_{sr} = 13$ ksi.

Example 4-6 presents the design of a two-angle tension member subjected to fluctuating loads, using Appendix K of the ASD Specification. Stress fluctuations and reversals are an everyday problem in the design of bridge structures. The 1989 AASHTO Specifications Article 10.3 provide allowable-stress ranges determined in a manner very similar to those of the ASD Specification.

■ Example 4-6

An 18-ft member is to consist of a pair of equal-leg angles with fillet-welded end connections. The dead-load working tensile force is 30 k, while it is estimated that the working live-load force may be applied 250,000 times and may vary from a compression of 12 k to a tension of 65 k. Select the angles, using A36 steel and the ASD Specification.

Solution. Referring to Appendix K of the ASD Specification and selecting the following values:

From Table A-K4.1—loading condition 2

From Fig. A-K4.1 and Table A-K4.2—illustrative example 17 and stress category E

Therefore, from Table A-K4.3—allowable stress range is 13 ksi

Range of loads:
For maximum tension:

$$T = +30 + 65 = +95 \text{ k}$$

For compression:

$$C = +30 - 12 = +18 \text{ k}$$

Therefore, member remains in tension.
Estimated section size:

$$F_t = 0.60F_y$$

$$A_{reqd} = \frac{95}{(0.60)(36)} = 4.40 \text{ in.}^2$$

Try 2 L4 × 4 × $\frac{5}{16}$s ($A = 4.80$ in.^{2})

$$\text{Max tension } f_t = \frac{30 + 65}{4.80} = 19.79 \text{ ksi}$$

$$\text{Min tension } f_t = \frac{30 - 12}{4.80} = 3.75 \text{ ksi}$$

Actual stress range = 16.04 ksi > 13 ksi   NG (no good)

Try 2 L4 × 4 × $\frac{1}{2}$s ($A = 7.50$ in.^{2})

$$\text{Max } f_t = \frac{30 + 65}{7.50} = 12.67 \text{ ksi}$$

$$\text{Min } f_t = \frac{30 - 12}{7.50} = \underline{2.40 \text{ ksi}}$$

Actual stress range $= 10.27$ ksi < 13 ksi   OK

Use 2 L4 $\times$ 4 $\times \frac{1}{2}$s   ■

PROBLEMS

4-1 Select the lightest W12 section available to support a tensile load of 420 k using A36 steel. The member is to be 25 ft long and is assumed to have two lines of holes for 1-in. bolts in each flange. There will be at least three bolts in each line. (*Ans.* W12 $\times$ 72)

4-2 Repeat Prob. 4-1 using A441 steel.

4-3 Select the lightest W14 available with A36 steel to support a tensile load of 300 k. Assume there are two lines of $\frac{7}{8}$-in. bolts in each flange (at least three bolts in each line). The member is to be 32 ft long. (*Ans.* W14 $\times$ 53)

4-4 Select the lightest American Standard channel that will safely support a tensile load of 175 k. The member is 15 ft long and is assumed to have one line of holes for 1-in. bolts in each flange. Use A36 steel and assume there are at least three holes in each line.

4-5 A welded tension member is to support a tensile load $T = 430$ k and is to consist of two channels placed 12 in. back-to-back with the flanges turned in. Select the lightest standard channels available using A36 steel. Assume $U = 0.87$. The member is to be 30 ft long. (*Ans.* 2C15 $\times$ 33.9s)

4-6 to 4-11   Select a section as specified, assuming there are two lines of bolts in each flange.

PROBLEM	SECTION	TENSILE LOAD (K)	F_y (KSI)	F_u (KSI)	LENGTH (FT)	BOLT DIAMETER (IN.)	BOLTS IN EACH LINE
4-6	W14 $\times$?	400	36	58	20	$\frac{7}{8}$	3
4-7	W12 $\times$?	300	36	58	18	$\frac{7}{8}$	3
4-8	W14 $\times$?	600	36	58	12	1	2
4-9	W10 $\times$?	250	50	70	16	$\frac{3}{4}$	3
4-10	W8 $\times$?	150	50	70	12	$\frac{3}{4}$	3
4-11	S $\times$?	200	36	58	14	$\frac{7}{8}$	3

(*Ans.* Prob. 4-7: W12 $\times$ 50 slightly overstressed, or W12 $\times$ 53; Prob. 4-9: W10 $\times$ 33; Prob. 4-11: S10 $\times$ 35 or S12 $\times$ 35)

4-12 A36 steel is to be used in selecting a single-angle member to support a tensile load of 170 k. The member is to be 18 ft long and is assumed to be connected with one line of four $\frac{7}{8}$-in. bolts in one leg.

4-13 Select a single-angle tension member to resist a tensile load of 60 k. The member is to be 15 ft long and is to be connected with one line of four $\frac{7}{8}$-in. bolts. Assume $F_y = 40$ ksi and $F_u = 60$ ksi. (*Ans.* L4 × 4 × $\frac{3}{8}$ or L5 × 3 × $\frac{3}{8}$)

4-14 Repeat Prob. 4-12 using a pair of angles and A36 steel. Assume angle legs are touching and assume one hole for a $\frac{7}{8}$-in. bolt is to be taken out of each angle. Also assume $U = 0.85$.

4-15 Select a W12 for member $L_2 L_3$ of the truss shown. Use A36 steel and assume there are two lines of three $\frac{3}{4}$-in. bolts in each flange. (*Ans.* W12 × 30)

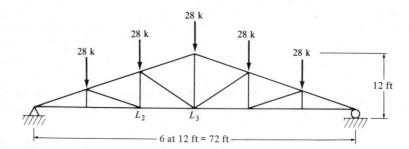

4-16 Select an ST section using A36 steel to serve as member $L_0 U_3$ in the truss shown. Assume the member is to be connected through its web with one line of three or more $\frac{7}{8}$-in. bolts.

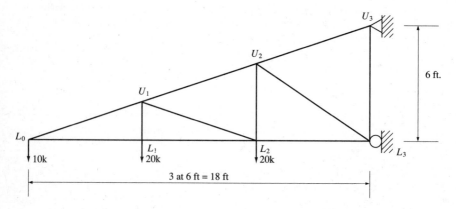

4-17 Select an ST for member $U_1 L_1$ of the truss of Prob. 4-16 using A36 steel and assuming the load is transmitted to the member by welds. Assume $U = 0.85$. (*Ans.* ST1.5 × 3.75)

4-18 Repeat Prob. 4-5 assuming that one line of $\frac{7}{8}$-in. bolts will be used in each flange with at least three bolts in each line. Also design tie plates. Assume distance or gage from back of each channel to center of line of bolts is 2 in. U is to be determined from ASD Specification B3.

4-19 A tension member is to consist of four equal-leg angles arranged as shown to support a tensile load of 560 k. The member is assumed to be 30 ft long and is to have one line of three $\frac{7}{8}$-in. bolts in each angle. Design the member with A36 steel including the necessary tie plates. (*Ans.* 4 L5 × 5 × $\frac{3}{8}$s; tie PL$\frac{1}{4}$ × 8 × 1 ft 3 in., 15 ft 0 in. on centers)

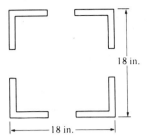

18 in.

18 in.

4-20 Select a threaded round rod as a hanger to resist a tensile load of 15 k using A36 steel.

4-21 Repeat Problem 4-20 if the tensile load is 120 k. (*Ans.* 3-in.-diameter rod with 4 threads per in.)

4-22 The horizontal thrust at the base of the three-hinged arch shown is to be resisted by a tie-rod of A36 steel. What size threaded round tie-rod should be used if the arch supports the working loads shown?

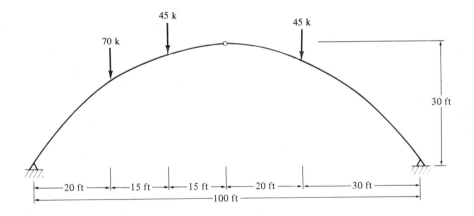

45 k

45 k

70 k

30 ft

20 ft — 15 ft — 15 ft — 20 ft — 30 ft

100 ft

4-23 Select a threaded steel rod for member *AB* using A36 steel. (*Ans.* $2\frac{3}{4}$-in.-diameter rod with 4 threads per in.)

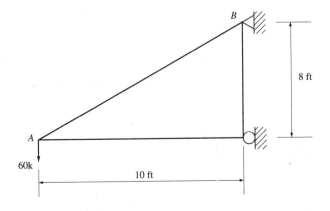

B

8 ft

A

60k

10 ft

4-24 The roof trusses for a particular industrial building are spaced 21 ft on centers, have a roof covering weighing an estimated 6 psf of roof surface, and have purlins spaced as shown and weighing an estimated 3 psf of roof surface. Design sag rods with A36 steel, assuming a snow load of 25 psf of horizontal roof surface. The sag rods are to be used at the one-third points.

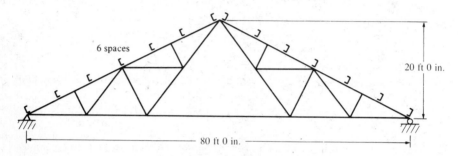

4-25 A W12 section consisting of A36 steel is to be selected to support a tensile dead load of 80 k and a live load that is estimated to vary from a tension of 60 k to a compression of 40 k. Assume that it can be applied 300,000 times. The loads are applied axially through full penetration groove welds. (*Ans.* W12 × 22)

4-26 A double-angle tension member is to be selected to support a tensile dead load of 30 k and a live load that is estimated to vary from 70 k tension to a compression of 30 k. The angles, which are to be separated by $\frac{3}{8}$ in., are to be $\frac{1}{2}$ in. thick, arranged with long legs back-to-back and connected with two lines of fully tensioned $\frac{7}{8}$-in bolts at least three in a line. Use A36 steel. Assume 750,000 loading cycles.

5

Introduction to Axially Loaded Compression Members

5-1 GENERAL

There are several types of compression members, the column being the best known. Among the other types are the top chords of trusses, bracing members, the compression flanges of rolled beams and built-up beam sections, and members that are subjected simultaneously to bending and compressive loads. Columns are usually thought of as being straight vertical members whose lengths are considerably greater than their thicknesses. Short vertical members subjected to compressive loads are often called struts or simply compression members; however, the terms *column* and *compression member* will be used interchangeably in the pages that follow.

There are two significant differences between tension and compression members.

1. Whereas tensile loads tend to hold a member straight, compressive loads tend to bend them out of the plane of the loads (a serious situation).

2. The presence of bolt or rivet holes in tension members reduces the areas available for resisting loads; but in compression members the bolts or rivets are assumed to fill the holes (although there may be

some very slight initial slippage until the bolts or rivets bear against the adjoining material), and the entire gross areas are available for resisting load.

Tests on all but the shortest columns show that they will fail at P/A stresses well below the elastic limit of the column material because of their tendency to buckle or bend laterally. For this reason their design stresses are reduced in some relation to the danger of buckling. The longer a column becomes for the same cross section, the greater its tendency to buckle and the smaller the load it will support. The tendency of a member to buckle is usually measured by its *slenderness ratio,* which has previously been defined as the ratio of the length of the member to its least radius of gyration. The greater the slenderness ratio the smaller will be the load-carrying ability of the column. Very slender columns become unstable at very small stresses. The tendency to buckle is also affected by such factors as the types of end connections, eccentricity of load application, imperfection of column material, initial crookedness of columns, residual stresses from manufacture, and so on.

The loads supported by a building column are applied by the column section above and by the connections of other members directly to the column. The ideal situation is for the loads to be applied uniformly across the column with the center of gravity of the loads coinciding with the center of gravity of the column. Furthermore, it is desirable for the column to have no flaws, to consist of a homogeneous material, and to be perfectly straight; but these situations are obviously impossible to achieve.

Loads that are exactly centered over a column are referred to as *axial* or *concentric loads.* The dead loads may or may not be concentrically placed over an interior building column, and the live loads may never be centered. For an outside column the load situation is probably even more eccentric, as the center of gravity of the loads will usually fall well on the inner side of the column. In other words, it is doubtful that a perfect axially loaded column will ever be encountered in practice.

The other desirable situations are also impossible to achieve because of imperfections of cross-sectional dimensions, residual stresses, holes punched for bolts or rivets, erection stresses, and transverse loads. It is difficult to take into account all of these imperfections in a formula.

Slight imperfections in tension members and beams can be safely disregarded, as they are of little consequence. On the other hand, slight defects in columns may be of major significance. A column that is slightly bent at the time it is put in place may have serious bending moments. Obviously a column is a more critical member in a structure than is a beam or tension member because minor imperfections in materials and dimensions mean a great deal. This fact can be illustrated by a bridge truss that has some of its members damaged by a truck. The bending of tension members probably will not be serious, as the tensile loads will tend to straighten those members; but the bending of any compression members is a serious matter, as compressive loads will tend to magnify the bending in those members.

The preceding discussion should clearly show that column imperfections cause bending, and the designer must consider stresses due to those moments

Two International Place, Boston, Mass. (Courtesy of Owen Steel Company, Inc.)

as well as due to axial loads. Chapters 5 to 7 are limited to a discussion of axially loaded columns, while members subjected to a combination of axial loads and bending loads are discussed in Chapter 10.

5-2 RESIDUAL STRESSES

Research at Lehigh University has shown that residual stresses and their distribution are very important factors affecting the behavior of axially loaded steel columns. These stresses are of particular importance for columns with slenderness ratios varying from approximately 40 to 120, a range that includes a very large percentage of practical columns. A major cause of residual stress is the uneven cooling of shapes after hot rolling. For instance, in a W shape the outer tips of the flanges and the middle of the web cool quickly while the areas at the intersection of the flange and web cool more slowly.

The quicker cooling parts of the sections when solidified resist further shortening, while those parts that are still hot tend to shorten further as they cool. The net result is that the areas which cooled more quickly have residual compressive stresses, and the slower cooling areas have residual tensile stresses. The magnitude of these stresses varies from about 10 to 15 ksi (69 to 103 MPa), although some values greater than 20 ksi (138 MPa) have been found.

When rolled-steel column sections with their residual stresses are tested, their proportional limits are reached at P/A values of only a little more than half of their yield stresses, and the stress-strain relationship is nonlinear from there up to the yield stress. Because of the early localized yielding occurring at some points of the column cross sections, buckling strengths are appreciably reduced. Reductions are greatest for columns with slenderness ratios varying from approximately 70 to 90 and may possibly be as high as 25 percent.[1]

As a column load is increased, some parts of the column will quickly reach the yield stress and go into the plastic range because of residual compression stresses. The stiffness of the column will be reduced and become a function of the part of the cross section that is still elastic. A column with residual stresses will behave as though it has a reduced cross section. This reduced section or elastic portion of the column will change as the applied stresses change. The buckling calculations for a particular column with residual stresses can be handled by using an effective moment of inertia I_e of the elastic portion of the cross section or by using a tangent modulus (that is, word stack the ratio of stress to strain at some point above the proportional limit). For the usual sections used as columns the two methods give fairly close results.

Residual stresses may also arise during fabrication when cambering is performed by cold bending and during cooling after welding. Cambering is the bending of a member in one direction so that it won't look so bad when the service loads bend it in the other direction. For instance, we may bend a beam initially upward ⌒ so that it will be approximately straight when its

normal gravity loads are applied. ⌣ Welding can produce severe residual stresses in columns which actually can approach the yield point in the vicinity of the weld. Another important fact is that columns may be appreciably bent by the welding process, decidedly affecting their load-carrying ability. Figure 5-1 shows the effect of residual stresses (due to cooling and fabrication) on the stress-strain diagram for a hot-rolled W shape.

---

[1] L. S. Beedle and L. Tall, "Basic Column Strength," *Proceedings ASCE* 86 (July 1960), pp. 139–173.

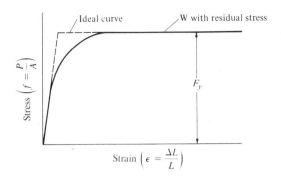

Figure 5-1 Effect of residual stresses on column stress-strain diagram.

5-3 SECTIONS USED FOR COLUMNS

Theoretically an endless number of shapes can be selected to safely resist a compressive load in a given structure. From a practical viewpoint, however, the number of possible solutions is severely limited by such considerations as sections available, connection problems, and type of structure in which the section is to be used. The paragraphs that follow are intended to give a brief résumé of the sections which have proved to be satisfactory for certain conditions. These sections are shown in Fig. 5-2, and the letters in parentheses in the paragraphs to follow refer to the parts of this figure.

The sections used for compression members are usually similar to those used for tension members, with certain exceptions. The exceptions are necessary because the strengths of compression members vary in some inverse rela-

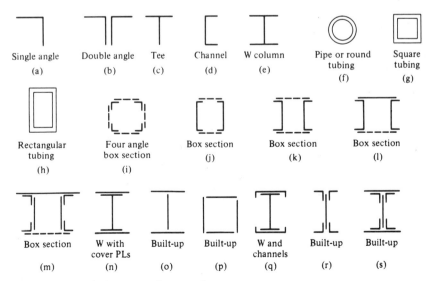

Figure 5-2 Types of compression members.

tion to the slenderness ratios, and stiff members are required. Individual rods, bars, and plates are usually too slender to make satisfactory compression members unless they are very short and lightly loaded.

Single-angle members (a) are satisfactory for bracing and compression members in light trusses. Equal-leg angles may be more economical than unequal-leg angles because their least r values are greater for the same area of steel. The top chord members of bolted roof trusses might consist of a pair of angles back-to-back (b). There will often be a space between them for the insertion of a gusset or connection plate at the joints necessary for connections to other members. An examination of this section will show that it is probably desirable to use unequal-leg angles with the long legs back-to-back to give a better balance between the r values about the x and y axes.

If roof trusses are welded, gusset plates may be unnecessary, and structural tees (c) might be used for the top chord compression members because the web members can be welded directly to the stems of the tees. Single channels (d) are not satisfactory for the average compression member because of their almost negligible r values about their web axes. They can be used if some method of providing extra lateral support in the weak direction is available. The W shapes (e) are the most common shapes used for building columns and for the compression members of highway bridges. Their r values, although far from being equal about the two axes, are much more nearly balanced than for channels.

Several famous bridges constructed during the nineteenth century (such as the Firth of Forth Bridge in Scotland and Ead's Bridge in St. Louis) made extensive use of tube-shaped members. Their use declined due to connection problems and manufacturing costs, but with the development of economical welded tubing their use is again increasing (although the tube shapes of today are very small compared with the giant ones used in those early steel bridges).

For small and medium loads, pipe sections or round tubing (f) are quite satisfactory. They are often used as columns in long series of windows, as short columns in warehouses, as columns for the roofs of covered walkways, in the basements and garages of residences, and so on. Pipe columns have the advantage of being equally rigid in all directions and are usually very economical unless moments are large. The ASD Manual furnishes the sizes of these sections and classifies them as being standard, extra strong, and double extra strong.

Square and rectangular tubing (g and h) have not been used to a great extent for columns until recently. In fact, for many years only a few steel mills manufactured steel tubing for structural uses. Perhaps the major reason why tubing was not used to a great extent was the difficulty of making connections with rivets or bolts. This problem has been fairly well eliminated, however, by the advent of modern welding. The use of tubing for structural purposes by architects and engineers in the years to come will probably be greatly increased for several reasons.

1. The most efficient compression member is one that has a constant radius of gyration about its centroid, a property available in round tubing. Square tubing is the next most efficient compression member.

Bents fabricated from tubing sections in New Jersey. (Courtesy of Bethlehem Steel Corporation.)

2. Their smooth surfaces permit easier painting.
3. They have excellent torsional resistance.
4. The surfaces of tubing are quite attractive.
5. When exposed, round tubing has a wind resistance only about two-thirds of that of flat surfaces of the same width.

A slight disadvantage in certain cases is that the ends of tubing may have to be sealed to protect their inaccessible inside surfaces from corrosion. Although making very attractive exposed members for beams, tubing is at a definite weight disadvantage as compared with W sections, which have so much larger resisting moments for the same weights.

Where compression members are designed for very large structures, it may be necessary to use built-up sections. Built-up sections are needed where the members are long and support very heavy loads and/or when there are connection advantages. Generally speaking, a single shape such as a W section is more economical than a built-up section having the same cross-sectional area. With heavy column loads, high-strength steels can frequently be used with very economical results particularly if their increased strength permits the use of W sections rather than built-up members.

When built-up sections are used, they must be connected on their open sides with some type of *lacing* (also called lattice bars) to hold the parts together in their proper positions and to assist them in acting together as a unit. The ends of these members are connected with tie plates (also called batten plates or stay plates). Several types of lacing for built-up compression members are shown in Fig. 6-6.

The dotted lines in Fig. 5-2 represent lacing or discontinuous parts, and the solid lines represent parts that are continuous for the full length of the members. Four angles are sometimes arranged as shown in (i) to produce large *r* values. This type of member may often be seen in towers and crane booms. A pair of channels (j) is sometimes used as a building column or as a web member in a large truss. It will be noted that there is a certain spacing for each pair of channels at which their *r* values about the *x* and *y* axes are equal. Sometimes the channels may be turned out as shown in (k).

A section well suited for the top chords of bridge trusses is a pair of channels with a cover plate on top (1) and with lacing on the bottom. The gusset plates at joints are conveniently connected to the insides of the channels and may also be used as splices. When the largest channels available will not produce a top chord member of sufficient strength, a built-up section of the type shown in (m) may be used.

When the rolled shapes do not have sufficient strength to resist the column loads in a building or the loads in a very large bridge truss, their areas may be increased by adding plates to the flange (n). In recent years it has been found that for welded construction a built-up column of the type shown in part (o) is a more satisfactory shape than a W with welded cover plates (n). It seems that in bending (as where a beam frames into the flange of a column) it is difficult to efficiently transfer tensile force through the cover plate to the column without pulling the plate away from the column. For very heavy column loads a welded box section of the type shown in (p) has proved to be quite satisfactory. Some other built-up sections are shown in (q) through (s). The built-up sections shown in (n) through (q) have an advantage over those shown in (i) through (m) in that they do not require the expense of the latticework necessary in the former. Lateral shearing forces are negligible for the single-column shapes and for the nonlatticed built-up sections, *but they are definitely not negligible for the built-up latticed columns.*

Today *composite columns* are being increasingly used. These columns usually consist of steel pipe and structural tubing filled with concrete or of W shapes encased in concrete usually square or rectangular in cross section.

5-4 DEVELOPMENT OF COLUMN FORMULAS

The use of columns goes back before the dawn of history, but it was not until 1729 that a paper was published on the subject by Pieter van Musschenbroek, a Dutch mathematician.[2] He presented an empirical column formula for estimating the strength of rectangular columns. In 1757 Leonhard Euler, a Swiss mathematician, wrote a paper of great value concerning the buckling of columns. He was probably the first person to realize the significance of buckling. The Euler formula, the most famous of all column expressions, which is derived in Section 5-5, marked the real beginning of theoretical and experimental investigation of columns.

Engineering literature is filled with formulas developed for ideal column conditions, but these conditions are not encountered in actual practice. Conse-

---

[2] L. S. Beedle et al., *Structural Steel Design* (New York: Ronald Press, 1964), p. 269.

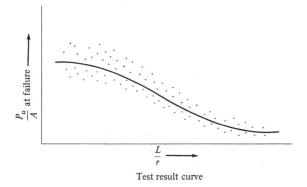

Test result curve

Figure 5-3 Column strength curve.

quently practical column design is based primarily on formulas that have been developed to fit with reasonable accuracy test-result curves. The reasoning behind this procedure is simply that the independent derivation of column expressions does not yield formulas that give results comparing closely with test-result curves for all slenderness ratios.

The testing of columns with various slenderness ratios results in a scattered range of values such as those shown by the broad band of dots in Fig. 5-3. The dots will not fall on a smooth curve even if all of the testing is done in the same laboratory, because of the difficulty of exactly centering the loads, lack of perfect uniformity of the materials, varying dimensions of the sections, residual stresses, end-restraint variations, and so on. The usual practice is to attempt to develop formulas which give results represented by an approximate average of the test results. The student should also realize that laboratory conditions are not field conditions and that column tests probably give the limiting values of column strengths.

The magnitudes of the yield stresses of the sections tested are quite important for short columns, as their failure stresses are close to those yield stresses. For columns with intermediate slenderness ratios the yield stresses are of lesser importance in their effect on failure stresses, and they are of no significance for long slender columns. For intermediate-range columns residual stresses have more effect on the results, while the failure stresses for long slender columns are very sensitive to end-support conditions. In addition to residual stresses and nonlinearity of material another dominant factor in column strength is member out-of-straightness.

5-5 DERIVATION OF THE EULER FORMULA

The Euler formula is derived in this section for a straight, concentrically loaded, homogeneous, long slender column with rounded ends. It is assumed that this perfect column has been laterally deflected by some means as shown in Fig. 5-4, and that if the concentric load P was removed, the column would straighten out completely.

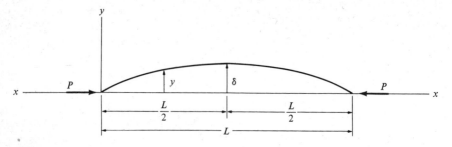

Figure 5-4

The x and y axes are located as shown in the figure. As the bending moment at any point in the column is $-Py$, the equation of the elastic curve can be written

$$EI\frac{d^2y}{dx^2} = -Py$$

For convenience in integration both sides of the equation are multiplied by $2dy$ and the integration is performed.

$$EI\ 2\frac{dy}{dx}d\frac{dy}{dx} = -2Pydy$$

$$EI\left(\frac{dy}{dx}\right)^2 = -Py^2 + C_1$$

When $y = \delta$, $dy/dx = 0$, and the value of C_1 will equal $P\delta^2$ and

$$EI\left(\frac{dy}{dx}\right)^2 = -Py^2 + P\delta^2$$

The preceding expression is arranged more conveniently as follows:

$$\left(\frac{dy}{dx}\right)^2 = \frac{P}{EI}(\delta^2 - y^2)$$

$$\frac{dy}{dx} = \sqrt{\frac{P}{EI}}\sqrt{\delta^2 - y^2}$$

$$\frac{dy}{\sqrt{\delta^2 - y^2}} = \sqrt{\frac{P}{EI}}dx$$

Integrating this expression, the result is

$$\text{arc sin}\frac{y}{\delta} = \sqrt{\frac{P}{EI}}x + C_2$$

When $x = 0$ and $y = 0$, $C_2 = 0$. The column is bent into the shape of a sine curve expressed by the equation

$$\text{arc sin}\frac{y}{\delta} = \sqrt{\frac{P}{EI}}x$$

When $x = L/2$, $y = \delta$, resulting in

$$\frac{\pi}{2} = \frac{L}{2}\sqrt{\frac{P}{EI}}$$

In this expression P is the *critical buckling load* or the maximum load which the column can support before it becomes unstable. Solving for P,

$$P = \frac{\pi^2 EI}{L^2}$$

This expression is the Euler formula, but it is usually written in a little different form involving the slenderness ratio. Since $r = \sqrt{I/A}$ and $r^2 = I/A$ and $I = r^2 A$, the Euler formula may be written as

$$\frac{P}{A} = \frac{\pi^2 E}{(L/r)^2}$$

The student should carefully note that the buckling load determined from the Euler equation is independent of the strength of the steel used. The equation is useful only if the end-support conditions are carefully considered. The results obtained by application of the formula to specific examples compare very well with test results for centrally loaded, long slender columns with rounded ends. Designers, however, do not encounter perfect columns of this type. The columns with which they work do not have rounded ends and are not free to rotate because their ends are bolted, riveted, or welded to other members. These practical columns have different amounts of restraint against rotation varying from slight restraint to almost fixed conditions. For the actual cases encountered where the ends are not free to rotate, different length values can be used in the formula and more realistic values will be obtained.

To use the Euler equation successfully for practical columns, the value of L should be the distance between points of inflection in the buckled shape. This distance is referred to as the *effective length* of the column. For a pinned-end column (whose ends can rotate but cannot translate) the points of inflection or zero moment are located at the ends a distance L apart. For columns with different end conditions the effective lengths may be entirely different. Effective lengths are described in detail in Section 5-9.

Example 5-1 illustrates the application of the Euler equation to a steel column. *If the $\frac{P}{A}$ value obtained exceeds the proportional limit, the elastic Euler equation is not applicable.*

■ Example 5-1

(a) A W10 × 22 is used as a 15-ft pin-connected column. Using the Euler formula determine the allowable axial compression load the member can support, using a factor of safety of 2. Assume the steel has a proportional limit of 36 ksi.

(b) Repeat part (a) for a length of 8 ft.

Solution.

(a) Using a W10 × 22 ($A = 6.49$ in.^{2}, $r_x = 4.27$ in., $r_y = 1.33$ in.)

Minimum $r = r_y = 1.33$ in.

$$\frac{L}{r} = \frac{(12)(15)}{1.33} = 135.34$$

Critical or buckling $\dfrac{P}{A} = \dfrac{(\pi^2)(29 \times 10^3)}{(135.34)^2} = 15.63$ ksi < 36 ksi

<u><u>OK</u></u> column is in elastic range

Allowable $\dfrac{P}{A} = \dfrac{15.63}{2} = 7.81$ ksi

Allowable $P = (7.81)(6.49) = \underline{\underline{50.7 \text{ k}}}$

(b) Using an 8-ft W10 × 22

$$\frac{L}{r} = \frac{(12)(8)}{1.33} = 72.18$$

Critical or buckling $\dfrac{P}{A} = \dfrac{(\pi^2)(29 \times 10^3)}{(72.18)^2} = 54.94$ ksi > 36 ksi

<u>NG column is in inelastic range and Euler equation not applicable</u> ■

The design of a column of a certain length for a particular load using the Euler formula (divided by a factor of safety) is a trial-and-error problem. We do not know the allowable stress until we have a section (with its r value), and we can't pick the section until we have an allowable stress. To handle the problem we may guess a column size, compute the load it can support according to the formula, and then, if it is overstressed or too much understressed, try another size and repeat the process. This is exactly the procedure described in the next few sections for other column formulas.

5-6 OBSOLETE COLUMN FORMULAS

As mentioned in Section 5-5, almost all practical steel column design is based on empirical formulas that have been developed through the years to fit as closely as possible test-result curves. Each of the resulting expressions is then divided by the desired factor of safety. Among the general types of formulas are the straight-line, parabolic, Gordon-Rankine, and secant. A few comments are made in this section about each of these types of equations to provide a general background for the use of all types of column equations.

Straight-line formulas can be devised that will approach test results for L/r values from about 50 to 120. These same formulas, however, give allowable stresses that are much too high when the slenderness ratio is below about 50 and much too low when it is above about 120. Thus a maxmum stress value is specified for columns with low slenderness ratios, and a maximum slenderness ratio above which the formula cannot be used is provided. The following is a typical straight-line column formula:

$$F_a = \text{allowable } \frac{P}{A} = 15{,}000 - 50\frac{L}{r} \text{ with a maximum value of 12,500 psi}$$

Parabolic formulas can be set up to represent test results fairly well for slenderness ratios varying from 0 to about 130 or 140. Above this range they give values that are much too conservative. Though the parabolic formula listed below is obsolete, most of the practical steel column formulas in use today are of the parabolic type.

$$F_a = 17{,}000 - \frac{1}{2}\left(\frac{L}{r}\right)^2$$

The Gordon-Rankine formulas have denominators larger than 1.0 and which increase with increasing slenderness ratios. Thus allowable stresses decrease as columns become more slender. These types of formulas can be made to correspond very well with test results for columns with L/r values from about 120 to 200. The following Gordon-Rankine formula was used for many years by the AISC for the design of small compression members with high slenderness ratios:

$$F_a = \frac{18{,}000}{1 + \left[\left(\dfrac{L}{r}\right)^2 \bigg/ 18{,}000\right]}$$

The *secant formula* can be made to approximate test results for all ranges of L/r. Many persons first looking at a secant formula say that it is too complicated for practical use. This is not true, however, because the results of all column formulas, no matter how complex, can be plotted on a graph or recorded in a table for any desired range of L/r. We will see in Chapter 6 that this is what the AISC does in the ASD Manual with their complex-looking column equations. In this secant formula,

$$F_a = \frac{F_y}{1 + \dfrac{ec}{r^2}\sec\left(\sqrt{\dfrac{P}{AE}}\dfrac{L}{2r}\right)}$$

F_y is the yield point of the steel, c is the distance from the centroid to the extreme fiber, and e is the eccentricity of load application. This formula has a definite advantage in that it provides specifically for some eccentricity of load application. On the basis of studies of test results a value for ec/r^2 (called the eccentric ratio) equal to 0.25 was generally recommended. This value supposedly provides for secondary stresses caused by bending and initial crookedness of columns.

Two very simple numerical examples are presented here to illustrate the use of these old column expressions. In Example 5-2 a parabolic column formula is used to determine the allowable load that a W section can carry.

Capstone House, University of South Carolina, Columbia, S.C. (Courtesy of Owen Steel Company, Inc.)

■ Example 5-2

Using the parabolic formula given, determine the allowable axial compression load that a 14-ft W12 × 96 can support.

$$F_a = 17,000 - \frac{1}{2}\left(\frac{L}{r}\right)^2$$

Solution. Properties of a W12 × 96 from ASD Manual ($A_g = 28.2$ in.^{2}, $r_y = 3.09$ in.):

$$\frac{L}{r} = \frac{(12)(14)}{3.09} = 54.37$$

$$F_a = 17,000 - \left(\frac{1}{2}\right)(54.37)^2 = 15,522 \text{ psi}$$

$$P = F_a A_g = (15.522)(28.2) = \underline{\underline{437.7 \text{ k}}} \qquad ■$$

The design applications for column equations involve a trial-and-error process. For a particular design the allowable stress is not known until a column size is selected and vice versa. Once a trial section is selected, the least r for that section is known and can be substituted into the column equation to determine the allowable stress.

The designer will usually assume an allowable stress, divide that stress into the column load to give an estimated column area, select a column, and check to see if the section selected is over- or understressed, and if appreciably so, try another size. Students may feel that they do not have a sufficient background or knowledge to make a reasonable stress assumption. If they read the information contained in the next paragraph, however, they will immediately be able to make excellent estimates.

The slenderness ratios for average columns of 10- to 15-ft lengths will generally fall in a range of about 40 to 60. If for a particular column a slenderness ratio somewhere in this range is assumed and substituted into the column equation, it will almost always provide a satisfactory allowable-stress estimate.

In Example 5-3 a 10-ft column is selected using the parabolic equation of Example 5-2. A slenderness ratio of 50 is assumed and substituted into the formula, and the resulting stress is divided into the column load to estimate the area required.

To estimate the slenderness ratio for a particular column the designer may pick a value a little higher than 40 to 60 if the column is appreciably longer than the 10- to 15-ft range and vice versa. A very heavy load, say larger than 500 k (2,224 kN) will yield a rather large column for which the least r will be rather large, and the designer may estimate a little smaller value of L/r. For lightly loaded bracing members he or she will estimate high slenderness ratios perhaps over 100.

■ Example 5-3

With the parabolic equation of Example 5-2 select a W14 section to support an axial load of 245 k. The member is to be 10 ft long.

Solution.   Assume an $\dfrac{L}{r} = 50$, resulting in an allowable stress $= 17,000 - (\tfrac{1}{2})(50)^2 = 15,750$ psi

$$A_{reqd} = \frac{245}{15.750} = 15.55 \text{ in.}^2$$

Try W14 × 53 ($A_g = 15.6$ in.^{2}, $r_y = 1.92$ in.)

$$\frac{L}{r} = \frac{(12)(10)}{1.92} = 62.5$$

$$F_a = 17,000 - \left(\frac{1}{2}\right)(62.5)^2 = 15,047 \text{ psi}$$

$$P = (15.047)(15.6) = 234.7 \text{ k} < 245 \text{ k} \qquad \text{NG}$$

Use W14 × 61 by examination   ■

5-7 LONG, SHORT, AND INTERMEDIATE COLUMNS

A column subject to an axial compression load will shorten in the direction of the load. If the load is increased until the column buckles, the shortening will stop, and the column will suddenly bend or deform laterally and may at the same time twist in a direction perpendicular to its longitudinal axis.

The strength of a column and the manner in which it fails are greatly dependent on its length. A very short stocky steel column may be loaded until the steel yields and perhaps on into the strain-hardening range. As a result it can support about the same load in compression that it can in tension.

As the length of a column increases, its buckling stress will decrease. If the length exceeds a certain value, the buckling stress will be less than the proportional limit of the steel. Columns in this range are said to fail *elastically*.

As shown in Section 5.5 very long steel columns will fail at loads which are proportional to the bending rigidity of the column (*EI*) and independent of the strength of the steel. *For instance, a long column constructed with a 36 ksi yield stress steel will fail at just about the same load as one constructed with a 100 ksi yield stress steel.*

Columns are sometimes classed as being long, short, or intermediate. A brief discussion of each of these classifications is presented in the paragraphs to follow.

Long Columns

The Euler formula predicts very well the strength of long columns where the axial buckling stress remains below the proportional limit. Such columns will buckle elastically.

Short Columns

For very short columns the failure stress will equal the yield stress and no buckling will occur. (For a column to fall into this class it would have to be so short as to have no practical application. Thus no further reference is made to them here.)

Intermediate Columns

For intermediate columns some of the fibers will reach the yield stress and some will not. The members will fail by both yielding and buckling, and their behavior is said to be *inelastic*. (For the Euler formula to be applicable for such columns it would have to be modified according to the reduced modulus concept or the tangent modulus concept to account for the presence of residual stresses.) The vast majority of columns fall into this class.

5-8 THE ASD AND AASHTO FORMULAS

The ASD expressions were developed to incorporate the latest research available concerning the behavior of steel columns. These formulas take into account the effect of residual stresses, the actual end-restraint conditions of the columns, and the varying strengths of different steels. The important effect of residual stresses on the stress-strain curves was discussed in Section 5-2 and illustrated in Fig. 5-1. Different types of end restraints cause entirely different effective lengths and column strengths. A discussion of this subject was presented in Section 5-5 and is continued here as it pertains to the ASD formulas.

The ASD formulas result in more logical and economical designs than those given by the older expressions. Design by many other column formulas gives members that are appreciably overdesigned in the lower L/r range, but the ASD formulas give fairly economical designs for all ranges of L/r. The expressions are rather complex to solve mathematically, but as tables are available in Part 3 of the ASD Manual for columns with $F_y = 36$ ksi and $F_y = 50$ ksi, designs can be made with little difficulty.

The ASD assumes that because of residual stresses the upper limit of elastic buckling is defined by an average stress equal to one-half of the yield point ($\frac{1}{2}F_y$). If this stress is equated to the Euler expression, the value of the slenderness ratio at this upper limit can be determined for a particular steel. This value is referred to as C_c, the slenderness ratio dividing elastic from inelastic buckling, and is determined as follows:

$$\frac{1}{2}F_y = \frac{\pi^2 E}{(L/r)^2} = \frac{\pi^2 E}{C_c^2}$$

$$C_c = \sqrt{\frac{2\pi^2 E}{F_y}}^*$$

The values of C_c can be computed with little difficulty, but the ASD Manual gives its values for each steel (126.1 for A36, 116.7 for the 42,000 psi yield point steels, etc.). For slenderness ratios less than C_c a parabolic formula, ASD Equation E2-1, is used. This is the Structural Stability Research Council equation for the ultimate strength of a centrally loaded column with a factor of safety applied.[3] In this expression F_a is the allowable axial stress (P/A), and K is the factor to be multiplied by the unsupported column length to give its estimated effective length (see Section 5-9).

$$F_a = \frac{\left[1 - \dfrac{(KL/r)^2}{2C_c^2}\right]F_y}{\dfrac{5}{3} + \dfrac{3(KL/r)}{8C_c} - \dfrac{(KL/r)^3}{8C_c^3}} \qquad \text{(ASD Equation E2-1)}$$

---

*For SI units $C_c = 1987/\sqrt{F_y}$ with F_y in MPa.

[3] Structural Stability Research Council, *Guide to Stability Design Criteria for Metal Structures,* 4th ed., T.V. Galambos, ed. (New York: Wiley, 1988).

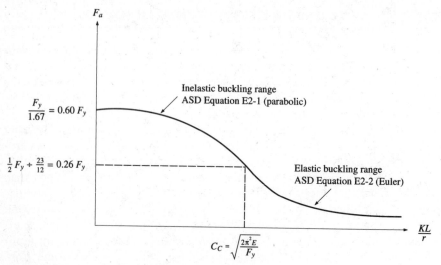

Figure 5-5 ASD column allowable-stress curve.

For values of KL/r greater than C_c the Euler formula is used. With a factor of safety of 1.92 (or 23/12) the expression becomes

$$F_a = \frac{12\pi^2 E}{23(KL/r)^2}$$ (ASD Equation E2-2)

Figure 5-5 shows the ranges in which the two ASD expressions are used.

The denominator of Equation E2-1 is the factor of safety to be used and usually gives a value not much greater than the one used for axially loaded tension members. Tests have shown that short columns are not greatly affected by small eccentricities permitting the use of lower factors of safety. As columns become slenderer, they become more sensitive to small imperfections, and the factor of safety is increased up to 15 percent. It might also be noticed that the medium-length columns are those in which residual stresses and initial column crookedness become quite important. The factor of safety (F.S.) expression is a quarter sine wave which equals $\frac{5}{3}$ when KL/r is equal to zero and increases up to $\frac{23}{12}$ when KL/r equals C_c.

The 1989 AASHTO Specifications has the same two types of column formulas as the ASD Specification. The two categories, as for the ASD, are for those columns that have low slenderness ratios and fail due to inelastic buckling and those columns that have high slenderness ratios and fail by elastic buckling. For the inelastic range the AASHTO requires the use of the parabolic formula

$$F_a = \frac{F_y}{\text{F.S.}}\left[1 - \frac{\left(\dfrac{KL}{r}\right)^2 F_y}{4\pi^2 E}\right]$$

and for the elastic range (that is, where $Kl/r > C_c$) the Euler formula is to be used:

$$F_a = \frac{\pi^2 E}{\text{F.S.}(KL/r)^2}$$

For both cases a constant F.S. equal to 2.12 is used.

Central Laboratory and Office Building, Baltimore, Md. (Courtesy of Owen Steel Company, Inc.)

A heavy column showing base plate with combination welded and high-strength bolted connection. (Courtesy of Bethlehem Steel Company.)

5-9 END RESTRAINT AND EFFECTIVE LENGTHS OF COLUMNS

End restraint and its effect on the load-carrying capacity of columns is a very important subject indeed. Columns with appreciable end restraint can support considerably more load than those with little end restraint, as at hinged ends.

The effective length of a column was defined in Section 5-5 as the distance between points of zero moment in the column, that is, the distance between its inflection points. In steel specifications the effective length of a column is referred to as *KL*, where *K* is the *effective length factor*. *K* is the number that must be multiplied by the length of the column to obtain its effective length. Its size depends on the rotational restraint supplied at the ends of the column and upon the resistance to lateral movement provided.

The concept of effective lengths is simply a mathematical method of taking a column whatever its end and bracing conditions and replacing it with an equivalent pinned-end braced column. A complex buckling analysis could be made for a frame to determine the critical stress in a particular column. The *K* factor is determined by finding the pinned-end column with an equivalent length that provides the same critical stress. The K factor procedure is a method of making simple solutions for complicated frame buckling problems.

Columns with different end conditions have entirely different effective lengths. For this initial discussion it is assumed that no sidesway or joint translation is possible. (Sidesway or joint translation means that one or both ends of a column can move laterally with respect to each other.) Should a column be connected with frictionless hinges as shown in Fig. 5-6(a), its effective length would be equal to the actual length of the column, and K would equal 1.0. If there were such a thing as a perfectly fixed-end column, its points of inflection (or points of zero moment) would occur at its one-fourth points, and its effective length would equal $L/2$ as shown in Fig. 5-6(b). As a result its K value would equal 0.50.

Obviously the smaller the effective length of a particular column, the smaller its danger of lateral buckling and the greater its load-carrying capacity. In Fig. 5-6(c) a column is shown with one end fixed and one end pinned. The K for this column is theoretically 0.70.

Actually there are no perfect pin connections nor any perfect fixed ends, and the usual column in a braced frame falls between the two extremes. This discussion would seem to indicate that column effective lengths always vary from an absolute minimum of $L/2$ to an absolute maximum of L, but there are exceptions to this rule. An example is given in Fig. 5-7(a) where a simple bent is shown. The base of each of the columns is pinned, and the other end is free to rotate and move laterally (sidesway). Examination of this figure will show that the effective length will exceed the actual length of the column, as the elastic curve will theoretically take the shape of the curve of a pinned-end column of twice its length and K will theoretically equal 2.0. Notice in part (b)

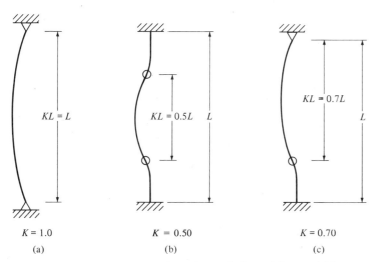

$KL = L$

$K = 1.0$

(a)

$KL = 0.5L$ L

$K = 0.50$

(b)

$KL = 0.7L$

L

$K = 0.70$

(c)

Figure 5-6 Effective lengths for columns in braced frames (sidesway prevented).

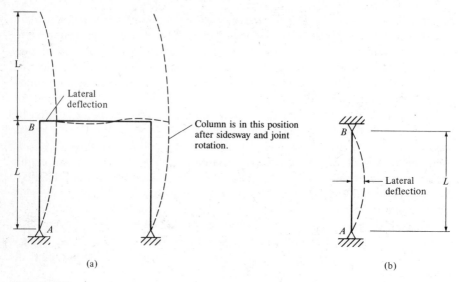

Figure 5-7

of the figure how much smaller the lateral deflection of column AB would be if it were pinned both top and bottom so as to prevent sidesway.

Structural steel columns serve as parts of frames, and these frames are sometimes *braced* and sometimes *unbraced*. In a braced frame sidesway or joint translation is prevented by means of bracing, shear walls, or lateral support from adjoining structures. An unbraced frame does not have any of these types of bracing and must depend on the bending stiffness of beams and columns rigidly connected to each other to prevent lateral buckling. For braced frames K values can never be greater than 1.0, but for unbraced frames the K values will always be greater than 1.0 because of sidesway.

Table C-C2.1 of the Commentary on the ASD Specification gives recommended effective length factors when ideal conditions are approximated. This table is reproduced here as Table 5-1(a). Two sets of K values are provided in the table, one being the theoretical values and the other being the recommended design values based on the fact that perfectly pinned and fixed conditions are not possible. If the ends of the column of Fig. 5-6(a) were not quite fixed, the column would be a little freer to bend laterally and its points of inflection would be farther apart. The recommended design K given is 0.65 while the theoretical value is 0.5. As no column ends are perfectly fixed or perfectly hinged, the designer may wish to interpolate between the values given in the table, the interpolation to be based on his or her judgment of the actual restraint conditions.

Effective lengths are discussed in greater detail in Section 7-1. The alignment charts there give supposedly more precise K factors than does Table 5-1.

TABLE 5-1 COLUMN EFFECTIVE LENGTHS

	(a)	(b)	(c)	(d)	(e)	(f)
Buckled shape of column shown by dashed line						
Theoretical K value	0.5	0.7	1.0	1.0	2.0	2.0
Recommended design value when ideal conditions are approximated	0.65	0.80	1.2	1.0	2.10	2.0

End condition code	
	Rotation fixed and translation fixed
	Rotation free and translation fixed
	Rotation fixed and translation free
	Rotation free and translation free

Source: American Institute of Steel Construction, *Manual of Steel Construction Allowable Stress Design,* 9th ed. (Chicago: AISC, 1989), Table C-C2.1, "Commentary on the Specification for Structural Steel Buildings Allowable Stress Design and Plastic Design," p. 5-135. Reprinted with the permission of AISC.

5-10 MAXIMUM SLENDERNESS RATIOS

The ASD Specification (B7) states that compression members *preferably* should be designed with KL/r ratios not exceeding 200. If $KL/r = 200$ for a compression member, its allowable stress will equal 3.73 ksi regardless of its yield stress. Should slenderness ratios larger than these be used, the allowable-stress values will be very small and will require the user to substitute into the column formulas provided in Section 5-8.

5-11 EXAMPLES

In this section three simple numerical column problems using the ASD Specification are presented. In each case the design strength of a column is calculated. In Example 5-4 the author determines the allowable strength of a W section. The value of K is determined from Table 5-1, the effective slenderness ratio is computed, and the allowable design stress F_a of the member is selected from Table C-36 of Part 3 of the Manual and multiplied by the cross-sectional area of the column.

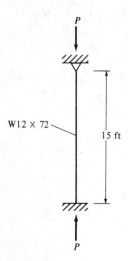

Figure 5-8

■ Example 5-4

Using the column design stress values shown in Table C-36 in the ASD Manual for A36 steel, determine the allowable compression load P that the axially loaded column shown in Fig. 5-8 can support.

Solution.   Using a W12 × 72 ($A = 21.1$ in.^{2}, $r_y = 3.04$ in.)

$$K = 0.80 \text{ from Table 5-1}$$

$$\frac{KL}{r} = \frac{(0.80)(12 \times 15)}{3.04} = 47.37$$

$$F_a = 18.58 \text{ ksi from Table C-36 of ASD Manual}$$

$$P = F_a A_g = (18.58)(21.1) = \underline{\underline{392 \text{ k}}} \qquad ■$$

In Example 5-5 the allowable axial compression load for a built-up column section is determined using the allowable ASD stresses for A36 steel. For this column it is necessary to locate the centroidal axes of the section and compute I_x and I_y and then the least r of the section. This r value is used in the column equation.

■ Example 5-5

Using the ASD allowable column stress table for A36 steel, determine the allowable column load that the built-up section of Fig. 5-9 can support. $KL = 19$ ft.

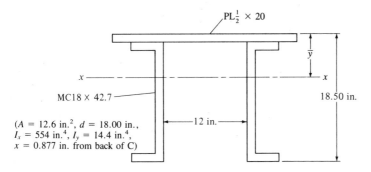

Figure 5-9

Solution.

$$A = (20)(\tfrac{1}{2}) + (2)(12.6) = 35.2 \text{ in.}^2$$

$$\bar{y} \text{ from top} = \frac{(10)(0.25) + (2)(12.6)(9.50)}{35.2} = 6.87 \text{ in.}$$

$$I_x = (2)(554) + (25.2)(2.63)^2 + (\tfrac{1}{12})(20)(\tfrac{1}{2})^3 + (10)(6.62)^2$$
$$= 1721 \text{ in.}^4$$

$$I_y = (2)(14.4) + (12.6)(6.877)^2(2) + (\tfrac{1}{12})(\tfrac{1}{2})(20)^3 = 1554 \text{ in.}^4$$

$$\text{Least } r = \sqrt{\frac{1554}{35.2}} = 6.64 \text{ in.}$$

$$\frac{KL}{r} = \frac{(12)(19)}{6.64} = 34.34$$

$$F_a = 19.63 \text{ ksi}$$

$$P = F_a A_g = (19.63)(35.2) = \underline{\underline{691 \text{ k}}}$$ ■

To determine the allowable design compression stress to be used for a particular column, it is theoretically necessary to compute both $(KL/r)_x$ and $(KL/r)_y$. The reader will notice, however, that for most of the steel sections used for columns, r_y will be much less than r_x. As a result only $(KL/r)_y$ is calculated for most columns and used in the applicable column formulas.

For some columns, particularly the long ones, bracing is supplied perpendicular to the weak axis, thus reducing the slenderness or the length free to buckle in that direction. This may be accomplished by framing braces or beams into the sides of a column. For instance, horizontal members called *girts* running parallel to the exterior walls of a building frame may be framed into the sides of columns. The result is stronger columns, for which the designer needs to calculate both $(KL/r)_x$ and $(KL/r)_y$. The larger ratio obtained for a particular column indicates the weaker direction and will be used for calculating the allowable design stress F_a for that member.

Bracing members must be capable of providing the necessary lateral forces without buckling themselves. The forces to be taken are quite small and are often conservatively estimated to equal 0.02 times the column design loads. These members can be selected as are other compression members. A bracing member must be connected to other members that can transfer the horizontal force by shear to the next restrained level. If this is not done, little lateral support will be provided for the original column in question.

If the lateral bracing were to consist of a single bar or rod (⊥) it would not prevent twisting and torsional buckling of the column (see Section 7-4 and Appendix). As torsional buckling is a difficult problem to handle, we should provide lateral bracing that prevents lateral movement and twist.[4]

Steel columns may also be built into substantial masonry walls in such a manner that they are substantially supported in the weaker direction. The designer should be quite careful in assuming complete lateral support parallel to the wall, however, because a poorly built wall will not provide 100 percent lateral support.

Example 5-6 illustrates the calculations necessary to determine the strength of a column with two unbraced lengths.

■ Example 5-6

Determine the allowable axial compression load that the A36 column of Fig. 5-10 can support, using the ASD Specification.

Solution.   Using a W14 × 145 ($A = 42.7$ in.², $r_x = 6.33$ in., $r_y = 3.98$ in.)

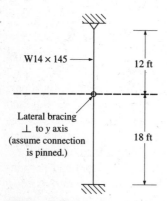

Figure 5-10

---

[4]J. A. Yura, "Elements for Teaching Load and Resistance Factor Design" (New York: AISC, July 1987), p. 20.

$$K_x L_x = (0.80)(30) = 24 \text{ ft}$$
$$K_y L_y = (1.0)(12) = 12 \text{ ft}$$
$$K_y L_y = (0.80)(18) = 14.4 \text{ ft} \quad \longleftarrow$$
$$\frac{K_x L_x}{r_x} = \frac{(12)(24)}{6.33} = 45.50 \quad \longleftarrow$$
$$\frac{K_y L_y}{r_y} = \frac{(12)(14.4)}{3.98} = 43.42$$

$F_a = 18.74$ ksi from Table C-36 in ASD Manual

Allowable $P = (18.74)(42.7) = \underline{\underline{800.2 \text{ k}}}$ ∎

PROBLEMS

Probs. 5-1 to 5-4   Determine the critical buckling load for each of the columns using the Euler equation. E = 29 × 10⁶ psi. Proportional limit = 30,000 psi. Assume simple ends and maximum permissible L/r = 200.

5-1 A solid square bar 1.0 in. × 1.0 in.
  (a) $L = 3$ ft 0 in. (*Ans.* 18.44 k)
  (b) $L = 4$ ft 6 in. (*Ans.* 8.20 k)

5-2 The pipe section shown.
  (a) $L = 36$ ft 0 in.
  (b) $L = 24$ ft 0 in.
  (c) $L = 12$ ft 0 in.

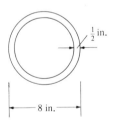

$\frac{1}{2}$ in.

8 in.

5-3 A W12 × 50. $L = 20$ ft 0 in. (*Ans.* 280.6 k)

5-4 The two C12 × 30s shown for $L = 36$ ft 0 in.

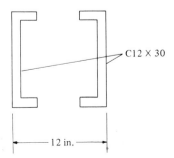

C12 × 30

12 in.

5-5 A W8 × 40 is to serve as an 18-ft pin-connected column. What is the maximum allowable load it can support? Use the Euler expression, a proportional limit of 36,000 psi, $E = 29 \times 10^6$ psi, and a safety factor of 2.5. (*Ans.* 119.5 k)

5-6 An 8-ft pin-connected column has the cross section shown. Using the Euler equation, determine the allowable load which the column can support if the factor of safety is 2. $E = 29 \times 10^6$ psi. Proportional limit = 36,000 psi.

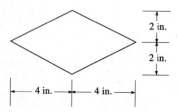

5-7 Using the Euler equation with a safety factor of 2, select a W8 section to support a 60-k load for a 20-ft pin-connected column. $E = 29 \times 10^6$ psi. Proportional limit = 36,000 psi. Maximum permissible $L/r = 200$. (*Ans.* W8 × 31)

5-8 A W10 × 49 is to be used as a 34-ft pin-ended column but is braced at middepth in the weak direction. Using a safety factor of 2, $E = 29 \times 10^6$ psi, and a proportional limit of 36,000 psi, determine the allowable total axial load which the column can support, using the Euler equation.

5-9 Using the straight-line formula $F_a = 15,000 - 50\, L/r$ with a maximum value of 12,500 psi and a maximum L/r of 120, determine the allowable axial compression load that each of these columns can support.
(a) A 12-ft W12 × 65 (*Ans.* 238.8 k)
(b) A 14-ft W14 × 82 (*Ans.* 279.9 k)
(c) An 8-ft C15 × 33.9 (*Ans.* 96.5 k)
(d) A 15-ft HP14 × 102 (*Ans.* 374.2 k)

5-10 Using the parabolic equation $F_a = 17,000 - \frac{1}{2}(L/r)^2$, determine the allowable axial compression load that each of these columns can support.
(a) A 10-ft W14 × 38
(b) An 8-ft W8 × 31
(c) A 12-ft S20 × 96
(d) An 11-ft MC13 × 50

5-11 Using the parabolic column equation of Prob. 5-10, determine the allowable loads for the columns shown. (*Ans.* (a) 654.0 k, (b) 228.3 k)

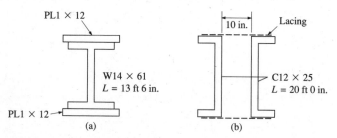

5-12 Using the parabolic equation of Prob. 5-10, find the allowable axial load for the column shown.

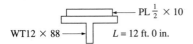

WT12 × 88 → PL $\frac{1}{2}$ × 10 L = 12 ft. 0 in.

5-13 Select the lightest available W12 section to support a 300-k axial compression load for an unsupported length of 15 ft, using the column formula of Prob. 5-9. (*Ans.* W12 × 87)

5-14 Select the lightest available W14 section to support a 500-k axial load for an unsupported length of 14 ft 0 in., using the column formula of Prob. 5-9.

5-15 Select the lightest W12 section that will support an axial compressive load of 350 k for an unsupported length of 16 ft 0 in., using the column equation of Prob. 5-10. (*Ans.* W12 × 87)

5-16 Using the ASD Specification, determine the allowable axial compressive loads that these columns can support.
  (a) A W8 × 35 with fixed ends, L = 16 ft 6 in, A36 steel
  (b) A W12 × 96 with pinned ends, L = 20 ft 0 in., A36 steel
  (c) A W10 × 68 with one end fixed and the other pinned, L = 24 ft 6 in., F_y = 50 ksi
  (d) A W14 × 193 with fixed ends, L = 22 ft 0 in., F_y = 50 ksi
  (e) A Pipe 10 Std. with pinned ends, L = 20 ft 0 in., A36 steel
  (f) Two L8 × 8 × $\frac{3}{4}$s separated $\frac{3}{8}$ in. (for gusset PL at ends), pinned ends, L = 24 ft 6 in., A36 steel

5-17 **(a)** Determine the allowable axial load that a W14 × 53 with an effective length of 16 ft can support. Use the ASD Specification and A36 steel. (*Ans.* 202.5 k)
  (b) Repeat the problem if a 1 × 10 plate is welded to one flange as shown. (*Ans.* 388.1 k)

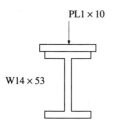

PL1 × 10

W14 × 53

Probs. 5-18 to 5-24   Determine the allowable column loads using A36 steel and the ASD Specification.

5-18

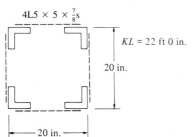

4L5 × 5 × $\frac{7}{8}$s

KL = 22 ft 0 in.

20 in.

20 in.

5-19 (*Ans.* 1052.4 k)

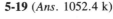

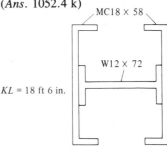

MC18 × 58

W12 × 72

KL = 18 ft 6 in.

5-20

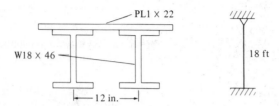

5-21 (*Ans.* 324.4 k)

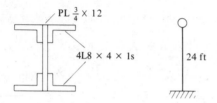

5-22   **5-23** (*Ans.* 585.6 k)

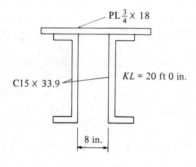   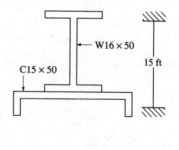

5-24

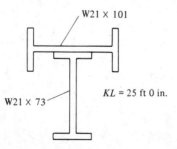

5-25 A W12 × 170 consisting of A36 steel is used as a 20-ft column. Using the ASD Specification, determine the total allowable compression load the column can support if lateral bracing is provided perpendicular to the y-axis at middepth. Assume all $Ks = 1.0$. (*Ans.* 952.5 k)

5-26 A 24-ft W10 × 112 consisting of a steel with $F_y = 50$ ksi is braced at its one-third points perpendicular to the y axis. Using the ASD specification, determine the total allowable compression load the member can support, assuming the top and bottom of the column are fixed and the lateral supports are pinned.

6

Design of Axially Loaded Compression Members

6-1 INTRODUCTION

In this chapter the designs of several axially loaded columns are presented. Included are the selections of single shapes, W sections with cover plates, and built-up sections constructed with channels. Also included are the designs of sections whose unbraced lengths in the x and y directions are different and the sizing of lacing and tie plates for built-up sections with open sides.

In Example 6-1(a) a column with $KL = 14$ ft is selected using the ASD equations. An effective slenderness ratio of 50 is assumed, the allowable stress for that value is determined from Table C-36 of the ASD Manual, and the resulting stress is divided into the column load to obtain an estimated column area. After a trial section is selected with approximately that area, its actual slenderness ratio and allowable strength are computed. The first estimated size, though quite close, is a little too small, and the next larger section in that series of shapes is tried and found to be satisfactory.

In Example 6-1(b) the problem is repeated using a higher-strength steel. It should be remembered that the yield points of high-strength steels vary for different thicknesses (unless specially alloyed to keep values equal for thicker members); therefore, when these steels are involved, it is necessary to carefully determine their yield stresses as described on pages 1-7 and 1-8 of the ASD Manual.

■ **Example 6-1**
  (a) Select a W14 section for the column and load shown in Fig. 6-1, using A36 steel.
  (b) Repeat part (a) using A441 steel.

Solution.

(a) Using A36 steel
  Assume $KL/r = 50$

$$F_a = 18.35 \text{ ksi from Table C-36 in ASD Manual}$$

$$A_{reqd} = \frac{375}{18.35} = 20.44 \text{ in.}^2$$

Try W14 × 74 ($A = 21.8 \text{ in.}^2$, $r_y = 2.48$ in.

$$\frac{KL}{r} = \frac{(1.0)(12 \times 14)}{2.48} = 67.74$$

$$F_a = 16.67 \text{ ksi}$$

Allowable $P = (16.67)(21.8) = 363 \text{ k} < 375 \text{ k}$ NG

Try W14 × 82 ($A = 24.1 \text{ in.}^2$, $r_y = 2.48$ in.)

$$\frac{KL}{r} = \frac{(1.0)(12 \times 14)}{2.48} = 67.74$$

$$F_a = 16.67 \text{ ksi}$$

Allowable $P = (16.67)(24.1) = 401.7 \text{ k} > 375 \text{ k}$ OK

Use W14 × 82

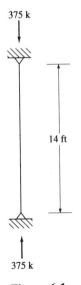

375 k

14 ft

375 k

Figure 6-1

(b) Using A441 steel

Assume $KL/r = 70$ and $F_y = 50$ ksi

$$F_a = 20.94 \text{ ksi from Table C-50 in ASD Manual}$$

$$A_{reqd} = \frac{375}{20.94} = 17.91 \text{ in.}^2$$

Try W14 × 61 ($A = 17.9$ in.^{2}, $r_y = 2.45$ in., Group 2 on page 1-8 in Manual $\therefore F_y = 50$ ksi, page 1-7)

$$\frac{KL}{r} = \frac{(1.0)(12 \times 14)}{2.45} = 68.57$$

$$F_a = 21.20 \text{ ksi}$$

$$\text{Allowable } P = (21.20)(17.9) = 379.5 \text{ k} > 375 \text{ k} \qquad \text{OK}$$

$$\text{Use W14} \times 61 \qquad \blacksquare$$

6-2 COLUMN DESIGN TABLES

The reader can see that it is quite a simple matter to compute the ASD allowable axial compression load for a particular column. For instance, in Example 6-1 we found that a W14 × 74 of A36 steel with $KL = 14$ ft has an allowable compression load of 363 k. In the same fashion we could compute the allowable loads for the same section for KL values of 10 ft, 11 ft, 12 ft, and other practical lengths. If we made the same calculations for all the other steel sections commonly used as columns, we could create a set of tables just like those given in Part 3 of the ASD Manual.

These tables tremendously expedite column calculations. They include the allowable column loads for each of the steel shapes normally used as columns (Ws, Ms, Ss, pipes, tubes, pairs of angles, and structural tees) for most of the commonly used effective lengths or KL values in feet. The column values are given with respect to the least radius of gyration for steels with $F_y = 36$ and 50 ksi (with the exception of square and rectangular tubes, which are only available in 46-ksi steel).

The resulting tables are very simple to use. The designer takes the KL value for the y direction in feet, enters the table in question from the left-hand side, and moves horizontally across the table. Under each section is listed the allowable design strength P for that KL and the steel yield stress. For example, assume that we have a 540-k load, $K_y L_y = 12$ ft, and we want to select the lightest available W14 section using A36 steel. We enter the tables with $KL = 12$ ft in the left column and read from left to right for A36 steel the numbers 4277, 3892, 3529 k, and so on until several pages later where the consecutive values 561 and 511 k are found. The 511-k value is not sufficient, and we go back to 561 k, which falls under the W14 × 99. A similar procedure can be followed for the other available shapes.

Example 6-2 illustrates the selection of W sections as well as pipes and tubes. It is possible to support a given column load with a standard pipe column; or with an extra strong pipe column (× strong) which has a smaller diameter but thicker walls, thus heavier and more expensive; or with a double extra strong pipe column (×× strong) which has an even smaller diameter and even thicker walls and heavier weight.

■ Example 6-2
Use the ASD column tables of Part 3 of the Manual.
(a) Select the lightest W section available for a load $P = 260$ k and $KL = 10$ ft. Use A36 steel.
(b) Select the lightest standard, extra strong, and double extra strong pipe columns for the situation of part (a) of this problem.
(c) Select the lightest square and rectangular tubes satisfactory for the situation of part (a), except use an F_y of 46 ksi.

Eversharp building at Milford, Conn. (Courtesy of Bethlehem Steel Corporation.)

Solution.

(a) Enter tables with $K_y L_y = 10$ ft and $P = 260$ k:

$$W14 \times 53 \ (P = 268 \text{ k})$$
$$W12 \times 53 \ (P = 288 \text{ k})$$
$$W10 \times 49 \ (P = 268 \text{ k}) \leftarrow$$
$$W8 \times 58 \ (P = 303 \text{ k})$$
$$\underline{\underline{\text{Use W10} \times 49}}$$

(b) Pipe columns:

$$12 \text{ pipe std } (P = 293 \text{ k}) \text{ wt} = 49.56 \text{ lb/ft} \leftarrow$$
$$\text{Pipe } 10\times \text{ strong } (P = 318 \text{ k}) \text{ wt} = 54.74 \text{ lb/ft}$$
$$\text{Pipe } 6\times\times \text{ strong } (P = 275 \text{ k}) \text{ wt} = 53.16 \text{ lb/ft}$$

(c) Square and rectangular tubing ($F_y = 46$ ksi):

$$9 \times 9 \times \frac{5}{16} \ (P = 262 \text{ k}) \text{ wt} = 36.10 \text{ lb/ft} \leftarrow$$

$$9 \times 7 \times \frac{1}{2} \ (P = 262 \text{ k}) \text{ wt} = 37.69 \text{ lb/ft} \qquad \blacksquare$$

6-3 COLUMN SPLICES

For multistory buildings, column splices are usually placed 2 or 3 ft above floor levels to keep from interfering with the beam and column connections. Typical column splices are shown in Fig. 6-2. The column ends are usually milled so they can be placed firmly in contact with each other for purposes of load transfer. When the contact surfaces are milled, a large part of the axial compression (if not all) can be transferred through the contacting areas. It is obvious that splice plates are necessary even though full contact is made between the columns and only axial loads are involved. They are even more necessary when consideration is given to the shears and moments existing in practical columns subjected to off-center loads, lateral forces, moments, and so on.

There is obviously a great deal of difference between tension splices and compression splices. In tension splices all load has to be transferred through the splice, whereas in splices for compression members a large part of the load can be transferred directly in bearing between the columns. The splice material is then needed to transfer only the remaining part of the load.

Fig. 6-2(a) shows a splice that may be used for columns with substantially the same nominal depths. The student will notice in the ASD Manual that wide-flange shapes of the same series (as W14) generally have the same inside distances between flanges although their total actual depths may vary as much

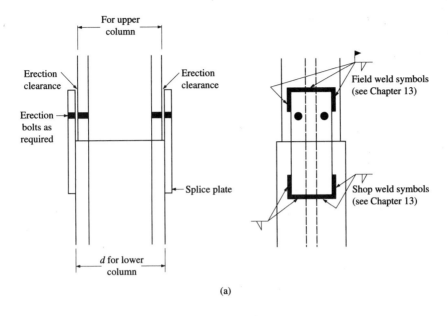

(a)

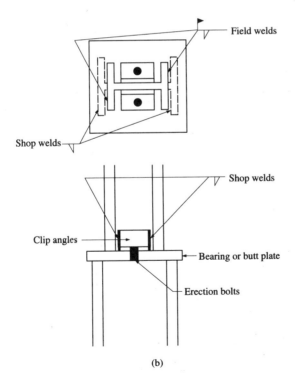

(b)

Figure 6-2 Column splices. (a) Columns from same W series with total depths close to each other ($d_{lower} < 2$ in. greater than d_{upper}). (b) Columns from different W series.

as several inches (8.68 in. difference from a W14 × 730 to a W14 × 22). It will be necessary to use filler plates between the splice plates and the upper column if the upper column has a total depth significantly less than that of the lower column.

Fig. 6.2(b) shows a type of splice which can be used for columns of equal or different nominal depths. For this type of splice the butt plate is shop-welded to the lower column, and the clip angles used for field erection are shop-welded to the upper column. In the field the erection bolts shown are installed, and the upper column is field-welded to the butt plate. The horizontal welds on this plate resist shears and moments in the columns.

Sometimes splices are used on all four sides of columns. The web splices are bolted in place in the field and field-welded to the column webs. The flange splices are shop-welded to the lower column and field-welded to the top column. The web plates may be referred to as *shear plates* and the flange plates as *moment plates*.

For multistory buildings the columns may be fabricated for one, two, or more stories. Theoretically, column sizes can be changed at each floor level so that the lightest total column weight is used. The splices needed at each floor will be quite expensive, however, and as a result it is usually more economical to use the same column sizes for at least two stories even though the total steel weight will be higher. Seldom are the same sizes used for as many as three stories because three-story columns are so difficult to erect. The two-story heights work out very well most of the time.

Welded column splice for Colorado State Building, Denver, Colo. (Courtesy of Lincoln Electric Company.)

It is most economical to use the simple splices of Fig. 6-2(a). This can easily be accomplished by using one series of shapes for as many stories of a building as possible. For instance, we may select a particular W14 column section for the top story or top two stories of a building and then keep selecting heavier and heavier W14s for that column as we come down in the building. We may also switch to higher strength steel columns as we move down in the building, thus enabling us to stay with the same W series for even more floors.

6-4 COLUMNS WITH DIFFERENT UNBRACED LENGTHS

An axially loaded column is laterally restrained in its weak direction in Fig. 6-3. Example 6-3 illustrates the design of such a column with its different unsupported lengths in the x and y directions. The student can easily solve this problem by trial and error. A trial section can be selected as described earlier in this chapter, the slenderness values $(KL/r)_x$ and $(KL/r)_y$ computed, F_a determined and multiplied by A_g to obtain P. Then, if necessary another size can be tried and so on.

For this discussion it is assumed that K is the same in both directions. Then if we are to have equal strengths about the x and y axes, the following relation must hold:

$$\frac{L_x}{r_x} = \frac{L_y}{r_y}$$

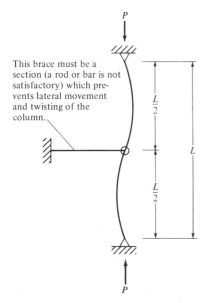

Figure 6-3 A column laterally restrained at middepth in its weaker direction

For L_y to be equivalent to L_x we would have

$$L_x = L_y \frac{r_x}{r_y}$$

If $L_y(r_x/r_y)$, is less than L_x, then L_x controls; if greater than L_x, then L_y controls.

Based on the preceding information the ASD Manual provides a method with which a section can be selected from its tables with little trial and error when the unbraced lengths are different. The designer enters the appropriate table with $K_y L_y$, selects a shape, takes the r_x/r_y value given in the table for that shape, and multiplies it by L_y. If the result is larger than $K_x L_x$, then $K_y L_y$ controls, and the shape initially selected is the correct one. If the result of the multiplication is less than $K_x L_x$, then $K_x L_x$ controls, and the designer will reenter the tables with a larger or effective $K_y L_y$ equal to $K_x L_x/(r_x/r_y)$ and select the final section.

■ Example 6-3

Select the lightest satisfactory W12 for the following conditions: A36 steel, $P = 465$ k, $K_x L_x = 26$ ft, and $K_y L_y = 13$ ft.
(a) By trial and error
(b) Using ASD tables

Solution.
(a) Using trial and error:

Assume $\dfrac{KL}{r} = 50$:

$$F_a = 18.35 \text{ ksi}$$

$$A_{reqd} = \frac{465}{18.35} = 25.34 \text{ in.}^2$$

Try W12 × 87 ($A = 25.6$ in.^{2}, $r_x = 5.38$ in., $r_y = 3.07$ in.

$$\left(\frac{KL}{r}\right)_x = \frac{(12)(26)}{5.38} = 57.99 \leftarrow$$

$$\left(\frac{KL}{r}\right)_y = \frac{(12)(13)}{3.07} = 50.81$$

$$F_a = 17.62 \text{ ksi}$$
$$P = (17.62)(25.6) = 451 \text{ k} < 465 \text{ k} \qquad\qquad \text{NG}$$

A subsequent check of the next W12 section (96 lb) shows it will work.

Use W12 × 96

(b) Using ASD tables:
    Enter tables with $K_y L_y = 13$ ft

$$\underline{\text{Try W12} \times 87} \left(\frac{r_x}{r_y} = 1.75 \right)$$

$$(K_y L_y)\left(\frac{r_x}{r_y} \right) = (13)(1.75) = 22.75 < K_x L_x$$

Therefore, $K_x L_x$ controls.

Georgia Railroad Bank and Trust Company Building, Atlanta, Ga. (Courtesy of Bethlehem Steel Corporation.)

$$\text{Reenter tables with new } K_y L_y = \frac{K_x L_x}{r_x/r_y} = \frac{26}{1.75} = 14.86$$

$$\underline{\underline{\text{Use W12} \times 96}} \qquad\qquad \blacksquare$$

In Example 6-4(a) a W section of A572 Grade 50 steel is selected for a large column load. Usually it is more economical to select a single W section to support a column load, but in Example 6-4(b) it is assumed that the desired W section is not available. As a result a smaller W section consisting of A36 steel is selected together with sufficiently large A36 cover plates to support the load. It is assumed that the plates are connected to the W section at its ends and at sufficient intermediate points so the parts act together. As this type of section is not shown in the column tables of the ASD Manual, it is necessary to use a trial-and-error procedure. An effective slenderness ratio is assumed; F_a for that KL/r is determined and divided into the column design load to estimate the total area required. The area of the W section is subtracted from the estimated total area to obtain the estimated cover plate area. A cover plate size is selected for each flange to provide the estimated area, and then F_a is calculated for the whole section, after which it may be necessary to revise the cover plate size and try again.

■ Example 6-4

(a) Select a W section for the load and column shown in Fig. 6-4 using A572 Grade 50 steel.
(b) Using A36 steel and a W14 × 311, select plates so the section will be satisfactory for the conditions of Fig. 6-4.

2700 k

15 ft

2700 k

Figure 6-4

Solution.
(a) Using A572 Grade 50 steel and ASD column tables

$$KL = (1.0)(15) = 15 \text{ ft}$$

Use W14 × 370

(b) Using a W14 × 311 with cover plates (A36 steel)

Assume $\dfrac{KL}{r} = 40$

$$F_a = 19.19 \text{ ksi}$$

$$A_{\text{reqd}} = \frac{2700}{19.19} = 140.70 \text{ in.}^2$$

Try W14 × 311 ($A = 91.4$ in.^{2}, $d = 17.12$ in., $I_x = 4330$ in.^{4}, $I_y = 1610$ in.^{4})
With 1 cover PL20 × $1\frac{1}{4}$ each flange $= 50.00$ in.^{2}

Total area $= 141.4$ in.^{2}

$$I_x = 4330 + (2)(25)(9.185)^2 = 8548 \text{ in.}^4$$

$$I_y = 1610 + (\tfrac{1}{12})(2.50)(20)^3 = 3277 \text{ in.}^4$$

$$r = \sqrt{\frac{3277}{141.4}} = 4.81 \text{ in.}$$

$$KL = (1.0)(15) = 15 \text{ ft}$$

$$\frac{KL}{r} = \frac{(12)(15)}{4.81} = 37.42$$

Allowable $F_a = 19.39$ ksi

Allowable $P = (19.39)(141.4) = 2742$ k > 2700 k

Use W14 × 311 with 1 cover PL20 × $1\frac{1}{4}$ each flange   ∎

6-5 BUILT-UP COLUMNS

Both tension and compression members may be constructed with two or more plates or shapes built up into one member. A pair of angles separated only by the thickness of the connection or gusset plates is one common type of member used in light trusses. When such members are used, the various parts must be fastened together at intervals so they will act as a unit. For tension members this process is often called *stitching*. Welds may be used at intervals (with a spacer between the parts), or the connections may be made with *stitch bolts* or *stitch rivets*. For these connections a washer or "ring fill" is placed between the parts to keep them at the proper spacing.

For built-up columns with their parts in continuous contact with each other or where the parts are separated by fillers, these requirements are listed in ASD Specification E-4:

1. When such columns consist of different components which are in contact with each other and which are bearing on base plates or milled surfaces, they must be connected at their ends with bolts or welds. If welds are used, the weld lengths must at least equal the maximum width of the member. If bolts or rivets are used, they may not be spaced longitudinally more than four diameters on center and the connection must extend for a distance at least equal to $1\frac{1}{2}$ times the maximum width of the member.

2. Throughout the length of built-up members the parts in contact with each other must be connected closely enough to provide for the transfer of calculated stresses. The spacing of bolts, rivets, or intermittent welds used for this purpose may not exceed 24 in.

3. When a built-up column that is painted or that is unpainted and not subject to corrosion has an outside plate, the ASD Specification provides certain maximum spacings for fasteners. If intermittent welds are used along the edges of the components or if bolts or rivets are provided along all gage lines at each section, their maximum spacing may not be greater than $127/\sqrt{F_y}$ times the thickness of the thinner outside plate or 12 in. Should these fasteners be staggered on each gage line, however, they may not be spaced farther apart on each gage line than $190/\sqrt{F_y}$ times the thickness of the thinner part or 18 in.

4. When built-up members are composed of two or more rolled shapes separated by intermittent fillers, the shapes must be connected at the fillers so that the KL/r values of the individual shapes do not exceed $\frac{3}{4}$ the governing slenderness ratio of the whole built-up member. (The governing slenderness ratio is the least KL/r for the whole member.) Furthermore, at least two intermediate connections must be used along the member.

Other spacing values are given in ASD Specification (J3.10) for built-up members made from unpainted weathering steel.

Example 6-5 presents the design of a double-angle compression member using the ASD column tables and A36 steel. The allowable loads for both double-angle and WT sections about their y axes are based on flexural-torsional buckling. This buckling, which has been accounted for in the ASD tables, is discussed in Chapter 7 and the Appendix.

It is assumed in this example that the 120-k column load includes the effect of wind forces. The AISC feels that severe wind and earthquake loads are of such short duration that allowable stresses may be increased by one-third (ASD specification A5.2). This is equivalent to multiplying the design loads by three-fourths.

■ Example 6-5

Using the ASD column tables and A36 steel, select a compression member consisting of two angles to support a 120-k axial load that includes wind effects. Assume the angles are spaced $\frac{3}{8}$ in. apart for end connections or gusset plates. The 12-ft-long member is to be braced at midlength perpendicular to the x axis. Assume $K = 1.0$ for all cases.

Solution.

$$\text{Reduced } P = \left(\frac{3}{4}\right)(120) = 90 \text{ k}$$

$$KL_x = 6 \text{ ft}$$
$$KL_y = 12 \text{ ft}$$

Possible selections from double-angle column tables (Part 3 of Manual):

2 L5 × 5 × $\frac{3}{8}$s (24.6 lb) will support 133 k with respect to x axis and 111 k with respect to y axis.

2 L4 × 4 × $\frac{1}{2}$s (25.6 lb) will support 131 k with respect to x axis and 115 k with respect to y axis.

Use 2 L5 × 5 × $\frac{3}{8}$s ■

The design of another built-up section is presented in Example 6-6.

■ Example 6-6

Select a pair of channels for the column and load shown in Fig. 6-5, using A36 steel and the ASD Specification. For connection purposes the back-to-back distance of the channels is to be 12 in.

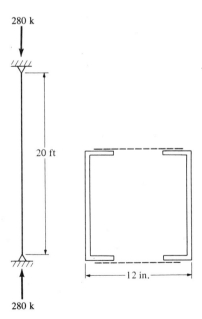

Figure 6-5

Solution.   Assume that the allowable F_a is 17.5 ksi and that the area required is $280/17.5 = 16.0$ in.^{2}

$$\text{Try 2 C12} \times 30\text{s} \ (A = 17.64 \text{ in.}^2, \ r_x = 4.29 \text{ in.})$$

$$I_y = (2)(5.14) + (17.64)(5.33)^2 = 511 \text{ in.}^4$$

$$r_y = \sqrt{\frac{511}{17.64}} = 5.38 \text{ in.}$$

$$KL = (1)(20) = 20 \text{ ft}$$

$$\frac{KL}{r} = \frac{(12)(20)}{4.29} = 56$$

Allowable $F_a = 17.81$ ksi

Allowable $P = (17.81)(17.64) = 314 \text{ k} > 280 \text{ k}$ OK

$$\underline{\text{Use 2 C12} \times 30\text{s}}$$ ∎

6-6 LACING AND TIE PLATES

The necessity for built-up compression members in large buildings and bridges was discussed in Chapter 5. When members are built up from more than one section, it is necessary to connect or lace them together across their open sides. The purpose of lacing or latticework is to hold the various parts parallel and the correct distance apart and to equalize the stress distribution among the various parts. The student will understand the necessity for lacing if he or she considers a built-up member consisting of several sections (such as the four-angle member of Fig. 5-2[i]) which supports a heavy compressive load. Each of the parts will tend to buckle laterally unless they are tied together to act as a unit in supporting the load. In addition to lacing it is necessary to have tie plates (also called stay plates or batten plates) as near the ends of the member as possible and at intermediate points if the lacing is interrupted. Fig. 6-6 (a and b) shows arrangements of tie plates and lacing. Other possibilities are shown in Fig. 6-6 (c and d).

The failure of several structures in the past has been attributed to inadequate lacing of built-up compression members. Perhaps the best-known example was the failure of the Quebec Bridge in 1907. Following its collapse the general opinion was that the latticework of the compression chords was too weak and resulted in failure. This disaster brought home to the engineering profession the importance of carefully designed lacing.

Dimensions of tie plates and lacing are usually controlled by the specifications. The ASD Specification (E4) states that tie plates shall have a thickness at least equal to one-fiftieth the distance between the connection lines of rivets, bolts, or welds and shall have a length parallel to the axis of the main member at least equal to the distance between the connection lines.

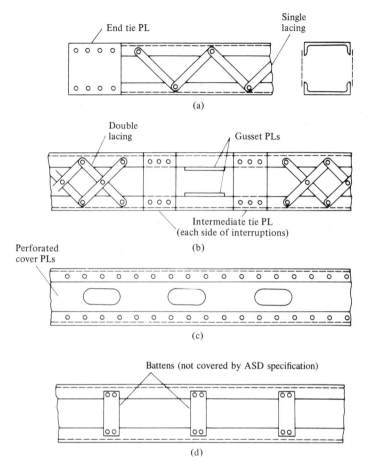

Figure 6-6 Lacing and tie plates.

Lacing usually consists of flat bars but may consist of angles, perforated cover plates, channels, or other rolled sections. These pieces must be so spaced that the individual parts being connected will not have L/r values between connections which exceed $\frac{3}{4}$ times the governing value for the entire built-up member. Lacing is assumed to be subjected to a shearing stress normal to the member and equal to not less than 2 percent of the total compression in the member. The ASD column formulas are used to design the lacing in the usual manner. Slenderness ratios are limited to 140 for single lacing and 200 for double lacing. Double lacing should be used if the distance between connection lines is greater than 15 in.

Rather than using tie plates and lacing bars it is permissible to use continuous cover plates over the open sides of the built-up sections. Access holes are necessary, and these plates are referred to as *perforated cover plates*. Stress concentrations and secondary bending stresses are usually neglected, but lateral shearing stresses must be checked as they must be for other types of latticework. Perforated cover plates are becoming popular with an increasing percentage of engineers because they possess these advantages:

1. They are easily fabricated with modern gas cutting methods.
2. Some specifications permit the inclusion of their net areas in the effective section of the main members, provided the holes are made in accordance with their empirical requirements, which have been developed on the basis of extensive research.
3. Painting of the members is probably simplified as compared to ordinary lacing bars.

Example 6-7 illustrates the design of lacing and end tie plates for the built-up column of Example 6-6. Bridge specifications are somewhat different in their lacing requirements from the ASD, but the design procedure is much the same.

■ Example 6-7
Using the ASD Specification, design bolted single lacing for the column of Example 6-6. Reference is made to Fig. 6-7.

Solution.   Distance between lines of bolts is 8.5 in. < 15 in.; therefore, single lacing OK.

Assume an inclination of 60° with axis of member. Length of channels between lacing connections is 8.5/cos 30° = 9.8 in., and L/r of one channel between connections is $9.8/0.763 = 12.9 < \frac{3}{4} \times \dfrac{KL}{r}$ of main member = $(\frac{3}{4})(56) = 42$.

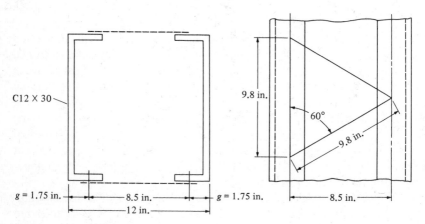

Figure 6-7

450 Lexington Ave., New York City. (Courtesy of Owen Steel Company, Inc.)

Force on lacing bar:

$$V = 0.02\,P = (0.02)(280) = 5.6\text{ k}$$
$$\tfrac{1}{2}V = 2.8\text{ k} \; - \; \text{shearing force on each plane of lacing}$$
$$\text{Force on bar} = (9.8/8.5)(2.8) = 3.23\text{ k}$$

Properties of a flat bar:

$$I = \tfrac{1}{12}bt^3$$
$$A = bt$$
$$r = \sqrt{\dfrac{\tfrac{1}{12}bt^3}{bt}}$$
$$r = 0.289t$$

Design of bar:

$$\text{Max } \frac{L}{r} = 140$$

$$\frac{9.8}{0.289t} = 140$$

$$t = 0.242 \text{ in.}$$

$$\underline{\text{try } \tfrac{1}{4}\text{-in. flat bar}}$$

$$\frac{L}{r} = \frac{9.8}{(0.289)(0.250)} = 136$$

Allowable $F_a = 8.07$ ksi

$$A_{\text{reqd}} = \frac{3.23}{8.07} = 0.400 \text{ in.}^2 \; (1.60 \times \tfrac{1}{4})$$

Minimum edge distance if $\tfrac{3}{4}$-in. bolt used $= 1\tfrac{1}{4}$ in.

$$\underline{\text{Use } \tfrac{1}{4} \times 2\tfrac{1}{2} \text{ bar}}$$

Design of end tie plates:

$$\text{Min length} = 8.5 \text{ in.}$$

With t not less than $\tfrac{1}{50}$ distance between bolt lines,

$$t = (\tfrac{1}{50})(8.5) = 0.17 \text{ in.}$$

$$\underline{\text{Use } \tfrac{3}{16} \times 8\tfrac{1}{2} \times 11\tfrac{1}{2} \text{ end tie PLs}}$$ ■

6-7 STIFFENED AND UNSTIFFENED ELEMENTS

Up to this point in the text the author has considered only the overall stability of members yet it is entirely possible for the thin flanges or webs of a column or beam to buckle locally in compression well before the calculated buckling strength of the whole member is reached. When thin plates are used to carry compressive stresses, they are particularly susceptible to buckling about their weak axes due to the small moments of inertia in those directions.

The ASD Specification (Table B5.1) provides limiting values for the width-thickness ratios of the individual parts of compression members and for the parts of beams in their compression regions. The student is well aware of the lack of stiffness of thin pieces of cardboard or metal or plastic with free edges. If, however, one of these elements is folded or restrained, its stiffness is appreciably increased. For this reason two categories are listed in the Manual, these being *stiffened elements* and *unstiffened elements*.

An unstiffened element is a projecting piece with one free edge parallel to the direction of the compression force, while a stiffened element is supported along the two edges in that direction. These two types of elements are illustrated in Fig. 6-8. In each case the width *b* and the thickness *t* of the elements in question are shown.

Depending on the ranges of different width-thickness ratios for compression elements and on whether the elements are stiffened or unstiffened, the elements will buckle at different stress situations.

For establishing width-thickness ratio limits for the elements of compression members, the ASD Specification divides members into three classifications: compact sections, noncompact sections, and slender compression elements. These classifications, which decidedly affect the design compression stresses to be used for columns, are discussed in the paragraphs to follow.

Compact Sections

A compact section has a sufficiently stocky profile so that it is capable of developing a fully plastic stress distribution before buckling. The term *plastic* means stressed throughout to the yield stress and is discussed at length in Chapter 18. For a compression member to be classified as compact, its flanges must be continuously connected to its web or webs and the width-thickness ratios of its compression elements may not be greater than the limiting ratios given in Table 6-1 (Table B5.1 in the ASD Specification).

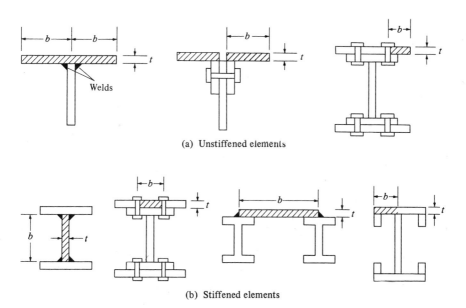

(a) Unstiffened elements

(b) Stiffened elements

Figure 6-8

TABLE 6-1   LIMITING WIDTH-THICKNESS RATIOS FOR COMPRESSION ELEMENTS

DESCRIPTION OF ELEMENT	WIDTH-THICK-NESS RATIO	LIMITING WIDTH-THICKNESS RATIOS	
		COMPACT	NONCOMPACT [a]
Flanges of I-shaped rolled beams and channels in flexure[b]	b/t	$65/\sqrt{F_y}$	$95/\sqrt{F_y}$
Flanges of I-shaped welded beams in flexure	b/t	$65/\sqrt{F_y}$	$95/\sqrt{F_{yf}/k_c}$ [c]
Outstanding legs of pairs of angles in continuous contact; angles or plates projecting from rolled beams or columns; stiffeners on plate girders	b/t	NA	$95/\sqrt{F_y}$
Angles or plates projecting from girders, built-up columns, or other compression members; compression flanges of plate girders	b/t	NA	$95/\sqrt{F_y/k_c}$
Stems of tees	d/t	NA	$127/\sqrt{F_y}$
Unstiffened elements simply supported along one edge, such as legs of single-angle struts, legs of double-angle struts with separators, and cross or star-shaped cross sections	b/t	NA	$76/\sqrt{F_y}$
Flanges of square and rectangular box and hollow structural sections of uniform thickness subject to bending or compression;[d] flange cover plates and diaphragm plates between lines of fasteners or welds	b/t	$190/\sqrt{F_y}$	$238/\sqrt{F_y}$
Unsupported width of cover plates perforated with a succession of access holes[e]	b/t	NA	$317/\sqrt{F_y}$
All other uniformly compressed stiffened elements, i.e., supported along two edges	b/t h/t_w	NA	$253/\sqrt{F_y}$
Webs in flexural compression[b]	d/t h/t_w	$640/\sqrt{F_y}$ —	— $760/\sqrt{F_b}$
Webs in combined flexural and axial compression	d/t_w	for $f_a/F_y \leq 0.16$ $\dfrac{640}{\sqrt{F_y}}\left(1 - 3.74\dfrac{f_a}{F_y}\right)$	

TABLE 6-1 *Continued*

DESCRIPTION OF ELEMENT	WIDTH-THICK-NESS RATIO	LIMITING WIDTH-THICKNESS RATIOS	
		COMPACT	NONCOMPACT [a]
		for $f_a/F_y > 0.16$ $257/\sqrt{F_y}$	—
	h/t_w	—	$760/\sqrt{F_b}$
Circular hollow sections in axial compression in flexure	D/t	$3300/F_y$ $3300/F_y$	— —

Source: American Institute of Steel Construction, *Manual of Steel Construction Allowable Stress Design,* 9th ed. (Chicago: AISC, 1989), Table B5.1, p. 5-36. Reprinted with the permission of AISC.

[a] For design of slender sections that exceed the noncompact limits, see ASD Manual, Appendix B5.

[b] For hybrid beams, use the yield strength of the flange F_{yf} instead of F_y.

[c] $k_c = \dfrac{4.05}{(h/t)^{0.46}}$ if $h/t > 70$, otherwise $k_c = 1.0$.

[d] See also ASD Manual, Section F3.1.

[e] Assumes net area of plate at widest hole.

Noncompact Sections

In a noncompact section the yield stress can be reached in some but not all of the compression elements before buckling occurs. Noncompact sections cannot reach a fully plastic stress distribution. They do not qualify as compact shapes but have width-thickness ratios that do not exceed the values given for noncompact shapes.

Slender Compression Elements

Slender compression elements have width-thickness ratios greater than those given for noncompact sections and will buckle elastically before the yield stress is reached in any part of the section. For such elements it is necessary to consider elastic buckling strengths.

A shape with a cross section that does not satisfy the width-thickness requirements of Table 6-1 can still be used as a column, but its allowable stress will have to be reduced. For most sections the limits can be met, and allowable-stress reductions are unnecessary. All of the W and tube sections listed in the ASD Manual with $F_y = 36$ and 50 ksi meet the requirements except the W14 × 43 in 50-ksi steel.

Should the width-thickness limits for noncompact sections be exceeded, Appendix B of the ASD Manual must be consulted for allowable-stress reductions. The formulas presented there are so complex and tedious to apply that we would be wise to almost never permit the use of members falling in this classification of slender compression elements. Slender Compression elements are discussed in detail in Section A-1 of the Appendix of this book and a numerical example (A-1) is presented.

6-8 BASE PLATES FOR AXIALLY LOADED COLUMNS

The allowable compressive stress in a concrete or other type of masonry footing is much smaller than it is in a steel column. When a steel column is supported by a footing, the column load must be spread over a sufficient area to keep the footing from being overstressed. Loads from a steel column are transferred through a steel base plate to a fairly large area of the footing below. (It will be noted that a footing performs a related function in that it spreads the load over an even larger area so that the underlying soil will not be overstressed.)

The base plates for steel columns can be welded directly to the columns or fastened by means of some type of bolted or welded lug angles. These connection methods are illustrated in Fig. 6-9. A base plate welded directly to the column is shown in part (a) of the figure. For small columns these plates are usually shop-welded to the columns, but for larger columns it may be necessary to ship the plates separately and set them to the correct elevations. The columns are then set and connected to the footing with anchor bolts which pass through the lug angles which have been shop-welded to the columns. This type of arrangement is shown in Fig. 6-9(b). Some designers like to use lug angles on both flanges and web.

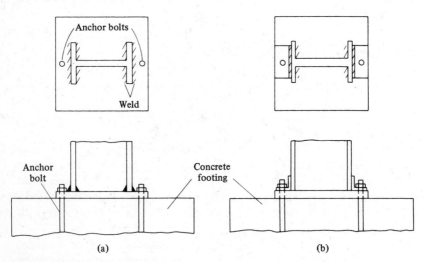

Figure 6-9 Column base plates.

To make sure that column loads are spread uniformly over their base plates, it is essential to have good contact between the two. To accomplish this objective it is necessary to straighten plates thicker than 2 in. up through 4 in. by pressing or milling. Plates thicker than 4 in. need to have their upper surfaces milled. The bottom of the plates will be in contact with cement grout and do not have to be milled. The top surfaces of base plates that are connected to columns with complete-penetration welds (see Chapter 13) do not have to be pressed or milled.

A critical phase in the erection of a steel building is the proper positioning of column base plates. If they are not located at their correct elevations, serious stress changes may occur in the beams and columns of the steel frame. One of the following three methods is used for preparing the site for the erection of a column to its proper elevation: leveling plates, leveling nuts, or preset base plates. An article by Ricker[1] describes these procedures in considerable detail.

For small-to-medium base plates (up to 20 to 22 in.), approximately $\frac{1}{4}$-in.-thick leveling plates with the same dimensions as the base plates (or a little larger) are shipped to the job and carefully grouted in place to the proper elevations. Then the columns with their attached base plates are set on the leveling plates.

For larger base plates up to about 36 in., some types of leveling nuts are used to adjust the base plates up or down.

If the base plates are larger than about 36 in., the columns with the attached base plates are so heavy and cumbersome that it is difficult to ship them together. For such cases the base plates are shipped to the job and placed in advance of the steel erection. They can be leveled with shims or wedges. Frequently bolts or threaded rods are used for leveling purposes.

For tremendously large base plates weighing several tons or more, angle frames may be built to support the plates. They are carefully leveled and filled with concrete, which is screeded off to the correct elevations and the base plates are set directly on the concrete.

A column transfers its load to the supporting pier or footing through the base plate. Should the supporting concrete area A_2 be larger than the plate area A_1, the allowable pressure will be higher. In that case the concrete surrounding the contact area supplies appreciable lateral support to the directly loaded part, with the result that the loaded concrete can support more load. This fact is reflected in the allowable stresses.

The ASD Specification (J9) provides two allowable pressures for concrete supports. These values are based on the compressive strength of the concrete f_c' and the percentage of the support area covered by the base plate.

If the entire concrete area A_2 is covered by the plate whose area is A_1,

$$F_p = 0.35 f_c' \tag{1}$$

[1]D. T. Ricker, "Some Practical Aspects of Column Bases," *Engineering Journal,* AISC, 26, no. 3 (3d quarter, 1989), pp. 81–89.

If A_1 is less than A_2,

$$F_p = 0.35 f_c' \sqrt{\frac{A_2}{A_1}} \leq 0.7 f_c' \tag{2}$$

For the usual footing, F_p will equal $0.7 f_c'$. For bearing on brick in cement mortar, a value of $F_p = 0.25$ ksi is specified, while 0.40 ksi is permitted by the ASD Specification (J9) for bearing on sandstone or limestone.

If we substitute $P/A_1 = F_p$ into Equation (2),

$$\frac{P}{A_1} = 0.35 f_c' \sqrt{\frac{A_2}{A_1}}$$

$$A_1 = \frac{1}{A_2}\left(\frac{P}{0.35 f_c'}\right)^2 \tag{3}$$

The smallest possible value of A_1 will occur if $F_p = 0.7 f_c'$.

$$A_1 = \frac{P}{0.7 f_c'} \tag{4}$$

Thus, A_1 is the largest value obtained from Equations (3) and (4).

The ASD Manual (pages 3-107 and 3-108) develops a formula that enables the designer to select the minimum area of a pedestal such that the optimum concrete bearing stress is obtained ($A_2 \geq P/0.175 f_c'$). Their Example 13 illustrates the use of this expression.

To analyze the base plate shown in Fig. 6-10, the column is assumed to apply a total load P to the base plate, and this load is assumed to be transmitted uniformly through the base plate to the footing below with a value of f_p psi (P/A_1). The footing will push back with a pressure of f_p psi and tend to curl up the cantilevered parts of the base plate outside the column, as shown. This pressure also tends to push up the part of the base plate between the flanges of the column.

With reference to Fig. 6-10 the ASD Manual suggests that maximum moments in a base plate occur at distances approximately $0.80 b_f$ and $0.95 d$ apart. The bending moment is calculated at each of these sections, and the larger value is used to determine the plate thickness needed. This method of analysis is only a rough approximation of the true conditions because the actual plate stresses are caused by a combination of bending in two directions.

From Fig. 6-10 the following moment expressions can be written for the two critical sections, considering in each case a 1-in. width of plate:

$$M = f_p n \frac{n}{2} = \frac{f_p n^2}{2}$$

$$M = f_p m \frac{m}{2} = \frac{f_p m^2}{2}$$

The section modulus of a 1-in. width of plate of t thickness is

$$S = \frac{I}{C} = \frac{(\frac{1}{12})(1)(t^3)}{t/2} = \frac{t^2}{6}$$

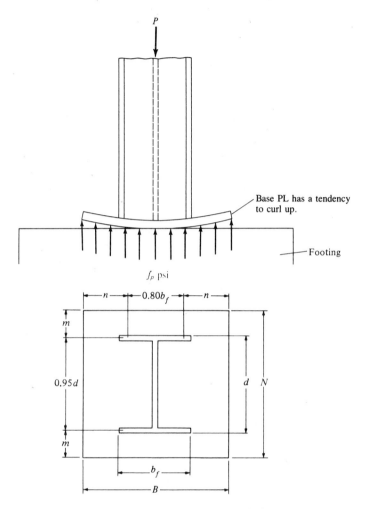

Figure 6-10 Dimensions used in base plate design.

Since the stress is $Mc/I = M/S$, the required thickness of the base plate can be determined.

$$F_b = \frac{M}{S} = \frac{f_p(m^2/2)}{t^2/6}$$

$$= \frac{3f_p m^2}{t^2}$$

$$t = \sqrt{\frac{3f_p m^2}{F_b}}$$

The ASD Specification (F2.1) provides an allowable bending stress in the plates, F_b, equal to $0.75F_y$. Substituting this value into the preceding expression results in

$$t = 2m\sqrt{\frac{f_p}{F_y}}$$

Similarly in the other direction,

$$t = 2n\sqrt{\frac{f_p}{F_y}}$$

Should the values of m and n be equal, the calculated moments will be equal, and the resulting plate thickness will be kept to a minimum. This condition can be approached if

$$N = \sqrt{A_1} + \Delta$$

where

$$A_1 = \text{area of plate} = BN$$

$$\Delta = 0.5(0.95d - 0.80b_f) \text{ and } B \approx \frac{A_1}{N}$$

If the column is supporting a very small load, the area of the plate, and thus m and n, will be very small and the computed plate thickness may be close to zero. For such situations the load is assumed to be distributed over the cross-hatched area shown in Fig. 6-11. This cross-hatched area can be determined with the following expression:

$$A = b_f d - \left(\frac{b_f - 2L}{2}\right)(2)(d - 2L)$$

$$= 2(d + b_f - 2L)L$$

The dimension L can be determined from the following quadratic expression. Two L values are obtained, but the smaller one is used in subsequent calculations.

$$F_p = \frac{P}{2(d + b_f - 2L)L}$$

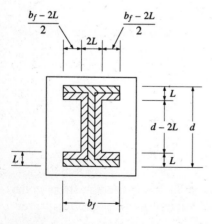

Figure 6-11 Distribution of small column load to the supporting concrete.

Assuming the plate is fixed at the edge of the web, the moment for a 1-in.-wide strip of plate is equated to the resisting moment of a plate of t thickness and 1-in. width. The resulting expression is solved for t. We must check our plate thickness with this expression as well as the earlier ones.

$$(f_p L)\left(\frac{L}{2}\right) = \left(\frac{1}{12}\right)(1)(t^3)F_b$$

$$t = L\sqrt{\frac{3 f_p}{F_b}}$$

It is possible for these formulas to yield some rather odd values for the plate thickness. As a result several papers[2] have suggested revisions. Thornton[3] recommends that a value $n' = \frac{1}{4}\sqrt{db_f}$ be used in place of the L value just described and that the third required plate thickness expression be

$$t_p = 2n'\sqrt{\frac{f_p}{F_y}}$$

Such a procedure is quite similar to the one recommended in the eighth edition of the ASD Manual. It very conservatively affects the thicknesses of small base plates.

Before the final dimensions of base plates are selected, reference should be made to the bars and plates section in Part 1 of the ASD Manual so that standard sizes are used. In this way the designer will be insured of both economy and promptness of delivery. Example 6-8 illustrates the design of a base plate by this procedure.

■ Example 6-8

Design a base plate with A36 steel for a W12 × 65 column ($d = 12.12$ in., $b_f = 12.00$ in.) and a load of 370 k. The column is to be supported by an 8 ft × 8 ft reinforced-concrete footing with $f_c' = 3$ ksi.

Solution.   Determine plate area:

$$A_1 = \frac{1}{A_2}\left(\frac{P}{0.35 f_c'}\right)^2 = \frac{1}{(12 \times 8)^2}\left(\frac{370}{0.35 \times 3}\right)^2 = 13.5 \text{ in.}^2$$

$$A_1 = \frac{P}{0.7 f_c'} = \frac{370}{(0.7)(3)} = 176.2 \text{ in.}^2 \leftarrow$$

---

[2] For example, S. Ahmed and R. R. Kreps, "Inconsistencies in Column Base Plate Design in the New AISC ASD Manual," *Engineering Journal*, AISC, 27, no. 3 (3rd quarter, 1990), pp. 106–107.

[3] W. A. Thornton, "Design of Small Base Plates for Wide Flange Columns," *Engineering Journal*, AISC, 27, no. 3 (3rd quarter, 1990), pp. 108–110.

Dimensions of plate:

$\Delta = 0.5(0.95d - 0.80b_f) = 0.5(0.95 \times 12.12 - 0.80 \times 12.00) = 0.96$ in.

$N \approx \sqrt{A_1} + \Delta = \sqrt{176.2} + 0.96 = 14.23$ in.　　(Say 14 in.)

$B = \dfrac{A_1}{N} = \dfrac{176.2}{14} = 12.59$ in.　　(Say 13 in.)

Pressure on concrete footing:

$$f_p = \frac{P}{B \times N} = \frac{370}{(13)(14)} = 2.03 \text{ ksi}$$

Compute m and n dimensions:

$$m = \frac{N - 0.95d}{2} = \frac{14 - 0.95 \times 12.12}{2} = 1.24 \text{ in.}$$

$$n = \frac{B - 0.80b_f}{2} = \frac{13 - 0.80 \times 12}{2} = 1.70 \text{ in.}$$

Determine L:

As plate does not cover all of concrete,

$$F_p = 0.35f'_c \sqrt{\frac{A_2}{A_1}} = (0.35)(3)\sqrt{\frac{(12 \times 8)^2}{(13)(14)}}$$

$$= 7.47 > 0.7f'_c = 2.1 \text{ ksi}$$

$$F_p = \frac{P}{2(d + b - 2L)L}$$

$$2.10 = \frac{370}{2(12.12 + 12.00 - 2L)L}$$

Solving preceding expression for L yields an imaginary number. Calculate plate thickness:

$$t_p = n\sqrt{\frac{f_p}{0.25F_y}} = 1.70\sqrt{\frac{2.03}{(0.25)(36)}} = 0.81 \text{ in. (Say } \tfrac{13}{16} \text{ in.)}$$

$$\underline{\underline{\text{Use PL } \tfrac{13}{16} \times 13 \times 1 \text{ ft 2 in.}}} \quad \blacksquare$$

Moment Resisting Column Bases

The designer will often be faced with the need for moment resisting column bases. Before such a topic is introduced, the author feels that the student needs to be familiar with the design of welds (Chapter 13) and moment resisting connections between members (Chapters 13 and 14). As a result, the subject of moment resisting base plates has been placed in the Appendix (Section A-5).

PROBLEMS

All of the columns for these problems are assumed to be in frames that are braced against sidesway. Use ASD specification.

6-1 to 6-3   Use a trial-and-error procedure in which a KL/r value is estimated, a value of F_a determined from the appropriate table (C-36 or C-50 in Part 3 of the Manual), the estimated column area determined, a trial section selected, its allowable P computed, and then another size tried if necessary, and so on.

6-1 Select the lightest available W12 section to support an axial compression load $P = 235$ k. $KL = 14$ ft, and A36 steel is used. (*Ans.* W12 × 53)

6-2 Select the lightest available W14 section to support an axial load $P = 500$ k. $KL = 12$ ft, and A36 steel is used.

6-3 Repeat Prob. 6-2 if $F_y = 50$ ksi. (*Ans.* W14 × 74)

6-4 to 6-20   Take advantage of all available column tables in the Manual, particularly those in Part 3.

6-4 Repeat Prob. 6-1.

6-5 Repeat Prob. 6-2. (*Ans.* W14 × 90)

6-6 Repeat Prob. 6-3.

6-7 Several building columns are to be designed using A36 steel and the ASD Specification. Select the lightest available W sections for these columns.
 (a) $P = 350$ k, $L = 12$ ft, pinned-end supports (*Ans.* W12 × 65)
 (b) $P = 305$ k, $L = 14$ ft, fixed-end supports (*Ans.* W8 × 58 or W12 × 58)
 (c) $P = 450$ k, $L = 16$ ft 6 in., fixed at bottom, pinned at top (*Ans.* W12 × 87)
 (d) $P = 1200$ k, $L = 15$ ft, pinned-end supports (*Ans.* W12 × 230)

6-8 (a) Select the lightest W section available in A36 steel to serve as a pinned-end column to support an axial load $= 700$ k. Assume KL is 14 ft.
 (b) Repeat with $F_y = 50$ ksi.

6-9 A W section is to be selected to support an axial compressive load $P = 610$ k. The member, which is to be 24 ft long and is to be pinned top and bottom, has lateral support (pinned) supplied in the weak direction at middepth. Select the lightest W12 or 14 section using A36 steel. (*Ans.* W12 × 120 or W14 × 120)

6-10 Repeat Prob. 6-9 if $F_y = 50$ ksi.

6-11 Repeat Prob. 6-9 if lateral support is provided in the weak direction at the one-third points and the total column length is changed to 33 ft. (*Ans.* W14 × 132)

6-12 A 27-ft column is laterally supported in the weak direction at its middepth. Select the lightest W section that can adequately support the axial gravity load $P = 250$ k, using A36 steel. Assume all K's are 1.0.

6-13 A 14-ft column is to be built into a wall in such a manner that it will be continuously braced in its weak direction but not in its strong direction. If the member is to consist of A36 steel and is assumed to have pinned ends, select the lightest satisfactory W12 section available using the ASD Specification. $P = 630$ k. (*Ans.* W12 × 120)

6-14 Repeat Prob. 6-13 if $F_y = 50$ ksi.

6-15 A W14 section of A36 steel is to be selected to support an axial compressive load $P = 525$ k. The member, which is to be 30 ft long, is to be fixed top and bottom and is to have lateral support at its one-third points perpendicular to the y axis (pinned). (*Ans.* W14 × 99)

6-16 Using A36 steel (except $F_y = 46$ ksi for square and rectangular tubing), select the lightest available rolled sections (W, M, S, HP, square, rectangular, or round tubing) that are adequate for the following situations:
(a) $P = 200$ k, $L = 14$ ft, pinned ends
(b) $P = 260$ k, $L = 15$ ft, fixed ends
(c) $P = 460$ k, $L = 20$ ft, one end pinned and the other fixed

6-17 Assuming axial loads only, select W sections for an interior column of the frame shown. Use A36 steel and the ASD Specification. Each column section can be used for one or two stories before it is spliced, whichever seems advisable. Miscellaneous data: concrete weighs 150 lb/ft³. Live load on roof = 40 psf (includes roofing). Live load on interior floors = 80 psf. Partition load on interior floors = 15 psf. All points pinned. Frames 30 ft on center. (*Ans.* Several possible solutions such as W14 × 61 top two stories and W14 × 132 bottom two stories; or W12 × 65 top two stories and W12 × 136 bottom two stories)

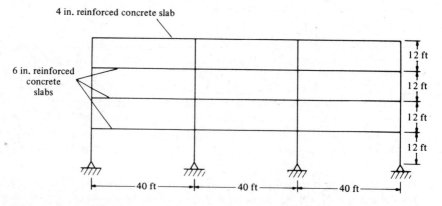

6-18 It is desired to design a column for $P = 1300$ k using A36 steel with $KL = 12$ ft. A W14 × 145 is on hand with a good supply of $\frac{3}{4}$-in. thick plates. Design cover plates to be welded to the flanges of the W section to enable the column to support the required load.

6-19 Repeat Prob. 6-18 if a steel with $F_y = 50$ ksi is used and $P = 1575$ k. (*Ans.* W14 × 145 with 1 PL $\frac{3}{4}$ × 12 each flange)

6-20 Select a pair of channels (Cs or MCs) to support an axial load of 320 k. The member is to be 24 ft long with both ends pinned and is to be arranged as shown. Use A36 steel.

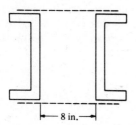

6-21 Four $4 \times 4 \times \frac{1}{2}$ angles are used to form the member shown. The member is 20 ft long, has pinned ends, and consists of A36 steel. Determine the allowable compressive strength of the member. Design single lacing and end tie plates, assuming connections are made to the angles with $\frac{3}{4}$-in. bolts. (*Ans.* P = 298.8 k. Use $\frac{1}{4} \times 13 \times 1$ ft 4 in. end tie plates and $\frac{15}{32} \times 2\frac{1}{2} \times 1$ ft $8\frac{7}{8}$ in. single lacing at 45°)

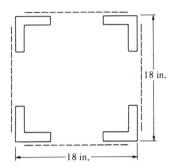

6-22 Select a pair of miscellaneous 12-in. channels to support an axial compressive load of $P = 450$ k. The member is to be 24 ft long with both ends pinned and is to be arranged as shown. Use A36 steel, and design single lacing and end tie plates, assuming $\frac{3}{4}$-in. bolts are to be used for connections. Assume bolts located $2\frac{1}{4}$ in. from back of channels.

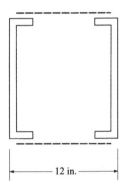

6-23 A W12 × 65 column supports an axial load of 350 k. Using A36 steel, design a base plate for the column if the supporting 9 ft × 9 ft reinforced-concrete footing has an $f'_c = 3$ ksi. (*Ans.* PL$\frac{5}{8} \times 12 \times 1$ ft 2 in.)

6-24 Repeat Prob. 6-23 if the column is supported by a 24 in. × 24 in. pedestal.

6-25 Repeat Prob. 6-23 if the load is 500 k and $F_y = 50$ ksi. (*Ans.* PL$1\frac{1}{8} \times 14 \times 1$ ft 5 in.)

6-26 Design a base plate with A36 steel for a W14 × 120 supporting an axial load of 600 k. The reinforced-concrete footing is 10 ft × 10 ft, and f'_c is 3.5 ksi.

6-27 Repeat Prob. 6-26 if the load is 650 k and f'_c is 3 ksi. (*Ans.* PL$1\frac{1}{4} \times 17 \times 1$ ft 7 in.)

6-28 A W8 × 24 column supporting a 75-k axial load is bearing on a block wall with an allowable bearing pressure of 600 psi. Design a base plate for the column using A36 steel.

6-29 A W14 × 74 column has an 2 × 20 × 1 ft 8 in. A36 base plate on a 10 ft × 10 ft reinforced-concrete footing with f_c' = 3 ksi. What is the maximum allowable axial load that this plate can support? (*Ans.* 404 k)

7

Design of Axially Loaded Compression Members Continued

7-1 FURTHER DISCUSSION OF EFFECTIVE LENGTHS

The subject of effective lengths was introduced in Chapter 5, and some suggested K factors were presented in Table 5-1. These factors were developed for columns with certain idealized conditions of end restraint which may be very different from practical design conditions. The table values are usually quite satisfactory for preliminary designs and for situations where sidesway is prevented by bracing. Should the columns be part of a continuous frame subject to sidesway, however, it is often advantageous to make a more detailed analysis as described in this section. To a lesser extent this is also desirable for columns in frames braced against sidesway.

Perhaps a few explanatory remarks should be made at this point, defining sidesway as it pertains to effective lengths. For this discussion sidesway refers to a type of buckling. In statically indeterminate structures sidesway occurs where the frames deflect laterally due to the presence of lateral loads or unsymmetrical vertical loads or where the frames themselves are unsymmetrical. Sidesway also occurs in columns whose ends can move transversely when they are loaded until buckling occurs.

Should frames with diagonal bracing or rigid shear walls be used, the columns will be prevented from sidesway and provided with some rotational

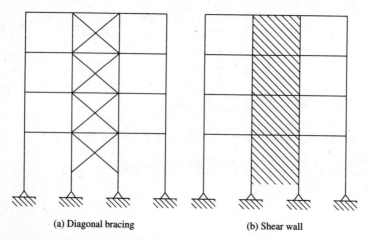

(a) Diagonal bracing   (b) Shear wall

Figure 7-1 Sidesway inhibited.

restraint at their ends. For these situations, pictured in Fig. 7-1, the K factors will fall somewhere between cases (a) and (d) of Table 5-1.

The ASD Specification (C2) states that $K = 1.0$ should be used for columns in frames with sidesway inhibited unless an analysis shows that a smaller value can be used. $K = 1.0$ is often quite conservative, and an analysis made as described herein may result in some savings.

The effective length of a column is a property of the whole structure of which the column is a part. In many existing buildings it is probable that the masonry walls provide sufficient lateral support to prevent sidesway. When light curtain walls are used, however, as they often are in modern buildings, there is probably little resistance to sidesway. Sidesway is also present in tall buildings in appreciable amounts unless a definite diagonal bracing system or shear walls are used. For these cases it seems logical to assume that resistance to sidesway is primarily provided by the lateral stiffness of the frame alone.

Theoretical mathematical analyses may be used to determine effective lengths, but such procedures are usually too lengthy and perhaps too difficult for the average designer. The usual procedure is to use either Table 5-1, interpolating between the idealized values as the designer feels appropriate, or the alignment charts or nomograms that are described in this section.

The charts shown in Fig. 7-2 present a practical method for estimating K values.[1,2] One chart was developed for columns braced against sidesway and one for columns subject to sidesway. Their use enables the designer to obtain

[1] O. G. Julian and L. S. Lawrence, "Notes on J and L Monograms for Determination of Effective Lengths" (1959).

[2] Structural Stability Research Council, *Guide to Stability Design Criteria for Metal Structures,* 4th ed. T. V. Galambos, ed. (New York: Wiley, 1988).

Shearson Lehman/American Express Information Services Center, New York City. (Courtesy of Owen Steel Company, Inc.)

good K values without struggling through lengthy trial-and-error procedures with the buckling equations.

To use the alignment charts it is necessary to have preliminary sizes for the girders and columns framing into the column in question before the K factor can be determined for that column. In other words, before the chart can be used, we have to either assume some member sizes or carry out a preliminary design.

The resistance to rotation furnished by the beams and girders meeting at one end of a column is dependent on the rotational stiffnesses of those members. The moment needed to produce a unit rotation at one end of a member if the other end of the member is fixed is referred to as its *rotational stiffness*. From our structural analysis studies this works out to be equal to $4EI/L$ for a homogeneous member of constant cross section. Based on the preceding we

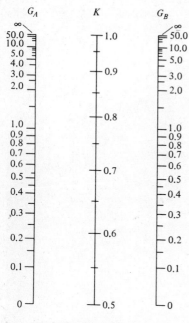

(a) Sidesway prevented

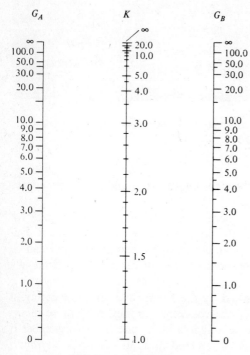

(b) Sidesway uninhibited

Figure 7-2 Column effective length chart. "The subscripts A and B refer to the joints at the two ends of the column section being considered. G is defined as

$$G = \frac{\sum \dfrac{I_c}{L_c}}{\sum \dfrac{I_g}{L_g}}$$

in which Σ indicates a summation of all members rigidly connected to that joint and lying in the plane in which buckling of the column is being considered, I_c is the moment of inertia and L_c the unsupported length of a column section, and I_g is the moment of inertia and L_g the unsupported length of a girder or other restraining member. I_c and I_g are taken about axes perpendicular to the plane of buckling being considered.

"For column ends supported by but not rigidly connected to a footing or foundation, G is theoretically infinity, but, unless actually designed as a true friction free pin, may be taken as '10' for practical designs. If the column end is rigidly attached to a properly designed footing, G may be taken as 1.0. Smaller values may be used if justified by analysis."

From American Institute of Steel Construction, *Manual of Steel Construction Allowable Stress Design*, 9th ed. (Chicago: AISC, 1989), pp. 3–5, 5–137. Reprinted with the permission of AISC.

can say that the rotational restraint at the end of a particular column is proportional to the ratio of the sum of the column stiffnesses to the girder stiffnesses meeting at that joint.[3]

$$G = \frac{\sum \dfrac{4EI}{L} \text{ for columns}}{\sum \dfrac{4EI}{L} \text{ for girders}} = \frac{\sum \dfrac{I_c}{L_c}}{\sum \dfrac{I_g}{L_g}}$$

To determine a K value for a particular column, the following steps are taken:

1. Select the appropriate chart (sidesway prevented or sidesway uninhibited).
2. Compute G at each end of the column and label the values G_A and G_B as desired.
3. Draw a straight line on the chart between the G_A and G_B values and read K where the line hits the center K scale.

The Structural Stability Research Council (SSRC) makes several recommendations concerning the use of the alignment charts.

1. For pinned columns G is theoretically infinite, as where a column is connected to a footing with a frictionless hinge. It is recommended that G be made equal to 10 where such nonrigid supports are used.

---

[3] W. T. Segui, *Fundamentals of Structural Steel Design* (Boston: PWS-Kent, 1989), pp. 81–82.

2. For rigid connections of columns to footings G theoretically approaches zero but a value of 1.0 is recommended.

3. If a beam or girder is rigidly attached to a column, its stiffness should be multiplied by the appropriate factor given in Table 7-1, depending on the condition at the far end of the member.

TABLE 7-1 MULTIPLIERS FOR RIGIDLY ATTACHED MEMBERS

CONDITION AT FAR END OF GIRDER	SIDESWAY PREVENTED, MULTIPLY BY:	SIDESWAY UNINHIBITED, MULTIPLY BY:
Pinned	1.5	0.5
Fixed against rotation	2.0	0.67

The effective lengths of each of the columns of a frame are estimated with the alignment charts in Example 7-1. (When sidesway is possible, it will be found that the effective lengths are always greater than the actual lengths as is illustrated in this example. When frames are braced in such a manner that sidesway is not possible, K will be less than 1.0.) An initial design has provided preliminary sizes for each of the members in the frame. After the effective lengths are determined, each column can be redesigned. Should the sizes change appreciably, new effective lengths can be determined, the column designs repeated, and so on. Several tables are used in the solution of this example. These should be self-explanatory after the clear directions given on the alignment chart are examined.

■ Example 7-1
Determine the effective lengths of each of the columns of the frame shown in Fig. 7-3 if the frame is braced against sidesway. Use the alignment charts of Fig. 7-2(b) and assume the far ends of the girders are fixed against rotation.

Solution.   Stiffness factors:

MEMBER	SHAPE	I	L	I/L
AB	W8 × 24	82.8	144	0.575
BC	W8 × 24	82.8	120	0.690
DE	W8 × 40	146	144	1.014
EF	W8 × 40	146	120	1.217
GH	W8 × 24	82.8	144	0.575
HI	W8 × 24	82.8	120	0.690
BE	W18 × 50	800	240	3.333
CF	W16 × 36	448	240	1.867
EH	W18 × 97	1750	360	4.861
FI	W16 × 57	758	360	2.106

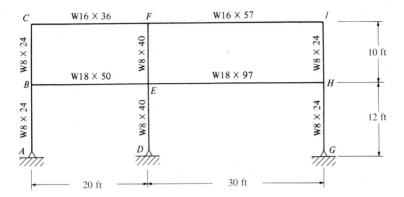

Figure 7-3

G factors for each joint:

JOINT	$\sum(I_c/L_c)/\sum(I_g/L_g)$	G
A	See Fig. 7-2(b)	10.0
B	$\dfrac{0.575 + 0.690}{(2.0)(3.333)}$	0.190
C	$\dfrac{0.690}{(2.0)(1.867)}$	0.185
D	See Fig. 7-2(b)	10.0
E	$\dfrac{1.014 + 1.217}{(2.0)(3.333 + 4.861)}$	0.136
F	$\dfrac{1.217}{(2.0)(1.867 + 2.106)}$	0.153
G	See Fig. 7-2(b)	10.0
H	$\dfrac{0.575 + 0.690}{(2.0)(4.861)}$	0.130
I	$\dfrac{0.690}{(2.0)(2.106)}$	0.164

Column K factors from chart (Fig. 7-2[b]):

COLUMN	G_A	G_B	K
AB	10.0	0.190	1.68
BC	0.190	0.185	1.04
DE	10.0	0.136	1.67
EF	0.136	0.153	1.05
GH	10.0	0.130	1.67
HI	0.130	0.164	1.05

For most buildings the values of K_x and K_y should be examined separately. The reason for such individual study lies in the different possible framing conditions in the two directions. Many multistory frames consist of rigid frames in one direction and conventionally connected frames with sway bracing in the other. In addition the points of lateral support may often be entirely different in the two planes.

The alignment chart of Fig. 7-2(b) for frames with sidesway uninhibited always indicates that $K \geq 1.0$. In fact calculated K factors of 2.0 to 3.0 are common, and even larger values are occasionally obtained. To many designers such large factors seem completely unreasonable. If the designer obtains seemingly high K factors, he or she should carefully review the numbers used to enter the chart (that is, the G values) as well as the basic assumptions used in preparing the charts. These assumptions are discussed in detail in Section 7-2. For example, column behavior was assumed to be elastic. This will be the case if a column has an effective slenderness ratio greater than C_c. If it is less than C_c (inelastic buckling), the K factors obtained will be overly conservative.

7-2 STIFFNESS REDUCTION FACTORS

The alignment charts were developed on the basis of a set of idealized conditions which are seldom if ever completely met in a real structure. A complete list of these assumptions is shown on page 5-136 of the ASD Manual. A few of them are: column behavior is purely elastic, all columns buckle simultaneously, all members have constant cross sections, all joints are rigid, and so on.

If the actual conditions are different from these assumptions, unrealistically high K factors may be obtained from the charts, and overconservative designs may result. A large percentage of columns will fall in the inelastic range, but the alignment charts were prepared assuming elastic failure. This situation, previously discussed in Chapter 5, is illustrated in Fig. 7-4. For such cases the chart K values are too conservative and should be corrected as described in this section.

In the elastic range the stiffness of a column is proportional to EI where $E = 29,000$ ksi, while in the inelastic range its stiffness is more accurately proportional to $E_T I$ where E_T is a reduced or tangent modulus.

The buckling strength of columns in framed structures was shown in the alignment charts to be related to

$$G = \frac{\text{column stiffness}}{\text{girder stiffness}} = \frac{\Sigma (EI/L) \text{ columns}}{\Sigma (EI/L) \text{ girders}}$$

If the columns behave elastically, the modulus of elasticity will be canceled from the preceding expression for G. If the column behavior is inelastic, however, the column stiffness factor will be smaller and will equal $E_T I/L$. As a result the G factor used to enter the alignment chart will be smaller, and the K factor selected from the chart will be smaller.

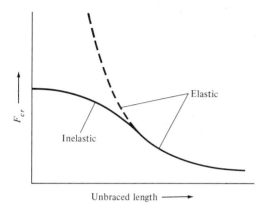

Figure 7-4

Though the alignment charts were developed for elastic column action, they may be used for an inelastic column situation if the G value is multiplied by a correction factor called the *stiffness reduction factor* (SRF). This reduction factor equals the tangent modulus over the elastic modulus (E_T/E) and is approximately equal to

$$\frac{F_a \times \text{inelastic factor of safety}}{F_e' \times \text{elastic factor of safety}} \approx \frac{F_a}{F_e'}$$

where F_a is the allowable compressive stress for an inelastic column with $KL/r < C_c$ and where F_e' is the allowable compressive stress for an elastic column with $KL/r > C_c$ (that is, the Euler buckling stress divided by the safety factor).

The reader should note that the SRF values are approximate because the inelastic factor of safety is variable while the elastic factor of safety is a constant $\frac{23}{12}$. Values of the correction are shown for various P/A or f_a values in Table A on page 3-8 in the ASD Manual.

The steps involved in using the SRF are

1. Calculate P and select a trial column size.
2. Calculate $f_a = P/A$ and pick the SRF from Table A in page 3–8 in the Manual. (If f_a is less than the values given in the table, the column is in the elastic range and no reduction needs to be made. For a large f_a value where the table shows a dash, SRF will equal 0.)
3. The value of G_{elastic} is computed and multiplied by the SRF, and K is picked from the chart.
4. The effective slenderness ratio KL/r is computed; F_a is obtained from the Manual and multiplied by the column area to obtain P. If this value is appreciably different from the value computed in step 1, another column size is tried and the four steps are repeated.

Example 7-2 illustrates these steps for the design of a column in a frame subject to sidesway. *It will be noticed in this example that the author has con-*

sidered only in-plane behavior and only bending about the x axis. As a result of inelastic behavior the effective length factor is appreciably reduced.

■ Example 7-2

Select a W12 section for column *AB* of the frame shown in Fig. 7-5,(a) assuming elastic column behavior and (b) assuming inelastic column behavior. $P = 880$ k and A36 steel is used. The columns above and below *AB* are assumed to be approximately the same size as *AB*. Consider only in-plane behavior and assume the frame is subject to sidesway.

Solution.

(a) Assuming column in elastic range:

Assume $\dfrac{KL}{r} = 50$

$$F_a = 18.35 \text{ ksi}$$

$$A_{\text{reqd}} = \frac{880}{18.35} = 47.96 \text{ in.}^2$$

Try W12 × 170 ($A = 50.0$ in.^{2}, $I_x = 1650$ in.^{4}, $r_x = 5.74$ in.)

$$G_A = G_B = \frac{\Sigma (I_c/L_c)}{\Sigma (I_g/L_g)} = \frac{(2)(1650/12)}{(2)(800/30)} = 5.16$$

$K = 2.27$ from Fig. 7-2(b) alignment chart

$$\frac{KL}{r} = \frac{(2.27)(12 \times 12)}{5.74} = 56.95$$

$$F_a = 17.72 \text{ ksi}$$

$$P = (17.72)(50.0) = 886 \text{ k} > 880 \text{ k} \qquad\qquad \text{OK}$$

$$\underline{\underline{\text{Use W12} \times 170}}$$

(b) Assuming column in inelastic range:

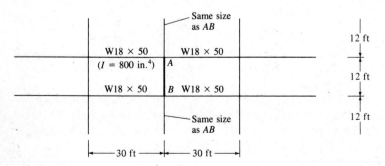

Figure 7-5

Try a lighter section W12 × 152 ($A = 44.7$ in.^{2}, $I_x = 1430$ in.^{4}, $r_x = 5.66$ in.)

$$\frac{P}{A} = \frac{880}{44.7} = 19.69 \text{ ksi}$$

SRF = 0.148 from page 3-8 in ASD Manual

∴ Column is in inelastic range.

$$G_A = G_B = \frac{\Sigma \ (I_c/L_c)}{\Sigma \ (I_g/L_g)} (SRF)$$

$$= \frac{(2)(1430/12)}{(2)(800/30)} (0.148) = 0.661$$

$K = 1.22$ from Fig. 7-2(b) alignment chart

$$\frac{KL}{r} = \frac{(1.22)(12 \times 12)}{5.66} = 31.04$$

$$F_a = 19.87 \text{ ksi}$$
$$P = (19.87)(44.7) = 888.2 \text{ k} > 880 \text{ k} \qquad \text{OK}$$

Use W12 × 152 ∎

7-3 COLUMNS LEANING ON EACH OTHER FOR IN-PLANE DESIGN

When we have an unbraced frame with beams rigidly attached to columns, it is safe to design each column individually using the sidesway uninhibited alignment chart to obtain the K factors (which will probably be appreciably larger than 1.0).

A column cannot buckle by sidesway unless all of the columns on that story buckle by sidesway. One of the assumptions on which the alignment chart of Fig. 7-2(b) for unbraced frames was prepared was that all of the columns on the story in question would buckle at the same time. If this assumption is correct, the columns cannot support or brace each other because, if one gets ready to buckle, they all supposedly are ready to buckle.

In some situations, however, certain columns in a frame have some excess buckling strength. If, for instance, the buckling loads of the exterior columns of the unbraced frame of Fig. 7-6 have not been reached when the buckling loads of the interior columns are reached, the frame will not buckle. The interior columns in effect will lean against the exterior columns; that is, the exterior columns will brace the interior ones. For this situation shear resistance is provided in the exterior columns which resists the sidesway tendency.[4]

---

[4] J. A. Yura, "The Effective Length of Columns in Unbraced Frames," *Engineering Journal*, AISC, 8, no. 2 (2d quarter, 1971), pp. 37–42.

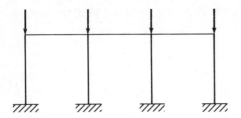

Figure 7-6

There are many practical situations where some columns have excessive buckling strength. This might happen when the designs of different columns on a particular story are controlled by different loading conditions. For such cases failure of the frame will occur only when the gravity loads are increased sufficiently to offset the extra strength of the lightly loaded columns. As a result the critical loads for the interior columns of Fig. 7-6 are increased, and in effect their effective lengths are decreased. In other words, if the exterior columns are bracing the interior ones against sidesway, the K factors for those interior columns are approaching 1.0. Yura says that the effective length of some of the columns in a frame subject to sidesway can be reduced to 1.0 in this type of situation even though there is no apparent bracing system present.

The net effect of the information presented here is that the total gravity load which an unbraced frame can support equals the sum of the strength of the individual columns. In other words, the total gravity load that will cause sidesway buckling in a frame can be split up among the columns in any proportion just so long as the maximum load applied to any one column does not exceed the load that column could support if it were braced against sidesway with $K = 1.0$.

For this discussion the unbraced frame of Fig. 7-7(a) is considered. It is assumed that each column has a $K = 2.0$ and will buckle under the loads shown.

When sidesway occurs, the frame will lean to one side as shown in Fig. 7-7(b), and $P\Delta$ moments equal to 200Δ and 700Δ will be developed.

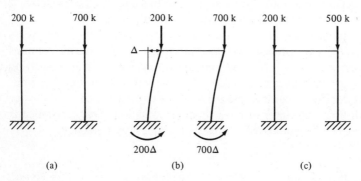

Figure 7-7

Suppose that we load the frame with 200 k on the left-hand column and 500 k on the right-hand column (or 200 k less than we had before). We know that for this situation, which is shown in part (c) of the figure the frame will not buckle by sidesway until we reach a moment of 700Δ at the right-hand column base. This means that our right-hand column can take an additional moment of 200Δ. Thus as Yura says, the right-hand column has a reserve of strength that can be used to brace the left-hand column and prevent its sidesway buckling.

Obviously the left-hand column is now braced against sidesway, and sidesway buckling will not occur until the moment at its base reaches 200Δ. Therefore, it can be designed with a K factor less than 2.0 and can support an additional load—but this load must not be greater than would be its capacity if it were braced against sidesway with $K = 1.0$. It should be mentioned that the total load the frame can carry is still 900 k as in Fig. 7-7(a).

The advantage of the frame behavior described here is illustrated in Example 7-3, where it is assumed that the interior columns of a frame are braced against sidesway by the exterior columns. As a result the interior columns are assumed to each have K factors equal to 1.0. They are designed for the loads shown (450 k each). Then the K factors are determined for the exterior columns with the sidesway uninhibited chart of Fig. 7-2(b), and they are each designed for column loads equal to $300 + 450 = 750$ k.

■ Example 7-3

For the frame of Fig. 7-8, which consists of A36 steel, beams are rigidly connected to the exterior columns while all other connections are simple. The columns are braced top and bottom against sidesway out of the plane of the frame so that $K = 1.0$ in that direction. Sidesway is possible in the plane of the frame. Design the interior columns, assuming $K = 1.0$ and the exterior columns with K as determined from the alignment chart and $P = 750$ k.

Solution.   Design of interior columns:
Assume $K = 1.0$, $KL = (1.0)(15) = 15$ ft, $P = 450$ k.

Use W14 × 90

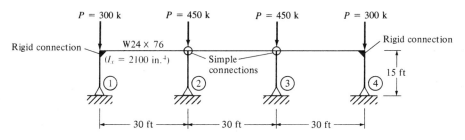

Figure 7-8

Design of exterior columns:

Out of plane $K_y = 1.0$, $P = 300$ k

In plane $P = 300 + 450 = 750$ k, K_x to be determined from alignment chart

Try W14 × 145 ($A = 42.7$ in.^{2}, $I_x = 1710$ in.^{4}, $r_x = 6.33$ in.)

$$G_{top} = \frac{1710/15}{2100/30 \times 1.5} = 1.09$$

Note that girder stiffness is multiplied by 1.5, since sidesway is permitted and far end of girder is hinged.

$$G_{bottom} = 10$$
$$K_x = 1.68 \text{ from alignment chart}$$
$$\frac{K_x L_x}{r_x} = \frac{(1.68)(12 \times 15)}{6.33} = 42.77$$
$$F_a = 18.55 \text{ ksi}$$
$$P = (18.55)(42.7) = 792.1 \text{ k} > 750 \text{ k} \qquad \text{OK}$$

Note that out of plane with $K_y L_y = (1)(15) = 15$ ft, P from Manual tables is 801 k > 300 k.

<u>Use W14 × 145</u>   ∎

It is rather frightening to think of additions to existing buildings and the leaner column theory. If we have a building (represented by the solid lines in Fig. 7-9) and we decide to add onto it (indicated by the dotted lines in the same figure), we may think that we can use the old frame to brace the new one and that we can keep expanding laterally with no effect on the existing building. Sadly, we may be in for quite a surprise. The leaning of the new columns may cause one of the old ones to fail.

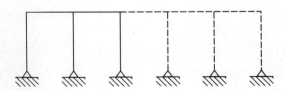

Figure 7-9

7-4 INTRODUCTORY REMARKS CONCERNING FLEXURAL-TORSIONAL BUCKLING OF COMPRESSION MEMBERS

Axially loaded compression members can theoretically fail in three different fashions: by flexural buckling, by torsional buckling, or by flexural-torsional buckling.

Flexural buckling (also called Euler buckling) is the situation considered up to this point in our column discussions where we have computed slenderness ratios for the principal column axes and determined F_a for the highest ratios so obtained. Doubly symmetrical column members (such as W sections) are subject only to flexural buckling and torsional buckling.

As torsional buckling can be very complex, it is very desirable to prevent its occurrence. This may be done by careful arrangements of the members and by providing bracing to prevent lateral movement and twisting. If sufficient end supports and intermediate lateral bracing are provided, flexural buckling will always control. The values given in the ASD column tables for W, M, S, tube, and pipe sections are based on flexural buckling.

Open sections such as Ws, Ms, and channels have little torsional strength, but box beams have a great deal. Thus if a torsional situation is encountered, it may be well to use box sections or to make box sections out of W sections by adding welded side plates ( ). Another way in which torsional problems can be reduced is to shorten the lengths of members that are subject to torsion.

For a singly symmetrical section such as a tee or double angle, Euler buckling may occur about the x or y axis. For equal-leg single angles Euler buckling may occur about the z axis. For all these sections flexural-torsional buckling is definitely a possibility and may control. (It will always control for unequal-leg single-angle columns.) The values given in the ASD column load tables for double-angle and structural tee sections were computed for buckling about the weaker of the x or y axis and for flexural-torsional buckling.

The average designer does not consider the torsional buckling of symmetrical shapes or the flexural-torsional buckling of unsymmetrical shapes. The feeling is that these conditions don't control the critical column loads or at least don't affect them very much. Should we have unsymmetrical columns or even symmetrical columns made up of thin plates, however, we will find that torsional buckling or flexural-torsional buckling may significantly reduce column capacities.

The reader may have noted that the author has not up to this point presented any designs for single-angle columns or for any other columns for which neither the x nor the y axis is an axis of symmetry. The ASD Manual includes tables that provide allowable compression loads for concentrically loaded W and S shapes, structural tees, pipes, square and rectangular tubes, and double angles but not for single angles. Yet single-angle struts may be used on many occasions.

It is stated in the ASD Manual that column tables for single angles are not included because of the difficulty of loading them concentrically. The Manual states that in practice actual eccentricities for single-angle members are rather large, and if they are neglected, some seriously underdesigned members may result.

The reader may think that he or she will select a single-angle strut by trial and error as we have in other sections in this chapter and the preceding one by assuming a KL/r value (it is KL/r_z here) and determining F_a from Table C-36 or C-50 in the Manual, and so on. Then the designer may just say, "I'll reduce these values by 20 or 25 percent." Such a procedure may be seriously in error on the unsafe side.

The plain fact is that the designer using single-angle struts and other similar members is going to have to wade through the flexural-torsional formulas (or get a computer program to do so) and make some allowances for eccentricities of loading and the resulting moments. An ASD flexural-torsional buckling example is presented in the Appendix (Example A-2).

PROBLEMS

7-1 Determine the effective length factors for each of the columns in the frame shown, using the appropriate alignment chart. Assume that the frame is not braced against sidesway and that the girders are rigidly attached to the exterior columns and simply connected to the interior one. (*Ans. AB* = 1.90, *DE* = 1.92, *HI* = 1.38)

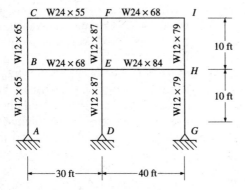

7-2 Repeat Prob. 7-1, assuming the frame is not braced against sidesway.

7-3 Select a W12 section for column *AB*. *P* = 750 k. Use A36 steel. The columns above and below *AB* are to be approximately the same size as *AB*. Consider only in-plane behavior. (a) Assume elastic column behavior. (b) Assume inelastic column behavior. (*Ans.* [a] W12 × 136 [b] W12 × 136)

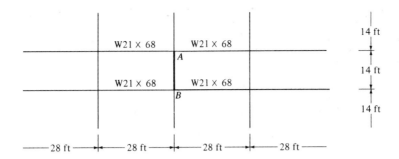

7-4 Repeat Prob. 7-3 using $F_y = 50$ ksi.

7-5 Repeat Prob. 7-3 using a W14. (*Ans.* [a] W14 × 132, [b] W14 × 132)

7-6 Select a W14 section for column *CD*. Consider only in-plane behavior. (a) Assume elastic column behavior. (b) Assume inelastic column behavior. $P = 800$ k and A36 steel is to be used. The columns above and below are assumed to be approximately the same size as *CD*.

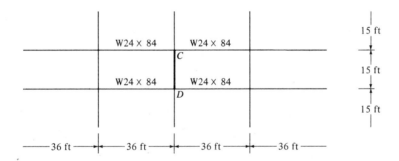

7-7 Design W12 columns for the bent shown, with A36 steel and using the inelastic *K*-factor procedure. The columns are braced top and bottom against sidesway out of the plane of the frame so that $K = 1.0$ in that direction. Sidesway is possible in the plane of the frame. Design the right-hand column using $K = 1.0$ and the left-hand column with *K* as determined from the alignment chart and $P = 1000$ k. The beam is rigidly connected to the left column but has only a simple connection to the right column (*Ans.* W12 × 96 right-hand column, W12 × 190 left-hand column)

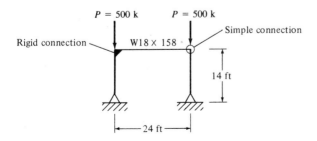

7-8 Repeat Prob. 7-7 with P loads 700 k each.

7-9 The columns for the frames shown are braced top and bottom against sidesway out of the plane of the frame so that $K = 1.0$ in that direction. Sidesway is possible in the plane of the frame. Design the interior column assuming $K = 1.0$ and the exterior columns with K as determined from the alignment chart and $P = 1200$ k. Use A36 steel and a W14 section. (*Ans*. W14 × 233 all columns).

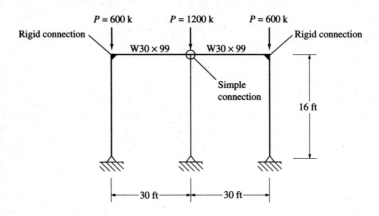

7-10 Repeat Prob. 7-9 using $F_y = 50$ ksi.

7-11 For the frame shown the beams are rigidly connected to the exterior columns while all other connections are simple. The columns are braced top and bottom against sidesway out of the plane of the frame so that $K = 1.0$ in that direction. Sidesway is possible in the plane of the frame. Design W14 interior columns of A36 steel, assuming $K = 1.0$ and W14 exterior columns with K as determined from the alignment chart and $P = 1050$ k. (*Ans*. W14 × 99 for interior columns and W14 × 211 for exterior columns)

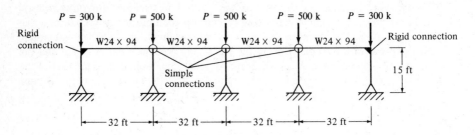

7-12 Repeat Prob. 7-11 using $F_y = 50$ ksi and assuming the beam is a W24 × 76.

8

Introduction to Beams

8-1 TYPES OF BEAMS

Beams are usually said to be members that support transverse loads. They are probably thought of as being used in horizontal positions and subjected to gravity or vertical loads; but there are frequent exceptions—rafters, for example.

Among the many types of beams are joists, lintels, spandrels, stringers, and floor beams. *Joists* are the closely spaced beams supporting the floors and roofs of buildings, while *lintels* are the beams over openings in masonry walls such as windows and doors. A *spandrel beam* supports the exterior walls of buildings and perhaps part of the floor and hallway loads. The discovery that steel beams as a part of a structural frame could support masonry walls (together with the development of passenger elevators) is said to have permitted the construction of today's skyscrapers. *Stringers* are the beams in bridge floors running parallel to the roadway, whereas *floor beams,* found in many bridges, are the larger beams perpendicular to the roadway of the bridge and used to transfer the floor loads from the stringers to the supporting girders or trusses. The term *girder* is rather loosely used but usually indicates a large beam and perhaps one into which smaller beams are framed. These and other types of beams are discussed in the sections to follow.

8-2 SECTIONS USED AS BEAMS

The W shapes will normally prove to be the most economical beam sections, and they have largely replaced channels and S sections for beam usage. Channels are sometimes used for beams subjected to light loads, such as purlins, and in places where the clearances available require narrow flanges. They have very little resistance to lateral forces and need to be braced as illustrated by the sag rod problem in Chapter 4. The W shapes have more steel concentrated in their flanges than do S beams and thus have larger moments of inertia and resisting moments for the same weights. They are relatively wide and have appreciable lateral stiffness. (The small amount of space devoted to S beams in the ASD Manual clearly shows how much their use has decreased from former years. They are today used primarily for special situations where narrow flange widths are desirable, or where shearing forces are very high, or where the greater flange thickness next to the web may be desirable where lateral bending occurs, perhaps with crane rails.)

Another common type of beam section is the open web joist or bar joist which is discussed at length in Chapter 15. This type of section which is commonly used to support floor and roof slabs is actually a light shop-fabricated parallel chord truss. It is particularly economical for long spans and light loads.

Harrison Avenue Bridge, Beaumont, Tex. (Courtesy of Bethlehem Steel Corporation.)

8-3 BENDING STRESSES

For an introduction to bending stresses the rectangular beam and stress diagrams of Fig. 8-1 are considered. (For this initial discussion the beam's compression flange is assumed to be fully braced against lateral buckling. Lateral buckling is discussed at length in Sections 8-7 and 8-8.) If the beam is subjected to some bending moment, the stress at any point may be computed with the flexure formula $f_b = Mc/I$. The value of I/c is a constant for a particular section and is known as the *section modulus* (S). The flexure formula may then be written

$$f_b = \frac{M}{S}$$

When the moment is first applied to the beam, the stress will vary linearly from the neutral axis to the extreme fibers. This situation is shown in Fig. 8-1(b). If the moment is increased, there will continue to be a linear variation of stress until the yield stress is reached in the outermost fibers as shown in (c). The *yield moment* of a cross section is defined as the moment that will just produce the yield stress in the outermost fiber of the section.

If the moment in a laterally braced ductile steel beam is increased beyond the yield moment, the outermost fibers that had previously been stressed to their yield point will continue to have the same stress but will yield, and the duty of providing the necessary additional resisting moment will fall on the fibers nearer to the neutral axis. This process will continue, with more and more parts of the beam cross section stressed to the yield point as shown by the stress diagrams of Fig. 8-1(d and e), until finally a full plastic distribution is approached as shown in (f). When the stress distribution has reached this stage, a *plastic hinge* is said to have formed because no additional moment can be resisted at the section. Any additional moment applied at the section will cause the beam to rotate with little increase in stress. The *plastic moment* is the moment that will produce full plasticity in a member cross section and create a plastic hinge. The ratio of the plastic moment M_p to the yield moment M_y is called the *shape factor*. The shape factor equals 1.50 for rectangular sections and varies from about 1.10 to 1.20 for standard rolled-beam sections.

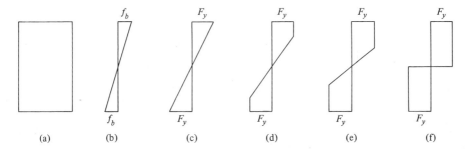

Figure 8-1 Stress diagrams.

8-4 DESIGN WITH THE FLEXURE FORMULA

Included in the items that need to be considered in beam design are moments, shears, crippling, buckling, lateral support, deflection, and perhaps fatigue. Beams will probably be selected which satisfactorily resist the bending moments and then checked to see if any of the other items are critical. To select a beam for a given situation the maximum moment is calculated for the assumed loading and a section having that much resisting moment is selected from the ASD Manual. It is usually more convenient, however, to work with the section modulus as described herein.

In the flexure formula f_b is the fiber stress in the outermost fiber at a distance c from the neutral axis, and I is the moment of inertia of the cross section. It should be remembered that this formula is limited to stress situations below the elastic limit because it is based on the usual elastic assumptions: a plane section before bending remains a plane section after bending; stress is proportional to strain; and so on.

If a beam is to be designed for a particular bending moment M and for a certain allowable stress F_b, the section modulus required to provide a beam of sufficient bending strength can be obtained from the flexure formula.

$$\frac{M}{F_b} = \frac{I}{c} = S$$

From the table in Part 2 of the ASD Manual entitled "Allowable Stress Design Selection Table," steel shapes having sufficient section moduli can be quickly selected. Two important items should be remembered in selecting shapes.

1. These steel sections cost so many cents per pound, and it is therefore desirable to select the lightest possible shape having the required section modulus (assuming that the resulting section will reasonably fit into the structure). The table has the sections arranged in various groups having certain ranges of section moduli. The heavily typed section at the top of each group is the lightest section in that group, and the others are arranged successively in the order of their section moduli. Normally the deeper sections will have the lightest weights giving the required section moduli, and they will usually be selected unless their depth causes a problem in obtaining the desired headroom, in which case a shallower but heavier section will be selected.

2. The section moduli values in the table are given about the horizontal axes for beams in their upright positions. If a beam is to be turned on its side, the proper section modulus can be found in the tables giving dimensions and properties of shapes in the ASD Manual. A W shape turned on its side may only be from 5 to 15 percent as strong as one in the upright position when subjected to gravity loads. In the same manner, the strength of a wood joist with the actual dimensions 2 × 10 in. turned on its side would only be 20 percent as strong as in the upright position (the percentage being based on the relative values of the section moduli).

The examples to follow illustrate the design of steel beams whose compression flanges have lateral support, thus permitting the use of the same allowable stresses in the tension and compression flanges. Beams without sufficient lateral support for the compression flanges are considered in Sections 8-7 and 8-8.

In each of these examples the weight of the beam to be selected must be included in the calculation of the bending moment to be resisted, as the beam must support itself as well as the external loads. The estimates of beam weight are very close here because the author was fortunately able to perform a little preliminary paperwork before making his estimate. Students are not expected to be able to glance at a problem and estimate exactly the weight of the beam required. Following the same procedure as did the author, however, they can do a little figuring on the side and make a very reasonable estimate. For instance, they could calculate the moment due to the external loads only, obtain the required section modulus, and select a beam having the required value. From this beam size a very good estimate of the weight of the final beam section can easily be made. It will often be a little larger than the trial beam size.

■ Example 8-1

Select a beam section for the span and loading shown in Fig. 8-2, assuming full lateral support is provided for the compression flange by the floor above. Allowable bending stresses are 24 ksi (165 MPa).

Solution.   Assume beam weight = 62 lb/ft

$$M = \frac{wL^2}{8} = \frac{(4.362)(21)^2}{8} = 240.46 \text{ ft-k}$$

$$\text{Required section modulus} = \frac{(12)(240.46)}{24} = 120.23 \text{ in.}^3$$

The possible solutions include (a) A W14 × 82 is the section that has the closest S on the safe side. (b) A W21 × 62 is the most economical solution. (c) Should depth be restricted, a W18 × 71 or even a W14 × 82 (very uneconomical) could be selected.

<u>Use W21 × 62</u>   ■

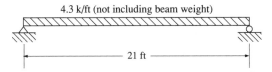

4.3 k/ft (not including beam weight)

21 ft

Figure 8-2

■ Example 8-2

A 5-in. reinforced-concrete slab is to be supported with steel beams 8 ft 0 in. on centers. The beams, which will span 20 ft, are assumed to be simply supported. If the concrete slab is designed to support a live load of 100 psf, determine the lightest steel section required to support the slab. The compression flange of the beam will be incorporated in the concrete slab and is thus laterally supported. The concrete weighs 150 lb/ft^3, and the allowable bending stress in the steel is 24 ksi.

Solution.

$$\text{Dead load: slab} = (8)\left(\frac{5}{12}\right)(150) = 500 \text{ lb/ft}$$

$$\text{Estimated beam weight} = 26$$
$$\text{Live load: } 8 \times 100 = \underline{800}$$
$$\text{Total uniform load} = 1326 \text{ lb/ft}$$

$$M = \frac{(1.33)(20)^2}{8} = 66.5 \text{ ft-k}$$

$$S_{\text{reqd.}} = \frac{(12)(66.5)}{24} = 33.2 \text{ in.}^3$$

$$\underline{\text{Use W12} \times 26 \ (S_x = 33.4 \text{ in.}^3)} \qquad ■$$

The limiting moment permitted by the allowable-stress method is the moment at which the stress in the outermost fibers first reaches the yield point. The true bending strength of a beam, however, is larger than this commonly used value because the beam will not fail at this condition. The outermost fibers will yield, and the stress in the inner fibers will increase until they reach the yield stress, and so on, until the whole section is plastified (see Section 18-3). It is shown in Chapter 18 that before local failure will occur, moments must be produced approximately 12 percent larger than those which will first produce the yield stress in the outermost fibers of W shapes.

The plastification process just mentioned is correct only if the beam remains stable in other ways; that is, it must have sufficient lateral support to prevent lateral buckling of the compression flange (see Section 8-7) and it must have a sufficiently stocky profile to prevent local buckling.

The ASD Specification (F1) gives different allowable bending stresses for different conditions. For most cases the allowable bending stress is

$$F_b = 0.66F_y$$

This expression can be used to determine the allowable bending stress in the extreme fibers of compact hot-rolled shapes and built-up members (not including members consisting of steels with $F_y > 65$ ksi or hybrid girders built up with different yield stress steels) which are symmetrical about and loaded in the plane of their minor axis and which meet the other requirements of Section

B5.1 of the ASD Specification for compact sections. One of these requirements is that the flange must be continuously attached to the web. A built-up section with its flanges intermittently welded to the web does not meet this requirement.

8-5 COMPACT SECTIONS

A *compact section* is capable of developing its plastic moment capacity before any local buckling occurs. To qualify as compact a section must meet the requirements of Section B5.1 of the ASD Specification. Nearly all W and S shapes of A36 steel and a large percentage of those shapes made from higher-strength steels are compact.

For noncompact laterally supported sections the ASD Specification requires a reduction in F_b below $0.66F_y$, while for laterally supported compact sections the allowable stress is equal to $0.66F_y$. Sections which contain *slender compression elements* with width-thickness ratios in excess of the noncompact values listed in Section B5.1 must be proportioned in accordance with the provisions of Appendix B5 of the ASD Specification. The proportions necessary for a section to be classed as compact are specified by the ASD and are summarized in the following paragraphs.

Flanges

Limitations are given by the ASD for the width-thickness ratios for both unstiffened and stiffened compression beam flanges. For the usual hot-rolled sections such as W sections the flanges are unstiffened, while they may very well be stiffened for certain built-up sections (see Fig. 6-8).

The ASD Specification requires that the width of an unstiffened projecting element of a compression flange divided by its thickness (that is, $b_f/2t_f$) not exceed $65/\sqrt{F_y}$. For stiffened elements the width-thickness ratio (b/t_f) may not be greater than $190/\sqrt{F_y}$, where b is the actual width of the stiffened element.

Webs

In addition to the flange requirements the depth-thickness ratios (d/t) of compact sections are not permitted to exceed certain values. These values are $640/\sqrt{F_y} [1 - 3.74(f_a/F_y)]$ when $f_a/F_y \leq 0.16$ and $257/\sqrt{F_y}$ when $f_a/F_y > 0.16$. The term f_a represents the stress caused by a concurrent axial load (if any).

The limitations of web and flange sizes are calculated for different yield stress values and tabulated in Table 8-1. These values are as given in Table 5 of the "Numerical Values" part of the ASD Specification immediately after their Appendix. Nearly all W and S sections are compact when made of A36 steel, while a large proportion of the same shapes are compact if F_y is 50 ksi.

TABLE 8-1 MAXIMUM WIDTH-THICKNESS RATIOS FOR COMPACT SECTIONS

YIELD STRESS F_y (KSI)		36	42	46	50	60	65
Unstiffened flanges $\dfrac{65}{\sqrt{F_y}}$		10.8	10.0	9.6	9.2	8.4	8.1
Stiffened flanges $\dfrac{190}{\sqrt{F_y}}$		31.7	29.3	28.0	26.9	24.5	23.6
Web	$\dfrac{640}{\sqrt{F_y}}$	106.7	98.8	94.4	90.5	82.6	79.4
	$\dfrac{257}{\sqrt{F_y}}$	42.8	39.7	37.9	36.3	33.2	31.9

It is obvious from the values shown in Table 8-1 that the higher the yield stress of a particular section the more likely it is to be noncompact. It is quite simple to determine the yield stress above which the flange of a particular section is noncompact as it is for the web. For instance, if the maximum width-thickness ratio of an unstiffened flange is equated to $b_f/2t_f$ and solved for F_y, the result, which is referred to as F_y', is

$$\frac{65}{\sqrt{F_y}} = \frac{b_f}{2t_f}$$

$$F_y = F_y' = \left(\frac{65}{b_f/2t_f}\right)^2$$

A similar derivation for the web when it is subject to combined bending and axial stress with $f_a/F_a > 0.16$ follows.

$$\frac{257}{\sqrt{F_y}} = \frac{d}{t_w}$$

$$F_y = F_y''' = \left(\frac{257t_w}{d}\right)^2$$

If the yield stress in question is $>F_y'$, the flange is noncompact, and if $>F_y'''$ the web is noncompact. The previously mentioned allowable-stress selection table has the noncompact shapes clearly indicated by showing the value of F_y' for each section. In Part I of the Manual values of F_y''' are tabulated (as are F_y' values) for W, M, S, and HP sections. The seventh edition of the ASD Manual contained F_y'' values computed for the depth-thickness ratio of beam webs when f_a was zero. The values of F_y'' are no longer shown because they are all higher than 70 ksi and plastic behavior is not recognized by the ASD Specification for such steels. If $F_y > 70$ ksi, the maximum value of F_b permitted is $0.60F_y$, which is the lower stress range for noncompact sections anyway.

If the web is noncompact, the maximum allowable bending stress permitted by the ASD is $0.60F_y$. If, however, the web is compact and the flange has a $b_f/2t_f$ value $>65/\sqrt{F_y}$ but less than $95/\sqrt{F_y}$, it is said to be *partially compact*. For partially compact sections a linear transition in F_b between $0.66F_y$ and $0.60F_y$ is provided by ASD Equation F1-3. The purpose of this formula is to avoid some of the abruptness of the transition from an allowable stress of $0.66F_y$ to $0.60F_y$. The transition does not apply to steels with $F_y > 65$ ksi or to hybrid girders. The partially compact section equation is

$$F_b = F_y\left[0.79 - 0.002\left(\frac{b_f}{2t_f}\right)\sqrt{F_y}\right] \quad \text{(ASD Equation F1-3)}$$

Should a doubly symmetric I or H shape be bent about its minor or y axis and should $b_f/2t_f > 65/\sqrt{F_y}$ but less than $95/\sqrt{F_y}$, the allowable bending stress is to be computed with the expression

$$F_b = F_y\left[1.075 - 0.005\left(\frac{b_f}{2t_f}\right)\sqrt{F_y}\right] \quad \text{(ASD Equation F2-3)}$$

Example 8-3 illustrates the calculations necessary to determine the allowable bending stress and the resisting moment of a noncompact section. It will be noted that the allowable-stress design selection table in the ASD Manual contains resisting moment values (M_R) for the commonly used beam sections with 36 and 50 ksi yield stress steels. Their values have been computed with the correct F_b values whether the sections are compact or noncompact. The values given in the other beam tables in the Manual have also accounted for reduced allowable bending stresses for noncompact sections.

■ Example 8-3

Compute the resisting moment of a W12 × 65 with (a) $F_y = 36$ ksi and (b) $F_y = 50$ ksi. Assume the section has full lateral support for its compression flange.

Solution.
(a) $F_y = 36$ ksi:
    Using a W12 × 65 ($d = 12.12$ in., $t_w = 0.390$ in., $b_f = 12.000$ in., $t_f = 0.605$ in., $S_x = 87.9$ in.^{3}) and checking "compact section" requirements

$$\frac{b_f}{2t_f} = \frac{12.000}{(2)(0.605)} = 9.92 < 10.8 \qquad \text{OK}$$

$$\frac{d}{t_w} = \frac{12.12}{0.390} = 31.08 < 106.7 \qquad \text{OK}$$

$$F_b = 0.66F_y$$
$$M_R = F_b S_x = (0.66)(36)(8.79) = 2089 \text{ in.-k} = \underline{174 \text{ ft}}$$

(b) $F_y = 50$ ksi:

Checking compact section requirements

$$\frac{b_f}{2t_f} = \frac{12.000}{(2)(0.605)} = 9.92 > 9.2$$

Therefore the flange is noncompact.

$$\frac{d}{t_w} = \frac{12.12}{0.390} = 31.08 < 90.5 \qquad\qquad \text{OK}$$

Applying ASD Equation F1-3

$$F_b = 50\left[0.79 - 0.002\left(\frac{12.000}{2 \times 0.605}\right)\sqrt{50}\right] = 32.49 \text{ ksi}$$

$$M_R = F_b S_x = (32.49)(87.9) = 2856 \text{ in.-k} = \underline{\underline{238 \text{ ft-k}}} \qquad\blacksquare$$

Note: The ASD Manual can be used to quickly determine if a section is compact as follows:

$$F_y' = 43.0 \text{ ksi} > 36 \text{ ksi but} <50 \text{ ksi}$$

Therefore the flange is noncompact for $F_y = 50$ ksi.

$$F_y''' = >70 \text{ ksi and is thus not listed}$$

Therefore the web is compact for both steels.

8-6 HOLES IN BEAMS

It is often necessary to have holes in steel beams. They are obviously required for the installation of bolts and rivets and sometimes for pipes, conduits, ducts, and so on. If at all possible, these latter types of holes should be completely avoided. When absolutely necessary they should be placed through the web if the moment is large and through the flange if the shear is large. Cutting a hole through the web of a beam does not reduce its section modulus greatly or its resisting moment; but, as will be described in Section 9-1, a large hole in the web tremendously reduces the shearing strength of a steel section. When large holes are put in beam webs, extra plates are sometimes connected to the webs around the holes to reinforce against possible web buckling. A sample design for such reinforcing is presented by Kussman and Cooper.[1]

The presence of holes of any type in a beam certainly does not make it stronger and in all probability weakens it somewhat. The effect of holes has been a subject which has been argued back and forth for many years. Two frequent questions are, Is the neutral axis affected by the presence of holes? and Is it necessary to subtract holes from the compression flange which are going to be plugged with rivets or bolts?

---

[1] R. L. Kussman and P. B. Cooper, "Design Example for Beams with Web Openings," *Engineering Journal*, AISC, 13, no. 2 (2d quarter, 1976), pp. 48–56.

The theory that the neutral axis might move from its normal position to the theoretical position of its net section when holes are present is quite questionable. A linear distribution of stress has been assumed in the preceding paragraphs of this chapter, but when holes are present, the situation is changed because there is considerable stress concentration around the holes.

Tests seem to show that flange holes for rivets or bolts do not appreciably change the location of the neutral axis. It is logical to assume that the location of the neutral axis will not follow the exact theoretical variation with its abrupt changes in position at rivet or bolt holes as shown in Fig. 8-3(b). A more reasonable change in neutral axis location is shown in Fig. 8-3(c), where it is assumed to have a more gradual variation in position.

It is interesting to note that flexure tests of steel beams seem to show that their failure is based on the strength of the compression flange even though there may be rivet or bolt holes in the tension flange. The presence of these holes does not seem to be as serious as might be thought, particularly as compared to holes in a pure tension member. These tests show little difference in the strengths of beams with no holes and in beams with the usual holes for bolts or rivets.

The ASD Specification does not require a reduction in the allowable strength of a beam with bolts or rivets in either flange as long as the tensile fracture strength of the net flange area divided by a factor of safety of 2.0 is at least as large as the tensile yield strength of the gross flange area divided by a factor of safety of 1.67. Expressed as an equation, no deduction for bolt or rivet holes in either flange is necessary if

$$0.5F_u A_{fn} \geq 0.6F_y A_{fg} \qquad \text{(ASD Equation B10-1)}$$

In this equation A_{fn} is the net flange area and A_{fg} is the gross flange area. Substituting into this expression we find that no deduction is necessary if the net flange area is equal to or greater than 75 percent of the gross flange area

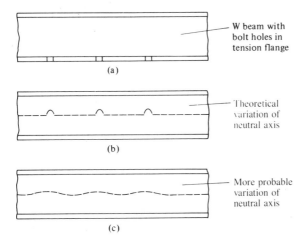

(a)   W beam with bolt holes in tension flange

(b)   Theoretical variation of neutral axis

(c)   More probable variation of neutral axis

Figure 8-3

TABLE 8-2 WHEN TO IGNORE BOLT AND RIVET HOLES IN BEAM AND GIRDER FLANGES

STEEL	NO REDUCTION IF $A_{fn}/A_{fg} \geq$
A36	0.75
A572 Grade 50	0.92
A588	0.86

for A36 steel or 92 percent for A572 Grade 50 steel. These values are shown in Table 8-2.

Should $0.5F_u A_{fn}$ be less than $0.6F_y A_{fg}$, the ASD Specification requires that the flexural properties of the section be based on an effective tension flange area A_{fe}, determined as follows:

$$A_{fe} = \frac{5}{6}\frac{F_u}{F_y}A_{fn} \qquad \text{(ASD Equation B10-3)}$$

In other words, to calculate the moment of inertia for the flanges, the transfer expression Ad^2 becomes $A_{fe}d^2$. If the bolt or rivet holes are in the tension flange, we assume they are in the compression flange as well. If they are only in the compression flange, they are ignored because it is felt that the fasteners can adequately transmit compression through the holes.

AASHTO and AREA require the calculation of two moments of inertia when holes are present. For compressive stresses the gross moment of inertia is to be used regardless of the presence of rivet or bolt holes. For tensile stresses the net moment of inertia is to be used. The neutral axis is assumed to remain at its normal position for both calculations. The effect of using the two different moments of inertia is to assume that rivet or bolt holes on the compression side of the beam have less effect than those on the tension side.

The usual practice is to subtract the same area of holes from both flanges whether they are present in the compression flange or not. For a section with two holes in the tension flange only, the properties of the section would be computed based on the subtraction of two holes from the tension flange and two holes from the compression flange. Example 8-4 illustrates this method. Again the author has made a few preliminary calculations in estimating the member size.

■ **Example 8-4**

Redesign the beam of Example 8-1, assuming that it will be necessary to punch holes for two $\frac{3}{4}$-in. bolts in the tension flange. The ASD reduction is not to be permitted in this beam. Figure 8-4 shows a sketch of the assumed section.

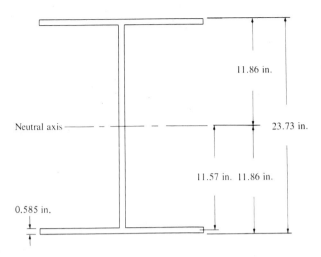

Figure 8-4

Solution. Assume beam weight = 68 lb/ft

$$M = \frac{(4.368)(21)^2}{8} = 240.8 \text{ ft-k}$$

$$\text{net } S_{reqd} = \frac{(12)(240.8)}{24} = 120.4 \text{ in.}^3$$

Try W24 × 68 ($S = 154$ in.^{3}, $d = 23.73$ in., $b_f = 8.965$ in., $t_f = 0.585$ in.)
Assuming two holes in each flange, I of holes about neutral axis is

$$(4)(\tfrac{7}{8})(0.585)(11.57)^2 = 274 \text{ in.}^4$$

S of holes is

$$\frac{274}{11.86} = 23.1 \text{ in.}^3$$

$$\text{Actual net } S = 154 - 23.1 = 130.9 \text{ in.}^3 > 120.4 \text{ in.}^3 \qquad \text{OK}$$

Use W24 × 68 ∎

A beam with holes in its tension flange is examined in Example 8-5 in accordance with the provisions of ASD Specification B10.

■ Example 8-5

A W16 × 31 beam section ($d = 15.88$ in., $b_f = 5.525$ in., $t_f = 0.440$ in.)
consisting of A572 Grade 50 steel has two holes for $\tfrac{7}{8}$-in. bolts in its tension
flange. Does the flange area need to be reduced for calculating the beam properties? If so, how much should I_x be reduced?

Solution.

$$A_{fg} = (5.525)(0.440) = 2.431 \text{ in.}^2$$
$$A_{fn} = 2.431 - (2)(1.0)(0.440) = 1.551 \text{ in.}^2$$

$$\frac{A_{fn}}{A_{fg}} = \frac{1.551}{2.431} = 0.638 < 0.92 \text{ for A572 steel}$$

$$\therefore \text{ A reduction is necessary}$$

$$A_{fe} = \frac{5}{6}\frac{F_u}{F_y}A_{fn} = \left(\frac{5}{6}\right)\left(\frac{65}{50}\right)(1.551) = 1.680 \text{ in.}^2$$

Reduction in flange area $= 2.431 - 1.680 = 0.751 \text{ in.}^2$

Reduction in $I_x = 2Ad^2$

$$= (2)(0.751)\left(\frac{15.88}{2} - \frac{0.440}{2}\right)^2 = \underline{\underline{89.5 \text{ in.}^4}} \quad\blacksquare$$

The ASD requirements (B10) apply to the design of hybrid beams and girders whose flanges consist of a stronger grade of steel than their webs. This is true as long as these members are not required to resist an axial force larger than 0.15 times the F_y of the flanges times their gross areas.

Should a hole be present in only one side of a flange of a W section, there will be no axis of symmetry for the net section of the shape. A correct theoretical elastic solution of the problem would involve the location of the principal axes, the calculation of the principal moments of inertia, and so on, or substitution into the lengthy generalized equations for unsymmetrical bending presented in Section 9-4. Rather than following such lengthy procedures over a minor point, it seems logical to assume holes in both sides of the flange. The results obtained will probably be just as satisfactory as those obtained by the more laborious methods mentioned.

8-7 LATERAL SUPPORT OF BEAMS

Probably the large majority of steel beams are used in such a manner that their compression flanges are restrained against lateral buckling. (Unfortunately, however, the percentage has not been quite as high as the design profession has assumed.) The upper flanges of beams used to support concrete building and bridge floors are often incorporated in these concrete floors. For situations of this type, where the compression flanges are restrained against lateral buckling, the allowable bending fiber stresses in the tension and compression flanges are considered to be equal.

Should the compression flange of a beam be without lateral support for some distance, it will have a stress situation similar to that existing in columns. As is well known the longer and slenderer a column becomes, the greater becomes the danger of its buckling for the same loading condition. When the compression flange of a beam is long enough and slender enough, it may quite possibly buckle unless lateral support is provided.

There are many factors affecting the amount of stress which will cause buckling in the compression flange of a beam. Some of these factors are the properties of the material, the spacing and types of lateral support provided, the types of end support or restraints, the loading conditions, and so on.

The tension in the other flange of a beam tends to keep that flange straight and restrain the compression flange from buckling; but as the bending moment is increased, the tendency to buckle may become large enough to overcome the tensile restraint. When the compression flange does begin to buckle, twisting or torsion will occur, and the smaller the torsional strength of the beam the more rapid will be the failure. The W, S, and channel shapes so frequently used for beam sections do not have a great deal of resistance to lateral buckling and the resulting torsion. Some other shapes, notably the built-up box shapes, are tremendously stronger. These types of members have a great deal more torsional resistance than the W, S, and plate girder sections. Tests have shown that they will not buckle laterally until the strains developed are well in the plastic range. If box girders meet the requirements for compact shapes given in ASD Specification F3, their allowable bending stresses are $0.66F_y$, and if they are not compact, their allowable bending stresses are equal to $0.60F_y$ unless their depths are greater than six times their widths. For such cases lateral support requirements should be specially investigated.

Some judgment needs to be used in deciding what does and what does not constitute satisfactory lateral support for a steel beam. Perhaps the most common question asked by practicing steel designers is, What is lateral support? A beam that is wholly encased in concrete or that has its compression flange incorporated in a concrete slab is certainly well supported laterally. When a concrete slab rests on the top flange of a beam, the engineer must study the situation carefully before counting on friction to provide full lateral support. Perhaps if the loads on the slab are fairly well fixed in position, they will contribute to the friction, and it may be reasonable to assume full lateral support. If on the other hand there is much movement of the loads and appreciable vibration, the friction may well be reduced and full lateral support should not be assumed. Such situations are caused in bridges by traffic, and in buildings by vibrating machinery such as printing presses.

Should lateral support of the compression flange not be provided by a floor slab, such support may be provided by connecting beams or by special members inserted for that purpose. Beams that frame into the sides of the beam or girder in question and are connected to the compression flange can usually be counted on to provide full lateral support at the connection. If the connection is made primarily to the tensile flange, little lateral support is provided to the compression flange. Before support is assumed from these beams, the designer should note if they themselves are prevented from moving. The series of beams represented with horizontal dotted lines in Fig. 8-5 provides questionable lateral support for the main beams between columns. For a situation of this type some system of x-bracing may be desirable in one of the bays. Such a

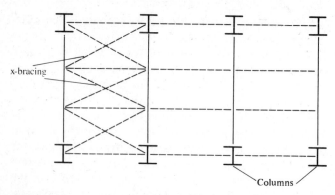

Figure 8-5 Lateral support for beams.

system is shown in Fig. 8-5. This one system will provide sufficient lateral support for the beams for several bays.

The corrugated sheet-metal roofs which are usually connected to the purlins with metal straps probably furnish only partial lateral support. A similar situation exists when wood flooring is bolted to supporting steel beams. At this time the student quite naturally asks, "If only partial support is available, what am I to consider to be the distance between points of lateral support?" The answer is to use his or her judgment. As an illustration, a wood floor is assumed to be bolted every 4 ft to the supporting steel beams in such a manner that it is thought only partial lateral support is provided at those points. After studying the situation the engineer might well decide that the equivalent of full lateral support at 8-ft intervals is provided. Such a decision seems to be within the spirit of the specifications.

8-8 LATERALLY UNBRACED BEAMS

The general practice through the years has been to reduce the allowable stress in the fibers of the compression flange of a beam with little lateral support. As an illustration, the 1989 AASHTO Specifications permit bending stresses of $0.55F_y$ or 20,000 psi for A36 steel if full lateral support is provided by embedment in concrete. If full lateral support is not provided, the allowable compressive stress is to be reduced in accordance with the formula

$$F_b = 20,000 - 7.5\left(\frac{L}{b_f}\right)^2$$

In this expression L is the distance in inches between points of lateral support, and b_f is the flange width in inches. For higher-strength steels the allowable values can be computed with similar expressions to be found in those specifications. Example 8-6 illustrates the use of the AASHTO expression for the design of a beam. The application of lateral-support formulas is quite similar to that required for the various column expressions, and a section that can adequately resist the moment can quickly be found. It will be noted, however, that the determination of the lightest section that can adeqately support the loads may involve quite a lengthy trial-and-error process, and for this reason curves such as those available in the ASD Manual and to be discussed later in this section are highly desirable.

■ Example 8-6

Select a steel section for the loads and span of Fig. 8-6. The beam is to be designed with A36 steel and the 1989 AASHTO Specifications. Lateral support is provided only at the beam ends.

Solution.   Assume beam weight = 70 lb/ft

$$M = (20)(10) + \frac{(0.070)(20)^2}{8} = 203.5 \text{ ft-k}$$

Assume allowable fiber stress = 15 ksi

$$S_{reqd} = \frac{12 \times 203.5}{15} = 163 \text{ in.}^3$$

Try W24 × 76 ($b = 8.990$ in., $S = 176$ in.^{3})

$$F_b = 20,000 - 7.5\left(\frac{12 \times 20}{8.990}\right)^2 = 14.65 \text{ ksi}$$

$$\text{Resisting moment} = F_b S = \frac{(14.65)(176)}{12} = 215 \text{ ft-k} > 203.5 \text{ ft-k}   \text{OK}$$

$$\underline{\underline{\text{Use W24} \times 76}}$$ ■

The ASD Specification presents three expressions (F1-6, F1-7, and F1-8) for determining the allowable bending fiber stresses in beams for which continuous lateral support is not provided. The expressions are applicable to rolled shapes, plate girders, and built-up members having an axis of symmetry in the plane of the web. Depending upon the proportions of the member and the unbraced length, the designer will substitute in Equations F1-6 and F1-8 or into F1-7 and F1-8 and use the larger value so obtained, provided the result is not greater than the maximum permissible value of $0.60F_y$.

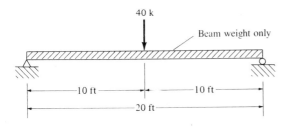

Figure 8-6

The lateral buckling strength of a beam can be estimated by taking into account the torsional resistance of the beam about its longitudinal axis and the lateral bending resistance of the beam plus the resistance of the flange to torsion.[2-5] The resulting expression is, however, too complicated for practical engineering use.

For shallow thick-walled sections the resistance to torsion about the longitudinal axis and the lateral buckling resistance are the most important factors. For these cases, ASD Equation F1-8 is considered to give a reasonable approximation of an allowable buckling stress. In the expressions that follow, L is the distance between points of lateral support, d is the beam depth, and A_f is the flange area. *Should the beam under consideration be a cantilever, the unsupported length can be conservatively assumed to equal the actual length.*[6]

The allowable bending compression stresses permitted by the ASD Specification for sections loaded in the plane of their webs and having an axis of symmetry in their webs equal the larger values computed by Equations F1-6 or F1-7 and F1-8, as described in the following paragraphs, except not more than $0.60F_y$. This procedure for determining allowable stresses also applies to compression on extreme fibers of channels bent about their major axes. When

$$\sqrt{\frac{102 \times 10^3 C_b}{F_y}} \le \frac{L}{r_T} \le \sqrt{\frac{510 \times 10^3 C_b}{F_y}}$$

we use

$$F_b = \left[\frac{2}{3} - \frac{F_y(L/r_T)^2}{1530 \times 10^3 C_b} \right] F_y \qquad \text{(ASD Equation F1-6)}$$

When

$$\frac{L}{r_T} \ge \sqrt{\frac{510 \times 10^3 C_b}{F_y}}$$

we use

$$F_b = \frac{170 \times 10^3 C_b}{(L/r_T)^2} \qquad \text{(ASD Equation F1-7)}$$

[2] Karl De Vries, "Strength of Beams as Determined by Lateral Buckling," *Transactions ASCE* **112** (1947), pp. 1245–1271.

[3] G. Winter et al., "Discussion of 'Strength of Beams as Determined by Lateral Buckling,'" *Transactions ASCE* **112** (1947), pp. 1272–1320.

[4] G. G. Kubo, B. G. Johnston, and W. J. Eney, "Nonuniform Torsion of Plate Girders," *Transactions ASCE* **121** (1956), pp. 759–785.

[5] K. Basler and B. Thürlimann, "Strength of Plate Girders In Bending," *Proceedings ASCE* **87**, no. 6 (August 1961), pp. 153–181.

[6] S. Timoshenko and J. M. Gere, *Theory of Elastic Stability*, 2d ed. (New York: McGraw-Hill, 1961), pp. 257–262.

and, when the compression flange is solid and approximately rectangular in cross section and its area is not less than that of the tension flange,

$$F_b = \frac{12 \times 10^3 C_b}{Ld/A_f} \qquad \text{(ASD Equation F1-8)}$$

In these expressions L is the unbraced length of the compression flange; r_T is the radius of gyration of the compression flange plus one-third of the compression web area taken about an axis in the plane of the web; and C_b is a moment coefficient that is included in the formulas to account for the effect of different moment gradients on lateral-torsional buckling. In other words, lateral buckling may be appreciably affected by the end restraint and loading conditions of the member.

If a section has a compression flange distinctly smaller than its tension flange, Equation F1-8 will often yield values that are much too conservative and should not be used. *Equation F1-8 is to be used for determining the allowable bending stresses for channels bent about their major axes.*

Equation F1-8 gives allowable values that are very reasonable for shallow thick-walled sections but are somewhat conservative for a few deep thin W sections and nearly all plate girders. The designer could use this formula and ignore the other ones. His or her designs would be perfectly safe although considerably overdesigned for the cases mentioned. For these members the resistance of the flange to torsion is the predominant factor, and ASD Equations F1-6 and F1-7 more clearly estimate the effect of this item.

In Fig. 8-7(a) the reader can see that the moment in the unbraced beam causes a worse compression flange situation than does the moment in the unbraced beam of part (b). For one reason the upper flange of the beam in part (a) is in compression for its entire length, while in (b) the length of the "column," that is, the length of the upper flange which is in compression, is much less (thus a much shorter "column"). It will also be noted that for the

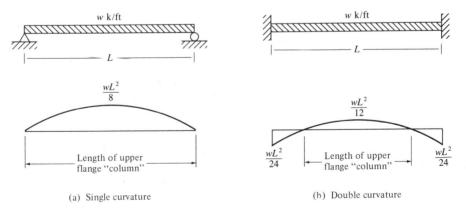

(a) Single curvature

(b) Double curvature

Figure 8-7

same span length and loading, the maximum moment in the second beam is substantially reduced.

For the simply supported beam in Fig. 8-7(a), C_b is taken as 1.0, while for the beam in part (b) it is taken as larger than 1.0. The allowable-stress expressions (F1-6 to F1-8) were developed for laterally unbraced beams subject to single curvature bending with $C_b = 1.0$. Frequently beams are not bent in single curvature and so can resist more moment. (It is theoretically possible under certain circumstances for a member with a moment diagram having reverse curvature to have an allowable bending strength as much as 2.3 times as large as the same member with a rectangular moment diagram.)

To handle this situation the ASD Specification provides moment or C_b coefficients larger than 1.0, which are to be multiplied by the allowable stresses for cases where $C_b = 1.0$. The results are higher moment capacities. The designer who conservatively says, "I'll always use $C_b = 1.0$," is missing out on the possibility of significant savings in steel weight for some situations. *When using C_b values the designer should clearly understand that the allowable stress obtained may not be larger than $0.60F_y$.*

The value of C_b is determined from the expression to follow in which M_1 is the smaller and M_2 the larger of the bending-moments at the ends of the unbraced length taken about the strong axis of the member. Should the moment at any point within the unbraced length be larger than the end moments, C_b shall be taken as 1.0. The ratio M_1/M_2 is considered positive if M_1 and M_2 have the same sign (reverse curvature bending) and negative if they have opposite signs (single curvature bending).

$$C_b = 1.75 + 1.05\left(\frac{M_1}{M_2}\right) + 0.3\left(\frac{M_1}{M_2}\right)^2 \leq 2.3$$

C_b is equal to 1.0 for unbraced cantilever beams and for beams that have a moment over an appreciable part of their unbraced span equal to or larger than the larger of the segment's end moments. Some typical values of C_b are shown in Fig. 8-8 for various beam and moment situations. The ASD Appendix (Table 6) presents calculated values of C_b for different M_1/M_2 ratios.

The ASD Manual provides information that greatly simplifies the application of these seemingly complex lateral-support equations. Of particular use are the L_c and L_u values which are provided in the beam and column sections of the Manual. In Section F1.1 of the ASD Specification it is stated that for flanged beams the distances between points of lateral bracing should not exceed $76b_f/\sqrt{F_y}$ or $20,000/[(d/A_f)F_y]$ if the members are to be assumed to have adequate lateral bracing. The least of these two values is referred to as L_c.

Should the distance between points of lateral bracing be greater than L_c, the ASD says that the allowable bending stress must be reduced from $0.66F_y$ with the appropriate formula, but in no case may it exceed $0.60F_y$. When these formulas are used, however, there is a range in which they give a value above $0.60F_y$. For each beam there is an unbraced length for which the controlling formula yields an allowable stress exactly equal to $0.60F_y$. This length is called L_u throughout the Manual. Based on this information it is possible to make the following simplifying statements:

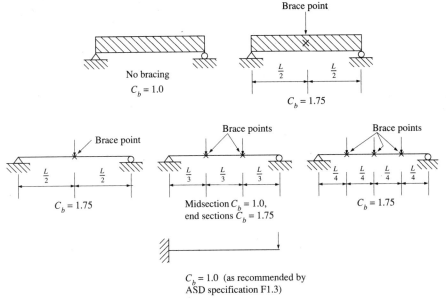

Figure 8-8 Sample C_b values.

1. If the unbraced length $\leq L_c$, F_b equals $0.66F_y$, assuming the other requirements of ASD Section F1.1 are met.
2. If the unbraced length is $>L_c$ but $\leq L_u$, F_b equals $0.60F_y$.
3. If the unbraced length is $>L_c$ and $>L_u$, F_b will be less than $0.60F_y$ and can be determined by the appropriate formulas from Section F1.3 of the ASD Specification. The designer will often find that the curves given in the beam part of the ASD Manual entitled "Allowable Moments in Beams with Unbraced Length Greater Than L_u" and discussed in Section 8-9 will be helpful in this regard.

Example 8-7 illustrates the calculations necessary to determine the allowable stresses in a W section that has different unbraced lengths.

■ **Example 8-7**

Determine the allowable bending stresses in a W33 × 130 (see Fig. 8-9) for simple spans of 10, 13, 20, and 30 ft without lateral support. Use A36 steel and the ASD Specification.

Solution. Computing properties of the section:

$$A_f + \frac{1}{6}A_w = (11.510)(0.855) + \left(\frac{1}{3}\right)\left(\frac{31.38}{2}\right)(0.580) = 12.87 \text{ in.}^2$$

$$r_T = \sqrt{\frac{\frac{1}{2} \times 218}{12.87}} = 2.91 \text{ in.}$$

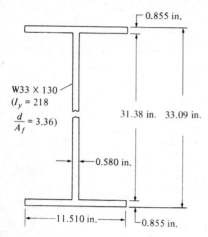

Figure 8-9

Approximate I_y for tee section $= \frac{1}{2}I_y$ for whole section. More exact values of r_T are given in the ASD Manual, with $r_T = 2.88$ in. for this section. $C_b = 1.0$ since moment in span exceeds value at both ends.

$L_{\text{unbraced}} = 10$ ft

$$L_c = 12.1 \text{ ft from Manual} > 10 \text{ ft}$$

Therefore $F_b = 0.66F_y = \underline{24 \text{ ksi}}$

$L_{\text{unbraced}} = 13$ ft

$$L_u = 13.8 \text{ ft from Manual}$$
$$L_c < L_{\text{unbraced}} < L_u$$

Therefore $F_b = 0.60F_y = \underline{22 \text{ ksi}}$

$L_{\text{unbraced}} = 20$ ft

$$L_{\text{unbraced}} > L_u$$

Therefore, one must use formulas

$$\sqrt{\frac{102 \times 10^3 \times 1.0}{36}} = 53 < \frac{L}{r_T} \text{ of } \frac{12 \times 20}{2.88} = 83.3 < \sqrt{\frac{510 \times 10^3 \times 1.0}{36}}$$

$$= 119$$

Use Equations F1-6 and F1-7

$$F_b = \left[\frac{2}{3} - \frac{(36)(83.3)^2}{(1530)(10^3)(1)}\right]36 = \underline{18.1 \text{ ksi}} \leftarrow$$

$$F_b = \frac{12 \times 10^3 \times 1}{12 \times 20 \times 3.36} = 14.9 \text{ ksi}$$

$L_{\text{unbraced}} = 30$ ft

$$\sqrt{\frac{102 \times 10^3 \times 1.0}{36}} = 53 < \frac{L}{r_T} \text{ of } \frac{12 \times 30}{2.88} = 125 > \sqrt{\frac{510 \times 10^3 \times 1.0}{36}}$$

$$= 119$$

Use Equations F1-7 and F1-8

$$F_b = \frac{170 \times 10^3 \times 1.0}{(125)^2} = \underline{\underline{10.88 \text{ ksi}}} \leftarrow$$

$$F_b = \frac{12 \times 10^3 \times 1.0}{12 \times 30 \times 3.36} = 9.92 \text{ ksi} \qquad \blacksquare$$

8-9 DESIGN OF LATERALLY UNBRACED BEAMS WITH C_b = 1.0

Example 8-8 illustrates the design of a beam without full lateral support, using the ASD Specification. The problem is very simple if the charts in the Manual entitled "Allowable Moments in Beams with Unbraced Lengths Greater Than L_u" are used. In these charts the resisting moments of the sections commonly used as beams are plotted for different unbraced lengths and with C_b assumed to equal 1.0. The unbraced length corresponding to L_c for each beam section is shown on the charts as a black dot, while the unbraced length corresponding to L_u is shown as an open circle.

The designer enters the chart with unbraced length on the bottom scale and bending moment on the vertical scale. He or she goes to the intersection of these two values and then moves up and to the right. Any section encountered in that direction will have a greater unbraced length and a greater resisting moment than needed. The first solid line encountered represents the most economical section available. Frequently a dashed line is encountered before a solid line is reached. The shape represented by the dashed line will have a sufficient resisting moment but will not be the most economical solution. The designer should continue moving up and to the right until a solid line is encountered.

■ Example 8-8
Select the lightest available steel section for the beam shown in Fig. 8-10, using the ASD Specification and A36 steel. Lateral support is provided at the ends only.

Solution. Assume beam weight = 100 lb/ft

$$M = \frac{(7.1)(20)^2}{8} = 355 \text{ ft-k}$$

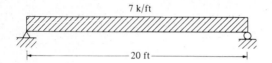

Figure 8-10

From chart, noting $C_b = 1.0$,

Use W30 × 99

8-10 DESIGN OF LATERALLY UNBRACED BEAMS WITH $C_b > 1.0$

Though the ASD charts were developed for cases where $C_b = 1.0$, they may easily be used for cases where C_b is larger than 1.0 by following the special procedure described in this section. In an excellent paper entitled "Fast Design of Beams with C_b Greater Than 1.0" a useful procedure is presented for handling this type of problem.[7] The following steps are suggested:

1. Compute an effective length ($L_{eff} = L_b/C_b$) and select a trial section with that length and the computed moment.
2. The value of F_b is limited to a maximum value of $0.60F_y$; our bending stress f_b may be computed and compared to this value, or the value of the allowable moment when $F_b = 0.60F_y$ may be selected from the curves. If this stress or the corresponding moment is exceeded, we try the next stronger section and repeat this step.
3. Next the allowable stress from ASD Equation F1-8 and the corresponding allowable moment are determined. If the result is ≥ our moment, the section is adequate and no further checking is necessary. If not, we may try the next stronger section and repeat these first three steps or follow Step 4.
4. We may have found that our section was unsatisfactory by ASD Equation F1-8 but Equation F1-6 may yet show it is satisfactory. We compute $L_e = L_b/\sqrt{C_b}$ and go to the beam curves with that value to obtain the allowable moment. If this value is ≥ our moment, the section is satisfactory.

Examples 8-9 and 8-10 present the design of two beams with the ASD charts for cases where C_b is larger than 1.0.

---

[7]L. B. Burgett and R. H. R. Tide, *Engineering Journal*, AISC, 17, no. 3 (3d quarter, 1980), pp. 74–78.

■ **Example 8-9**
Using A36 steel and the ASD Specification, select the lightest available W section for a moment of 150 ft-k and an unbraced length of 24 ft, if $C_b = 1.75$.

Solution.

$$L_{\text{eff}} = \frac{24}{1.75} = 13.7 \text{ ft}$$

From charts try W18 × 55 ($S_x = 98.3$ in³, $\dfrac{d}{A_f} = 3.82$ in.⁻¹)

Moment at L_u on chart = 180.2 ft-k > 150 ft-k   OK

Check trial section with Equation F1-8:

$$F_b = \frac{(12)(10)^3(1.75)}{(12)(24)(3.82)} = 19.09 \text{ ksi}$$

$$M = \frac{(19.09)(98.3)}{12} = 156.4 \text{ ft-k} > 150 \text{ ft-k}$$ OK

Use W18 × 55   ■

■ **Example 8-10**
Using a steel with $F_y = 36$ ksi and the ASD Specification, select the lightest available W section for a moment of 580 ft-k and an unbraced length of 20 ft, if $C_b = 2.0$.

Solution.

$$L_{\text{eff}} = \frac{20}{2.0} = 10.0 \text{ ft}$$

From charts try W30 × 108 ($S_x = 299$ in.³, $\dfrac{d}{A_f} = 3.75$ in.⁻¹)

Moment at L_u on chart = 548 ft-k < 580 ft-k   NG

Try next stronger section on chart, W30 × 116 ($S_x = 329$ in³, $\dfrac{d}{A_f} = 3.36$ in.⁻¹)

Moment at L_u = 603 ft-k > 580 ft-k   OK

Check trial section with Equation F1-8:

$$F_b = \frac{(12)(10)^3(2.0)}{(12)(20)(3.36)} = 29.76$$

Use $0.60F_y = 22$ ksi

$$M = \frac{(22)(329)}{12} = 603 \text{ ft-k} > 580 \text{ ft-k} \qquad \text{OK}$$

$$\underline{\underline{\text{Use W30} \times 116}} \qquad\qquad \blacksquare$$

Should it be necessary to design a beam without full lateral support when the charts cannot be used (as where charts are not available for the steel being used, or where the particular conditions do not fall on the chart, etc.), it will be necessary to use a trial-and-error solution. An allowable stress can be assumed, the required section modulus computed, a trial beam selected, the allowable stress determined for the trial beam size, and so on. There are several variables in the ASD equations, however, and although it's not difficult to find a section that will adequately support the load, it is quite difficult to find the absolutely lightest section by trial and error.

The student may quite logically ask, "What do I do if the beam in question has no compression flange?" Such a situation might very well occur in the bottom chord member of a truss subject to an intermediate load which causes bending in addition to the normal axial stress. Should the member consist of a pair of angles or a structural tee with the flanges on the tension side, there will be no compression flange, only a web.

The formulas presented in this section are for members that are symmetrical about both x and y axes, as W or S shapes. For other shapes more complicated expressions are needed for estimating the allowable stresses. For reference the student is referred to the *Guide to Stability Design Criteria for Metal Structures*.[8] A reasonable and very conservative practice is to assume $F_b = 0.60F_y$ for all such situations, as long as the local buckling requirements of ASD Section B5 are satisfied.

8-11 DESIGN OF CONTINUOUS MEMBERS

The ASD Specification for the elastic design of continuous members leans definitely toward the plastic design theories. Both theory and tests show clearly that continuous ductile steel members meeting the requirements for compact sections have the desirable ability of redistributing moments caused by overloads. (The student is again referred to the introductory paragraphs of Chapter 18 for a detailed discussion of this subject.)

The magnitude of the load applied to the beam in Fig. 8-11 can be increased until it reaches a value above which there will be no increase in the support moments. (These moments, which are defined as plastic moments in Chapter 18, occur when the steel has been stressed to its yield point all the way

---

[8] Structural Stability Research Council, *Guide to Stability Design Criteria for Metal Structures*, 4th ed., T. V. Galambos, ed. (New York: Wiley, 1988).

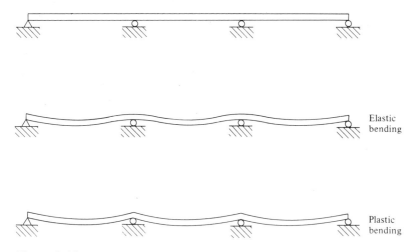

Figure 8-11

through the section from top to bottom at some point.) Should the load be increased further, the beam will act as though there is no continuity over the interior supports. These points will act as though they are hinges so far as load increases go but will continue to transfer the plastic moments. Increases in load will cause increases in the positive moments out in the "simple beams" between supports but no increases at the supports. The result is that the negative and positive moments in the beam tend to equalize as the load is increased. The beam may be said to have a reserve of strength and will not fail until the load is decidedly increased above the value at which the outermost fibers of the steel were first stressed to its yield point.

To take some advantage of this plastic overload capacity the ASD Specification (F1.1) says that for continuous compact sections the design may be made on the basis of nine-tenths of the maximum negative moments caused by gravity loads if the positive moments are increased by one-tenth of the average negative moments at the adjacent supports. (The 0.9 factor is applicable only to gravity loads and not to lateral loads such as those caused by wind and earthquake. The factor can also be applied to columns which have axial stresses of less than $0.15F_y$.) This moment reduction does not apply to members consisting of A514 steel, to hybrid girders or to moments produced by loading on cantilevers. Example 8-11(a) illustrates the design of a two-span beam with full lateral support. Part (b) of this example illustrates the design of the same beam if continuous lateral bracing is not provided.

■ Example 8-11

(a) A W18 × 55 consisting of a steel with F_y = 45 ksi is used for the span and loads of Fig. 8-12. Is this section satisfactory according to the ASD Specification if continuous lateral support is provided by a reinforced-concrete slab? The beam dimensions are shown in Fig. 8-13.

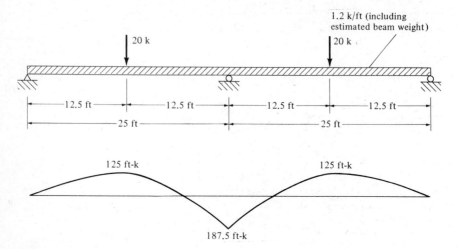

Figure 8-12

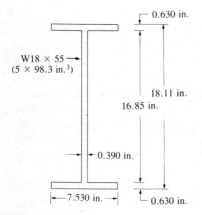

Figure 8-13

(b) Is the section satisfactory if lateral support is provided only at beam supports and midspans?

Solution.
(a) Continuous lateral support:

$$F_b = 0.66F_y = 29.7 \text{ ksi}$$

Max negative M for design $= (0.9)(187.5) = 168.8$ ft-k

Max positive M for design $= 125 + (0.10)\left(\dfrac{0 + 187.5}{2}\right) = 134.4$ ft-k

$$f_b = \frac{12 \times 168.8}{98.3} = 20.6 \text{ ksi} < 29.7 \text{ ksi} \qquad \text{OK}$$

(b) Lateral support at 12.5-ft intervals:
  Considering the 12.5 ft at ends of beam

$$C_b = 1.75 + (1.05)\left(\frac{0}{125}\right) + (0.3)\left(\frac{0}{125}\right)^2 = 1.75$$

$r_T = 1.95$ from Manual

$$\sqrt{\frac{102 \times 10^3 \times 1.75}{45}} = 63.0 < \frac{12 \times 12.5}{1.95} \text{ of}$$

$$76.9 < \sqrt{\frac{510 \times 10^3 \times 1.75}{45}} \qquad \text{of } 140.8$$

Use Equations F1-6 and F1-8

$$F_b = \left[\frac{2}{3} - \frac{45(76.9)^2}{1530 \times 10^3 \times 1.75}\right]45 = 25.5 \text{ ksi}$$

$$F_b = \frac{12 \times 10^3 \times 1.75}{12 \times 12.5 \times 3.82} = 36.6 \text{ ksi}\leftarrow$$

Use $0.60F_y = 27$ ksi

$$f_b = \frac{12 \times 125}{98.3} = 15.30 \text{ ksi} < 27 \text{ ksi} \qquad \text{OK}$$

Considering the interior 12.5-ft section

$$C_b = 1.75 + 1.05\left(+\frac{125}{187.5}\right) + 0.3\left(+\frac{125}{187.5}\right)^2 = 2.58 > 2.30$$

$C_b = 2.30$

$$\sqrt{\frac{102 \times 10^3 \times 2.30}{45}} = 72.2 < \frac{12 \times 12.5}{1.95} \text{ of}$$

$$76.9 < \sqrt{\frac{510 \times 10^3 \times 2.30}{45}} \text{ of } 161.5$$

Use Equations F1-6 and F1-8

$$F_b = \left[\frac{2}{3} - \frac{(45)(76.9)^2}{1530 \times 10^3 \times 2.30}\right]45 = 26.6 \text{ ksi}$$

$$F_b = \frac{12 \times 10^3 \times 2.30}{12 \times 12.5 \times 3.82} = 48.2 \text{ ksi}\leftarrow$$

Use $0.60F_y = 27$ ksi

$$f_b = \frac{12 \times 168.8}{98.3} = 20.6 \text{ ksi} < 27 \text{ ksi} \qquad \text{OK}$$

<u>Section is satisfactory</u>

∎

PROBLEMS

8-1 Select the most economical section available for a 30-ft simple span if the beam is to support a uniform load of 4k/ft. The beam is assumed to have full lateral support and an allowable bending stress of 24,000 psi. (*Ans.* W30 × 90)

8-2 to 8-9   Select the most economical section available for each of the beams shown, assuming full lateral support is provided for the compression flanges. Use A36 steel and the ASD Specification.

8-2

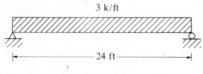

3 k/ft

|— 24 ft —|

8-3 (*Ans.* W30 × 99)

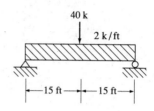

40 k

2 k/ft

|—15 ft —|— 15 ft—|

8-4

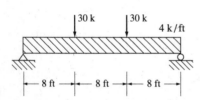

30 k   30 k

4 k/ft

|— 8 ft —|— 8 ft —|— 8 ft —|

8-5 (*Ans.* W30 × 90)

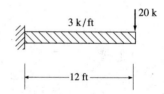

3 k/ft   20 k

|————— 12 ft —————|

8-6

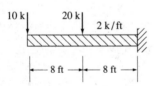

10 k   20 k

2 k/ft

|— 8 ft —|— 8 ft —|

8-7 (*Ans.* W12 × 14)

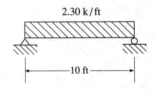

2.30 k/ft

|————— 10 ft —————|

8-8

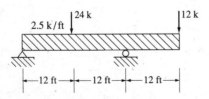

24 k   12 k

2.5 k/ft

|—12 ft —|— 12 ft —|— 12 ft—|

8-9 Repeat Prob. 8-4 with the 30-k loads doubled. (*Ans.* W33 × 130)

8-10 A36 steel beams 10 ft on center with simple spans of 32 ft are to support a 4-in. reinforced-concrete slab (which is assumed to supply full lateral bracing for the beams). If the live load is 100 lb/ft² and the concrete weighs 150 lb/ft³, design the beams.

8-11 Select the most economical section for the beam shown using a steel with an allowable bending stress of 30,000 psi. Full lateral support is assumed for the entire span. (*Ans.* W24 × 84)

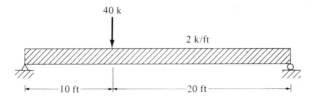

40 k

2 k/ft

| 10 ft | 20 ft |

8-12 Select the most economical section available for a 32-ft simple span if the beam is to support a uniform load of 10 k/ft. Use A242 steel and the ASD Specification, and assume that full lateral support is provided.

8-13 The illustration shows the arrangement of beams and girders which are used to support a 6-in. reinforced-concrete floor for a small industrial building. Using A36 steel and the ASD Specification, design the beams and girders, assuming they are simply supported. Assume full lateral support and a live load of 150 psf. (*Ans.* Beams W33 × 118, girders W40 × 268)

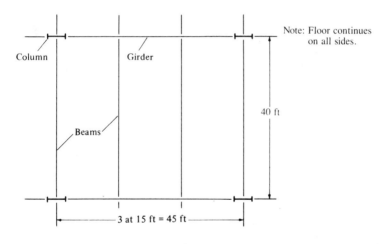

Column Girder

Note: Floor continues on all sides.

40 ft

Beams

3 at 15 ft = 45 ft

8-14 A W24 × 68 consisting of A36 steel is used for a simple span of 30 ft. Using the ASD Specification and assuming full lateral support, determine the maximum allowable uniform load the beam can support in addition to its own weight.

8-15 A beam consists of a W16 × 40 with a $\frac{1}{2}$ × 12 cover plate welded to each flange. If the allowable bending stress is 24,000 psi, determine the allowable uniform load it can support in addition to its own weight for a 30-ft simple span. (*Ans.* 2.711 k/ft)

8-16 Using the ASD Specification and A36 steel and assuming full lateral support, determine the maximum allowable uniform load the beam shown can support in addition to its own weight for a 24-ft simple span.

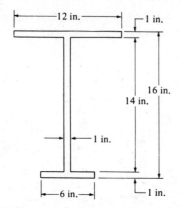

8-17 A 40-ft simple beam is to support two movable 30-k loads a distance of 15 ft apart. If the beam is to have full lateral support, select a section to resist the largest possible bending moment. Use ASD Specifications and A36 steel. (*Ans.* W27 × 84)

8-18 to 8-19   Using A36 steel and assuming full lateral bracing, design the beams. Be sure to consider different live-load placements so as to cause maximum moments.

8-18

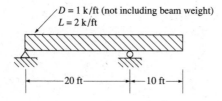

8-19 (*Ans.* W24 × 55)

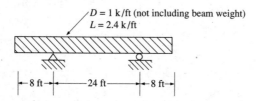

8-20 A W27 × 94 with full lateral support is used for a 30-ft simple span and supports a 4 k/ft uniform load. Assume two holes for 1-in. bolts in each flange. Compute the maximum compressive stress using the gross section properties, and the maximum tensile stress using the net section properties. Do not use ASD Specification B10.

8-21 Repeat Prob. 8-20, assuming four holes for 1-in. bolts in the tension flange. Comply with ASD Specification B10. (*Ans.* −22.75 ksi, +28.60 ksi)

8-22 The section shown has two $\frac{7}{8}$-in. bolts passing through each flange. Find the allowable uniform load this section can support in addition to its own weight for a

40-ft simple span. Use A572 Grade 50 steel and comply with ASD Specification B10.

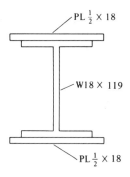

PL $\frac{1}{2}$ × 18

W18 × 119

PL $\frac{1}{2}$ × 18

8-23 Rework Prob. 8-1, assuming two $\frac{7}{8}$-in. bolts pass through each flange of the section at the point of maximum moment. Use ASD Specification B10 and A588 steel. (*Ans.* W24 × 84)

8-24 It is desired to select a section to support a 2.5 k/ft load for an 18-ft simple span. If full lateral support is assumed and two $\frac{3}{4}$-in. bolts are needed in each flange, select a W section using A36 steel and the ASD Specification.

8-25 Select a section for the span and loads shown. Use an allowable F_b = 24 ksi and assume there are to be two holes for 1-in. bolts in the tension flange. Use the ASD Specification. (*Ans.* W27 × 84)

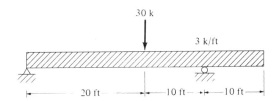

30 k

3 k/ft

20 ft   10 ft   10 ft

8-26 Compute the allowable bending stresses in a W24 × 68 for simple spans of 6, 12, and 25 ft if lateral support is provided at ends only. Use A36 steel and the ASD Specification.

8-27 A W36 × 150 consisting of A36 steel is used for a simple span of 24 ft and has lateral support at its ends only. What is the largest concentrated load that can be placed at the beam centerline? Use the ASD Specification. (*Ans.* 111.2 k)

8-28 Using a steel with F_y = 65 ksi and the ASD Specification, what uniformly distributed load can a simply supported W33 × 241 carry for a span of 40 ft when (a) the compression flange is braced laterally and (b) the compression flange has lateral support only at its ends?

8-29 What is the allowable uniform load that can be placed on a W30 × 99 (A242 steel) which has lateral support provided at its ends only? The beam is simply supported and has a span of 30 ft. Use the ASD Specification. (*Ans.* 1.627 k/ft)

8-30 to 8-34   The beams shown have lateral bracing provided for their compression flanges only at the supports and at other clearly marked points. Using A36 steel and the

ASD Manual charts for beams with unbraced lengths greater than L_u, select the most economical sections.

8-30

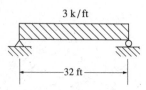

8-31 *(Ans.* W18 × 60)

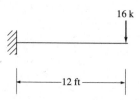

8-32

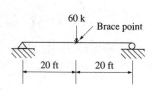

8-33 *(Ans.* W40 × 149)

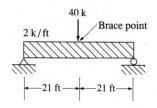

8-34

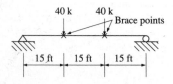

8-35 A W33 × 130 is used to support the loads and moment shown. Using A36 steel and the ASD Specification and neglecting the dead weight of the beam, find if the beam is overloaded
    (a) if full lateral support is provided. *(Ans.* Beam is satisfactory, $M = 340$ ft-k, resisting moment = 812 ft-k)
    (b) if lateral support is provided at the ends only. *(Ans.* Beam is satisfactory, $M = 340$ ft-k, resisting moment = 368.1 ft-k)

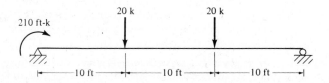

8-36 The W30 × 99 shown has lateral support supplied at the 25-ft points only (beam ends and centerline). If F_y is 65 ksi and the ASD Specification is used, determine the maximum permissible value of the concentrated load P. Neglect beam weight.

8-37 A W30 × 99 with lateral support supplied only at its ends supports a concentrated load at the center of its 30-ft simple span. Using A242 steel and the ASD Specification, determine the maximum permissible value of the load. *(Ans.* 24.4 k)

Figure for Problem 8-36

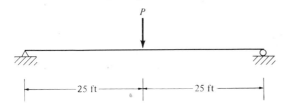

8-38 to 8-42   The members shown are assumed to have full lateral bracing for their compression flanges. Neglecting beam weight in each case and considering moment only, select the lightest section permissible by the ASD Specification if A36 steel is used.

8-38

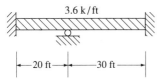

8-39   (*Ans.* W24 × 55)

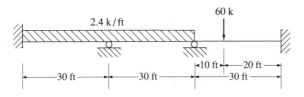

8-40

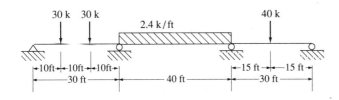

8-41   (*Ans.* W24 × 55)

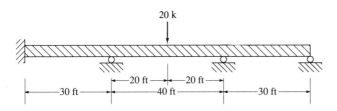

8-42

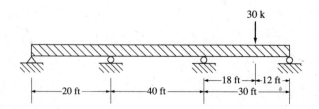

8-43 Three methods of supporting a roof are shown. Design the beam for each case, using A36 steel and the ASD Specification, if a 2-k/ft uniform load is to be supported. Assume full lateral support and consider moment only. (*Ans.* [a] W24 × 76, [b] W21 × 62, [c] W21 × 62)

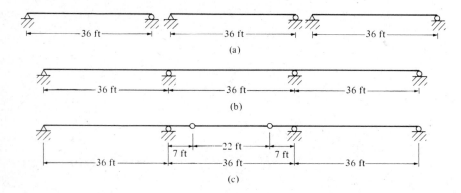

9

Design of Beams Continued

9-1 SHEAR

The beam supporting transverse loads, shown in Fig. 9-1(a), is sometimes said to be subjected to two types of shear—*transverse* and *longitudinal*. The first of these two shears is illustrated in Fig. 9-1(b), where there is a tendency of the

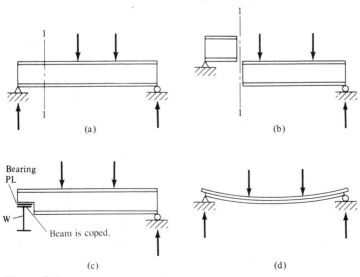

(a)

(b)

Bearing PL

W

Beam is coped.

(c)

(d)

Figure 9-1

part of the beam to the left of section 1-1 to slide upward with respect to the part of the beam to the right of the section. Actually, this type of shear failure will not occur in a regular steel beam because web crippling (discussed in Section 9-3) will occur first. Transverse shear, however, can cause failure directly if the beam has been deeply coped, as shown in Fig. 9-1(c).

The second type of shear is illustrated in Fig. 9-1(d) and occurs due to the bending of the member, which causes changes in lengths of the longitudinal fibers. For positive bending the lower fibers are stretched and the upper fibers are shortened while somewhere in between there is a neutral axis where the fibers do not change in length. Due to these varying deformations, a particular fiber has a tendency to slip on the fiber above or below. The largest value of longitudinal or horizontal shear occurs at the neutral axis.

If a wooden beam was made by stacking boards on top of each other and not connecting them, they would tend to take the shape shown in Fig. 9-1(d). The student may have observed short heavily loaded timber beams with large transverse shears which split along horizontal planes.

This presentation may be entirely misleading in seeming to completely separate horizontal and vertical shears. In reality horizontal and vertical shears at any point are the same and may not be separated. Furthermore, one cannot occur without the other.

The tendency to slip is resisted by the shearing strength of the material. Although steel beam sizes are rarely controlled by shear, it is wise to check them, particularly if they are short and heavily loaded. Maximum external shear usually occurs near the supports, but shear is present throughout the beam. The average transverse shearing stress on the cross section of a beam at a certain point in the span equals the external shear divided by the cross-sectional area of the beam. The longitudinal shearing stress formula, however, shows that the shearing stress is not constant across a beam cross section but is zero at the outermost fibers and has its largest value at the neutral axis. The familiar formula, to follow, can be used to calculate the unit shearing stress at any point. (This formula applies to beams with open cross sections which are not subjected to torsion.)

$$f_v = \frac{VQ}{bI}$$

where

V = external shear at the section in question

Q = statical moment of that portion of the section lying outside (either above or below) the line on which f_v is desired, taken about the neutral axis

I = moment of inertia of the entire section about the neutral axis

b = width of the section where the unit shearing stress is desired

When this expression is used to calculate the shear across the face of a W, M, or S section, the resulting values are very small in the flanges and quite large in the web. The values in the web are fairly uniform from top to bottom.

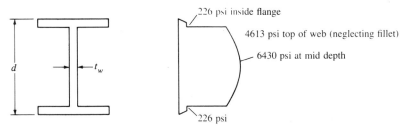

226 psi inside flange

4613 psi top of web (neglecting fillet)

6430 psi at mid depth

226 psi

Figure 9-2 Variation of shear stress across a W24 × 76 subjected to an external shear of 60,000 lbs.

Fig. 9-2 shows the variation of shear on the cross section of a W24 × 76 which is resisting an external shear of 60,000 lb.

The common practice of assuming a uniform shearing stress variation in the web from top to bottom (using the full depth of the section) appears to be reasonable after examining Fig. 9-2. The shearing stress obtained in accordance with this assumption is $60,000/(23.92)(0.440) = 5,700$ psi, which is not much smaller than the maximum theoretical value of 6,430 psi. The truth of the matter is that the allowable shearing stresses given by most specifications have been developed on the assumption that uniformly distributed shear calculations will be made. For most W, S, and channel sections having fairly large flanges and thin webs, the results obtained from this assumption are reasonable. For other shapes a few calculations may be necessary to see if the error percentage is appreciable.

A more conservative approach followed by a few engineers is to use only the web depth in figuring the average shear. This method would give a result of $60,000/(22.56)(0.440) = 6,044$ psi in the case being considered. Although some designers use the total beam depth and others only the web depth or even the depth between the toes of the fillets for rolled sections, they nearly all agree to use only the web depth for plate girders.

The ASD Specification (F4) states that the allowable shear stress F_v is $0.40F_y$ if $h/t_w \leq 380/\sqrt{F_y}$. In this expression h is the clear distance between the beam or girder flanges. Almost all rolled sections fall into this class, and the cross-sectional area effective in resisting shear is considered to be the overall depth of the member times its web thickness. For such cases the shear stress is calculated as follows:

$$f_v = \frac{V}{dt_w} \leq 0.40F_y$$

Should h/t_w be greater than $380/\sqrt{F_y}$, the allowable shear stress F_v becomes $F_y/2.89(C_v) \leq 0.40F_y$, where C_v is given in Section F4 of the Specification and is the ratio of the "critical" web stress to the shear yield stress of the web. (C_v is discussed in Chapter 17.) This allowable shear stress is applicable to the clear distance between the flanges times the web thickness.

$$f_v = \frac{V}{ht_w}$$

The maximum permissible external shear which a particular beam can resist can be calculated by substituting the allowable shear stress F_v into the preceding expression and solving for V.

$$V = F_v \, dt_w \quad \text{or} \quad V = F_v h t_w$$

In the beam tables of the ASD Manual this value V is tabulated for each of the sections normally used as beams with yield stresses of 36 and 50 ksi. With a glance at these tables the designer can check the shear in a particular beam. If it is too high he or she can quickly select another section which has a satisfactory allowable V. On some occasions when short spans and high shears are present, S beams with their rather heavy webs may prove to be economical.

Generally shear is not a problem in steel beams because the webs of rolled shapes are capable of resisting quite large shearing stresses. Perhaps it is well, however, to list here the most common situations where shear might be excessive.

1. Should large concentrated loads be placed near beam supports, they will cause large external shears without corresponding increases in bending moments. A fairly common example of this type occurs in tall buildings where on a particular floor the upper columns are offset with respect to the columns below. The loads from the upper columns applied to the beams on the floor level in question will be quite large if there are many stories above.

Combined welded and bolted joint, Transamerica Pyramid, San Francisco, Calif. (Courtesy of Kaiser Steel Corporation.)

2. Probably the most common shearing stress problem occurs where two members (as a beam and a column) are rigidly connected so their webs lie in a common plane. In Section E6 of the *Commentary on the ASD Specification,* formulas are given which show when the webs are overstressed in shear and how much they need to be thickened or reinforced when overstressed. This situation, which commonly occurs at the junction of columns and beams (or rafters) in rigid frame structures, is discussed in Section 19-4 as it relates to plastic design.

3. Where beams are notched or coped as was shown in Fig. 9-1(c), shear can be a problem. For this case shear stresses can be calculated for the remaining beam depth. A similar discussion can be made where holes are cut in beam webs for ductwork or other items.

4. Theoretically, very heavily loaded short beams can have excessive shears, but practically this does not occur too often unless case 1 applies.

5. Shear may very well be a problem even for ordinary loadings when very thin webs are used as in plate girders or in light-gage cold-formed steel members.

Should calculated shearing stresses exceed allowable values *doubler plates* may be welded to the column web, or stiffeners may be connected to the webs in the zones of high shear.

Example 9-1 presents a brief illustration of a shear stress calculation.

■ Example 9-1

Select a W section for the load and span shown in Fig. 9-3. Use A36 steel and assume full lateral bracing for the compression flange. Also check shear.

Solution.   Assume beam weight = 76 lb/ft

$$M = \frac{(2.076)(24)^2}{8} + (30)(8) = 389.5 \text{ ft-k}$$

$$S_{reqd} = \frac{(12)(389.5)}{24} = 194.7 \text{ in}^3.$$

Try W24 × 76 ($d = 23.92$ in., $t_w = 0.440$ in.)
Check shear:

$$\frac{d}{t_w} = \frac{23.92}{0.440} = 54.36 < \frac{380}{\sqrt{36}} = 63.33$$

$\therefore F_v = 0.40 \, F_y = 14.4$ ksi

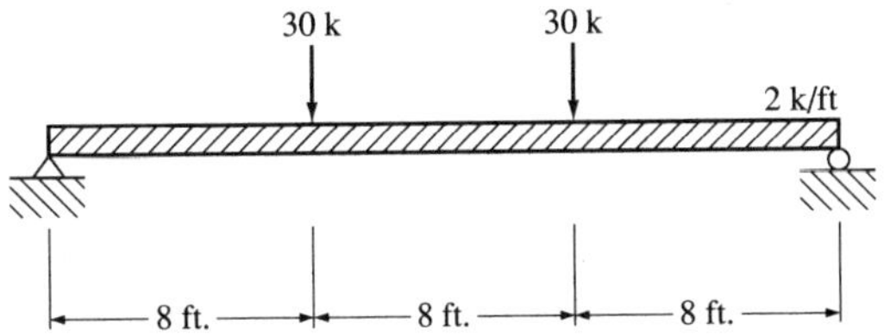

Figure 9-3

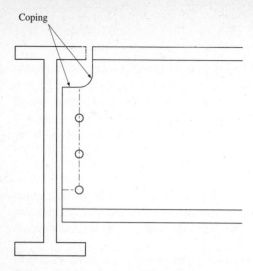

Coping

Figure 9-4 Block shear failure possible along dashed line.

$$\text{Max } V = (2.076)(12) + 30 = 54.91 \text{ k}$$

$$f_v = \frac{54.91}{(23.92)(0.440)} = 5.22 \text{ ksi} < 14.4 \text{ ksi} \qquad \text{OK}$$

Or from Part 2 tables of Manual for a W24 × 76 of A36 steel

$$\text{Max } V = 152 \text{ k} > 54.91 \text{ k} \qquad \text{OK}$$

$$\underline{\underline{\text{Use W24 × 76}}} \qquad \blacksquare$$

Sometimes a connection is made to only a small portion or depth of the web of a beam. The $0.40F_y$ allowable shear stress, however, is based on the assumption that the shear stress is averaged over the whole web (V/dt_w). For such cases the designer may decide to assume the shear is spread over a lesser depth.

When beams that have their top flanges at the same elevations (the usual situation) are connected to each other, it is frequently necessary to cope one of them, as shown in Fig. 9-4. For such cases there is a distinct possibility of a block shear failure along the dotted lines shown. This situation is discussed in Chapter 14.

9-2 DEFLECTIONS

The deflections of steel beams are usually limited to certain maximum values. Among the several excellent reasons for deflection limitations are the following:

1. Excessive deflections may damage other materials attached to or supported by the beam in question. Plaster cracks caused by large ceiling joist deflections are one example.

2. The appearance of structures is often damaged by excessive deflections.
3. Extreme deflections do not inspire confidence in the persons using a structure although it may be completely safe from a strength standpoint.
4. It may be necessary for several beams supporting the same loads to deflect equal amounts.

Standard American practice for buildings has been to limit service live-load deflections to approximately 1/360 of the span length. This deflection is supposedly the largest value that ceiling joists can deflect without causing cracks in underlying plaster. The 1/360 deflection is only one of many maximum deflection values in use because of different loading situations, different engineers, and different specifications. For situations where precise and delicate machinery is supported, maximum deflections may be limited to 1/1500 or 1/2000 of the span lengths. The 1989 AASHTO Specifications limit deflections in steel beams and girders due to live load and impact to 1/800 of the span. (For bridges in urban areas which are partly used by pedestrians, the AASHTO *recommends* a maximum value equal to 1/1000 of the span lengths.)

The ASD Specification does not specify exact maximum permissible deflections. There are so many different materials, types of structures, and loadings that no one set of deflection limitations is acceptable for all cases. Thus limitations must be set by the individual designer on the basis of his or her experience and judgment.

Before substituting blindly into a formula that will give the deflection of a beam for a certain loading condition, the student should thoroughly understand the theoretical methods of calculating deflections. These methods include the moment area, conjugate beam, and virtual-work procedures. From these methods various expressions can be determined, such as the common one given at the end of this paragraph for the centerline deflection of a uniformly loaded simple beam.

$$\Delta = \frac{5wL^4}{384EI}$$

To use deflection expressions such as this one the reader must be very careful to use consistent units. Example 9-2 illustrates the application of the preceeding expression. The author has changed all units to pounds and inches. Thus, the uniform load given in the problem as so many k/ft is changed to so many lb/in.

■ **Example 9-2**
In Example 8-1 a W21 × 62 ($I_x = 1330$ in.^{4}) was selected to support a total uniform load of 4.362 k/ft, including its own weight for a 21-ft span. Compute the total deflection at the centerline of the beam.

Solution.

$$\Delta = \frac{5wL^4}{384\ EI} = \frac{(5)\left(\dfrac{4362}{12}\right)(12 \times 21)^4}{(384)(29 \times 10^6)(1330)} = \underline{\underline{0.495 \text{ in.}}}$$ ∎

In the beam tables of Part 2 of the ASD Manual a useful deflection value is given for each beam in the table, loaded to its maximum permissible uniform load. For this discussion the total uniform load W on the beam of Example 9-2 is $wL = (4.362)(21) = 91.60$ k. In the tables the maximum W for a 21-ft W21 × 62 of A36 steel is 96 k and its centerline deflection for that load is 0.52 in. By proportions our centerline deflection will be $(91.60/96)(0.52) = 0.496$ in.

Some specifications handle the deflection problem by requiring certain minimum depth-span ratios. For example, the AASHTO suggests the depth-span ratio be limited to a minimum value of 1/25. A shallower section is permitted, but it should have sufficient stiffness to prevent a deflection greater than would have occurred if the 1/25 ratio had been used.

In a similar fashion Section L3.1 of the ASD Commentary provides a suggested minimum depth equal to $(F_y/800)$ (span length) for fully stressed beams and girders in floor systems. Should shallower members be used, it is suggested that their allowable bending stresses be reduced in the same ratio as their depths are in proportion to the recommended value. The Commentary further recommends that the depth of fully stressed roof purlins should desirably be at least equal to $(F_y/1000)$ (span length) except for flat roofs.

A steel beam can be cold-bent or cambered an amount equal to the deflection caused by dead load or the deflection caused by dead load plus some percentage of the live load. Approximately 25 percent of the camber so produced is elastic and will disappear when the cambering operation is completed. Detailed information for particular shapes is given in Part 1 of the ASD Manual in the section entitled "Standard Mill Practice." It should be remembered that a beam which is bent upward looks much stronger and safer than one which sags downward (even a very small distance).

A camber requirement is quite common for longer steel beams. In fact a very large percentage of the beams used in composite construction (see Chapter 16) today are cambered.

Cambering costs more and is something of a nuisance to steel fabricators. One common rule of thumb is that it takes about one extra work-hour to camber each beam. Another problem with cambering is that fabricators may have to adjust their detail dimensions somewhat to achieve the proper fitting of the members. As a result of this problem and the extra cost to the fabricator (and thus the owner), it is often more economical to use a larger beam section that will reduce deflections sufficiently.

If we can move up one section in weight, thus reducing deflections so cambering is not needed, we may have a very desirable solution. Similarly, if a higher-strength steel is being used, it may be be desirable to switch the beams that need cambering to A36 steel. The results will be larger beams but smaller deflections and perhaps some economy if cambering can be eliminated.

Deflections may very well control the sizes of beams for longer spans or for short ones where deflection limitations are severe. To assist the designer in selecting sections where deflections may control, the ASD Manual includes in Part 2 a table entitled "Moment of Inertia Selection Table," in which the I_x values are given in numerically descending order for all of the sections normally used as beams. In this table the sections are arranged in groups, with the lightest section in each group printed in bold type. Example 9-3 presents the design of a beam where deflection controls the design.

■ Example 9-3

Select the lightest available section with $F_y = 36$ ksi to support a total load of 4.2 k/ft for a 30-ft simple span if the maximum permissible deflection is 1/1500 the span length. The section is to be fully braced laterally.

Solution.   Assume beam weight = 194 lb/ft (after some initial scratch work)

$$M = \frac{(4.394)(30)^2}{8} = 494.32 \text{ ft-k}$$

$$S_{reqd} = \frac{(12)(494.32)}{24} = 247.2 \text{ in.}^3$$

<u>Try W30 × 99 (I_x = 3990 in.^{4})</u>

$$\text{Maximum permissible } \Delta = \left(\frac{1}{1500}\right)(12 \times 30) = 0.24 \text{ in.}$$

$$\text{Actual } \Delta \text{ at centerline} = \frac{(5)\left(\dfrac{4394}{12}\right)(12 \times 30)^4}{(384)(29 \times 10^6)(3990)}$$

$$= 0.692 \text{ in.} > 0.24 \text{ in.} \qquad\qquad \text{NG}$$

I_x required to limit Δ to 0.24 in.

$$I_x = \left(\frac{0.692}{0.24}\right)(3990) = 11,504 \text{ in.}^4$$

From "Moment of Inertia Selection Table":

<u>Use W40 × 167</u>   ■

Vibrations

Though steel members may be selected which are satisfactory as to moment, shear, deflections and so on, some very annoying floor vibrations may still occur. This is perhaps the most common serviceability problem faced by designers. Objectionable vibrations will frequently occur where long spans and large open floors without partitions or other items which might provide suitable damping are used. The reader may often have noticed this situation in the floors of large malls.

Damping of vibrations may be achieved by using framed in place partitions each attached to the floor system in at least 3 places or by installing "false" sheetrock partitions between ceilings and the underside of floor slabs. A better procedure is to control the stiffness of the structural system.[1,2] (A rather common practice in the past has been to try to limit vibrations by selecting beams no shallower than $\frac{1}{20}$th of spans.)

Ponding

If water on a flat roof accumulates faster than it runs off, the increased load causes the roof to deflect into a dish shape that can hold more water, which causes greater deflections and so on. This process of *ponding* continues until equilibrium is reached or until collapse occurs. Ponding is a serious matter, as illustrated by the large annual number of flat roof failures in the United States.

Ponding will occur on almost any flat roof to a certain degree even though roof drains are present. Drains may be inadequate during severe storms or they may become stopped up, and, furthermore, they are often placed along the beam lines, which are actually the high points of the roof. The best method of preventing ponding is to have an appreciable slope on the roof ($\frac{1}{4}$ in./ft or more) together with good drainage facilities. It has been estimated that probably two-thirds of the flat roofs in the United States have slopes less than this value, which is the minimum recommended by the National Roofing Contractors Association (NRCA). It costs approximately 3 to 6 percent more to construct a roof with this desired slope than building with no slope.[3]

When a very large flat roof (perhaps an acre or more) is being considered, the effect of wind on water depth may be quite important. A heavy rainstorm will frequently be accompanied by heavy winds. When a large quantity of water is present on the roof, a strong wind may very well push a great deal of water to one end, creating a dangerous depth of water as regards the load in pounds per square foot applied to the roof. For such situations *scuppers* are sometimes used. These are large holes or tubes in the walls or parapets which enable water above a certain depth to quickly drain off the roof.

Ponding failures will be prevented if the roof system (consisting of the roof deck and supporting beams and girders) has sufficient stiffness. The ASD Specification (K2) describes a minimum stiffness to be achieved if ponding

---

[1] T. M. Murray, "Controlling Floor Movement" Modern Steel Construction (AISC, Chicago, Il., June 1991) pp. 17–19.

[2] T. M. Murray, "Acceptability Criterion for Occupant-Induced Floor Vibrations," *Engineering Journal,* AISC, 18, 2 (2nd quarter, 1981), pp. 62–70.

[3] Gray Van Ryzin, "Roof Design: Avoid Ponding by Sloping to Drain," *Civil Engineering* (New York: ASCE, January 1980), pp. 77–81.

failures are to be prevented. If this minimum stiffness is not provided, it is necessary to make other investigations to be sure that a ponding failure is not possible.

Theoretical calculations for ponding are very complicated. The ASD requirements are based on work by F. J. Marino[4] in which he considered the interaction of a two-way system of secondary or submembers supported by a primary system of main members or girders. A numerical ponding example (A-4) is presented in the Appendix.

9-3 WEBS AND FLANGES WITH CONCENTRATED LOADS

When steel members have concentrated loads perpendicular to one flange and symmetric to the web, their flanges and webs must have sufficient design strength in the areas of flange bending, web yielding, web crippling, and sidesway web buckling. Should a member have concentrated loads applied to both flanges, it must have a sufficient web design strength in the areas of web yielding, web crippling, and column web buckling. In this section formulas for determining strengths in these areas are presented.

Local Flange Buckling

In Fig. 9-5 it is assumed that the horizontal member shown is rigidly attached to the flanges of the vertical member. As a result it may be necessary to place transverse stiffeners in the vertical member (as shown by the dashed lines) if the bending stress in the flange of the supporting member is too large.

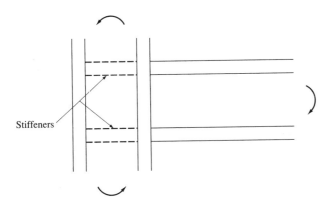

Figure 9-5 Column stiffeners.

---

[4] F. J. Marino. "Ponding of Two-Way Roof System," *Engineering Journal*, AISC, 3, no. 3, (3rd quarter 1966), pp. 93–100.

The bending stress is in effect limited by ASD Equation K1-1. In this equation, F_{yc} is the column yield stress and P_{bf} is $\frac{5}{3}$ times the computed force in kips applied from the flange or moment connection if that force is due to dead and live loads only. The multiplier becomes $\frac{4}{3}$ where the force is due to dead and live load combined with wind or earthquake forces. Sometimes this latter multiplier is different in other codes.

Stiffeners are required if the thickness of the flange member is less than

$$0.4\sqrt{\frac{P_{bf}}{F_{yc}}} \qquad \text{(ASD Equation K1-1)}$$

It is not necessary to check this formula if the length of loading across the beam flange is less than 0.15 times the flange width b_f. The required areas for stiffeners are not provided by the ASD Specification. Section K1.8 does, however, provide minimum stiffener dimensions. Furthermore, stiffeners used must meet the required width-thickness ratios of Section B5.

Equation K1-1 may be equated to the flange thickness t_f and solved for P_{bf} as follows:

$$P_{bf} = \frac{F_{yc}t_f^2}{0.16} \le P_{fb}$$

If this value, which is tabulated in the column tables of the Manual, is exceeded, stiffeners will be necessary.

Local Web Yielding

Beams that support heavy concentrated loads sometimes fail by web yielding unless the web is stiffened near the loads. Web yielding is due to the stress concentrations at the junction of the flange and web, where the beam is trying to transfer compression from the relatively wide flange to the narrow web.

As shown in Fig. 9-6 the loads are assumed to be spread longitudinally along the beam from the load application point for a distance = 2.5 k for loads

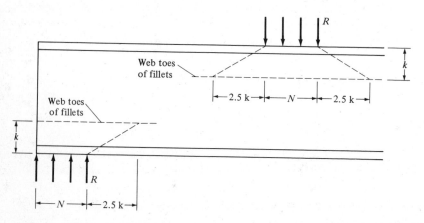

Figure 9-6 Local web yielding.

applied at or near the member end and for 2.5 k on each side of a load applied at a distance from the member end greater than the member depth d. The toe of the fillet is the most dangerous location for failure because the resulting area has its smallest value there. The ASD Specification (K1.3) does not permit the compression at this point to exceed $0.66F_y$ in beams without web stiffeners. From this information you can clearly see how the ASD expressions to follow were obtained. The stress is computed by dividing the concentrated load or re-action by an area equal to the web thickness t_w times the length over which the force is assumed to be spread.

If the force is a concentrated load or reaction that causes tension or com-pression and is applied at a distance greater than the member depth from the end of the member, use

$$\frac{R}{t_w(N + 5\ k)} \le 0.66F_y \qquad \text{(ASD Equation K1-2)}$$

If the force is a concentrated load or reaction applied at or near the end of the member, use

$$\frac{R}{t_w(N + 2.5\ k)} \le 0.66F_y \qquad \text{(ASD Equation K1-3)}$$

In the beam tables of Part 2 of the ASD Manual, numerical values have been substituted into parts of Equation K1-3 for the shapes normally used as beams with $F_y = 36$ ksi and 50 ksi and labeled R_1 and R_2. These terms are

$$R_1 = 1.65\ k\ F_{yw}t_w \qquad \text{(first part of equation)}$$
$$R_2 = 0.66F_{yw}t_w \qquad \begin{array}{l}\text{(second part of equation}\\ \text{except for } N)\end{array}$$

Then the whole equation may be written as

$$R = R_1 + R_2 N$$

The values of R_1 and R_2 can be selected from the beam tables and used in the above equation to determine the allowable R from the standpoint of web yielding for a certain value of N; or the equation can be solved for N, which is the minimum bearing length for a concentrated load or reaction at or near the end of a member, so that local web yielding will not be a problem.

$$N = \frac{R - R_1}{R_2}$$

Web Crippling

The ASD Specification (K1.4) says that bearing stiffeners are needed for the webs of a member if a concentrated force is larger than a certain value. This limiting value is determined by the appropriate equation of the two that follow (in which F_{yw} is the specified yield stress of the beam web). If web stiffeners are provided and extend for at least half of the web depth, web crippling does not need to be checked.

If the concentrated load is applied at a distance not less than $d/2$ from the end of the member, use

$$R = 67.5\, t_w^2 \left[1 + 3\left(\frac{N}{d}\right)\left(\frac{t_w}{t_f}\right)^{1.5} \right] \sqrt{\frac{F_{yw} t_f}{t_w}}$$ (ASD Equation K1-4)

If the concentrated load is applied at a distance less than $d/2$ from the end of the member, use

$$R = 34\, t_w^2 \left[1 + 3\left(\frac{N}{d}\right)\left(\frac{t_w}{t_f}\right)^{1.5} \right] \sqrt{\frac{F_{yw} t_f}{t_w}}$$ (ASD Equation K1-5)

Numerical values have been substituted into parts of Equation K1-5 and labeled R_3 and R_4 in the beam tables of the Manual.

$$R_3 = 34 t_w^2 \sqrt{\frac{F_{yw} t_f}{t_w}}$$ (first part of equation)

$$R_4 = 34 t_w^2 \left[3\left(\frac{1}{d}\right)\left(\frac{t_w}{t_f}\right)^{1.5} \right] \sqrt{\frac{F_{yw} t_f}{t_w}}$$ (second part of equation except for N)

Then the whole equation may be written as

$$R = R_3 + R_4 N$$

The values of R_3 and R_4 may be obtained from the beam tables and the above equation solved for N, which is the minimum bearing length for concentrated loads or reactions located at a distance less than $d/2$ from the end of a member so that web crippling is not a problem.

$$N = \frac{R - R_3}{R_4}$$

The minimum bearing length required at the end of a beam for web yielding and web crippling is determined in Example 9-4. In addition the 6-in. length over which a pair of interior concentrated loads are spread is checked.

For many beams the theoretical bearing length determined is too small to be practical and may even be negative. Should this be the case, the designer selects a reasonable value that will fit in with the construction requirements. For steel beams bearing on masonry a minimum bearing length of probably $3\frac{1}{2}$ or 4 in. (90 or 100 mm) should be used. It is to be remembered that a fairly large percentage of the structural failures which have occurred, particularly during erection, were due to insufficient bearing. Another fact to keep in mind is that the bearing area should be large enough to keep the bearing stress from exceeding the allowable value of the supporting material. This subject is considered in Section 9-7.

■ Example 9-4

A W33 × 130 consisting of A36 steel has been selected for the loading and span of Fig. 9-7. (a) Determine the minimum bearing length required at the

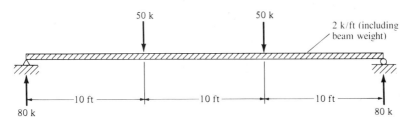

Figure 9-7

reactions. (b) The 50-k concentrated loads are applied to the beam over a width of 6 in. Is this sufficient?

Solution.   Using a W33 × 130 (t_w = 0.580 in. t_f = 0.855 in., k = $1\frac{11}{16}$ in. = 1.688 in., d = 33.09 in., R_1 = 58.1 k, R_2 = 13.8 k/in., R_3 = 83.3 k, and R_4 = 4.22 k/in.)

(a) Minimum bearing length required at the reactions:

$$\text{For web yielding } N = \frac{R - R_1}{R_2} = \frac{80 - 58.1}{13.8} = 1.59 \text{ in.}$$

$$\text{For web crippling } N = \frac{R - R_3}{R_4} = \frac{80 - 83.3}{4.22} = \text{negative number}$$

$$\underline{\text{Use } 3\frac{1}{2} \text{ in.}}$$

(b) Checking bearing length under 50-k loads:

$$\frac{R}{t_w(N + 5 k)} = \frac{80}{(0.580)(6 + 5 \times 1.688)} = 9.55 \text{ ksi} < 24 \text{ ksi} \qquad \text{OK}$$

$$R = 67.5t_w^2\left[1 + 3\left(\frac{N}{d}\right)\left(\frac{t_w}{t_f}\right)^{1.5}\right]\sqrt{\frac{F_{yw}t_f}{t_w}}$$

$$= (67.5)(0.580)^2\left[1 + 3\left(\frac{6}{33.09}\right)\left(\frac{0.580}{0.855}\right)^{1.5}\right]\sqrt{\frac{(36)(0.855)}{0.580}}$$

$$= 215.7 \text{ k} > 80 \text{ k} \qquad\qquad \text{OK} \quad \blacksquare$$

Sidesway Web Buckling

Should compressive loads be applied to braced compression flanges, the web will be put in compression and the tension flange may buckle as shown in Fig. 9-8.

It has been found that sidesway web buckling will not occur if the flanges are restrained against rotation with $(d_c/t_w)/L/b_f) > 2.3$ or if $(d_c/t_w)/(L/b_f) > 1.7$ when flange rotation is not restrained. In these expressions d_c is the web depth between the web toes of the fillets, that is, $d - 2k$, and L is the largest laterally unbraced length along either flange at the point of the load.

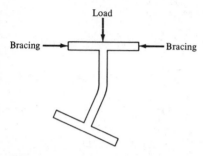

Figure 9-8 Sidesway web buckling.

It is also possible to prevent sidesway web buckling with properly designed lateral bracing or stiffeners at the load. The ASD Commentary (K1.5) suggests that local bracing for *both* flanges be designed for 1 percent of the magnitude of the concentrated load applied at the point. If stiffeners are used, they must extend from the point of load for at least one-half of the member depth and should be designed to carry the full load. Flange rotation must be prevented if the stiffeners are to be effective.

Should the flanges of members not be restrained against relative movement by stiffeners or lateral bracing and be subject to concentrated compressive loads, stiffeners may be required if the loads exceed the following limits:

When the loaded flange is braced against rotation and $(d_c/t_w)/(L/b_f)$ is less than 2.3, use

$$R = \frac{6800t_w^3}{h}\left[1 + 0.4\left(\frac{d_c/t_w}{L/b_f}\right)^3\right] \quad \text{(ASD Equation K1-6)}$$

When the loaded flange is not restrained against rotation and $(d_c/t_w)/L/b_f)$ is less than 1.7, use

$$R = \frac{6800t_w^3}{h}\left[0.4\left(\frac{d_c/t_w}{L/b_f}\right)^3\right] \quad \text{(ASD Equation K1-7)}$$

9-4 UNSYMMETRICAL BENDING

From mechanics of materials it is remembered that each beam cross section has a pair of mutually perpendicular axes known as the principal axes for which the product of inertia is zero. Bending that occurs about any axis other than one of the principal axes is said to be unsymmetrical bending. When the external loads are not in a plane with either of the principal axes or when loads are simultaneously applied to the beam from two or more directions, unsymmetrical bending is the result.

If a load is not perpendicular to one of the principal axes, it may be broken into components which are perpendicular to those axes and the moments about each axis, M_x and M_y, determined as shown in Fig. 9-9.

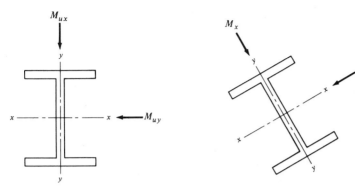

Figure 9-9

When the external loads are not in the principal plane, the stresses can be determined by breaking the loads into components perpendicular to the principal axes, calculating the moments about each axis, and determining the maximum stress caused by a combination of the two moments. The following expression can be written for the stress at any point in a beam subjected to unsymmetrical bending:

$$f_b = \frac{M_x y}{I_x} \pm \frac{M_y x}{I_y} = \frac{M_x}{S_x} \pm \frac{M_y}{S_y}$$

It might be noted that the longitudinal shearing stresses for a beam of this type can be calculated with a similar expression.

$$f_v = \frac{V_x Q_x}{b I_x} \pm \frac{V_y Q_y}{b I_y}$$

When a section has one axis of symmetry, that axis is one of the principal axes and the calculations necessary for determining stress are quite simple. For this reason unsymmetrical bending is not difficult to handle in the usual beam section, which is probably a W, S, M, or C. Each of these sections has at least one axis of symmetry, and the calculations are appreciably reduced. A further simplifying factor is that the loads are usually gravity loads and probably perpendicular to the x axis. For a case of this type the formula for stress is simply $f_b = Mc/I$.

Among the beams that must resist unsymmetrical bending are crane girders in industrial buildings and purlins for ordinary roof trusses. The x axes of purlins are parallel to the sloping roof surfaces, while the large percentage of their loads (roofing, snow, etc.) are gravity loads. These loads do not lie in a plane with either of the principal axes of the inclined purlins, and the result is unsymmetrical bending. Wind loads are generally considered to act perpendicular to the roof surface and thus perpendicular to the x axes of the purlins, with the result that they are not considered to cause unsymmetrical bending. The x axes of crane girders are usually horizontal, but the girders are subjected to lateral thrust loads from the moving cranes as well as to gravity loads.

Should a compact, laterally supported beam be bent about its x axis only, the allowable stress F_b will be $0.66F_y$ using the ASD Specification. Should a similar shape be bent about its y axis only, F_b equals $0.75F_y$.

This increase in allowable stress for compact sections and solid rectangular beams bent about their weak axes is based upon the high shape factor for such bending and upon the fact that these sections are not subject to lateral-torsional buckling when bent about their weak axes. The increase in allowable stress is not nearly as large as the increase in the shape factor, because it is desired to ensure elastic behavior. (Shape factors, briefly introduced in Section 8-3, are discussed in detail in Chapter 18.)

Which allowable stress is to be used when both types of bending occur simultaneously? The ASD Specification provides an allowable stress which is in effect a combination of the two allowable values. This is done by using the interaction expression to follow, which is ASD Equation H1-3 with $f_a = 0$. (Note that the actual stress cannot exceed the allowable for either axis even if the total is ≤ 1.0.)

$$\frac{f_{bx}}{F_{bx}} + \frac{f_{by}}{F_{by}} < 1.0$$

For compact laterally supported shapes the equation becomes

$$\frac{f_{bx}}{0.66F_y} + \frac{f_{by}}{0.75F_y} < 1.0$$

Examples 9-5 and 9-6 illustrate the design of beams subjected to unsymmetrical bending. To illustrate the trial-and-error nature of the problem, the author did not do quite as much scratch work as in the first examples. The first design problems of this type which the student attempts may quite well take several trials. Consideration needs to be given to the question of lateral support for the compression flange. Should the lateral support be of questionable nature, the engineer should reduce the allowable compressive stresses by means of one of the expressions previously given for that purpose.

■ Example 9-5

A certain beam is estimated to have a vertical bending moment of 120 ft-k and a lateral bending moment of 25 ft-k. These moments include the effect of the estimated beam weight. The loads are assumed to pass through the centroid of the section, and the entire section modulus value is therefore available about each axis. Select a W shape that can resist these moments. Use A36 steel and the ASD Specification and assume full lateral support for the compression flange.

Solution. First trial: The S required for vertical bending moments alone is $12 \times 120/24 = 60$ in.^{3} Try a section with a larger section modulus.

Try W24 × 62 (S_x = 131 in.³, S_y = 9.80 in.³)

$$f_{bx} = \frac{12 \times 120}{131} = 10.99 \text{ ksi}$$

$$F_{bx} = 0.66F_y = 24 \text{ ksi}$$

$$f_{by} = \frac{12 \times 25}{9.80} = 30.61 \text{ ksi}$$

$$F_{by} = 0.75F_y = 27 \text{ ksi}$$

$$\frac{f_{bx}}{F_{bx}} + \frac{f_{by}}{F_{by}} = \frac{10.99}{24} + \frac{30.61}{27} = 1.592 > 1.00 \qquad \text{NG}$$

Final trial:

Try W14 × 74 (S_x = 112 in.³, S_y = 26.6 in.³)

$$f_{bx} = \frac{12 \times 120}{112} = 12.86 \text{ ksi}$$

$$f_{by} = \frac{12 \times 25}{26.6} = 11.28 \text{ ksi}$$

$$\frac{f_{bx}}{F_{bx}} + \frac{f_{by}}{F_{by}} = \frac{12.86}{24} + \frac{11.28}{27} = 0.954 < 1.00 \qquad \text{OK}$$

$$\underline{\text{Use W14} \times 74} \qquad \blacksquare$$

It will be noticed in the solution for Example 9-5 that, though the procedure used will yield a section that will adequately support the moments, the selection of the absolutely lightest satisfactory section listed in the ASD Manual could be quite lengthy. In the steel textbooks by Gaylord and Gaylord[5] and by Salmon and Johnson[6] practical methods are presented to assist the designer in selecting I-shaped sections for biaxial bending.

A further complication in the design of beams subject to unsymmetrical bending is that the loads often do not pass through the centroid of the section. For instance, the loads applied to a purlin are generally applied to its top flange, and the result is a twisting moment or torsion in the inclined purlin. Rather than become involved in torsion calculations, a common design practice is to assume the lateral loads are carried only by the top flange of the beam. Only one-half of the section modulus about the y axis is considered effective, and the value of f_{by} becomes $M_y / \frac{1}{2} S_y$. This very inaccurate but very conservative practive is used in Example 9-6.

---

[5] E. H. Gaylord, Jr., and C. N. Gaylord, *Design of Steel Structures*, 2d ed. (New York: McGraw-Hill, 1972), Chapter 5.

[6] C. G. Salmon and J. E. Johnson, *Steel Structures: Design and Behavior*, 3d. ed. (New York: Harper & Row, 1990), Chapter 12.

■ Example 9-6

Redesign the beam of Example 9-5 if the lateral loads are assumed to be applied to the top flange of the beam and thus do not pass through the centroid of the section. Reduce the effective modulus for the y axis by 50 percent.

Solution.   Try W14 × 90 (S_x = 143 in.^{3}, S_y = 49.9 in.^{3})

$$f_{bx} = \frac{12 \times 120}{143} = 10.07 \text{ ksi}$$

$$f_{by} = \frac{12 \times 25}{\frac{1}{2} \times 49.9} = 12.02 \text{ ksi}$$

$$\frac{f_{bx}}{F_{bx}} + \frac{f_{by}}{F_{by}} = \frac{10.07}{24} + \frac{12.02}{27} = 0.865 < 1.0 \qquad\qquad \text{OK}$$

$$\underline{\text{Use W14} \times 90} \qquad\qquad ■$$

In this example the lateral moment was estimated to cause a stress of 12.02 ksi. Of this amount 6.01 ksi was added to approximate the torsion effect. (Is this a precise engineering calculation?) This subject of torsion is continued in the Appendix.

When a section does not have an axis of symmetry, the work required to locate the neutral axis, to calculate the principal moments of inertia, and to determine the necessary distances for a particular point are quite tedious. The expressions (from mechanics of materials) at the end of this paragraph can be used to locate the principal axes and calculate the principal moments of inertia. In these expressions I_{xy} is the product of inertia, while the various other symbols used are shown in Fig. 9-10.

$$\tan 2\phi = \frac{2I_{xy}}{I_y - I_x}$$

$$I_{x_1} = \cos^2 \phi I_x + \sin^2 \phi I_y - 2 \sin \phi \cos \phi I_{xy}$$
$$I_{y_1} = \cos^2 \phi I_y + \sin^2 \phi I_x + 2 \sin \phi \cos \phi I_{xy}$$

After the principal moments of inertia are determined, the stresses can be obtained by breaking the loads into components perpendicular to each of the principal axes, calculating the moments about each, and substituting in the unsymmetrical bending expression. It will be noted, however, that the remaining calculations are still tedious, particularly those involved in computing the perpendicular distances to each point (x and y in Fig. 9-10) from the principal axes.

The equation to follow may be more useful when the section in question has no axis of symmetry.

$$f = \frac{M_y I_x - M_x I_{xy}}{I_x I_y - I_{xy}^2} x + \frac{M_x I_y - M_y I_{xy}}{I_x I_y - I_{xy}^2} y$$

This general expression is applicable to any straight beam of constant cross section regardless of the shape and regardless of whether the section is open or

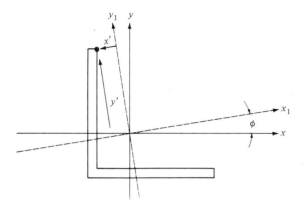

Figure 9-10

closed. The x and y axes can be assumed in any convenient direction as long as they are perpendicular to each other and pass through the centroid of the cross section.

Example 9-7 illustrates the application of this expression. For substitution in the formula (as well as in the calculations for I_{xy}), x is considered to be positive for points to the right of the y axis and minus to the left, while y is positive for points above the x axis and minus below. A plus sign for the resulting stress indicates compression, and a minus sign indicates tension.

■ Example 9-7

Determine the stress due to gravity loads which cause a moment of 60,000 in-lbs at points A, B, and C in the $6 \times 4 \times \frac{3}{4}$-in. angle shown in Fig. 9-11.

Solution.   Properties of section:

$$I_x = 24.5 \text{ in.}^4$$
$$I_y = 8.7 \text{ in.}^4$$

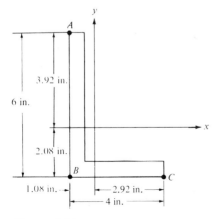

Figure 9-11

$$I_{xy} = (3.00)(+0.92)(-1.71) + (3.96)(-0.71)(+1.30)$$
$$= -8.37 \text{ in.}^4$$

Stress at A:

$$f = \frac{M_y I_x - M_x I_{xy}}{I_x I_y - I_{xy}^2} x + \frac{M_x I_y - M_y I_{xy}}{I_x I_y - I_{xy}^2} y$$

$$f_A = \frac{0 - (60,000)(-8.37)}{(24.5)(8.7) - (-8.37)^2}(-1.08) + \frac{(60,000)(8.7) - 0}{(24.5)(8.7) - (-8.37)^2}(+3.92)$$

$$= \underline{\underline{+10,510 \text{ psi}}} \text{ (compression)}$$

Stress at B:

$$f_B = \frac{0 - (60,000)(-8.37)}{(24.5)(8.7) - (-8.37)^2}(-1.08) + \frac{(60,000)(8.7) - 0}{(24.5)(8.7) - (-8.37)^2}(-2.08)$$

$$= \underline{\underline{-11,378 \text{ psi}}} \text{ (tension)}$$

Stress at C:

$$f_C = \frac{0 - (60,000)(-8.37)}{(24.5)(8.7) - (-8.37)^2}(+2.92) + \frac{(60,000)(8.7) - 0}{(24.5)(8.7) - (-8.37)^2}(-2.08)$$

$$= \underline{\underline{+2660 \text{ psi}}} \text{ (compression)} \qquad\blacksquare$$

9-5 DESIGN OF PURLINS

To avoid bending in the top chords of roof trusses, it is theoretically desirable to place purlins only at panel points. For large trusses, however, it is more economical to space them at closer intervals. If this practice is not followed for large trusses, the purlin sizes may become so large as to be impractical. When intermediate purlins are used, the top chords of the truss should be designed for bending as well as axial stress. Purlins are usually spaced from 2 to 6 ft apart depending on loading conditions, while their most desirable depth-span ratios are probably in the neighborhood of 1/24. Channels or S sections are the most frequently used sections, but on some occasions other shapes may be convenient.

As described in Chapter 5, the channel and S sections are very weak about their web axes, and sag rods are often necessary to reduce the span lengths for bending about those axes. Sag rods, in effect, make the purlins continuous sections for their y axes, and the moments about these axes are greatly reduced, as shown in Fig. 9-12. These moment diagrams are developed on the assumption that the changes in length of the sag rods are negligible. It is further assumed that the purlins are simply supported at the trusses. This assumption is on the conservative side, since purlins are often continuous over two or more trusses and appreciable continuity may be achieved at their splices. The student can easily reproduce these diagrams from his or her knowledge of mo-

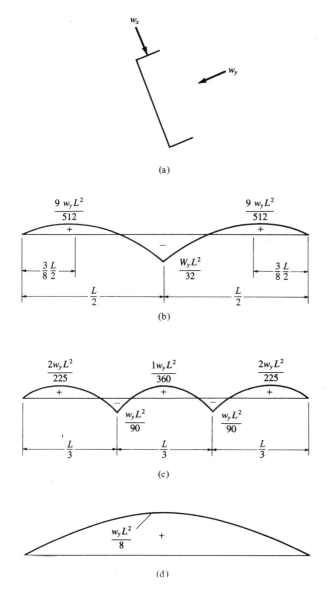

Figure 9-12 (a) Channel purlin. (b) Moment about web axis of purlins—sag rods at midspan. (c) Moment about web axis of purlins—sag rods at one-third points. (d) Moment about x axis of purlin.

ment distribution or other analysis methods. In the diagrams L is the distance between trusses, w_y is the load component *perpendicular* to the web axis of the purlin, and w_x is the load component *parallel* to the web axis.

If sag rods were not used, the maximum moment about the web axis of a purlin would be $w_y L^2/8$. When sag rods are used at midspan, this moment is reduced to a maximum of $w_y L^2/32$ (a 75 percent reduction) and when used at one-third points is reduced to a maximum of $w_y L^2/90$ (a 91 percent reduction). In Example 9-8 sag rods are used at the midpoints, and the purlins are designed for a moment of $w_x L^2/8$ parallel to the web axis and $w_y L^2/32$ perpendicular to the web axis.

In addition to being of advantage in reducing moments about the web axes of purlins, sag rods can serve other useful purposes. First, they can provide lateral support for the purlins; second, they are useful in keeping the purlins in proper alignment during erection until the roof deck is installed and connected to the purlins.

■ Example 9-8

Select a W6 purlin for the roof shown in Fig. 9-13. The trusses are 15 ft 0 in. on center, and sag rods are used at the midpoints between trusses. Full lateral support is assumed to be supplied from the roof above. Use A36 steel and the ASD Specification and assume there is no torsion. Thus the full value of S_y is to be used. Loads are as follows in terms of pounds per square foot of roof surface.

$$\text{Snow} = 30 \text{ psf}$$
$$\text{Roofing} = 6 \text{ psf}$$
$$\text{Estimated purlin weight} = \underline{3 \text{ psf}}$$
$$\text{Total} = 39 \text{ psf}$$
$$\text{Wind pressure} = 15 \text{ psf} \perp \text{ to roof surface}$$

Solution.

$$w_{\text{gravity}} = (4.42)(39) = 172.4 \text{ lb/ft}$$
$$w_{\text{wind}} = (4.42)(15) = 66.3 \text{ lb/ft}$$
$$w_x = 66.3 + \left(\frac{2}{\sqrt{5}}\right)(172.4) = 220.5 \text{ lb/ft}$$

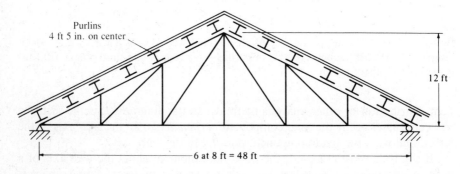

Purlins
4 ft 5 in. on center

12 ft

6 at 8 ft = 48 ft

Figure 9-13

$$w_y = \left(\frac{1}{\sqrt{5}}\right)(172.4) = 77.1 \text{ lb/ft}$$

$$M_x = \frac{(0.2205)(15)^2}{8} = 6.20 \text{ ft-k}$$

$$M_y = \frac{(0.0771)(15)^2}{32} \times 0.542 \text{ ft-k}$$

Try W6 × 9 ($S_x = 5.56$ in.^{3}, $S_y = 1.11$ in.^{3})

$$f_{bx} = \frac{(12)(6.20)}{5.56} = 13.38 \text{ ksi}$$

$$f_{by} = \frac{(12)(0.542)}{1.11} = 5.86 \text{ ksi}$$

$$\frac{f_{bx}}{F_{bx}} + \frac{f_{by}}{F_{by}} = \frac{13.38}{24} + \frac{5.86}{27} = 0.775 < 1.0 \qquad \text{OK}$$

Use $W6 \times 9$ ∎

9-6 THE SHEAR CENTER

The shear center is defined as the point on the cross section of a beam through which the resultant of the transverse loads must pass so that the stresses in the beam may be calculated only from the theories of pure bending and transverse shear. Should the resultant pass through this point, it is unnecessary to analyze the beam for torsional moments. If the loads do not pass through the shear center, both flexural and torsional stresses will occur. If possible, the dimensions of the structure and/or the connection should be changed to eliminate the eccentricity.

For a beam with two axes of symmetry the shear center will fall at the intersection of the two axes, thus coinciding with the centroid of the section. For a beam with one axis of symmetry the shear center will fall somewhere on that axis but not necessarily at the centroid of the section. This surprising statement means that, to avoid torsion in some beams, the lines of action of the loads should not pass through the centroids of the sections.

The shear center is of particular importance for beams whose cross sections are composed of thin parts which provide considerable bending resistance but little resistance to torsion. Many common structural members such as the W, S, and C sections, angles, and various beams made up of thin plates (as in aircraft construction) fall into this class, and the problem has wide application.

The locations of shear centers for several open sections are shown in Fig. 9-14. Sections such as these are relatively weak in torsion, and for them and similar shapes the location of the resultant of the external loads can be a very serious matter. A previous discussion has indicated that the addition of one or

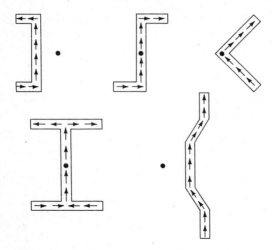

Figure 9-14 Shear centers of open sections.

more webs to these sections so they are changed into box shapes greatly increases their torsional resistance.

The average designer probably does not take the time to go through the sometimes tedious computations involved in locating the shear center and calculating the effect of twisting. Instead he or she may just ignore the situation or may make a rough estimate of the effect of torsion on the bending stresses. One very poor estimate used on many occasions is that of reducing S_y by 50 percent as previously described for use in the interaction equation.

Shear centers can be located quickly for beams with open cross sections and relatively thin webs. For other beams the shear centers can probably be found but only with considerable difficulty. The term *shear flow* is often used when speaking of thin-wall members, although there is really no flowing involved. It refers to the shear per inch of the cross section and equals the unit shearing stress times the thickness of the member. (The unit shearing stress has been determined by the expression VQ/bI, and the shear flow can be determined by VQ/I if the shearing stress is assumed to be constant across the thickness of the section.) The shear flow acts parallel to the sides of each element of a member.

The channel section of Fig. 9-15(a) will be considered for this discussion. In this figure the shear flow is shown with the small arrows, and in part (b) the values are totaled for each component of the shape and labeled H and V. The two H values are in equilibrium horizontally, and the internal V value balances the external shear at the section. Although the horizontal and vertical forces are in equilibrium, the same cannot be said for the moment forces unless the lines of action of the resultant of the external forces pass through a certain point called the shear center. The horizontal H forces in part (b) of the figure can be seen to form a couple. The moment produced by this couple must be opposed by an equal and opposite moment which can only be produced by the two V values. The location of the shear center is a problem in equilibrium;

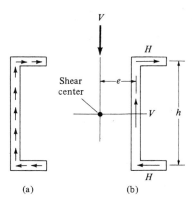

Figure 9-15

therefore moments should be taken about a point that eliminates the largest number of forces possible.

From this information the following equation can be used to locate the shear center:

$$Ve = Hh \quad \text{(moments taken about center of gravity of web)}$$

Examples 9-9 and 9-10 illustrate the calculations involved in locating the shear center for two shapes. It will be noted that the location of the shear center is independent of the value of the external shear.

■ Example 9-9

The channel section in Fig. 9-16(a) is subjected to an external shear of V. Locate the shear center.

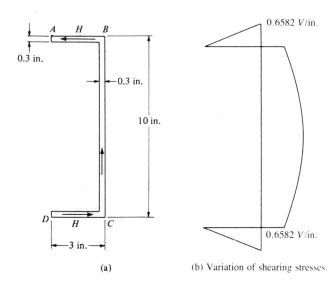

(a)

(b) Variation of shearing stresses

Figure 9-16

Solution.   Properties of section:

$$I_x = (\tfrac{1}{12})(3)(10)^3 - (\tfrac{1}{12})(2.7)(9.4)^3 = 63 \text{ in.}^4$$

$$f_v \text{ at } B = \frac{(V)(2.85 \times 0.3 \times 4.85)}{63} = 0.06582\ V/\text{in.}$$

$$\text{Total } H = (\tfrac{1}{2})(2.85)(0.06582V) = 0.0938V$$

Location of shear center:

$$Ve = Hh$$
$$Ve = (0.0938V)(9.7)$$
$$\underline{\underline{e = 0.91 \text{ in. from centerline of web}}}$$ ■

The student should clearly understand that shear stress variation across the corners, where the webs and flanges join, cannot be determined correctly with the mechanics of materials expression (VQ/bI or VQ/I for shear flow) and cannot be determined too well even after a complicated study with the theory of elasticity. The author as an approximation in the two examples presented here has assumed that shear flow continues up to the middle of the corners on the same pattern (straight-line variation for horizontal members and parabolic for others). The values of Q are computed for the corresponding dimensions. Other assumptions could have been made, such as that shear flow variation continues vertically for the full depth of webs and only for the protruding parts of flanges horizontally, or vice versa. It is rather disturbing to find that, whichever assumption is used, the values do not check out perfectly. For instance, in Example 9-10 the sum of the vertical shear flow values does not check out very well with the external shear.

■ Example 9-10

The open section of Fig. 9-17, is subjected to an external shear of V. Locate the shear center.

Solution.   Properties of section:

$$I_x = (\tfrac{1}{12})(0.25)(16)^3 + (2)(3.75 \times 0.25)(4.87)^2 = 129.8 \text{ in.}^4$$

Values of shear flow (labeled q):

$$q_A = 0$$

$$q_B = \frac{(V)(3.12 \times 0.25 \times 6.44)}{129.8} = 0.0387V/\text{in.}$$

$$q_C = q_B + \frac{(V)(3.75 \times 0.25 \times 4.87)}{129.8} = 0.0739V/\text{in.}$$

$$q_{\mathbb{C}} = q_C + \frac{(V)(4.87 \times 0.25 \times 2.44)}{129.8} = 0.0968V/\text{in.}$$

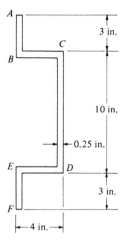

Figure 9-17

These shear flow values are shown in Fig. 9-18(a) and the summation for each part of the member is given in Fig. 9-18(b).

Taking moments about the center line of *CD*

$$-(0.211V)(9.75) + (2)(0.0807V)(3.75) + Ve = 0$$
$$e = \underline{\underline{1.45 \text{ in.}}} \qquad \blacksquare$$

The theory of the shear center is very useful in design, but it has certain limitations which should be clearly understood. For instance, the approximate analysis given in this section is valid only for thin sections. In addition steel

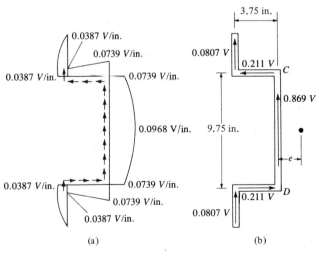

(a)   (b)

Figure 9-18

beams often have variable cross sections along their spans, with the result that the loci of the shear centers are not straight lines along those spans. If the resultant of the loads passes through the shear center at one cross section it might very well not do so at other cross sections.

When designers are faced with the application of loads to thin-wall sections such that twisting of those sections will be a problem, they will usually provide some means by which the twisting can be constrained. They may specify special bracing at close intervals, or attachments to flooring or roofing, or other similar devices. Should such solutions not be feasible, the designer will probably consider selecting sections with greater torsional stiffnesses.[7,8]

9-7 BEAM-BEARING PLATES

When the ends of beams are supported by direct bearing on concrete or other masonry construction, it is frequently necessary to distribute the beam reactions over the masonry by means of beam-bearing plates. The reaction is assumed to be spread uniformly through the bearing plate to the masonry, and the masonry is assumed to push up against the plate with a uniform pressure equal to the reaction R over the area of the plate A_1. This pressure tends to curl up the plate and the bottom flange of the beam. The ASD Manual recommends that the bearing plate be considered to take the entire bending moment produced and that the critical section for moment be assumed to be a distance k from the center line of the beam (see Fig. 9-19). The distance k is the same as the distance from the outer face of the flange to the web toe of the fillet given in the tables for each section (or it equals the flange thickness plus the fillet radius).

The determination of the true pressure distribution in a beam-bearing plate is a very formidable task, and the uniform pressure distribution assumption is usually made. This assumption is probably on the conservative side, as the pressure is usually larger at the center of the beam than at the edges. The outer edges of the plate and flange tend to bend upward, and the center of the beam tends to go down, concentrating the pressure there.

The required thickness of a 1-in.-wide strip of plate can be determined as follows:

$$f_p = \frac{R}{A}$$

$$M = f_p n \frac{n}{2} = \frac{f_p n^2}{2}$$

[7] C. G. Salmon and J. E. Johnson, *Steel Structures: Design and Behavior*, 3d ed. (New York: Harper & Row, 1990), pp. 767–768.

[8] AISC Engineering Staff, "Torsional Analysis of Steel Members" (Chicago: AISC, 1983).

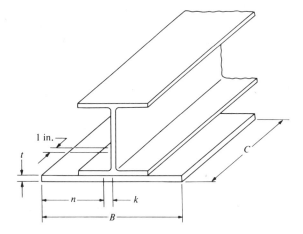

Figure 9-19

$$S = \frac{I}{C} = \frac{(\frac{1}{12})(1)(t)^3}{\frac{t}{2}} = \frac{t^2}{6}$$

$$\frac{M}{f} = S$$

$$\frac{f_p n^2}{2f} = \frac{t^2}{6}$$

$$t = \sqrt{\frac{3 f_p n^2}{F_b}}$$

If the plate extends for the full width of the wall or other support parallel to the beam, the area of the plate A_1 is determined by dividing the reaction R by the permissible pressure on the concrete or other masonry beneath the plate. For concrete this would normally be $0.35 f_c'$ where f_c' is the 28-day compressive strength of the concrete in ksi.

$$A_1 = \frac{R}{0.35 f_c'}$$

If the plate does not extend for the full width of the support, A_1 is to be determined by the following expression, according to Section J9 of the ASD Specification, where A_2 is the maximum area of the portion of the supporting surface that is geometrically similar to and concentric with the loaded area.

$$A_1 = \frac{1}{A_2}\left(\frac{R}{0.35 f_c'}\right)^2$$

After A_1 is determined, its length (parallel to the beam) and its width are selected. The length may not be less than the N required to prevent local web yielding or web crippling of the beam nor may it be less than about $3\frac{1}{2}$ or 4 in. for practical construction reasons. It may not be greater than the thickness of the wall or other support, and it may actually have to be less than that thickness, particularly at exterior walls, to prevent the steel from being exposed.

Example 9-11 illustrates the calculations involved in designing a beam-bearing plate. Notice that the width and the length of the plate are desirably taken to the nearest full inch.

■ Example 9-11

A W18 × 71 beam ($d = 18.47$ in., $t_w = 0.495$ in., $b_f = 7.635$ in., $t_f = 0.810$ in. and $k = 1\frac{1}{2}$ in.) has one of its ends supported by a reinforced-concrete wall with $f'_c = 3$ ksi. Design a bearing plate for the beam with A36 steel. The end reaction R is 70 k. Use the entire width of the wall (8.0 in.) for the bearing length N.

Solution. From the beam tables:

$$R_1 = 44.1 \text{ k}$$
$$R_2 = 11.8 \text{ k/in.}$$
$$R_3 = 63.9 \text{ k}$$
$$R_4 = 4.96 \text{ k/in.}$$

Minimum bearing length required:

$$N = \frac{R - R_1}{R_2} = \frac{70 - 44.1}{11.8} = 2.19 \text{ in.} < 8.0 \text{ in.} \qquad \text{OK}$$

$$N = \frac{R - R_3}{R_4} = \frac{70 - 63.9}{4.96} = 1.23 \text{ in.} < 8.0 \text{ in.} \qquad \text{OK}$$

Selecting plate size:

$$A_{1_{\text{reqd}}} = \frac{R}{0.35 f'_c} = \frac{70}{(0.35)(3)} = 66.67 \text{ in.}^2$$

$$B = \frac{A_1}{N} = \frac{66.67}{8.0} = 8.33 \text{ in.} \qquad \underline{\text{Use 9 in.}}$$

$$A_1 = (8.0)(9.0) = 72.0 \text{ in.}^2 > 66.67 \text{ in.}^2 \qquad \text{OK}$$

$$n = \frac{9}{2} - \frac{1.50}{2} = 3.75 \text{ in.}$$

$$f_p = \frac{70}{72.0} = 0.972 \text{ ksi}$$

$$t = \sqrt{\frac{3 f_p n^2}{F_b}} = \sqrt{\frac{(3)(0.972)(3.75)^2}{27}} = 1.23 \text{ in.}$$

$$\underline{\text{Use PL}1\frac{1}{4} \times 8 \times 0\text{ft } 9 \text{ in.}} \qquad ■$$

On some occasions the beam flanges alone probably provide sufficient bearing area, but bearing plates are nevertheless recommended, as they are useful in erection and ensure an even bearing surface for the beam. They can be placed separately from the beams and carefully leveled to the proper elevations. When the ends of steel beams are enclosed by the concrete or masonry walls, it is considered desirable to use some type of wall anchor to prevent the beam from moving longitudinally with respect to the wall. The usual anchor consists of a bent steel bar passing through the web of the beam and running parallel to the wall. These are called *government anchors,* and details of their sizes are given on page 4-136 of the ASD Manual. Occasionally clip angles attached to the web are used instead of government anchors. Should longitudinal loads of considerable size be anticipated, regular vertical anchor bolts may be used at the beam ends.

PROBLEMS

9-1 A W21 × 62 has a maximum external shear of 70 k.
  (a) Calculate the maximum theoretical shearing stress in the section (*Ans.* 9.36 ksi)
  (b) Calculate the average shearing stress in the section using the full depth of the section. (*Ans.* 8.34 ksi)

9-2 A W18 × 50 has a maximum end shear of 60 k. Calculate the average shearing stress in the section (using the total depth of the section) and the actual maximum theoretical shearing stress.

Probs. 9-3 to 9-4   Using A36 steel select the lightest available section for the span and loading shown. Consider moment and shear only and neglect beam weight in all calculations. The member is assumed to have full lateral bracing.

9-3 (*Ans.* W30 × 90)

9-4

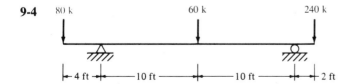

9-5 Repeat Prob. 9-4 using $F_y = 50$ ksi. (*Ans.* W27 × 84)

9-6 If a fully braced W14 × 53 section consisting of A36 steel is used for a simple span of 6 ft 6 in., determine the maximum uniform load w that it can support in addition to its own weight. Consider shear and moment only.

9-7 A W18 × 76 consisting of A36 steel is used as a simple beam for a span of 6 ft. If it has full lateral support, determine the maximum uniform load w it can support in addition to its own weight. Consider shear and moment only. (*Ans.* 37.074 k/ft)

9-8 A fully braced W36 × 135 consisting of A36 steel is used as a simple beam for a span of 16 ft. Considering moment and shear only, determine the maximum uniform load w that it can support in addition to its own weight.

9-9 A 30-ft simply supported beam is to support a moving concentrated load $P = 50$ k. Using A36 steel, select the most economical section considering moment and shear only. Neglect beam weight. (*Ans.* W24 × 84)

9-10 A 40-ft simple beam which supports a concentrated load $P = 40$ k at midspan is laterally unbraced except at its ends and centerline. If the maximum permissible centerline deflection under service loads equals 1/1000 of the span, select the most economical W section of A36 steel considering moment, shear, and deflection. Neglect beam weight.

9-11 Design a beam for a 24-ft simple span to support a uniform load of 3k/ft (includes beam weight). The maximum permissible deflection under working loads is 1/1200 of the span. Use A36 steel and consider moment, shear, and deflection. The beam is to be braced laterally for its full length. (*Ans.* W30 × 90)

9-12 Select the lightest available W section of A36 steel for the span and service loads shown. The beam will have full lateral support for its compression flange. Its maximum service load centerline deflection may not exceed 1/1500 of the span. Consider moment, shear, and deflection only.

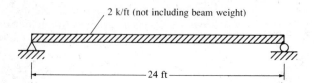

2 k/ft (not including beam weight)

24 ft

9-13 Select the lightest section available if $F_y = 50$ ksi for the span and working loads shown, if the section is to be fully braced laterally and have a maximum service load deflection of 1/1000 of its span length. Neglect beam weight in all calculations. (*Ans.* W36 × 135)

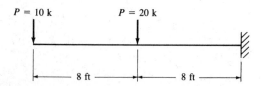

$P = 10$ k $\qquad$ $P = 20$ k

8 ft $\qquad$ 8 ft

9-14 If the maximum permissible service load deflection for the fully braced beam shown is 1/1200 of the span, select the lightest available section using A36 steel. Consider moment, shear, and deflections.

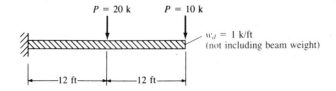

$P = 20$ k $\qquad P = 10$ k

$w_d = 1$ k/ft
(not including beam weight)

|←——12 ft——→|←——12 ft——→|

9-15 A simply supported W27 × 114 consisting of A36 steel is supporting a 300-k load as shown. If the length of bearing at the left support is 8 in. and at the concentrated load is 12 in., check the beam for shear, web yielding, and web crippling. Neglect beam weight in all calculations. (*Ans.* NG in web yielding and web crippling at both points)

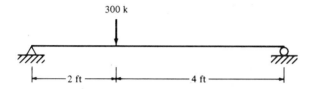

300 k

|←—— 2 ft ——→|←———— 4 ft ————→|

9-16 A 32-ft beam with full lateral support for its compression flange is supporting a moving concentrated load of 40 k. Using A36 steel, select the lightest section available for moment. Then check to see if the section is satisfactory in shear and compute the minimum length of bearing required at the supports from the standpoint of web yielding and web crippling.

9-17 The 20-ft simple beam shown has full lateral support for its compression flange and consists of A36 steel. The beam supports a 5K/ft uniform load in addition to its own weight. The loads are assumed to act through the center of gravity of the section. Select the lightest available W36 section. (*Ans.* W36 × 170)

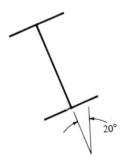

20°

9-18 Select the lightest available W30 section consisting of A36 steel to resist a gravity moment $M_x = 400$ ft-k and a lateral bending moment of $M_y = 100$ ft-k. The section is assumed to have full lateral support.

9-19 Repeat Prob. 9-18 for $F_y = 50$ ksi and $M_x = 500$ ft-k. (*Ans.* W30 × 173)

9-20 Select a W12 section to serve as a purlin between roof trusses 24 ft on centers. The roof is assumed to support a dead load of 20 psf of roof surface and a snow load of 20 psf of horizontal roof surface projection. The slope of the roof truss is

1 vertically to 2 horizontally, and the purlins are to be spaced 12 ft on centers. Use A36 steel and the ASD Specification and assume loads pass through the center of gravity of the section. Sag rods are to be placed halfway between trusses.

9-21 to 9-23 Find the shear centers for the section.

9-21 (*Ans. e* = 0.93 in. from center of web)

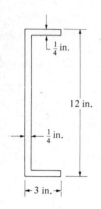

9-22

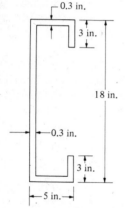

9-23 (*Ans. e* = 1.03 in. out from web centerline)

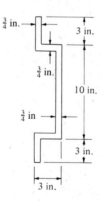

9-24 Design a steel bearing plate from A36 steel for a W27 × 94 beam with an end reaction R = 80 k. The beam will bear on a reinforced-concrete wall with f'_c = 4 ksi. In a direction perpendicular to the wall the bearing plate may not be longer than 8 in.

9-25 Design a steel bearing plate of A36 steel for a W30 × 124 beam supported by a reinforced-concrete wall with f'_c = 3 ksi. The maximum beam reaction R is 90 k. Assume the width of the plate perpendicular to the wall is 8 in. (*Ans.* PL1$\frac{1}{2}$ × 8 × 1 ft 0 in.)

9-26 Repeat Prob. 9-25 if R = 110 k and F_y = 50 ksi.

9-27 A W33 × 221 is required for a certain span but, due to a strike in the steel mills, cannot be obtained on time; however, an extra W33 × 118 of sufficient length is available along with a good supply of plates. Select $\frac{7}{8}$-in.-thick cover plates to be welded to the flanges of the W33 × 118 to provide a satisfactory substitute for the original beam. Use A36 steel and the ASD Specification. (*Ans.* PL$\frac{7}{8}$ × 16 each flange)

9-28 A W30 × 173 section has been specified for use on a certain job. By mistake, a W30 × 124 section was shipped to the field. This beam must be erected today. Assuming that plates are obtainable immediately, select $\frac{1}{2}$-in. cover plates to be welded to the flanges to obtain the necessary section. Use A36 steel and the ASD Specification.

9-29 Repeat Prob. 9-27 if the original beam was to have been of a steel with $F_y =$ 50 ksi and the substitute material is available only in A36 steel. Assume also that only $\frac{15}{16}$-in. plates are available. (*Ans.* W33 × 118 with PL$\frac{15}{16}$ × 24 each flange)

9-30 Design a beam of A36 steel with a depth no greater than 12.00 in. to support a uniform load of 8 k/ft for a 20-ft simple span. Use ASD Specification.

10

Bending and Axial Stress

10-1 OCCURRENCE

Structural members that are subjected to a combination of bending and axial stress are far more common than the student may realize. This section is devoted to listing a few of the more obvious cases. Columns that are part of a steel building frame must nearly always resist sizable bending moments in addition to the usual compressive loads. It is almost impossible to erect and center loads exactly on columns even in a testing lab, and in an actual building the student can see that it is even less possible. Even if building loads could be perfectly centered at one time, they would not stay in one place. Furthermore, columns may be initially crooked or have other flaws with the result that lateral bending is produced. The beams framing into columns are quite commonly supported with framing angles or brackets on the sides of the columns. These eccentrically applied loads produce moments. Wind and other lateral loads cause columns to bend laterally, and the columns in rigid frame buildings are subjected to moments even when the frame is supporting gravity loads alone. The members of bridge portals must resist combined stresses as do building columns. Among the causes of the combined stresses are heavy lateral wind loads, vertical traffic loads whether symmetrical or not, and the centrifugal effect of traffic on curved bridges.

Inryco Building, Chicago, Ill. (Courtesy of I. R. Construction Products Co.)

The previous experience of the student has probably been to assume that truss members are axially loaded only. Purlins for roof trusses, however, are frequently placed in between truss joints, causing the top chords to bend. Similarly the bottom chords may be bent by the hanging of light fixtures, ductwork, and other items between the truss joints. All horizontal and inclined truss members have moments caused by their own weights, while all truss members whether vertical or not are subjected to secondary bending stresses. Secondary stresses are developed because the members are not connected with frictionless pins as assumed in the usual analysis, the member centers of gravities or those of their connectors do not exactly coincide at the joints, and so on.

Moments in tension members are not as serious as those in compression members because tension tends to reduce lateral deflections, while compression increases them. Increased lateral deflection in turn results in larger moments, which result in larger lateral deflections, and so on. It is hoped that members in such situations are quite stiff so as to keep the additional lateral deflections from becoming excessive.

10-2 CALCULATION OF STRESSES

Sometimes members are encountered which have as their most important loadings transverse bending moments but which are also subjected to axial loads. However, this is not nearly as common a situation as where the major loading is axial with some transverse bending occurring at the same time. The name *beam-column* is often given to members that have appreciable amounts of both axial compression and bending stresses. The stresses in members subjected to a combination of axial load and bending are difficult to obtain exactly, and the stresses calculated in this chapter are truly approximate.

The stress at any point in a member subject to bending and direct stress is usually obtained from the familiar expression to follow. (This expression is said to be approximate because it does not include the effect of increased lateral deflections caused by moments and their subsequent effect on the moments and thus the stresses.)

$$f = \frac{P}{A} \pm \frac{Mc}{I}$$

Frequently bending occurs about some axis other than the x or y axis; that is, it occurs about both axes simultaneously. The corner columns of buildings may fall into this class. Stresses for members subjected to axial loads and bending about both axes are usually determined by

$$f = \frac{P}{A} \pm \frac{M_x y}{I_x} \pm \frac{M_y x}{I_y}$$

Example 10-1 shows the calculations involved in computing the total stress in a member of a bridge truss due to its computed axial stress plus the flexure stress caused by its own weight. The calculations here are quite straightforward, but a few comments will probably be of value. First, it will be noted that, although this is a bolted member, the gross area of the section is used, because the bolt holes are at the ends of the member while the maximum bending is assumed to occur at the centerline. Second, the flexure stresses due to the weight of the member in this example are not very large, and as a matter of fact they are not very large in the average truss member. As truss members become unusually long and heavy, the moments due to their own weights may be of appreciable magnitude.

The engineers who prepare the usual specifications used for the design of truss members have reduced the allowable axial stresses by approximately one-fourth to take into account the so-called secondary stresses. A similar discussion can be made for the magnitude of column flexure stresses due to column imperfections, slightly off-center loads and moments due to lateral deflections. This discussion shows that if the engineer went overboard theoretically, he or she would design all members, whether columns or beams or truss members, for axial load and bending. It is probable, however, that the decreased allowable stresses provide a sufficient margin to cover the usual case of off-center loads or otherwise imperfect columns and secondary truss member stresses. Unless the situation is severe, the designer will probably not include the moments in his or her design. Specifications rarely mention the matter, and it is up to the judgment of the individual designer. This discussion is not intended to apply to truss members that have transverse loads applied to them between the truss joints or to columns that have appreciable moments applied as a part of a rigid frame structure.

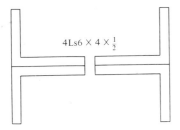

Figure 10-1

■ Example 10-1

A bottom chord of a bolted bridge truss is 22 ft long, has a maximum tensile force of 310 k, and is made up of 4 L6 × 4 × $\frac{1}{2}$s, arranged as shown in Fig. 10-1. Determine the maximum stress in the member due to the axial tension force plus the flexure stress due to the member's own weight.

Solution.

$$A_{\text{gross}} = 19 \text{ in.}^2, \ I_x = 2[12.5 + 9.50(0.987)^2] = 43.5 \text{ in.}^4$$

$$\text{Weight of L} = \left(\frac{19}{144}\right)(490) = 64.8 \text{ lb/ft}$$

$$M = \frac{wl^2}{8} = \frac{(64.8)(22)^2}{8} = 3920 \text{ ft-lb}$$

$$f = \frac{P}{A} + \frac{Mc}{I} = \frac{310,000}{19} + \frac{(12)(3,920)(4.00)}{43.5}$$

$$f = 16,315 + 4,325 = \underline{20,640 \text{ psi}}$$

$$\text{Percentage of stress due to member's weight} = \frac{4,325}{20,640} = 21\% \qquad ■$$

10-3 SPECIFICATIONS FOR COMBINED STRESSES

The preceding paragraphs have shown how stresses due to flexure and axial load can be combined. The calculations were quite simple, but the problem of establishing an allowable combined stress is much more difficult. There are allowable stresses for pure bending and other allowable stresses for pure axial loads, but the two values given in one particular specification are probably appreciably different. What is the value to be used when the two types of stresses occur simultaneously? Most organizations use an allowable stress which is some combination of the two individual allowables. Expressions of this type are referred to as *interaction equations*. One interaction equation used by many specifications is

$$\frac{f_a}{F_a} + \frac{f_b}{F_b} \leq 1.0$$

In this expression f_a is the axial stress (P/A), F_a is the allowable stress if only axial stresses occurred, f_b is the flexure stress (Mc/I), and F_b is the allowable flexure stress if only bending stresses are present. For many years the AISC also used this expression for members subjected to a combination of the two types of stresses. Today the ASD specification permits its use only for certain conditions (to be described in Section 10-6). This expression might be thought of as a percentage formula. For instance, if 60 percent of the allowable axial stress is used up by f_a/F_a, only 40 percent of the allowable flexure stress remains for the Mc/I stress. This expression has the effect of giving an allowable combined stress which will fall proportionately between the allowable individual values as each type of stress is to its allowable value. When the flexure stress is large with respect to the axial stress, the allowable combined stress will very nearly approach the allowable flexure stress. Similarly, when the axial stress is large in comparison to the flexure stress, the allowable combined stress will be close to the allowable axial stress. Example 10-2 presents an illustration of the combination of the two types of stresses according to this commonly used expression.

Should bending occur about both axes, the following expression is used to consider the combined stress situation:

$$\frac{f_a}{F_a} + \frac{f_{bx}}{F_{bx}} + \frac{f_{by}}{F_{by}} \le 1.0$$

■ Example 10-2

A W10 × 49 ($A = 14.40$ in.^{2}, $S_x = 54.6$ in.^{3}, $r_y = 2.54$ in.) is subjected to a moment of 40 ft-k and an axial compressive load of 100 k. The member is 15 ft long and has an allowable bending stress of 18 ksi and an allowable axial compressive stress to be determined from the formula $F_a = 15{,}000 - \frac{1}{4}(L/r)^2$. Is the member overstressed according to the $f_a/F_a + f_b/F_b \le 1$ expression?

Solution.

$$f_a = \frac{100}{14.4} = 6.94 \text{ ksi}$$

$$F_a = 15{,}000 - \frac{1}{4}\left(\frac{12 \times 15}{2.54}\right)^2 = 13.75 \text{ ksi}$$

$$f_b = \frac{12 \times 40}{54.6} = 8.8 \text{ ksi}$$

$$F_b = 18 \text{ ksi}$$

$$\frac{f_a}{F_a} + \frac{f_b}{F_b} = \frac{6.94}{13.75} + \frac{8.8}{18} = 0.994 < 1 \qquad \text{OK}$$

Member is satisfactory ■

10-4 DESIGN FOR AXIAL COMPRESSION AND BENDING

The average designer faced with the problem of selecting a member to resist combined stresses will probably estimate the size member required, and then check its combined stress in accordance with the requirements of the specification being used. Example 10-3 illustrates the design of a member of this type. To arrive at the assumed section in this example, it was assumed that roughly one-half of each allowable stress (F_a and F_b) was caused by the loads. The resulting axial stress value was divided into the total axial load to estimate the cross-sectional area required, and the flexure stress value was divided into the bending moment to estimate the required section modulus. A section was selected which had approximately these proportions. Although the section selected by the method was satisfactory in this example, it is usually necessary to make another trial or two to obtain an economical solution. The division of stresses was purely an estimate, and some other division might be better. For instance, if the axial load seems quite large in proportion to the bending moment, the designer might decide to estimate that 75 percent of the allowable axial stress is used and 25 percent of the allowable bending stress, or some other division. (Much more convenient methods are available for designing beam-columns, such as the one presented in Section 10-7 as it applies to the ASD Specification.)

■ Example 10-3

Select a W10 beam-column to support an axial load of 120 k and a bending moment of 50 ft-k. The member has an unsupported length of 18 ft, an allowable flexure stress of 18 ksi, and an allowable compressive stress of $F_a =$ $15,000 - \frac{1}{4}(L/r)^2$. The $f_a/F_a + f_b/F_b \leq 1.0$ expression is to be used in the design.

Solution. Assume $f_a/F_a = 0.5$ and estimate $F_a = 14$ ksi:

$$\frac{f_a}{F_a} = \frac{\dfrac{P}{A}}{14} = \frac{\dfrac{120}{A}}{14} = 0.5$$

Estimated $A = 17.1$ in.^{2}
Assume $f_b/F_b = 0.5$, $F_b = 18$ ksi:

$$\frac{f_b}{F_b} = \frac{\dfrac{M}{S}}{18} = \frac{\dfrac{(12)(50)}{S}}{18} = 0.5$$

Estimated $S = 66.7$ in.^{3}
Try W10 × 60($A = 17.6$ in.^{2}, $S_x = 66.7$ in.^{3}, $r = 2.57$ in.^{3})

$$f_a = \frac{120,000}{17.6} = 6,818 \text{ psi}$$

$$F_a = 15,000 - \frac{1}{4}\left(\frac{12 \times 18}{2.57}\right)^2 = 13,230 \text{ psi}$$

$$f_b = \frac{(12)(50,000)}{66.7} = 8,996 \text{ psi}$$

$$F_b = 18,000 \text{ psi}$$

$$\frac{f_a}{F_a} + \frac{f_b}{F_b} = \frac{6,818}{13,230} + \frac{8,996}{18,000}$$

$$= 0.515 + 0.500 = 1.015 \quad \text{(slightly overstressed)} \quad \text{OK}$$

<u>Use W10 × 60</u> ∎

10-5 MOMENT AMPLIFICATION AND MODIFICATION

When a beam-column is subjected to moment along its unbraced length, it will be displaced laterally in the plane of bending. The result is an increased or secondary moment caused by the axial load times the lateral displacement or eccentricity. In Fig. 10-2 the load P causes the column moment to be increased by an amount Pd. This moment will cause additional lateral deflection, which will cause a larger column moment, and so on until equilibrium is reached.

We could take the column moments, compute the lateral deflection, increase the moment at middepth by P times the deflection, recalculate the lateral deflection and the increased moment, and so on. Although about two cy-

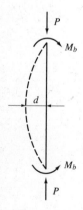

Figure 10-2 Moment amplification.

cles would be sufficient, this would still be a tedious and impractical procedure.

To account for the moment increase, the ASD Specification requires that the bending stress be increased by multiplying it by the *amplification factor* $(1/[1 - f_a/F_e'])$, which is larger than 1.0.

This factor may cause the secondary stress to be overestimated for some situations. As a result the amplified stress is multiplied by a reduction or *modification factor* C_m which is equal to 1.0 or less. The basic ASD interaction equation then becomes

$$\frac{f_a}{F_a} + \frac{C_m f_b}{\left(1 - \frac{f_a}{F_e'}\right) F_b} \leq 1.0$$

Amplification Factor

To derive the amplification factor the beam-column is assumed to deflect laterally in a certain shape (as a sine function), and then with the approximate differential equation of the member $(d^2y/dx^2 = -M/EI)$, an expression is derived for the lateral deflection at middepth.

The total bending moment at middepth M_b equals the initial moment plus Pd, the increase in moment due to the lateral deflection. From this expression for total moment comes the following expression for bending stress. It includes the terms P, which is the axial load, and P_E, which is the Euler buckling load.

$$f = f_b \left(\frac{1}{1 - \frac{P}{P_E}}\right)$$

To put everything in terms of stress, P and P_E are divided by the member's cross-sectional area A. The following expression is the result in which f_e is the critical Euler buckling stress:

$$f_{bm} = f_b \left(\frac{1}{1 - \frac{f_a}{f_e}}\right)$$

To compute the bending stress at the working or service load conditions, the Euler stress is divided by a factor of safety of $23/12$ and the expression becomes.

$$f_{bm} = f_b \left(\frac{1}{1 - \frac{f_a}{F_e'}}\right)$$

where $(1/[1 + f_a/F_e'])$ is the amplification factor and F_e' is the Euler buckling stress divided by the factor of safety.

Modification Factors

For some moment distributions in a beam-column the amplification provides flexural stresses that are too large. One such situation occurs when the moment at one end of the member is zero. For this situation the lateral deflection is about half of the deflection in effect provided by the amplification factor. Should we have approximately equal end moments that are causing reverse curvature bending, the deflection at middepth and the flexural stress are close to zero. As a result of these and other situations, the ASD Specification provides modification factors C_m, which are to be multiplied by f_b. These are discussed in detail in Section 10-6.

10-6 ASD EQUATIONS

The ASD Specification (H1) includes three equations for members subject to axial load and bending. These equations include bending about both the x and y axes. Should bending occur about one axis only, the appropriate term for the other axis will drop out.

The first equation is applicable to the middepth of the members. This equation is used to check the overall stability of the member.

$$\frac{f_a}{F_a} + \frac{C_{mx}f_{bx}}{\left(1 - \frac{f_a}{F'_{ex}}\right)F_{bx}} + \frac{C_{my}f_{by}}{\left(1 - \frac{f_a}{F'_{ey}}\right)F_{by}} \leq 1.0 \qquad \text{(ASD Equation H1-1)}$$

The equation that follows is applicable only at the ends of beam-columns and is used to check stress conditions at those points. It was prepared to account for cases where maximum moments occur at member ends. Stability is not a problem at a support, and thus, as shown in the first term of the equation, the allowable compression stress F_a is $0.60 F_y$. In addition F_{bx} is $0.66 F_y$ at the supports as lateral bracing is supplied. This equation will usually govern for members braced against sidesway and subjected to reverse curvature bending.

$$\frac{f_a}{0.60F_y} + \frac{f_{bx}}{F_{bx}} + \frac{f_{by}}{F_{by}} \leq 1.0 \qquad \text{(ASD Equation H1-2)}$$

When the axial load is relatively small, it does not cause appreciable moment amplification. As a result the ASD Specification provides a third beam-column equation that applies if f_a/F_a is equal to 0.15 or less.

$$\frac{f_a}{F_a} + \frac{f_{bx}}{F_{bx}} + \frac{f_{by}}{F_{by}} \leq 1.0 \qquad \text{(ASD Equation H1-3)}$$

In these expressions f_a, f_b, F_a, and F_b are the same values that have been previously defined. F'_e is the Euler stress divided by a safety factor of $\frac{23}{12}$. Its value is given by the expression to follow in which L_b is the actual unsupported length in the plane of bending, r_b is the corresponding radius of gyration, and K is the effective length factor in the plane of bending.

$$F_e' = \frac{12\pi^2 E}{23(KL_b/r_b)^2}$$

The value of F_e' can be increased by one-third for wind and seismic stresses according to the ASD Specification, provided the section used is not overstressed (not counting the one-third increase) by the gravity dead, live, and impact loads.

There are three separate categories of C_m as described in Section H1 of the ASD Specification.

1. In Category 1 the columns are considered to be parts of frames that depend upon the bending stiffnesses of their members for lateral stiffness. These members are subject to joint translation or sidesway, and C_m is considered to equal 0.85.

2. In Category 2 the members are assumed to be restrained against rotation and prevented from joint translation or sidesway, and they are not subject to transverse loading between their ends. For such members the modification factor is to be determined as follows:

$$C_m = 0.6 - 0.4\frac{M_1}{M_2}$$

In this expression M_1/M_2 is the ratio of the smaller moment to the larger moment at the ends of the unbraced length. The ratio is negative if the moments cause the member to bend in single curvature (or) and positive if they bend the member in reverse curvature (or). It should be obvious that a member in single curvature has larger lateral deflections than a similar member bent in reverse curvature. With larger lateral deflections the moments due to the axial loads and thus the stresses will be greater. When a column is bent in single curvature, substitution in the ASD Equation will, therefore, give smaller allowable stresses.

3. Category 3 applies to members which are subjected to transverse loading between the joints and which are braced against joint translation or sidesway in the plane of loading. The compression chord of a truss with a purlin load between its joints is a typical example of this category. The ASD Specification states that the value of C_m for this category may be taken as follows:

(a) For members with restrained ends $C_m = 0.85$;
(b) For members with unrestrained ends $C_m = 1.0$.

These two values are sufficiently accurate and safe for almost all cases. It is very difficult to obtain more precise C_m values, and if we do obtain them, they are of rather questionable value because we can't estimate the applied loads too well. Nevertheless, the ASD Commentary provides a supposedly more refined expression for estimating C_m for this category:

$$C_m = 1 + \psi \frac{f_a}{F_e'} \qquad \text{(ASD Equation C-H1-6)}$$

$$\psi = \frac{\pi^2 \delta_0 EI}{M_0 L} - 1 \qquad \text{(ASD Equation C-H1-7)}$$

where δ_0 is the maximum deflection due to transverse loading and M_0 is the maximum moment between the supports due to transverse loads. ψ has been computed for several common conditions of transverse loading and end restraint, and the results are shown in Table 10-1.

Most of the members encountered in actual practice that are subject to appreciable amounts of combined bending and axial stress are parts of a rigid frame structure, and the other members rigidly connected to the member in question have some appreciable effect on that member. This means that, to determine the allowable axial stress, the effective length of the member needs to be determined as previously described. From that previous discussion it is remembered that, if sidesway is possible, the effective length can be greater than

TABLE 10-1   AMPLIFICATION FACTORS ψ AND C_m

Case	ψ	C_m
(a)	0	1.0
(b)	-0.4	$1 - 0.4\frac{f_a}{F_e'}$
(c)	-0.4	$1 - 0.4\frac{f_a}{F_e'}$
(d)	-0.2	$1 - 0.2\frac{f_a}{F_e'}$
(e)	-0.3	$1 - 0.3\frac{f_a}{F_e'}$
(f)	-0.2	$1 - 0.2\frac{f_a}{F_e'}$

Source: American Institute of Steel Construction, *Manual of Steel Construction Allowable Stress Design*, 9th ed. (Chicago: AISC, 1989), Table C-H1.1, p. 5-154. Reprinted with the permission of AISC.

the actual length, but if sidesway is prevented as by lateral bracing, the effective length will be less than the actual length.

In category 1 the effective lengths of the members are used in calculating F_a, and it can never be less than the unbraced length and may even be greater. The effective length in the direction of bending is used in calculating F'_e. For the moment computations the actual unbraced length is used. In category 2 the columns have no sidesway or transverse loading, and the effective length is used for calculating F_a. It cannot be greater than the unbraced length and may be less. Again the effective length in the direction of bending is used in calculating F'_e. For the moment calculations the actual unbraced length is used.

Examples 10-4 to 10-6 illustrate the analysis of members subjected to combined bending and axial stress in accordance with the ASD Specification. It will be noted in these examples that f_b is the bending stress at the point of the member being considered. When there are no transverse loads, the stress will be computed for the largest of the moments at the ends of the unbraced length. When a transverse load is applied, the largest moment between the points of lateral support is used to compute f_b for substitution in the first of the two ASD expressions. For substitution in the second expression the largest moment at either of the supported points is used.

Particularly important is the check made in each example to see if $C_m/(1 - f_a/F'_e)$ is ≥ 1.0. It seems unreasonable to modify a moment more than we magnify it. Therefore, if the ratio is less than 1.0, a value of 1.0 will be used.

The joints of a truss are restrained from translation. For this reason it might seem reasonable to use an effective length for the compression members of a truss somewhat less than the actual lengths. Section C-2 of the ASD Commentary suggests, however, that the use of $K = 1.0$ is wise when the ultimate load situation is considered. Should all of the members of a truss reach their ultimate load capacity at the same time, the restraints against translation mentioned would be drastically reduced or eliminated.

Though web compactness is not included in these examples, the reader should realize that on some occasions it is necessary to check the web compactness expression of ASD Table B5.1 for beam-columns. If the web is noncompact, F_b cannot be larger than $0.60F_y$.

■ Example 10-4

Is a W14 × 159 satisfactory to support the loads and moments shown in Fig. 10-3 if A36 steel and the ASD Specification are used? In the plane of loading there is no bracing, and the column is subject to sidesway. An analysis of effective lengths has resulted in $K_x = 1.92$. In the perpendicular plane there is bracing, and K_y is estimated to equal 1.0.

Solution. Using a W14 × 159 ($A = 46.7$ in.^{2}, $S_x = 254$ in.^{3}, $r_x = 6.38$ in., $r_y = 4.00$ in., $L_c = 16.4$ ft, $L_u = 57.2$ ft, $F'_{ex}(K_xL_x)^2/10^2 = 422$ k)

$$f_a = \frac{400}{46.7} = 8.57 \text{ ksi}$$

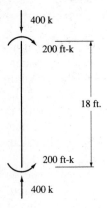

Figure 10-3

$$\frac{K_x L_x}{r_x} = \frac{(1.92)(12)(18)}{6.38} = 65.00 \leftarrow$$

$$\frac{K_y L_y}{r_y} = \frac{(1.00)(12)(18)}{4.00} = 54.00$$

$$F_a = 16.94 \text{ ksi}$$

$$\frac{f_a}{F_a} = \frac{8.57}{16.94} = 0.506 > 0.15$$

$\therefore$ Must use Equations H1-1 and H1-2

$$f_b = \frac{(12)(200)}{254} = 9.45 \text{ ksi}$$

F_b at middepth $= 0.60\, F_y = 22$ ksi since $L_{unbraced} > L_c < L_u$
F_b at member ends $= 0.66\, F_y = 24$ ksi

$$F'_{ex} = \frac{(422)(10)^2}{(1.92 \times 18)^2} = 35.33 \text{ ksi}$$

Member falls in category 1 for C_m

$$C_m = 0.85$$

Checking:

$$\frac{C_m}{1 - \dfrac{f_a}{F'_e}} = \frac{0.85}{1 - \dfrac{8.57}{35.33}} = 1.12 > 1.00 \qquad\qquad \text{OK}$$

Substituting into ASD equations:

$$\frac{f_a}{F_a} + \frac{C_m f_b}{\left(1 - \frac{f_a}{F_e'}\right)F_b} = \frac{8.57}{16.94} + \frac{(0.85)(9.45)}{\left(1 - \frac{8.57}{35.33}\right)22} = 0.987 < 1.0 \qquad \text{OK}$$

$$\frac{f_a}{0.6F_y} + \frac{f_b}{F_b} = \frac{8.57}{22} + \frac{9.45}{24} = 0.783 < 1.0 \qquad \text{OK}$$

<u>Section is satisfactory</u> ∎

■ Example 10-5

A 12-ft W12 × 170 consisting of A36 steel is braced against sidesway and is used to support an axial load $P = 400$ k and the moments $M_x = 200$ ft-k and $M_y = 60$ ft-k. Is the member satisfactory if it is bent in reverse curvature about both axes with equal end moments? There are no transverse loads, and K_x and K_y are assumed to be 1.0.

Solution. Using a W12 × 170 $\Big(A = 50.0$ in.^{2}, $r_x = 5.74$ in., $r_y = 3.22$ in.,

$S_x = 235$ in.^{3}, $\quad S_y = 82.3$ in.^{3}, $\quad L_c = 13.3$ ft, $\quad \dfrac{F_{ex}'(K_x L_x)^2}{10^2} = 342$ k,

$\dfrac{F_{ey}'(K_y L_y)^2}{10^2} = 108$ k $\Big)$

$$f_a = \frac{400}{50} = 8.00 \text{ ksi}$$

$$\frac{K_x L_x}{r_x} = \frac{(1.0)(12 \times 12)}{5.74} = 25.09$$

$$\frac{K_y L_y}{r_y} = \frac{(1.0)(12 \times 12)}{3.22} = 44.72 \leftarrow$$

$$F_a = 18.80 \text{ ksi}$$

$$\frac{f_a}{F_a} = \frac{8.00}{18.80} = 0.426 > 0.15$$

Therefore, check Equations H1-1 and H1-2

$$f_{bx} = \frac{(12)(200)}{235} = 10.21 \text{ ksi}$$

$$f_{by} = \frac{(12)(60)}{82.3} = 8.75 \text{ ksi}$$

$F_{bx} = 0.66F_y$ at middepth $= 24$ ksi since $L_{\text{unbraced}} < L_c$

$\qquad F_{by} = 0.75F_y = 27$ ksi

$$C_{mx} = C_{my} = 0.6 - 0.4\frac{M_1}{M_2} = 0.6 - 0.4(+1.0) = 0.2$$

$$F'_{ex} = \frac{(342)(10)^2}{(12)^2} = 237.5 \text{ ksi}$$

$$F'_{ey} = \frac{(108)(10)^2}{(12)^2} = 75 \text{ ksi}$$

Checking:

$$\frac{C_{mx}}{1 - \dfrac{f_a}{F'_{ex}}} = \frac{0.2}{1 - \dfrac{8.00}{237.5}} = 0.207 < 1.0 \qquad \underline{\text{Use 1.0}}$$

$$\frac{C_{my}}{1 - \dfrac{f_a}{F'_{ey}}} = \frac{0.2}{1 - \dfrac{8.00}{75}} = 0.224 < 1.0 \qquad \underline{\text{Use 1.0}}$$

Applying ASD Equations H1-1 and H1-2:

$$\frac{8.00}{18.80} + \frac{(1.0)(10.21)}{24} + \frac{(1.0)(8.75)}{27} = 1.174 > 1.0 \qquad \text{NG}$$

$$\frac{8.00}{22} + \frac{10.21}{24} + \frac{8.75}{27} = 1.113 > 1.0 \qquad \text{NG}$$

$$\underline{\text{Section is not satisfactory}} \qquad \blacksquare$$

When the young engineer begins to design actual structures, he or she is often distressed because they seldom fall into the exact textbook situations learned in school. As a result he or she must frequently make assumptions or interpolate between handbook values. For instance: Is a certain connection fixed or is it simple or is it really somewhere in between? The decision made may decidedly affect the values of moments, K values, C_m values, and so on.

The author likes to think that Example 10-6 is a very practical problem in that these types of assumptions have to be made. In this case the author assumed that the end conditions of the member being checked are somewhere in between ⊥⊥ and ⊥⫰. He then computed moments and C_m values for the two cases and averaged them. It is thought that this procedure is within the spirit of the ASD Specification.

■ Example 10-6

For the truss shown in Fig. 10-4(a) a W8 × 31 is used as a continuous top chord member from joint L_0 to joint U_3. If the member consists of A36 steel, does it have sufficient strength to resist the loads shown in Fig. 10-4 (b)? Part (b) shows the portion of the chord from L_0 to U_1, and the 12-k load represents the effect of a purlin. It is assumed that lateral support is provided for this member at its ends and centerline.

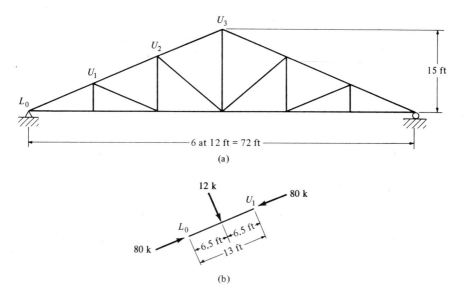

(a)

(b)

Figure 10-4

Solution. Using a W8 × 31 $\left(A = 9.13 \text{ in.}^2,\ S_x = 27.5 \text{ in.}^3,\ r_x = 3.47 \text{ in.,} \right.$

$r_y = 2.02 \text{ in.},\ L_c = 8.4 \text{ ft},\ \dfrac{F'_{ex}(K_x L_x)^2}{10^2} = 125 \text{ k} \left. \right)$

$$f_a = \frac{80}{9.13} = 8.76 \text{ ksi}$$

For a frame braced against sidesway $K = 1.0$

$$\frac{K_x L_x}{r_x} = \frac{(1.0)(12 \times 13)}{3.47} = 44.96 \leftarrow$$

$$\frac{K_y L_y}{r_y} = \frac{(1.0)(12 \times 6.5)}{2.02} = 38.61$$

$$F_a = 18.78 \text{ ksi}$$

$$\frac{f_a}{F_a} = \frac{8.76}{18.78} = 0.466 > 0.15$$

∴ Must use Equations H1-1 and H1-2

$$F'_{ex} = \frac{(125)(10)^2}{(13)^2} = 73.96 \text{ ksi}$$

From Table 10-1

For $C_m = 1 - 0.2\left(\frac{8.76}{73.96}\right) = 0.976$

For $C_m = 1 - 0.3\left(\dfrac{8.76}{73.96}\right) = 0.964$

Average $C_m = 0.97$

For $M_{\text{centerline}} = \dfrac{(12)(13)}{4} = 39$ ft-k

For $M_{\text{centerline}} = \dfrac{(5)(12)(13)}{32} = 24.38$ ft-k

Average $M_{\text{centerline}} = 31.69$ ft-k

$$f_b \text{ at centerline} = \frac{(12)(31.69)}{27.5} = 13.83 \text{ ksi}$$

Average M at end in same manner $= 14.62$ ft-k

$$f_b \text{ at end} = \frac{(12)(14.62)}{27.5} = 6.38 \text{ ksi}$$

Applying Equations H1-1 and H1-2:

$$\frac{C_m}{1 - \dfrac{f_a}{F_e'}} = \frac{0.97}{1 - \dfrac{8.76}{73.96}} = 1.10 > 1.0 \qquad \text{OK}$$

$$\frac{8.76}{18.78} + \frac{(1.10)(13.83)}{24} = 1.10 > 1.0 \qquad \text{NG}$$

$$\frac{8.76}{22} + \frac{6.38}{24} = 0.664 < 1.0 \qquad \text{OK}$$

Section is unsatisfactory, as both
equations must be satisfied ∎

10-7 DESIGN OF BEAM-COLUMNS

The design of beam-columns involves a trial-and-error procedure. A trial section is selected by some process and is then checked with the appropriate interaction equations. If the section does not satisfy the equations or if it's too much on the safe side (that is, if it is overdesigned), another section is selected and the interaction equations are applied again. Probably the first thought of the reader is "I sure hope that we can select a good section the first time and not have to go through all of that rigamarole more than once." We certainly can make a good initial estimate, as shown in this section.

A common method used for selecting sections to resist both moments and axial loads is the *equivalent axial load method*. In this method the eccentric load is replaced with a concentric load of such a magnitude that the maximum stress it causes is the same as the maximum stress produced by the eccentric load.

It is assumed here that it is desired to select the most economical section to resist both a moment and an axial load (say 140 ft-k and 400 k, respectively). By a trial-and-error procedure it is possible eventually to select the lightest section. Somewhere, however, there is a fictitious axial load which will require the same section size as that required for the actual moment and the actual axial load. This fictitious load is the *equivalent axial load*.

Equations are used to convert the bending moment into an estimated equivalent axial load P' which is added to the axial load P. The total of $P + P'$ is the equivalent or effective axial load P_{eq} or P_{eff}, and it is used to enter the concentric column tables for choice of a trial section.

Space is not taken here to show the complete derivation of the equivalent axial load expressions to be used in place of ASD Equations H1-1, H1-2, and H1-3. These equivalent expressions, referred to as Modified Equations H1-1, H1-2, and H1-3 by the ASD Manual, can be derived by the procedure described in the several paragraphs to follow.

If bending is present about the x axis only, Equation H1-1 can be written

$$\frac{f_a}{F_a} + \frac{C_m f_b}{(1 - f_a/F_e')F_b} \leq 1.0$$

Replacing f_a with P/A and f_b with M/S the formula becomes

$$\frac{P}{AF_a} + \frac{M}{SF_b}\left(\frac{C_m}{1 - P/AF_e'}\right) = 1.0$$

Multiplying both sides of the equation by AF_a gives

$$P + \left(\frac{MA}{S}\right)\left(\frac{F_a}{F_b}\right)\left(\frac{C_m}{1 - P/AF_e'}\right) = AF_a$$

If both numerator and denominator of the $C_m/(1 - P/AF_e')$ part of the formula are multiplied by F_e', the result is

$$\frac{C_m}{1 - P/AF_e'}\left(\frac{F_e'}{F_e'}\right) = \frac{C_m F_e'}{F_e' - P/A}$$

Substituting into this expression the value of $F_e' = 12\pi^2 E/23(KL_b/r_f)^2$ including $E = 29,000$ ksi, the expression becomes

$$\frac{C_m F_e'}{F_e' - P/A} = \frac{C_m 149,000 A r^2}{149,000 A r^2 - P(KL)^2}$$

Then this value is put into the expression for bending about the x axis only; A/S is replaced with B, the so-called bending factor; and $149,000 A r_x^2$ is replaced with the letter a_x. The values of B and a can easily be computed for each shape, and they are given in the concentric column tables of the ASD

Manual. The resulting equation is

$$P_{eq} = P + \left[BMC_m \left(\frac{F_a}{F_b}\right)\left(\frac{a_x}{a_x - P(KL)^2}\right) \right]$$

In a similar fashion the equivalent or modified formulas can be written for all three of the ASD Equations including bending about both x and y axes with the following results

$$P + P_x' + P_y' = P + \left[B_x M_x C_{mx}\left(\frac{F_a}{F_{bx}}\right)\left(\frac{a_x}{a_x - P(KL)^2}\right) \right]$$

$$+ \left[B_y M_y C_{my}\left(\frac{F_a}{F_{by}}\right)\left(\frac{a_y}{a_y - P(KL)^2}\right) \right] \quad \text{(Modified Equation H1-1)}$$

$$P + P_x' + P_y' = P\left(\frac{F_a}{0.6F_y}\right) + \left[B_x M_x\left(\frac{F_a}{F_{bx}}\right) \right]$$

$$+ \left[B_y M_y\left(\frac{F_a}{F_{by}}\right) \right] \quad \text{(Modified Equation H1-2)}$$

ASD Equation H1-3 is to be used in lieu of Equations H1-1 and H1-2 when

$$\frac{f_a}{F_a} \leq 0.15$$

$$P + P_x' + P_y' = P + \left[B_x M_x\left(\frac{F_a}{F_{bx}}\right) \right]$$

$$+ \left[B_y M_y\left(\frac{F_a}{F_{by}}\right) \right] \quad \text{(Modified Equation H1-3)}$$

These equations, which are very useful for preliminary design, yield columns that are on the conservative side. Should bending occur about one axis only, the equations are reduced by deleting the appropriate terms.

In these expressions B_x and B_y are the bending factors which are respectively equal to A/S_x and A/S_y. Their values, which are tabulated in the ASD Manual, vary from 0.168 to 1.133 for I-shaped sections for B_x and from 0.408 to 4.722 for B_y. The larger values occur with the smaller sections. a_x is a component of the amplification factor when bending is about the x axis and equals $0.149Ar_x^2 \times 10^6$. Similarly a_y is $0.149Ar_y^2 \times 10^6$, and both a_x and a_y are given for sections normally used as columns in the column section of the ASD Manual. In these equations the moments M_x and M_y must be used in in.-k.

Rather than using the above modified equations to determine the equivalent load, it is much quicker and simpler to use the following approximation developed from those equations.[1]

$$P_{eff} = P_{eq} = P_0 + M_x m + M_y mU$$

[1] L. B. Burgett, "Selection of a 'Trial' Column Section," *Engineering Journal*, AISC, 15, no. 2 (2d quarter, 1973), pp. 54–59.

TABLE 10-2 VALUES OF m

	$F_y = 36$ KSI							$F_y = 50$ KSI						
KL (ft)	10	12	14	16	18	20	22 & over	10	12	14	16	18	20	22 & over
FIRST APPROXIMATION														
All shapes	2.4	2.3	2.2	2.2	2.1	2.0	1.9	2.4	2.3	2.2	2.0	1.9	1.8	1.7
SUBSEQUENT APPROXIMATIONS														
W, S 4	3.6	2.6	1.9	1.6	—	—	—	2.7	1.9	1.6	1.6	—	—	—
W, S 5	3.9	3.2	2.4	1.9	1.5	1.4	—	3.3	2.4	1.8	1.6	1.4	1.4	—
W, S 6	3.2	2.7	2.3	2.0	1.9	1.6	1.5	3.0	2.5	2.2	1.9	1.8	1.6	1.5
W 8	3.0	2.9	2.8	2.6	2.3	2.0	2.0	3.0	2.8	2.5	2.2	1.9	1.6	1.6
W 10	2.6	2.5	2.5	2.4	2.3	2.1	2.0	2.5	2.5	2.4	2.3	2.1	1.9	1.7
W 12	2.1	2.1	2.0	2.0	2.0	2.0	2.0	2.0	2.0	2.0	1.9	1.9	1.8	1.7
W 14	1.8	1.7	1.7	1.7	1.7	1.7	1.7	1.8	1.7	1.7	1.7	1.7	1.7	1.7

Source: American Institute of Steel Construction, *Manual of Steel Construction Allowable Stress Design*, 9th ed. (Chicago: AISC, 1989), Table B, p. 3–10. Reprinted with the permission of AISC.

Note: Values of m are for $C_m = 0.85$. When C_m is other than 0.85, multiply the tabular value of m by $C_m/0.85$.

An approximate equivalent axial load is calculated with this expression in which P_0 is the actual axial load, m is a factor given in Table 10-2, and U is a factor provided in the column tables. *In applying the expression, M_x and M_y must be used in ft-k.*

To apply the expression a value of m is taken from the first approximation section of Table 10-2, and a value of U is assumed equal to 3. The equation is solved for P_{eff}, and a column is selected from the column tables for that load. Then the equation for P_{eff} is solved again using a revised value of m from the subsequent approximations part of Table 10-2, and the value of U is taken from the column tables for the column initially selected. Another shape is selected, and the process is continued until m and U stabilize, that is, until the shape selected does not change. Equations H1-1, H1-2 and H1-3 (as applicable) are then used to check the adequacy of the selected section. (An equivalent procedure is to substitute into the modified equations to obtain the final equivalent axial load and to compare that value with the allowable axial load given in the column tables for the section in question.)

Examples 10-7 to 10-9 illustrate the design of beam-columns using the equivalent axial load procedure.

■ **Example 10-7**

Using Table 10-2 select the lightest satisfactory W section to support an axial compression load P of 240k and the moments $M_x = 50$ ft-k and $M_y = 30$ ft-k. Assume $KL = 16$ ft, $C_{mx} = C_{my} = 0.85$, and use A36 steel.

Solution. From Table 10-2 the first approximation of m is 2.2 and U is initially assumed to equal 3.

$$P_{eff} = 240 + (50)(2.2) + (30)(2.2)(3) = 548 \text{ k}$$

Possible sections from column tables

$$W14 \times 109$$
$$W12 \times 120$$

Try W14 × 109 ($U = 2.49$ from column tables and m from Table 10-2 for subsequent approximations = 1.7)

$$P_{eff} = 240 + (50)(1.7) + (30)(1.7)(2.49) = 452 \text{ k}$$

Try a smaller section a W14 × 90.
Checking a W14 × 90 ($A = 26.5$ in.^{2}, $S_x = 143$ in.^{3}, $S_y = 49.9$ in.^{3}, $r_y = 3.70$ in., $L_c = 15.3$ ft, $L_u = 34.0$ ft, $\dfrac{F'_{ex}(K_x L_x)^2}{(10)^2} = 391$ k, $\dfrac{F'_{ey}(K_y L_y)^2}{(10)^2} = 142$ k):

$$f_a = \frac{240}{26.5} = 9.06 \text{ ksi}$$

$$\frac{K_y L_y}{r_y} = \frac{(12)(16)}{3.70} = 51.89$$

$$F_a = 18.18 \text{ ksi}$$

$$\frac{f_a}{F_a} = \frac{9.06}{18.18} = 0.498 > 0.15$$

∴ Must use Equations H1-1 and H1-2

$$f_{bx} = \frac{(12)(50)}{143} = 4.20 \text{ ksi}$$

$$f_{by} = \frac{(12)(30)}{49.9} = 7.21 \text{ ksi}$$

$F_{bx} = 0.60F_y$ at middepth $= 22$ ksi since $L_{\text{unbraced}} > L_e < L_u$ and $0.66F_y = 24$ ksi at column ends

$$F_{by} = 0.75F_y = 27 \text{ ksi}$$
$$C_{mx} = C_{my} = 0.85$$

$$F'_{ex} = \frac{(391)(10)^2}{(16)^2} = 152.73 \text{ ksi}$$

$$F'_{ey} = \frac{(142)(10)^2}{(16)^2} = 55.47 \text{ ksi}$$

$$\frac{C_{mx}}{1 - \dfrac{f_a}{F'_{ex}}} = \frac{0.85}{1 - \dfrac{9.06}{152.73}} = 0.904 < 1.0 \qquad \underline{\text{Use } 1.0}$$

$$\frac{C_{my}}{1 - \dfrac{f_a}{F'_{ey}}} = \frac{0.85}{1 - \dfrac{9.06}{55.47}} = 1.016 < 1.0 \qquad \text{OK}$$

Applying Equations H1-1 and H1-2:

$$\frac{9.06}{18.18} + \frac{(1.0)(4.20)}{22} + \frac{(1.016)(7.21)}{27} = 0.961 < 1.0 \qquad \text{OK}$$

$$\frac{9.06}{22} + \frac{4.20}{24} + \frac{7.21}{27} = 0.854 < 1.0 \qquad \text{OK}$$

$$\underline{\underline{\text{Use W14} \times 90}} \qquad \blacksquare$$

■ Example 10-8

Using a steel with $F_y = 50$ ksi, select the lightest satisfactory W12 section for the situation shown in Fig. 10-5, assuming $C_m = 1.0$. Lateral bracing is supplied only at the ends.

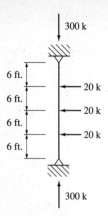

300 k

6 ft.

← 20 k

6 ft.

← 20 k

6 ft.

← 20 k

6 ft.

300 k

Figure 10-5

Solution. As the numbers in Table 10-2 are based on a C_m of 0.85, the m values obtained must be multiplied by $\dfrac{1.0}{0.85}$. Knowing we are to select a W12, we omit the first approximation and proceed to subsequent approximations.

$$KL = 24 \text{ ft}$$

$$m = (1.7)\left(\frac{1}{0.85}\right) = 2.0$$

$$M_x = (30)(12) - (20)(6) = 240 \text{ ft-k}$$

$$P_{\text{eff}} = 300 + (240)(2.0) = 780 \text{ k}$$

Try W12 × 170 (check interaction equations [omitted here])

Use W12 × 170 ■

■ Example 10-9

An analysis has been made of the frame shown in Fig. 10-6, and the beams have been designed. The beam is unbraced in the plane of the frame. It is braced in the other or y direction. Select a W section for column AB, assuming

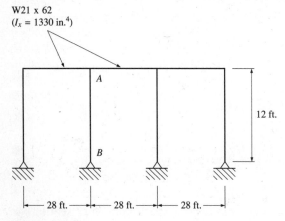

W21 x 62
($I_x = 1330$ in.^{4})

A

B

12 ft.

|← 28 ft. →|← 28 ft. →|← 28 ft. →|

Figure 10-6

it is A36 steel, supports an axial load of 120 k, and has an applied moment of 60 ft-k at its upper end.

Solution.   First trial (Assume $k = 1.0$ and $m = 2.3$):

$$P_{\text{eff}} = 120 + (60)(2.3) = 258 \text{ k}$$

Possible sections are W14 × 61, W12 × 53, and W10 × 54.
Second trial with W12 ($m = 2.1$):

$$P_{\text{eff}} = 120 + (60)(2.1) = 246 \text{ k}$$

Try W12 × 53 ($A = 15.6$ in.^{2}, $S_x = 70.6$ in.^{3}, $I_x = 425$ in.^{4}, $r_x = 5.25$ in.,
$r_y = 2.48$ in., $L_c = 10.6$ ft, $L_u = 22.0$ ft, $\dfrac{F'_{ex}(K_x L_x)^2}{(10)^2} = 284$ k)

Checking K_x

$$G_A = \frac{\sum \dfrac{I_c}{L_c}}{\sum \dfrac{I}{L}} = \frac{\dfrac{425}{12}}{\dfrac{(2)(1330)}{28}} = 0.373$$

$$G_B = 10.0$$

$K_x = 1.76$ from Fig. 7-2 sidesway uninhibited chart
Checking interaction equations (omitted here)

<u>Use a W12 × 53</u>   ■

10-8 PIPE COLUMNS AND STRUCTURAL TUBES

Although pipe columns and structural tubes are frequently used for columns in small buildings, Table B of the ASD Manual does not provide m values for them, nor do the column tables provide U values for them. However, these same column tables do provide B_x and B_y factors for pipe and tube sections. A rough value of P_{eff} can be obtained by substituting these values into the following equation, which is a simplified version of the ASD modified equations discussed in Section 10-7. *In this substitution note that M_x and M_y have the units in.-k.*

$$P_{\text{eff}} = P + B_x M_x + B_y M_y$$

Though this approximate equation does not include any moment magnification for lateral deflection or any moment modification, it does provide a fairly good first approximation for a trial-and-error design using the interaction equations. Example 10-10 presents the design of a square beam-column tube.

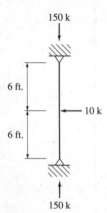

Figure 10-7

■ Example 10-10

Select a square structural tube for the situation shown in Fig. 10-7. $F_y =$ 50 ksi and $K = 1.0$. Lateral bracing is applied only at column ends.

Solution. Enter tables with $P_0 = 150$ k and 0 moment
Try $6 \times 6 \times \frac{3}{8}$ tube $(B = 0.583)$

$$M = \frac{(10)(12)}{4} = 30 \text{ ft-k}$$

$$P_{\text{eff}} = 150 + (0.583)(12 \times 30) = 360 \text{ k}$$

Try $8 \times 8 \times \frac{9}{16}$ $(B = 0.445)$

$$P_{\text{eff}} = 150 + (0.445)(12 \times 30) = 310.2 \text{ k}$$

Try $8 \times 8 \times \frac{1}{2}$ $(B = 0.437)$

$$P_{\text{eff}} = 150 + (0.437)(12 \times 30) = 307.7 \text{ k}$$

Checking interaction equations (omitted here)

$$\underline{\text{Use } 8 \times 8 \times \frac{1}{2} \text{ tube}}$$

Note: Frequently tubes, including this one, are noncompact (see ASD Table B5.1), and maximum F_b is $0.60F_y$. ■

In the ASD Manual sample designs are presented for beam-columns in which lateral wind loads are involved. In these examples the lateral loads are multiplied by three-fourths to take into account the provision of Section A5.2 of the ASD Specification which permits an increase in allowable stresses of one-third for wind or seismic loads or for wind or seismic loads acting in combination with the design dead and live loads. This reduction is permissible only if the resulting sections are as large as they would have been if the member were selected for the design dead and live loads and impact without the stress increase and without the wind or seismic loads applied.

10-9 LIMITATIONS OF ASD TABLES

The tables in the ASD Manual which have been described here work very well unless the moment becomes quite large in proportion to the axial load. For such cases the members selected with the tables will be capable of supporting the loads and moments but may not provide a very economical solution. The tables are limited to the W14s, W12s, and shallower sections, but when the moment is large in proportion to the axial load, there will often be a much deeper section, such as a W27 or W30, which may be more economical.

Example 10-11 illustrates this situation. The method presented in Example 10-3 is used for estimating the size, and then the regular ASD expressions are used to check the section. It will be noted that in this problem the estimated area required is 47.78 in.^{2} (which corresponds to about a W30 $\times$ 173) and the estimated section modulus required is 273 in.^{3} (which corresponds to about a W30 $\times$ 99). The author interpolated between these two sections and tried a W30 $\times$ 124, which proved to be satisfactory. Subsequent checks were made on W33s, W27s, and W24s. The W30 $\times$ 124 was the lightest satisfactory section. If the column tables in the ASD Manual were used together with the equivalent axial load method, appreciably heavier sections would be the result, such as the W14 $\times$ 159 or the W12 $\times$ 170.

■ Example 10-11

Using the ASD Specification and A36 steel, select a W section to support an axial load $P = 430$ k and a bending moment about the x axis $M_x = 250$ ft-k. The member, which has pinned ends, is to be 18 ft long and is to be laterally braced (pinned) in the weak direction at middepth. Assume $C_m = 0.85$.

Solution.   Assume.

$$\frac{f_a}{F_a} + \frac{f_b}{F_b} = 0.5 + 0.5 = 1.0$$

Assume:

$$F_a = 18 \text{ ksi}$$

$$\frac{f_a}{F_a} = \frac{430/A}{18} = 0.5$$

Approximate A required = 47.78 in.^{2}
Assume:

$$F_b = 22 \text{ ksi}$$

$$\frac{f_b}{F_b} = \frac{(12)(250)/S_x}{22} = 0.5$$

Approximate S_x required = 273 in.^{3}

From this information several possible estimated deep sections, including the W30 $\times$ 124, W27 $\times$ 146, W24 $\times$ 131, and others, may be tried.

Try W30 × 124 (checking this section with the interaction equations, it is satisfactory despite the fact that its flange is slightly noncompact.)

According to Table 10.2 and the tables for the usual column sections, a much heavier member is required.

Use a W14 × 159 or W12 × 170   ∎

10-10 COMBINED AXIAL TENSION AND BENDING

The foregoing paragraphs and examples have all pertained to members subjected to a combination of axial compression and bending. A less critical case but one that can occasionally be quite important occurs when a member is subjected to simultaneous axial tension and bending. Various specifications have different design requirements. For instance, the AREA says that members subject to both axial tension and bending stress shall be so proportioned that the total axial tension plus the bending tension shall not exceed the allowable tension for axially loaded tensile members. (In addition it says that the compression caused by such bending may not exceed the allowable value for axially loaded compression members.)

The ASD Specification requires a member subject to a combination of axial tension and bending to satisfy the expression which follows. In this expression f_b is the maximum tensile bending stress, and F_b is the allowable bending tensile stress.

$$\frac{f_a}{0.60F_y} + \frac{f_{bx}}{F_{bx}} + \frac{f_{by}}{F_{by}} \le 1.0$$

A further requirement is that the computed bending compressive stress may not exceed the applicable value for allowable bending compressive stress.

PROBLEMS

10-1 Determine the maximum stresses in the compression chord of a roof truss caused by its own weight. The member is 24 ft 0 in. long and consists of two C12 × 20.7s with a top cover plate $\frac{1}{2}$ × 12 in. Assume the lacing and tie plates weigh 200 lb. (*Ans.* $f_t = +1.18$ ksi, $f_c = -0.648$ ksi)

10-2 The top chord of a roof truss is shown. Determine the maximum stresses at the extreme fibers of this member if it is subjected to a moment of 90 ft-k and an axial compression of 450 k.

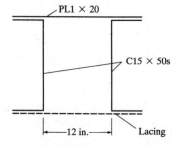

10-3 An 18-ft W14 × 90 is used to support an axial load of 300 k and a bending moment about its major axis of 60 ft-k. Is it adequate according to the following interaction formula and allowable stress values?

$$F_a = 15,000 - \frac{1}{4}\left(\frac{L}{r}\right)^2$$

$$F_b = 18,000 \text{ psi}$$

$$\frac{f_a}{F_a} + \frac{f_b}{F_b} \leq 1.0$$

(*Ans.* Section unsatisfactory; interaction equation gives 1.08.)

10-4 A column has a laterally unsupported length of 20 ft, an axial compression of 150 k, and a bending moment of 60 ft-k. Select a W section for this member using the interaction formula and allowable stress values of Prob. 10-3.

10-5 A 12-ft column in a rigid frame building is subjected to an axial load of 300 k and to a moment of 80 ft-k. Using the interaction formula and the allowable stresses of Example 10-3, select a W shape. (*Ans.* W14 × 109)

10-6 The column shown is to be designed to support an eccentric load of 150 k, located as shown in the figure. Select the lightest available W section using the design information of Prob. 10-3. Length = 18 ft.

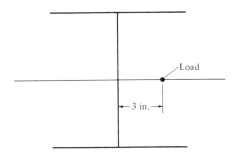

10-7 Repeat Prob. 10-6 with a load of 100 k applied as shown. (*Ans.* W12 × 72)

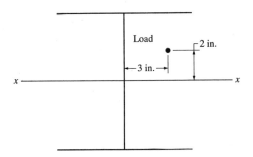

10-8 A column has a laterally unsupported length of 16 ft, an axial compression of 350 k, and a bending moment of 100 ft-k. Using the design formulas given in Prob. 10-3, select the most economical W section.

10-9 A column with an unsupported length of 28 ft must support an axial compressive load of 65 k and a moment of 65 ft-k. Select a W section using the design formulas of Prob. 10-3. (*Ans.* W12 × 53 slightly overstressed)

10-10 A W10 × 49 pin-connected beam-column, which is subjected to sidesway, is 15 ft long. A 150-k load is applied to the column at its upper end with an eccentricity of 2 in. so as to cause bending about the major axis of the section. Check the adequacy of the member with the ASD Specification if it consists of A36 steel.

10-11 Sidesway is prevented for the beam-column shown. If bending is about the major axis, is the member satisfactory? Use A36 steel and the ASD Specification. (*Ans.* Interaction equations yield 0.935 and of 0.774, therefore section is satisfactory.)

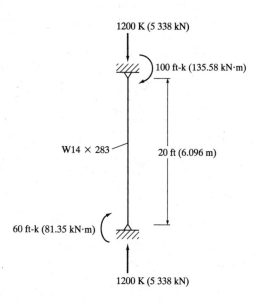

10-12 A pin-connected W10 × 45 consisting of A36 steel is used as a beam-column. If sidesway is possible and a load of 170 k is applied 1.50 in. off-center at the upper end so as to cause bending about the *y* axis, is the member satisfactory according to the ASD Specification? Length = 12 ft.

10-13 Is the member shown adequate? It is bent about its major axis and consists of A36 steel. Use the ASD Specification and assume sidesway is prevented. (*Ans.* Interaction equations yield 1.14 and 0.733, Section is not satisfactory.)

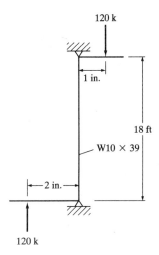

10-14 Select a W14 for the situation shown. Use A36 steel and the ASD Specification, and assume sidesway occurs.

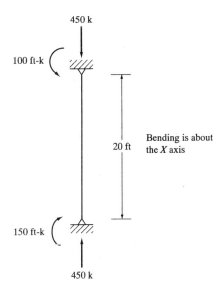

10-15 Repeat Prob. 10-8 using the ASD Specification and A36 steel. The column is assumed to be pinned at both ends, to have no sidesway, and to have no transverse loading applied. The moments are maximum at the column ends and tend to bend the column in reverse curvature. Assume L = 16 ft. (*Ans.* W14 × 90)

10-16 Repeat Prob. 10-9 using the ASD Specification, A36 steel and using a length of 14 ft. The member is fixed at its ends and has no sidesway or transverse loading. The moments are maximum at the column ends and tend to bend the column in single curvature.

10-17 A 16-ft pinned-end column is subjected to a moment of 100 ft-k at one end and 120 ft-k at the other end such that it is bent in single curvature. There are no transverse loads, and joint translation is prevented. If the axial load is 150 k, select a W14 section using the ASD Specification and A36 steel. (*Ans.* W14 × 82, W14 × 74 slightly overstressed)

10-18 A 15-ft pinned-end column is subject to a bending moment of 300 ft-k and an axial load of 300 k. If A36 steel is used and sidesway is not prevented, select the lightest satisfactory W section.

10-19 The accompanying diagram represents a highway sign board. Is a W10 × 45 of A36 steel satisfactory for the column labeled *AB* in the figure if the ASD Specification is used? Neglect member weights in all calculations. (*Ans.* Section is satisfactory.)

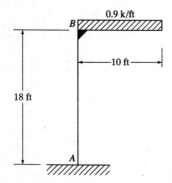

10-20 The W12 × 58 beam column shown is subject to an axial load of 170 k and to a lateral wind load perpendicular to the *x* axis of the member. Using A36 steel and the ASD Specification, determine the maximum value of *w* in ft.

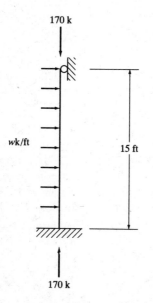

10-21 Using the ASD Specification and A36 steel, select a WT8 for a 10-ft horizontal truss member with a 65-k axial compression load and a transverse uniform load of 75 lb/ft. The ends of the member are assumed to be pinned, and joint translation is prevented. (*Ans.* WT8 × 22.5)

10-22 Repeat Prob. 10-21, assuming the member ends are fixed.

10-23 Repeat Prob. 10-21, assuming both ends are pinned. (*Ans.* WT8 × 22.5)
(*Ans.* WT8 × 22.5)

10-24 Repeat Prob. 10-7 using A36 steel and the ASD Specification. Assume the member is fixed at its ends and has no sidesway or transverse loading. The moments are maximum at the column ends and tend to bend it in single curvature.

10-25 For the frame shown, which is subject to sidesway, select tentative beam and column sizes assuming moments at beam ends of $wL^2/10$ and the same values in the columns. Use the ASD Specification and A36 steel. The roof is to be designed to support a built-up roof weighing 6 psf plus a live load of 30 psf, and the interior floor slab is to support a live load of 125 psf. Using the member sizes selected, determine the column K values from the chart given in Fig. 7-2. (*Ans.* W16 × 31 roof beams, W24 × 55 floor beams and W14 × 90 columns)

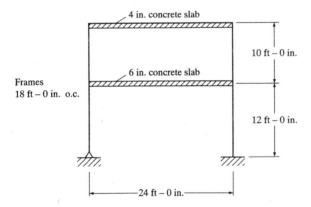

10-26 Select a W14 beam-column of A36 steel for an axial load of 200 k, M_x = 50 ft-k, and M_y = 30 ft-k. Use ASD Specification and assume KL_x = KL_y = 18 ft and C_{mx} and C_{my} = 0.85. Sidesway is uninhibited. Do not modify G factors for conditions at far ends of beam.

11

Bolted Connections

11-1 INTRODUCTION

For quite a few decades riveting was the accepted method used for connecting the members of steel structures. For the last few decades, however, bolting and welding have been the methods used for making structural steel connections, and riveting is almost extinct. This chapter and the next are almost entirely devoted to bolted connections, although some brief remarks at the end of Chapter 12 concern rivets.

Bolting of steel structures is a very rapid field erection process which requires less-skilled labor than does riveting or welding. This gives bolting a distinct economic advantage over the other connection methods in the United States, where labor costs are so very high. Even though the purchase price of a high-strength bolt is several times that of a rivet, the overall cost for bolted construction is cheaper than that for riveted construction because of reduced labor and equipment costs and the smaller number of bolts required to resist the same loads.

11-2 TYPES OF BOLTS

Several types of bolts can be used for connecting steel members.

Unfinished bolts are also called ordinary or common bolts. They are classified by the ASTM as A307 bolts and are made from carbon steels with stress-strain characteristics very similar to those of A36 steel. They are available in diameters from $\frac{5}{8}$ to $1\frac{1}{2}$ in. in $\frac{1}{8}$-in. increments.

A307 bolts generally have square heads and nuts to reduce costs, but hexagonal heads are sometimes used because they are a little more attractive, are easier to turn and easier to hold with the wrenches, and require less turning space. As A307 bolts have relatively large tolerances in shank and thread dimensions, their design strengths are appreciably smaller than those for rivets or high-strength bolts. They are primarily used in light structures subjected to static loads and for secondary members (such as purlins, girts, bracing, platforms, small trusses, and so forth).

Designers are often guilty of specifying high-strength bolts for connections when common bolts would be satisfactory. *The strength and advantages of common bolts have usually been greatly underrated in the past.* The analysis and design of A307 bolted connections are handled exactly as are riveted connections in every way except that the allowable stresses are different.

High-strength bolts are made from medium-carbon heat-treated steel and from alloy steel and have tensile strengths two or more times those of ordinary bolts. There are two basic types, the A325 bolts (made from a medium-carbon heat-treated steel) and the higher strength A490 bolts (also heat-treated but made from an alloy steel). High-strength bolts are used for all types of structures, from small buildings to skyscrapers and monumental bridges. These bolts were developed to overcome the weaknesses of rivets—primarily insufficient tension in their shanks after cooling. The resulting rivet tensions may not be large enough to hold them in place during the application of severe impactive and vibrating loads. The result is that they may become loose and vibrate and eventually have to be replaced. High-strength bolts may be tightened until they have very high tensile stresses so that the connected parts are clamped tightly together between the bolt and nut heads, permitting loads to be transferred primarily by friction.

Sometimes high-strength bolts are made from A449 steel in sizes larger than the $1\frac{1}{2}$-in. maximum diameter A325 and A490 bolts. These larger bolts may also be used as high-strength anchor bolts and for threaded rods of many different diameters.

11-3 HISTORY OF HIGH-STRENGTH BOLTS

The joints obtained using high-strength bolts are superior to riveted joints in performance and economy, and they are the leading field method of fastening structural steel members. C. Batho and E. H. Bateman first claimed in 1934

High-strength bolt. (Courtesy of Bethlehem Steel Corporation.)

that high-strength bolts could satisfactorily be used for the assembly of steel
structures,[1] but it was not until 1947 that the Research Council on Riveted and
Bolted Structural Joints of the Engineering Foundation was established in the
United States. This group issued their first specifications in 1951, and high-
strength bolts were adopted with amazing speed by building and bridge engineers for both static and dynamic loadings. They not only quickly became the
leading method of making field connections, but they also were found to have
many applications for shop connections. The construction of the Mackinac
Bridge in Michigan involved the use of more than one million high-strength
bolts.

Connections that were formerly made with ordinary bolts and nuts were
not too satisfactory when they were subjected to vibratory loads, because the
nuts frequently became loose. For many years this problem was dealt with by
using some type of locknut, but modern high-strength bolts furnish a far superior solution.

11-4 ADVANTAGES OF HIGH-STRENGTH BOLTS

Among the many advantages of high-strength bolts, partly explaining their
great success, are the following:

1. Smaller crews are involved as compared with riveting. Two two-
   person bolting crews can easily turn out in a day over twice as many
   bolts as the number of rivets driven by the standard four-person riveting crew. The result is quicker steel erection.
2. In comparison with rivets, fewer bolts are needed to provide the same
   strength.
3. Good bolted joints can be made by people with a great deal less train-
   ing and experience than is necessary to produce welded and riveted

---

[1] C. Batho and E. H. Bateman, "Investigations on Bolts and Bolted Joints" (London: H. M. Sta-
tionery Office, 1934).

450 Lexington Ave., New York City. (Courtesy of Owen Steel Company, Inc.)

connections of equal quality. The proper installation of high-strength bolts can be learned in a matter of hours.

4. No erection bolts, which may have to be later removed (depending on specifications), are required as in welded joints.
5. Though quite noisy, bolting is not nearly as bad as riveting.
6. Cheaper equipment is used to make bolted connections.
7. Bolts present no fire hazard and no danger like the tossing of hot rivets.
8. Tests on riveted joints and fully tensioned bolted joints under identical conditions definitely show that bolted joints have a higher fatigue strength. Their fatigue strength is also equal to or greater than that obtained with equivalent welded joints.
9. Where structures are to be later altered or disassembled, changes in connections are quite simple because of the ease of bolt removal.

11-5 SNUG-TIGHT AND FULLY TENSIONED BOLTS

Not all high-strength bolts have to be fully tensioned. Such a process is expensive, as is inspection to see that they are fully tensioned. Thus the ASD Specification requires that those bolts which have to be fully tensioned must be clearly identified on drawings. These will be the bolts used in slip-critical connections and in connections subject to direct tension. Slip-critical connections will be required when the working loads cause a large number of stress changes possibly causing fatigue problems. In Section J1.12 of the ASD Specification a detailed list of connections which must be made with fully tensioned bolts is presented. Included are connections for supports of running machinery or for live loads producing impact and stress reversal; column splices in all tier structures 200 ft or more in height; connections of all beams and girders to columns and other beams or girders on which the bracing of the columns is dependent for structures over 125 ft in height; and so on.

Other bolts need only be tightened to a *snug-tight* condition, when all the plies of a connection are in firm contact with each other. It usually means the tightness obtained by the full effort of a worker with a spud wrench or the tightness obtained after a few impacts of an impact wrench. Obviously there is some variation in the degrees of tightness obtained under these conditions. Snug-tight bolts must be clearly identified on both design and erection drawings.

Table 11-1 presents the fastener tensions required for slip-resistant connections and for connections subject to direct tension. To be fully tensioned both A325 and A490 bolts are tightened to at least 70 percent of their specified minimum tensile strengths.

The quality-control provisions specified in the manufacture of the A325 and A490 bolts are more stringent than those for the A449 bolts. As a result,

TABLE 11-1 FASTENER TENSION REQUIRED FOR SLIP-CRITICAL CONNECTIONS AND CONNECTIONS SUBJECT TO DIRECT TENSION, KIPS

BOLT SIZE (IN.)	A325 BOLTS	A490 BOLTS
$\frac{1}{2}$	12	15
$\frac{5}{8}$	19	24
$\frac{3}{4}$	28	35
$\frac{7}{8}$	39	49
1	51	64
$1\frac{1}{8}$	56	80
$1\frac{1}{4}$	71	102
$1\frac{3}{8}$	85	121
$1\frac{1}{2}$	103	148

Source: American Institute of Steel Construction, *Manual of Steel Construction Allowable Stress Design,* 9th ed. (Chicago: AISC, 1989), Table J3.7, p. 5-77. Reprinted with the permission of AISC.

Note: Equal to 0.70 of minimum tensile strength of bolts, rounded off to nearest kip, as specified in ASTM specifications for A325 and A490 bolts with UNC threads.

despite the method of tightening, the A449 bolts may not be used in slip-critical connections.

Although many engineers felt that there would be some slippage as compared with rivets (because the hot-driven rivets more nearly filled the holes), the results show that there is less slippage in fully tensioned high-strength bolted joints than in riveted joints under similar conditions.

It is interesting to note that the nuts used for fully tensioned high-strength bolts need no special provisions for locking. Once these bolts are installed and sufficiently tightened to produce the tension required, there is almost no tendency for the nuts to come loose. There are, however, a few situations where they will work loose under heavy vibrating loads. What do we do then? Some steel erectors have replaced the offending bolts with longer ones with two fully tightened nuts. Others have welded the nuts onto the bolts. Apparently the results have been somewhat successful.

11-6 METHODS FOR FULLY TENSIONING HIGH-STRENGTH BOLTS

We have already commented on the tightening required for snug-tight bolts. For fully tensioned bolts several methods of tightening are available. These methods, including the turn-of-the-nut method, the calibrated wrench method, and the use of alternate design bolts and direct tension indicators, are permitted without preference by the ASD Specification.

Turn-of-the-Nut Method

The bolts are brought to a snug-tight condition, and then they are given from one-third to one full turn, depending on their length and the slope of the surfaces under their heads and nuts. (The amount of turn given can easily be controlled by marking the snug-tight position with paint or crayon.)

Calibrated Wrench Method

With this method the bolts are tightened with an impact wrench that is adjusted to stall at that certain torque which is theoretically necessary to tension a bolt of that diameter and ASTM classification to the desired tension. Wrenches must be calibrated daily, and hardened washers must be used. Particular care needs to be given to protecting the bolts from dirt and moisture at the job site.

Direct Tension Indicator

The direct tension indicator is a hardened washer which has protrusions on one face in the form of small arches. The arches will be flattened as a bolt is tightened. The amount of gap at any one time is a measure of the bolt tension. For fully tensioned bolts the gaps should measure about 0.015 in. or less.

Alternate Design Fasteners

In addition to the preceding methods there are some alternate design fasteners which can be tensioned quite satisfactorily. Bolts with splined ends which extend beyond the threaded portion of the bolts are one example. Special wrench chucks are used to tighten the nuts until the splined ends shear off.

A maximum bolt tension is not specified in any of the preceding tightening methods. This means that the bolt can be tightened to the highest load that will not break it; the bolt will still do the job. Should the bolt break, another one is put in with no damage done. It might be noted that the nuts are stronger than the bolt, and the bolt will break before the nut strips.

For fatigue situations where members are subjected to constantly fluctuating loads, the slip-resistant connection is very desirable. If the force to be carried is less than the frictional resistance and thus no forces are applied to the bolts, how could we ever have a fatigue failure of the bolts? For a slip-critical joint the loads are not permitted to exceed the permissible frictional resistance.

Other situations where slip-critical connections are desirable include joints where bolts are used in oversized holes, joints where bolts are used in slotted holes where the loads are applied parallel or nearly so to the slots, joints that

Torquing the nut for a high-strength bolt with an air-driven impact wrench. (Courtesy of Bethlehem Steel Corporation.)

are subjected to significant force reversals, and joints where bolts and welds resist shear together on a common *faying surface* (the contact or shear area between the members).

11-7 SLIP-CRITICAL CONNECTIONS AND BEARING-TYPE CONNECTIONS

When high-strength bolts are fully tensioned, they clamp the parts being connected tightly together. The result is a considerable resistance to slipping on the faying surface. This resistance is equal to the clamping force times the coefficient of friction.

If the shearing load is less than the permissible frictional resistance, the connection is referred to as a slip-critical one. If the load exceeds the frictional resistance, the members will slip on each other and will tend to shear off the bolts, and at the same time the connected parts will push or bear against the bolts as shown in Fig. 11-1.

The surfaces of joints including the area adjacent to washers need to be free of loose scale, dirt, burrs, and other defects which might prevent the parts from solid seating. It is necessary for the surface of the parts to be connected to have slopes of not more than 1 to 20 with respect to the bolt heads and nuts unless beveled washers are used. For slip-critical joints the faying surfaces must also be free from oil, paint, and lacquer.

If the faying surfaces are galvanized, the slip factor will be reduced to almost half of its value for clean mill scale surfaces. The slip factor, however, may be significantly improved if the surfaces are subjected to hand wire brushing or to "brush off" grit blasting. However, such treatments do not seem to provide increased slip resistance for sustained loadings where there seems to be a creeplike behavior.[2]

The 1988 AASHTO Specifications permit hot-dip galvanization if the coated surfaces are scored with wire brushes or sandblasted after galvanization and before steel erection.

The ASTM Specification permits the galvanization of the A325 bolts themselves but not the A490 bolts. There is a danger of embrittlement of this higher-strength steel during galvanization because hydrogen may be introduced into the steel in the pickling operation.

If special faying surface conditions (such as blast-cleaned surfaces or blast-cleaned surfaces with special slip-resistant coatings applied) are used to increase the slip resistance, the designer may increase the values used here to the ones given by the Research Council on Structural Joints. These values are shown in the "Specification for Structural Joints using ASTM 325 or A490 Bolts" in Part 5 of the ASD Manual.

---

[2] G. L. Kulak, J. W. Fisher, and J. H. A. Struik, *Guide to Design Criteria for Bolted and Riveted Joints*, 2d ed. (New York: John Wiley & Sons, 1987).

11-8 MIXED JOINTS

Bolts may on occasion be used in combination with welds and on other occasions with rivets (as where they are added to old riveted connections to enable them to carry increased loads). The ASD Specification contains some specific rules for these situations.

Bolts in Combination with Welds

For new work neither A307 common bolts nor high-strength bolts designed for bearing-type connections may be considered to share the load with welds. (Before the connection's ultimate strength is reached the bolts will slip, with the result that the welds will carry a larger proportion of the load—the actual proportion being difficult to determine.) For such circumstances welds will have to be proportioned to resist the entire loads.

If high-strength bolts are designed for slip-critical conditions, they may be allowed to share the load with welds. Should we be making welded alterations to a structure that was designed for slip resistance, the welds need only be proportioned for the additional strength needed.

High-Strength Bolts in Combination with Rivets

High-strength bolts may be considered to share loads with rivets for new work or for alterations of existing connections that were designed as slip-critical. (The ductility of the rivets allows the capacity of both sets of fasteners to act together.)

11-9 SIZES OF HOLES FOR BOLTS AND RIVETS

In addition to the standard-size bolt and rivet holes which are $\frac{1}{16}$ in. larger in diameter than the bolts or rivets, there are three types of enlarged holes: oversized, short-slotted, and long-slotted. Oversized holes will on occasion be very useful in speeding up steel erection. In addition they give some latitude for adjustments in plumbing up frames during erection. The use of nonstandard holes requires the approval of the designer and is subject to the requirements of Section J3.7 of the ASD Specification. Table 11-2 provides the nominal dimensions of the various kinds of enlarged holes permitted for the different bolt sizes.

Oversized holes may be used in all plies of connections as long as the applied load does not exceed the permissible slip resistance. They may not be used in bearing-type connections. It is necessary for hardened washers to be used over oversized holes that are located in outer plies.

Short-slotted holes may be used regardless of the direction of the applied load if the permissible slip resistance is larger than the applied force. Should

TABLE 11-2 NOMINAL HOLE DIMENSIONS

BOLT DIAMETERS	HOLE DIMENSIONS			
	STANDARD (DIAMETERS)	OVERSIZE (DIAMETERS)	SHORT SLOT (WIDTH × LENGTH)	LONG SLOT (WIDTH × LENGTH)
$\frac{1}{2}$	$\frac{9}{16}$	$\frac{5}{8}$	$\frac{9}{16} \times \frac{11}{16}$	$\frac{9}{16} \times 1\frac{1}{4}$
$\frac{5}{8}$	$\frac{11}{16}$	$\frac{13}{16}$	$\frac{11}{16} \times \frac{7}{8}$	$\frac{11}{16} \times 1\frac{9}{16}$
$\frac{3}{4}$	$\frac{13}{16}$	$\frac{15}{16}$	$\frac{13}{16} \times 1$	$\frac{13}{16} \times 1\frac{7}{8}$
$\frac{7}{8}$	$\frac{15}{16}$	$1\frac{1}{16}$	$\frac{15}{16} \times 1\frac{1}{8}$	$\frac{15}{16} \times 2\frac{3}{16}$
1	$1\frac{1}{16}$	$1\frac{1}{4}$	$1\frac{1}{16} \times 1\frac{5}{16}$	$1\frac{1}{16} \times 2\frac{1}{2}$
$\geq 1\frac{1}{8}$	$d + \frac{1}{16}$	$d + \frac{5}{16}$	$(d + \frac{1}{16}) \times (d + \frac{3}{8})$	$(d + \frac{1}{16}) \times (2.5 \times d)$

Source: American Institute of Steel Construction, *Manual of Steel Construction Allowable Stress Design,* 9th ed. (Chicago: AISC, 1989), Table J3.1, p. 5–71. Reprinted with the permission of AISC.

the load be applied in a direction approximately normal (between 80° and 100°) to the slot, these holes may be used in any or all plies of connections for bearing-type connections. It is necessary to use washers (hardened if high-strength bolts are being used) over short-slotted holes in an outer ply.

Long-slotted holes may be used in *only one* of the connected parts of slip-critical or bearing-type connections at any one faying surface. For slip-critical joints these holes may be used in any direction, but for bearing-type connections the loads must be normal (between 80° and 100°) to axes of the slotted holes. If long-slotted holes are used in an outer ply, they will need to be covered with plate washers or a continuous bar. For high-strength bolted connections the washer or bar does not have to be hardened, but it must be made of structural-grade material and may not be less than $\frac{5}{16}$ in. thick.

11-10 LOAD TRANSFER AND TYPES OF JOINTS

The following paragraphs present a few of the elementary types of bolted or riveted joints subjected to axial forces (that is, the loads are assumed to pass through the centers of gravities of the groups of connectors). For each of these joint types some comments are made about the methods of load transfer. Eccentrically loaded connections are discussed in Chapter 12.

For this initial discussion the reader is referred to Fig. 11-1(a). It is assumed that the plates shown are connected with a group of snug-tight bolts. In other words, the bolts are not tightened enough to significantly squeeze the plates together. If there is assumed to be little friction between the plates, they will slip a little due to the applied loads shown. As a result the loads in the plates will tend to shear the connectors off on the plane between the plates and press or bear against the sides of the bolts as shown in the figure. These connectors are said to be in *single shear and bearing* (also called unenclosed bearing). They must have sufficient strength to satisfactorily resist these forces, and

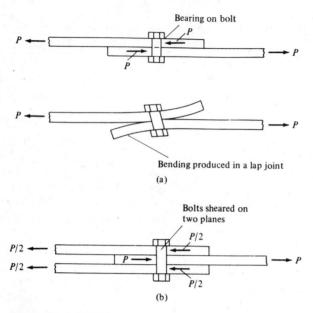

Figure 11-1 (a) Lap joint. (b) Butt joint.

the members forming the joint must be sufficiently strong to prevent the connectors from tearing through.

Should rivets be used instead of the snug-tight bolts, the situation is somewhat different because hot-driven rivets cool and shrink and squeeze or clamp
the connected pieces together with sizable forces which greatly increase the
friction between the pieces. As a result a large portion of the loads being transferred between the members is transferred by friction. The clamping forces
produced in riveted joints, however, are generally not considered to be dependable, and as a result specifications normally consider such connections to
be snug-tight with no frictional resistance. The same assumption is made for
A307 common bolts which are not tightened so as to have large dependable
tensions.

Fully tensioned high-strength bolts are in a different class altogether. Using the tightening methods previously described, a very dependable tension is
obtained in the bolts, resulting in large clamping forces and large dependable
amounts of frictional resistance to slipping. Unless the loads to be transferred
are larger than the frictional resistance, the entire forces are resisted by friction and the bolts are not really placed in shear or bearing. If the load exceeds
the frictional resistance, there will be slippage with the result that the bolts
will be placed in shear and bearing.

The Lap Joint

The joint shown in Fig. 11-1(a) is referred to as a *lap joint*. This type of joint
has a disadvantage in that the center of gravity of the force in one member is

not in line with the center of gravity of the force in the other member. A couple is present, which causes an undesirable bending in the connection as shown in the figure. For this reason the lap joint, which is desirably used only for minor connections, should be designed with at least two fasteners in each line parallel to the length of the member, to minimize the possibility of a bending failure.

The Butt Joint

A *butt joint* is formed when three members are connected as shown in Fig. 11-1(b). If the slip resistance between the members is negligible, the members will slip a little and tend to shear off the bolts simultaneously on the two planes of contact between the members. Again the members are bearing against the bolts, and the bolts are said to be in *double shear and bearing* (also called enclosed bearing). The butt joint is more desirable than the lap joint for two main reasons.

1. The members are arranged so that the total shearing force P is split into two parts, causing the force on each plane to be only about one-half of what it would be on a single plane if a lap joint were used. From a shear standpoint, therefore, the load-carrying ability of a group of bolts in double shear is theoretically twice as great as the same number of bolts in single shear.
2. A more symmetrical loading condition is provided. (In fact the butt joint does provide a symmetrical situation if the outside members are the same thickness and resist the same forces. The result is a reduction or elimination of the bending described for a lap joint.)

Double-Plane Connections

In the *double-plane connection* the bolts are subjected to single shear and bearing but bending moment is prevented. This type of connection, which is shown for a hanger in Fig. 11-2(a), subjects the bolts to single shear on two different planes.

Miscellaneous

Bolted connections generally consist of lap or butt joints or some combination of them, but there are other cases. For instance, there are occasionally joints in which more than three members are connected and the bolts are in multiple shear as shown in Fig. 11-2(b). Although the bolts in this connection are being sheared on more than two planes, the usual practice is to consider no more than double shear for strength calculations. It does not seem physically possible for shear failures to occur simultaneously on three or more planes. Several other types of bolted connections are discussed in this chapter and the next. These include bolts in tension, bolts in shear and tension, and so on.

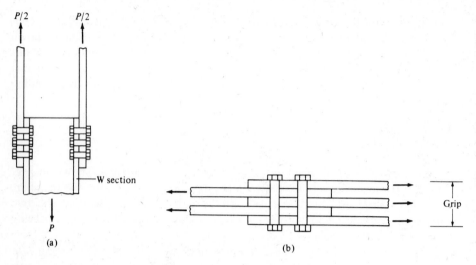

Figure 11-2 (a) Hanger connection. (b) Bolts in multiple shear.

11-11 FAILURE OF BOLTED JOINTS

Fig. 11-3 shows several ways in which failure of bolted joints can occur. To design bolted joints satisfactorily, it is necessary to understand these possibilities. These are described as follows:

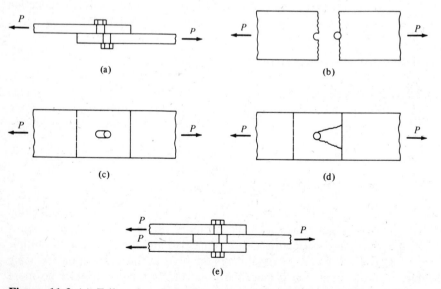

Figure 11-3 (a) Failure by single shearing of bolt. (b) Tension failure of plate. (c) Crushing failure of plate. (d) Shear failure of plate behind bolt. (e) Double shear failure of a butt joint.

1. The possibility of failure in a lap joint by shearing of the bolt on the plane between the members (single shear) is shown in (a).
2. The possibility of a tension failure of one of the plates through a bolt hole is shown in (b).
3. A possible failure of the bolts and/or plates by bearing between the two is shown in (c).
4. Another possibility is the shearing out of part of the member as shown in (d).
5. A butt joint with the possibility of a shear failure of the bolts along two planes (double shear) is shown in (e).

11-12 SPACING AND EDGE DISTANCES OF BOLTS

Before minimum spacings and edge distances can be discussed it is necessary for a few terms to be explained. The following definitions are given for a group of bolts in a connection and are shown in Fig. 11-4:

Pitch is the center-to-center distance of bolts in a direction parallel to the axis of the member.

Gage is the center-to-center distance of bolt lines perpendicular to the axis of the member.

The *edge distance* is the distance from the center of a bolt to the adjacent edge of a member.

The *distance between bolts* is the shortest distance between fasteners on the same or different gage lines.

Minimum Spacings

Bolts should be placed a sufficient distance apart to permit efficient installation and to prevent tension failures of the members between fasteners. The ASD Specification (J3.8) provides a minimum center-to-center distance for stan-

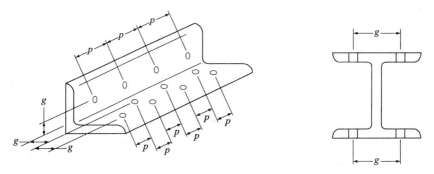

Figure 11-4 Bolt holes in connections. p = pitch, g = gage.

dard, oversized, or slotted fastener holes equal to not less than $2\frac{2}{3}$ diameters (with 3 diameters being preferred). If we are measuring along a line of transmitted force, this distance may have to be increased to prevent a bearing failure between the bolts in the direction of the applied force (ASD Specification J3.8). For such cases the distance may not be less than 3 diameters if the allowable bearing stress F_p is to be determined by ASD Equation J3-1 (for standard or short-slotted holes with two or more bolts in the direction of the line of force) or by Equation J3-2 (for long-slotted holes perpendicular to the direction of the force and with two or more bolts in the line of the force). These F_p values are discussed in the next section of this chapter. Otherwise the minimum center-to-center distance of standard holes is to be determined by the expression to follow, in which P is the force transmitted by one fastener to the critical part, F_u is the specified minimum tensile strength of the critical connected part, t is its thickness, and d is the diameter of the standard-size hole.

$$\text{Minimum distance center-to-center} = \frac{2P}{F_u t} + \frac{d}{2} \qquad \text{(ASD Equation J3-5)}$$

Should the holes be oversized or slotted, the minimum center-to-center distance is as determined above plus the applicable increment C_1 from Table 11-3. The clear distance between these enlarged holes may never be less than one bolt diameter.

The maximum edge distances and spacings of bolts used for weathering steel are smaller than they are for regular painted steel subject to corrosion or for regular unpainted steel not subject to corrosion. One of the requirements for using weathering steel is that it must not be allowed to be constantly in contact with water. As a result the ASD Specification tries to insure that the parts of a built-up weathering steel member are connected tightly together at frequent intervals to prevent forming of pockets that might catch water. The ASD Specification (J3.10) states that the maximum spacing center-to-center

TABLE 11-3 VALUES OF SPACING INCREMENT C_1 FOR DETERMINING MINIMUM SPACING OF ENLARGED HOLES

		SLOTTED HOLES		
NOMINAL DIAMETER OF FASTENER	OVERSIZE HOLES	PERPENDICULAR TO LINE OF FORCE	PARALLEL TO LINE OF FORCE	
			SHORT SLOTS	LONG SLOTS[a]
$\leq \frac{7}{8}$	$\frac{1}{8}$	0	$\frac{3}{16}$	$1\frac{1}{2}d - \frac{1}{16}$
1	$\frac{3}{16}$	0	$\frac{1}{4}$	$1\frac{7}{16}$
$\geq 1\frac{1}{8}$	$\frac{1}{4}$	0	$\frac{5}{16}$	$1\frac{1}{2}d - \frac{1}{16}$

Source: American Institute of Steel Construction, *Manual of Steel Construction Allowable Stress Design*, 9th ed. (Chicago: AISC, 1989), Table J3.4, p. 5–76. Reprinted with the permission of AISC.

[a] When length of slot is less than maximum allowed in Table 11-2, C_1 may be reduced by the difference between the maximum and actual slot lengths.

for painted members or for unpainted members not subject to corrosion is 24 times the thickness, not to exceed 12 in. For unpainted members consisting of weathering steel subject to atmospheric corrosion, the values are 14 times the thickness, not to exceed 7 in.

Minimum Edge Distances

Bolts should not be placed too near the edges of a member for two major reasons. The punching of holes too close to the edges may cause the steel opposite the hole to bulge out or even crack. The second reason applies to the ends of members where there is danger of the fastener tearing through the metal. The usual practice is to place the fastener a minimum distance from the edge of the plates equal to about 1.5 to 2.0 times the fastener diameter, so the metal there will have a shearing strength at least equal to that of the fasteners. For more exact information it is necessary to refer to the specification being used. The ASD Specification (J3.9) states that the distance from the center of a standard hole to the edge of a connected part may not be less than the applicable value given in Table 11-4 nor the value obtained from substituting, when applicable, into the formula to be described in the next paragraph.

In the direction of transmitted force the ASD Specification states that the minimum edge distance may not be less than $1\frac{1}{2}d$ when the allowable bearing

TABLE 11-4   MINIMUM EDGE DISTANCE FOR STANDARD HOLES

NOMINAL RIVET OR BOLT DIAMETER (IN.)	AT SHEARED EDGES	AT ROLLED EDGES OF PLATES, SHAPES, OR BARS EDGES, GAS CUT OR SAW-CUT EDGES[a]
$\frac{1}{2}$	$\frac{7}{8}$	$\frac{3}{4}$
$\frac{5}{8}$	$1\frac{1}{8}$	$\frac{7}{8}$
$\frac{3}{4}$	$1\frac{1}{4}$	1
$\frac{7}{8}$	$1\frac{1}{2}$[b]	$1\frac{1}{8}$
1	$1\frac{3}{4}$[b]	$1\frac{1}{4}$
$1\frac{1}{8}$	2	$1\frac{1}{2}$
$1\frac{1}{4}$	$2\frac{1}{4}$	$1\frac{5}{8}$
$>1\frac{1}{4}$	$1\frac{3}{4} \times$ diameter	$1\frac{1}{4} \times$ diameter

Source: American Institute of Steel Construction, *Manual of Steel Construction Allowable Stress Design,* 9th ed. (Chicago: AISC, 1989), Table J3.5, p. 5–76. Reprinted with the permission of AISC.

Note: Center of standard hole to edge of connected part. For oversized or slotted holes, see Table 11-5.

[a] All edge distances in this column may be reduced $\frac{1}{8}$ in. when the hole is at a point where stress does not exceed 25 percent of the maximum design strength in the element.

[b] These may be $1\frac{1}{4}$ in. at the ends of beam connection angles.

stress F_p is determined from ASD Equation J3-1 or J3-2 as for minimum center-to-center spacing. These bearing values are discussed in the next section. Otherwise the minimum edge distance is to be determined by the formula.

$$\text{Minimum edge distance, direction of transmitted force} = \frac{2P}{F_u t} \quad \text{(ASD Equation J3-6)}$$

Should the holes be oversized or slotted, the minimum edge distance may not be less than the value required for a standard hole plus an increment C_2 obtained from Table 11-5. Another minimum edge distance value is provided by the ASD Specification for end connections bolted to beam webs and designed for beam shear reactions only.

Maximum Edge Distances

Many specifications give maximum distances bolts can be placed from the edge of a connection. The ASD maximum (Specification J3.10) is 12 times the plate thickness but not to exceed 6 in. If bolts are too far from the edges, openings may develop between the members being connected. Special pitch and edge distance limitations are given in the ASD Specification (E4) for painted members and unpainted members not subject to corrosion. Maximum spacings for bolts may also be given for compression members so that buckling will not occur between bolts.

The maximum edge distances for bolts used for unpainted built-up members made of weathering steel are given as 8 times the thickness, not to exceed 5 in. in ASD Specification J3.10.

TABLE 11-5 VALUES OF MAXIMUM EDGE DISTANCE INCREMENT C_2 FOR ENLARGED HOLES

| | | SLOTTED HOLES | | |
| NOMINAL DIAMETER OF FASTENER (IN.) | OVERSIZED HOLES | PERPENDICULAR TO EDGE | | PARALLEL TO EDGE |
		SHORT SLOTS	LONG SLOTS[a]	
$\leq \frac{7}{8}$	$\frac{1}{16}$	$\frac{1}{8}$		
1	$\frac{1}{8}$	$\frac{1}{8}$	$\frac{3}{4}d$	0
$\geq 1\frac{1}{8}$	$\frac{1}{8}$	$\frac{3}{16}$		

Source: American Institute of Steel Construction, *Manual of Steel Construction Allowable Stress Design,* 9th ed. (Chicago: AISC, 1989), Table J3.6, p. 5–76. Reprinted with the permission of AISC.

[a] When length of slot is less than maximum allowable (see Table 11-2), C_2 may be reduced by one-half the difference between the maximum and actual slot lengths.

11-13 BEARING-TYPE CONNECTIONS—LOADS PASSING THROUGH CENTER OF GRAVITY OF CONNECTIONS

Shearing Strength

In bearing-type connections it is assumed that the loads to be transferred are larger than the frictional resistance caused by tightening the bolts, with the result that the members slip a little on each other putting the bolts in shear and bearing. The allowable design strength of a bolt in single shear equals the allowable shearing stress of the bolt in ksi times its cross-sectional area.

The allowable shearing stresses for high-strength bolts and for rivets are presented in Table 11-6. On observing these values you will note that the allowable shearing stress is reduced when the bolt threads are included in the shear planes because of the reduced cross-sectional area through the threads. For example, the allowable shear stress for A325 bolts for a bearing-type connection and standard-size holes is 21 ksi if threads are not excluded from the shear plane and 30 ksi if threads are excluded. Should a bolt be in double shear, its shearing strength is considered to be twice its single shear value.

The student may very well wonder what is done in design practice concerning threads excluded or not excluded from the shear planes. If normal bolt and member sizes are used, the threads will almost always be excluded from the shear plane. It is true that some extremely conservative individuals always assume the threads are not excluded from the shear plane.

Tightening high-strength bolts. (Courtesy of Bethlehem Steel Company.)

TABLE 11-6 ALLOWABLE STRESS ON FASTENERS, KSI

DESCRIPTION OF FASTENERS	ALLOWABLE TENSION[a] (F_t)	ALLOWABLE SHEAR[a] (F_v)					BEARING-TYPE CONNECTIONS[c]
		SLIP-CRITICAL CONNECTIONS[b,c]					
		STANDARD-SIZE HOLES	OVERSIZED AND SHORT-SLOTTED HOLES	LONG-SLOTTED HOLES			
				TRANSVERSE[d] LOAD	PARALLEL[d] LOAD		
A502, gr. 1, hot-driven rivets	23.0[e]						17.5[f]
A502, gr. 2 and 3, hot-driven rivets	29.0[e]						22.0[f]
A307 bolts	20.0[e]						10.0[f,g]
Threaded parts meeting the requirements of Sects. A3.1 and A3.4 and A449 bolts meeting the requirements of Sect. A3.4, when threads are not excluded from shear planes	$0.33F_u$[e,h,i]						$0.17F_u$[i]
Threaded parts meeting the requirements of Sects. A3.1 and A3.4, and A449 bolts meeting the requirements of Sect. A3.4, when threads are excluded from shear planes	$0.33F_u$[e,i]						$0.22F_u$[i]
A325 bolts, when threads are not excluded from shear planes	44.0[j]	17.0	15.0	12.0	10.0		21.0[f]
A325 bolts, when threads are excluded from shear planes	44.0[j]	17.0	15.0	12.0	10.0		30.0[f]

DESCRIPTION OF FASTENERS	ALLOW-ABLE TENSION[a] (F_t)	SLIP-CRITICAL CONNECTIONS[b,c]				BEARING-TYPE CONNEC-TIONS[e]
		STANDARD-SIZE HOLES	OVERSIZED AND SHORT-SLOTTED HOLES	LONG-SLOTTED HOLES		
				TRANSVERSE[d] LOAD	PARALLEL[d] LOAD	
A490 bolts, when threads are not excluded from shear planes	54.0[j]	21.0	18.0	15.0	13.0	28.0[f]
A490 bolts, when threads are excluded from shear planes	54.0[j]	21.0	18.0	15.0	13.0	40.0[f]

ALLOWABLE SHEAR[a] (F_v)

Source: American Institute of Steel Construction, Manual of Steel Construction Allowable Stress Design, 9th ed. (Chicago: AISC, 1989), Table J3.2, p. 5–73. Reprinted with the permission of AISC.

[a] See Sect. A5.2.

[b] Class A (slip coefficient 0.33). Clean mill scale and blast-cleaned surfaces with Class A coatings. When specified by the designer, the allowable shear stress, F_v, for slip-critical connections having special faying surface conditions may be increased to the applicable value given in the RCSC Specification.

[c] For limitations on use of oversized and slotted holes, see Sect. J3.2.

[d] Direction of load application relative to long axis of slot.

[e] Static loading only.

[f] When bearing-type connections used to splice tension members have a fastener pattern whose length, measured parallel to the line of force, exceeds 50 in., tabulated values shall be reduced by 20 percent.

[g] Threads permitted in shear plane.

[h] The tensile capacity of the threaded portion of an upset rod, based upon the cross-sectional area at its major thread diameter A_b, shall be larger than the nominal body area of the rod before upsetting times $0.60F_y$.

[i] See Table 2, Numerical Values Section for values for specific ASTM steel specifications.

[j] For A325 and A490 bolts subject to tensile fatigue loading, see Appendix K4.3.

Bearing Strength

The allowable design strength of a bolt in bearing equals the allowable bearing stress of the connected part in kips per square inch times the diameter of the bolt times the thickness of the member that bears against the bolt. (For countersunk bolts and rivets, one-half of the depth of the countersink should be deducted, says ASD Specification J3.3.) When the distance L_e in the direction of the force from the center of a regular or oversized hole (or from the center of the end of a slotted hole) to the edge of a connected part is not less than $1\frac{1}{2}$ times the bolt diameter d and the distance s center-to-center of the holes is not less than $3d$ and two or more bolts are used in the direction of the line of force, the bearing strength is

$F_p = 1.2F_u$ for standard or short-slotted holes   (ASD Equation J3-1)

$F_p = 1.0F_u$ for long-slotted holes perpendicular to the load

                                                 (ASD Equation J3-2)

Should deformations around the hole not be a design consideration, the two preceding expressions may be replaced with

$$F_p = 1.5F_u \qquad \text{(ASD Equation J3-4)}$$

For the examples and home problems in this chapter, F_p is assumed to equal $1.2F_u$. This is also the value used for the examples and tables of Part 4 of the ASD Manual.

Though the allowable bearing stress of a bolt with a small end distance is reduced *the allowable values of the other bolts in the connection are not reduced*. The value of F_p for a single bolt or for two or more bolts in the line *each with an end distance less than $1\frac{1}{2}d$* is to be determined with the following expression:

$$F_p = \frac{L_e F_u}{2d} \leq 1.2F_u \qquad \text{(ASD Equation J3-3)}$$

The allowable load on the edge bolt is the lesser of the allowable shear strength of the bolt on its cross-sectional area and the allowable bearing strength against the side of the hole. Should the allowable bearing strength control, we can increase the allowable strength of the connection by increasing the thickness of the parts being connected and by increasing the edge distance (if edge distance causes a reduction in F_p).

Tests of bolted joints have shown that neither the bolts nor the metal in contact with the bolts actually fail in bearing. However, these tests have shown that the efficiency of the connected parts in tension and compression is affected by the magnitude of the bearing stress. Therefore, the allowable bearing stresses given by the ASD Specification are values above which they feel the strength of the connected parts is impaired. In other words these apparently very high allowable bearing stresses are not really allowable bearing stresses at all but rather indexes of the efficiencies of the connected parts. If allowable bearing stresses larger than the values given are permitted, the holes seem to elongate more than about $\frac{1}{4}$ in. and impair the strength of the connections.

Minimum Connection Strength

The ASD Specification (J1.6) states that except for lacing, sag rods, and girts, connections must be provided with design strengths sufficient to support factored loads of at least 6 k.

Example 11-1 illustrates the calculations involved in determining the strength of a bearing-type connection. Using a similar procedure the number of bolts required for a certain loading condition is calculated in Example 11-2.

The values given for the strength of bolts in this chapter, whether bearing-type or slip-critical, can be obtained from the tables entitled "Bolts, Threaded Parts, and Rivets Shear Allowable Load in Kips" of Part 4 of the ASD Manual.

■ Example 11-1

Determine the allowable design strength P of the bearing-type connection shown in Fig. 11-5. The steel is A36, the bolts are $\frac{7}{8}$-in. A325, the holes are standard sizes, the threads are excluded from the shear plane, edge distances are $>1\frac{1}{2}d$, and the distance center-to-center of the holes is $>3d$. $F_p = 1.2F_u$.

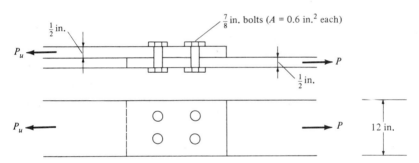

Figure 11-5

Solution.   Allowable tensile force:

$$A_g = (\tfrac{1}{2})(12) = 6.0 \text{ in.}^2$$
$$A_n = 6.00 - (2)(1.0)(\tfrac{1}{2}) = 5.0 \text{ in.}^2 = A_e$$
$$P = (0.60)(36)(6.0) = 129.6 \text{ k}$$
$$P = (0.50)(58)(5.0) = 145 \text{ k}$$

Bolts in single shear and bearing on $\frac{1}{2}$ in.:

$$P \text{ for single shear} = (0.6)(30)(4) = 72 \text{ k} \leftarrow$$
$$P \text{ for bearing} = (0.5)(\tfrac{7}{8})(1.2 \times 58)(4) = 121.8 \text{ k}$$

$$\underline{\text{Design } P = 72 \text{ k}} \qquad\qquad ■$$

■ Example 11-2

How many $\frac{3}{4}$-in. A325 bolts in standard-size holes with threads excluded from the shear plane are required for the bearing-type connection shown in Fig.

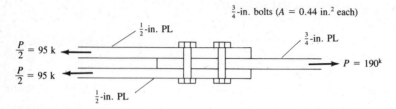

Figure 11-6

11-6? Use A36 steel and assume edge distances and spacing requirements are met, including those for $F_p = 1.2F_u$.

Solution. Bolts in double shear and bearing on $\frac{3}{4}$ in.:

Design shear strength per bolt $= (2)(0.44)(30) = 26.4$ k $\leftarrow$

Design bearing strength per bolt $= (\frac{3}{4})(\frac{3}{4})(1.2 \times 58) = 39.1$ k

Number of bolts required $= \dfrac{190}{26.4} = 7^+$

Use 8 or 9 bolts (depending on arrangement) ∎

Where cover plates are bolted to the flanges of W sections, the bolts must carry the longitudinal shear on the plane between the plates and the flanges. With reference to the cover-plated beam of Fig. 11-7 the unit longitudinal shearing stress to be resisted between a cover plate and the W flange can be determined with the expression $f_v = VQ/Ib$. The total shear force across the flange for a 1-in. length of the beam equals $(b)(1.0)(VQ/Ib) = VQ/I$.

The ASD Specification (E4) provides a maximum permissible spacing for bolts used in the outside plates of built-up members. It equals the thinner outside plate thickness times $127/\sqrt{F_y}$ or 12 in., whichever is smaller.

The spacing of pairs of bolts in Fig. 11-7 can be determined by dividing the design strength of two bolts by the shear force to be taken per inch at a particular section. The theoretical spacings will vary as the external shear varies along the span. Example 11-3 illustrates the calculations involved in determining bolt spacing for a cover-plated beam.

∎ Example 11-3

At a certain section in the cover-plated beam of Fig. 11-7 the external shear V is 190 k. Determine the spacing required for $\frac{7}{8}$-in. A325 bolts used in a bearing-type connection. Assume that all edge and center-to-center distance requirements are met and that the bolt threads are excluded from the shear plane. The steel is A36 and is painted for atmospheric corrosion resistance.

Solution.

$$I = 3630 + (2)(16 \times \tfrac{3}{4})(11.405)^2 = 6752 \text{ in.}^4$$

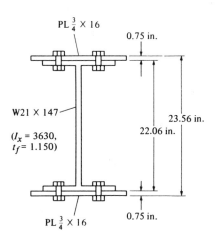

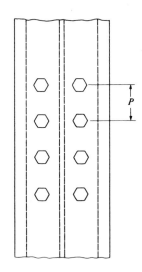

Figure 11-7

Shear to be taken per inch $= \dfrac{VQ}{I}$

$$= \frac{(190)(16 \times \frac{3}{4} \times 11.405)}{6752} = 3.851 \text{ k/in}$$

Bolts in single shear and bearing on 0.75 in.:

Allowable shear for 2 bolts $= (2)(0.6)(30) = 36$ k $\leftarrow$

Allowable bearing for 2 bolts $= (2)(0.75)(\frac{7}{8})(1.2 \times 58) = 91.3$ k

$p = \dfrac{36}{3.851} = 9.35$ in.   (say 9 in. on center)

Maximum $p = (24)(0.75) = 18$ in. but not > 12 in.

Place bolts 9 in. on center   ∎

The assumption has been made that the loads applied to a bearing-type connection are equally divided between the bolts. For this distribution to be correct the plates must be perfectly rigid and the bolts perfectly elastic, but actually the plates being connected are elastic too and have deformations which decidedly affect the bolt stresses. The effect of these deformations is to cause a very complex distribution of load in the elastic range.

Should the plates be assumed to be completely rigid and nondeforming, all bolts would be deformed equally and have equal stresses. This situation is shown in Fig. 11-8(a). Actually the loads resisted by the bolts of a group are probably never equal (in the elastic range) when there are more than two bolts in a line. Should the plates be deformable, the plate stresses and thus the deformations will decrease from the ends of the connection to the middle as shown

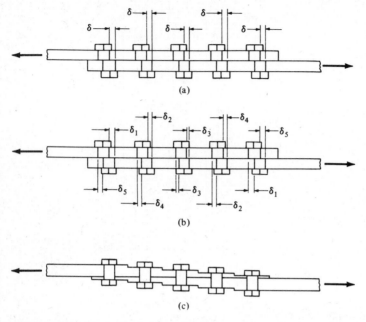

Figure 11-8 (a) Assuming nondeforming plates. (b) Assuming deformable plates. (c) Stepped joint (impractical).

in Fig. 11-8(b). The result is that the highest-stressed elements of the top plate will be over the lowest-stressed elements of the lower plate, and vice versa. The slip will be greatest at the end bolts and smallest at the middle bolts. The bolts at the ends will then have stresses much greater than those in the inside bolts.

The greater the spacing of bolts in a connection the greater will be the variation in bolt stresses due to plate deformation; therefore, the use of compact joints is very desirable, as they will tend to reduce the variation in bolt stresses. It might be interesting to consider a theoretical (although not practical) method of roughly equalizing bolt stresses. The theory would involve the reduction of the thickness of the plate toward its end in proportion to the reduced stresses by stepping. This procedure, which is shown in Fig. 11-8(c), would tend to equalize the deformations of the plate and thus the bolt stresses. A similar procedure would be to scarf the overlapping plates.

The calculation of the theoretically correct elastic stresses in a bolted group based on plate deformations is a tedious problem and is rarely handled in the design office. On the other hand the analysis of a bolted joint based on the plastic theory is a very simple problem. In this theory the end bolts are assumed to be stressed to their yield point. Should the total load on the connection be increased, the end bolts will deform without resisting additional load, the next bolts in the line will have their stresses increased until they too are at the yield point, and so forth. Plastic analysis seems to justify to a certain extent the assumption of rigid plates and equal bolt stresses which is usually

Three high-strength bolted structures in Constitution Plaza Complex, Hartford, Conn., using approximately 195,000 bolts. (Courtesy of Bethlehem Steel Corporation.)

made in design practice. This assumption is used in the examples in this chapter.

When there are only a few bolts in a line, the plastic theory of equal stresses seems to be borne out very well, but when there are a large number of bolts in a line, the situation changes. Tests have clearly shown that the end bolts will fail before the full redistribution takes place.[3]

For load-carrying bolted joints it is common for specifications to require a minimum of two or three fasteners. The feeling is that a single connector may fail to live up to its specified strength because of improper installation, material weakness, and so on, but if several fasteners are used, the effects of one bad fastener in the group will be overcome.

11-14 SLIP-CRITICAL CONNECTIONS—LOADS PASSING THROUGH CENTER OF GRAVITY OF CONNECTIONS

The situations where slip-critical connections are desirable were described in Section 11-5. These connections are particularly useful for fatigue-type situations where members are subjected to constantly fluctuating loads.

---

[3] *Transactions ASCE* 105 (1940), p. 1193.

If bolts are tightened to their required tensions for slip-critical connections (see Table 11-1), there is very little chance of their bearing against the plates which they are connecting. In fact tests show that there is very little chance of slip occurring unless there is a calculated shear of at least 50 percent of the total bolt tension. This means as we have said all along that slip-critical bolts are not stressed in shear; however, the ASD Specification provides allowable shear strengths (they are really permissible friction values on the faying surfaces) so that the designer can handle the connections in just about the same manner he or she uses for bearing-type connections. The ASD Specification assumes the bolts are in "shear" and no bearing, and the allowable shear stresses for high-strength bolts are given in Table 11-6.

The preceding discussion concerning slip-critical joints does not present the whole story, because during erection the joints may be assembled with bolts, and as the members are erected their weights will often push the bolts against the side of the holes before they are tightened and put them in some bearing and shear.

Example 11-4 presents the design of a slip-critical connection for a lap joint.

Bridge over Allegheny River at Kittaning, Pa. (Courtesy of American Bridge Company.)

■ Example 11-4

It is desired to design a slip-critical connection for the plates shown in Fig. 11-9 to resist an 80-k load using 1-in. A325 high-strength bolts with threads excluded from the shear plane and all edge and center-to-center distance requirements met.

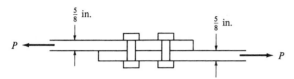

Figure 11-9

Solution. Slip-critical design, bolts in "single shear" and no bearing:

$$\text{"Single shear"} = (A_{bolt})(17.0) = (0.785)(17.0) = 13.35 \text{ k}$$

$$\text{Number of bolts required} = \frac{80}{13.35} = 5.99$$

<u>Use 6 bolts</u> ■

■ Example 11-5

The connection shown in Fig. 11-10 is made with $\frac{7}{8}$-in. A325 bearing-type bolts in standard-size holes with the threads excluded from the shear planes.

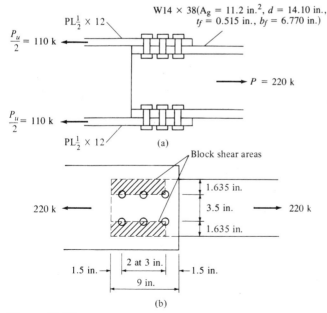

Figure 11-10

The beam and gusset plates consist of A36 steel. Check the following items: (a) the tensile strengths of the W section and the gusset plates, (b) the strength of the bolts in single shear and bearing, and (c) the block shear strength of the W section crosshatched areas shown in Fig. 11-10(b).

Solution.

(a) Tensile design strength of W section:

$$P = 0.6F_y A_g = (0.6)(36)(11.2) = 241.9 \text{ k} > 220 \text{ k} \qquad \text{OK}$$
$$A_n = 11.2 - (4)(1)(0.515) = 9.14 \text{ in.}^2$$
$$U = 0.85 \text{ since } b_f < \tfrac{2}{3}d$$
$$A_e = (0.85)(9.14) = 7.77 \text{ in.}^2$$
$$P = 0.5F_u A_e = (0.5)(58)(7.77) = 225.3 \text{ k} > 220 \text{ k} \qquad \text{OK}$$

Tensile design strength of gusset plates:

$$P = (0.6)(36)(2)(\tfrac{1}{2} \times 12) = 259.2 \text{ k} > 220 \text{ k} \qquad \text{OK}$$
$$A_n \text{ of 2 plates} = [(\tfrac{1}{2})(12) - (2)(1)(\tfrac{1}{2})]2 = 10 \text{ in.}^2$$
$$0.85 \, A_g = (0.85)(\tfrac{1}{2})(12)(2) = 10.2 \text{ in.}^2$$
$$P = (0.50)(58)(10.0) = 290 \text{ k} > 220 \text{ k} \qquad \text{OK}$$

<u>Tensile design strength of W section and plates is satisfactory</u>

(b) Bolts in single shear and bearing on $\tfrac{1}{2}$ in.:

Single shear design strength of bolts

$$= (0.6)(30)(12) = 216 \text{ k} < 220 \text{ k} \qquad \text{NG}$$

Bearing design strength of bolts

$$= (0.50)(\tfrac{7}{8})(1.2 \times 58)(12) = 365.4 \text{ k} > 220 \text{ k} \qquad \text{OK}$$

<u>Bolt strength is satisfactory</u>

(c) Block shear strength for the W section:

Block shear

$$= 0.30A_v F_u + 0.50A_t F_u$$
$$= [(0.30)(7.5 - 2\tfrac{1}{2} \times 1)(\tfrac{1}{2})(58) + (0.50)(1.635 - \tfrac{1}{2} \times 1)(\tfrac{1}{2})(58)]4$$
$$= 239.8 \text{ k} > 220 \text{ k} \qquad \text{NG}$$

<u>Block shear strength of W section is not satisfactory</u>   ■

The calculations for riveted connections and for connections made with A307 common bolts are made almost exactly as are the ones for high-strength bolts in bearing-type connections. The only differences are that the shear strength values for these connections are much smaller and the ASD Specification does not permit the design of slip-critical joints using rivets or common bolts. Rivet and common bolt examples are presented in Chapter 12.

PROBLEMS

Unless otherwise indicated, standard-size holes are used and spacing and edge distance requirements of the ASD Specification are satisfied so that $F_p = 1.2F_u$.

11-1 to 11-5   Determine the allowable tensile capacity for the member and the connection shown, assuming a bearing-type connection.

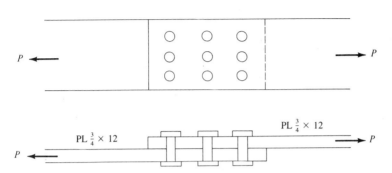

11-1 A36 steel, $\frac{7}{8}$-in. A325 bolts, threads excluded from shear plane. (*Ans.* 162.4 k)

11-2 A36 steel, $\frac{3}{4}$-in. A325 bolts, threads excluded from shear plane.

11-3 A36 steel, 1-in. A325 bolts, threads not excluded from shear plane. (*Ans.* 148.4 k)

11-4 A572, grade 50 steel; 1-in. A490 bolts; threads excluded from shear plane.

11-5 A572, grade 50 steel; $\frac{3}{4}$-in. A490 bolts; threads not excluded from shear plane. (*Ans.* 111.3 k)

11-6 to 11-10   Determine the allowable tensile capacity of the member and the connection shown, assuming a bearing-type connection.

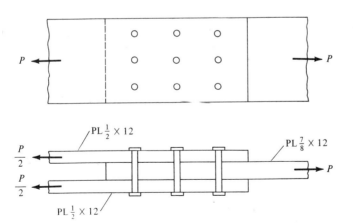

11-6 A36 steel, $\frac{7}{8}$-in. A325 bolts, threads not excluded from shear plane.

11-7 A36 steel, $\frac{7}{8}$-in. A325 bolts, threads excluded from shear plane. (*Ans.* 226.8 k)

11-8 A36 steel, $\frac{3}{4}$-in. A325 bolts, threads not excluded from shear plane.

11-9 $F_y = 50$ ksi, $F_u = 70$ ksi, 1-in. A490 bolts, threads excluded from shear plane. (*Ans.* 264.2 k)

11-10 $F_y = 50$ ksi, $F_u = 70$ ksi, $\frac{3}{4}$-in. A490 bolts, threads excluded from shear plane.

11-11 to 11-13   How many bolts are required for the bearing-type connection shown?

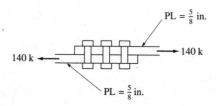

11-11 A36 steel, $\frac{3}{4}$-in. A325 bolts, threads excluded from shear plane. (*Ans.* 10.6, say 12)

11-12 $F_y = 50$ ksi, $F_u = 70$ ksi, $\frac{7}{8}$-in. A325 bolts, threads excluded from shear plane.

11-13 A36 steel, 1-in. A325 bolts, threads not excluded from shear plane. (*Ans.* 8.5, say 9 or 10)

11-14 to 11-16   How many bolts are required for the bearing-type connection shown?

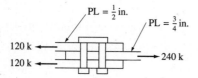

11-14 A36 steel, $\frac{7}{8}$-in. A325 bolts, threads excluded from shear plane.

11-15 A36 steel, $\frac{3}{4}$-in. A325 bolts, threads not excluded from shear plane. (*Ans.* 12.93, say 14 or 15)

11-16 $F_y = 50$ ksi, $F_u = 65$ ksi, $\frac{7}{8}$-in. A325 bolts, threads excluded from shear plane.

11-17 How many $\frac{3}{4}$-in. A325 bolts (threads excluded from the shear plane) in a bearing-type connection are required to develop the full capacity of the member shown? Assume A36 steel is used and there are two lines of bolts in each flange (at least three in a line). (*Ans.* 36.35, say 36 or 40)

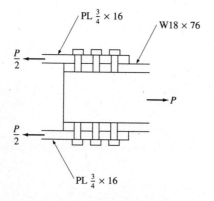

11-18 Repeat Prob. 11-17 using $\frac{7}{8}$-in. A490 bolts and A572, grade 50, steel.

11-19 For the beam shown what is the required spacing of $\frac{7}{8}$-in. A325 bolts in a bearing-type connection at a section where the external shear is 205 k? Use A36 steel. Threads excluded from shear plane. (*Ans.* 7.58 in., say $7\frac{1}{2}$ in.)

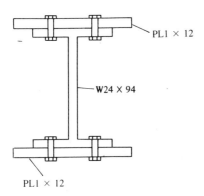

11-20 The cover-plated section shown is used to support a 14.3 k/ft uniform load (including beam weight) for an 18-ft simple span. If $\frac{7}{8}$-in. A325 bearing-type bolts are used as shown, work out a spacing diagram for the entire span. Assume the cover plates extend for the full span. Use A36 steel. Threads excluded from shear plane.

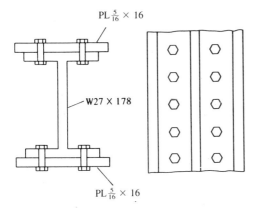

11-21 For the section shown, determine the required spacing of $\frac{3}{4}$-in. A490 bearing-type bolts if the member consists of A572, grade 60, steel. $V = 215$ k. Threads excluded from shear plane. (*Ans.* 9.17 in. use 7 in. maximum spacing as required by ASD J3.10.)

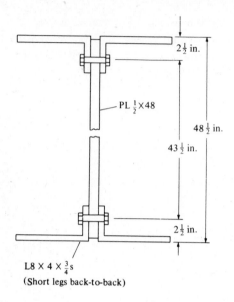

PL $\frac{1}{2}$×48

$2\frac{1}{2}$ in.

$48\frac{1}{2}$ in.

$43\frac{1}{2}$ in.

$2\frac{1}{2}$ in.

L8 × 4 × $\frac{3}{4}$ s
(Short legs back-to-back)

11-22 For an external shear of 380 k determine the spacing required for $\frac{3}{4}$-in. A325 web bolts used in a bearing-type connection for the built-up section shown. Use A36 steel. Threads excluded from shear plane.

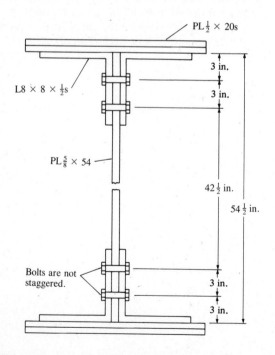

PL $\frac{1}{2}$ × 20s

3 in.

3 in.

L8 × 8 × $\frac{1}{2}$ s

PL $\frac{5}{8}$ × 54

$42\frac{1}{2}$ in.

$54\frac{1}{2}$ in.

Bolts are not staggered.

3 in.

3 in.

11-23 to 11-30 Repeat these problems, using slip-critical connections.

11-23 Prob. 11-1. (*Ans.* 92.0 k)

11-24 Prob. 11-7.

11-25 Prob. 11-9. (*Ans.* 264.2 k)

11-26 Prob. 11-11.

11-27 Prob. 11-13. (*Ans.* 10.49, say 12)

11-28 Prob. 11-17.

11-29 Prob. 11-19. (*Ans.* 4.29 in., say $4\frac{1}{4}$ in.)

11-30 Prob. 11-20.

11-31 The truss member shown consists of two C12 × 25s (A36 steel) connected to a 1-in. gusset plate. How many $\frac{7}{8}$-in. A325 bolts are required to develop the full tensile capacity of the member when it is used (a) as a slip-critical connection and (b) as a bearing-type connection? Threads excluded from shear plane. (*Ans.* (a) 15.53, say 16; (b) 8.80, say 10)

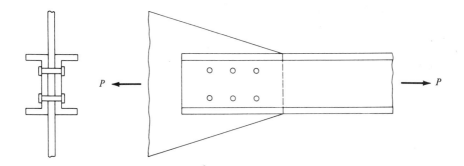

11-32 Rework Prob. 11-31 using 1-in. A490 bolts and A242 steel.

11-33 For the bearing-type connection shown, *P* is 500 k. Determine the number of 1-in. A325 bolts (threads excluded from shear plane) required. Use A36 steel. (*Ans.* 10.61, say 12)

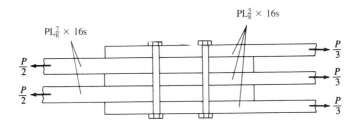

11-34 Repeat Prob. 11-33 using a slip-critical connection.

11-35 Repeat Prob. 11-33 using $\frac{7}{8}$-in. A490 bolts and A572, grade 50, steel. (*Ans.* 10.40, say 12)

11-36 Determine the allowable load that the connection shown can support if A36 steel and eight $\frac{7}{8}$-in. A325 bearing-type bolts (threads excluded from shear plane) are used in each flange. Include block shear in your calculations.

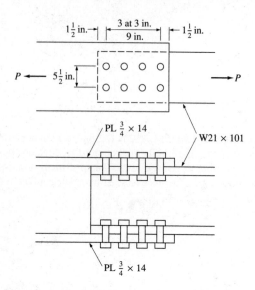

11-37 Repeat Prob. 11-36 using A572, grade 50 steel and 1-in. A325 bearing-type bolts. (*Ans.* 377 k)

12

Bolted Connections Continued and Historical Notes on Rivets

12-1 BOLTS SUBJECTED TO ECCENTRIC SHEAR

Eccentrically loaded bolt groups are subjected to shears and bending moments. The student may feel such situations are rare, but the truth is that they are much more common than he or she suspects. For instance, in a truss it is desirable to have the center of gravity of a member lined up exactly with the center of gravity of the bolts at its end connections. This feat is not quite as easy to accomplish as it may seem, and connections are often subjected to moments.

Eccentricity is quite obvious in Fig. 12-1(a), where a beam is connected to a column with a plate. In part (b) another beam is connected to a column

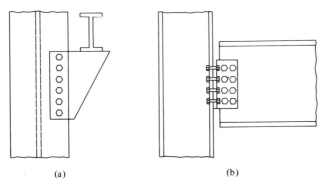

(a)   (b)

Figure 12-1 Eccentrically loaded connections.

with a pair of web angles. It is obvious that this connection must resist some moment because the center of gravity of the load from the beam does not coincide with the reaction from the column.

In general, specifications for bolts, rivets, and welds clearly state that the center of gravity of the connection shall coincide with the center of gravity of the member unless the eccentricity is accounted for in the calculations. However, Section J1.9 of the ASD Specification provides some exceptions to this rule. It states that the rule is not applicable to the end connections of statically loaded single angles, double angles, and similar members. In other words, the eccentricities between the centers of gravity of these members and the centers of gravity of the connections may be ignored unless fatigue loadings are involved. *Furthermore, the eccentricity between the gravity axes and the gage lines of bolted or riveted members may be neglected for statically loaded members.*

The ASD Specification presents values for computing the design strengths of individual bolts or rivets but does not specify a method for computing the forces on these fasteners when they are eccentrically loaded. As a result the method of analysis to be used is left up to the designer.

Three general approaches for the analysis of eccentrically loaded connections have been developed through the years. The first of the methods is the very conservative *elastic method* in which friction or slip resistance between the connected parts is neglected. In addition these connected parts are assumed to be perfectly rigid. This type of analysis has been commonly used since at least 1870.[1,2]

Tests have shown that the elastic method usually provides very conservative results. As a consequence, various *reduced* or *effective eccentricity methods* have been proposed.[3] The analysis is handled just as it is in the elastic method, except that smaller eccentricities and thus smaller moments are used in the calculations.

The third method, the *ultimate strength method,* provides the most realistic values as compared with test results but is extremely tedious to apply, at least with hand-held calculators. The tables in the ASD Manual for eccentrically loaded connections are based on the ultimate strength method and enable us to solve most of these types of problems quite easily as long as the bolt and rivet patterns are symmetrical. The remainder of this section is devoted to these three analysis methods.

---

[1] W. McGuire, *Steel Structures* (Englewood Cliffs, N.J.: Prentice Hall, 1968), p. 813.

[2] C. Reilly, "Studies of Iron Girder Bridges," *Proc. Inst. Civil Engrs.* 29 (London, 1870).

[3] T. R. Higgins, "New Formulas for Fasteners Loaded Off Center," *Engineering News-Record* (May 21, 1964).

Elastic Analysis

For this discussion the bolts of Fig. 12-2(a) are assumed to be subjected to a load P which has an eccentricity of e from the center of gravity (c.g.) of the bolt group. To consider the force situation in the bolts, an upward and down-ward force, each equal to P, is assumed to act at the center of gravity of the bolt group. This situation, shown in part (b), in no way changes the bolt forces. The force in a particular bolt will, therefore, equal P divided by the number of bolts in the group as seen in part (c), plus the force due to the moment caused by the couple shown in part (d).

The magnitude of the forces in the bolts due to the moment Pe will now be considered. The distances of each bolt from the center of gravity of the group are represented by the values d_1, d_2, and so on, in Fig. 12-3. The moment produced by the couple is assumed to cause the plate to rotate about the center of gravity of the bolt connection, with the amount of rotation or strain at a par-ticular bolt being proportional to its distance from the center of gravity. (For this derivation the gusset plates are again assumed to be perfectly rigid and the bolts are assumed to be perfectly elastic.) Rotation is greatest at the bolt that is the greatest distance from the center of gravity, as will be the stress since stress is proportional to strain in the elastic range.

The rotation is assumed to produce forces of r_1, r_2, r_3, and r_4 on the bolts in the figure. The moment transferred to the bolts must be balanced by resist-ing moments of the bolts as follows:

$$M_{c.g.} = Pe = r_1d_1 + r_2d_2 + r_3d_3 + r_4d_4 \tag{1}$$

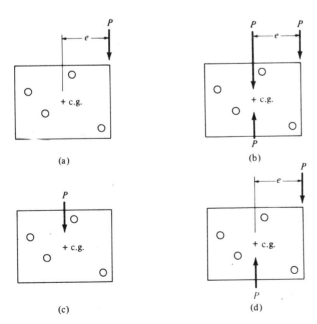

(a)

(b)

(c)

(d)

Figure 12-2

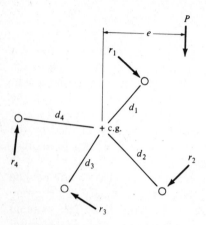

Figure 12-3

As the force on each bolt is assumed to be directly proportional to the distance from the center of gravity, the expression can be written

$$\frac{r_1}{d_1} = \frac{r_2}{d_2} = \frac{r_3}{d_3} = \frac{r_4}{d_4}$$

and writing each r value in terms of r_1 and d_1,

$$r_1 = \frac{r_1 d_1}{d_1} \qquad r_2 = \frac{r_1 d_2}{d_1} \qquad r_3 = \frac{r_1 d_3}{d_1} \qquad r_4 = \frac{r_1 d_4}{d_1}$$

Substituting these values into equation (1) and simplifying.

$$M = \frac{r_1 d_1^2}{d_1} + \frac{r_1 d_2^2}{d_1} + \frac{r_1 d_3^2}{d_1} + \frac{r_1 d_4^2}{d_1}$$

$$= \frac{r_1}{d_1}(d_1^2 + d_2^2 + d_3^2 + d_4^2)$$

Therefore,

$$M = \frac{r_1 \Sigma d^2}{d_1}$$

The force on each bolt can now be written

$$r_1 = \frac{M d_1}{\Sigma d^2} \qquad r_2 = \frac{d_2}{d_1} r_1 = \frac{M d_2}{\Sigma d^2} \qquad r_3 = \frac{M d_3}{\Sigma d^2} \qquad r_4 = \frac{M d_4}{\Sigma d^2}$$

Each value of r is perpendicular to the line drawn from the center of gravity to the particular bolt. It is usually more convenient to break these down into vertical and horizontal components. (see Fig 12-4.)

The horizontal and vertical components of the distance d_1 are represented by h and v respectively, and the horizontal and vertical components of force r_1

Bridge over New River Gorge in Fayette County, W. Va. near Charleston. (Courtesy of American Bridge Company.)

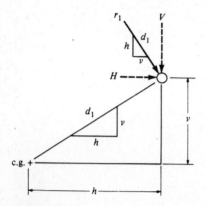

Figure 12-4

are represented by H and V respectively in this figure. It is now possible to write the ratio from which H can be obtained:

$$\frac{r_1}{d_1} = \frac{H}{v}$$

$$H = \frac{r_1 v}{d_1} = \left(\frac{Md_1}{\Sigma d^2}\right)\left(\frac{v}{d_1}\right)$$

Therefore,

$$H = \frac{Mv}{\Sigma d^2}$$

By a similar procedure V is found to equal

$$V = \frac{Mh}{\Sigma d^2}$$

■ Example 12-1

Determine the force in the most stressed bolt of the group shown in Fig. 12-5, using the elastic analysis method.

Solution. Each bolt and the forces applied to it by the direct load and the clockwise moment are shown in Fig. 12-6. From this sketch the student will see that the upper right-hand bolt and the lower right-hand bolt are the most stressed and have equal stresses.

$$e = 6 + 1.5 = 7.5 \text{ in.}$$
$$M = Pe = (30)(7.5) = 225 \text{ in.-k}$$
$$\Sigma d^2 = \Sigma h^2 + \Sigma v^2$$
$$\Sigma d^2 = (8)(1.5)^2 + (4)(1.5^2 + 4.5^2) = 108$$

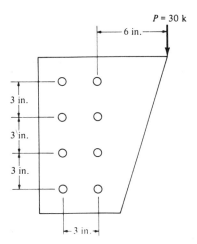

Figure 12-5

$$H = \frac{Mv}{\Sigma d^2} = \frac{(225)(4.5)}{108} = 9.38 \text{ k}$$

$$V = \frac{Mh}{\Sigma d^2} = \frac{(225)(1.5)}{108} = 3.13 \text{ k}$$

$$\frac{P}{8} = \frac{30}{8} = 3.75 \text{ k}$$

$$\downarrow 3.13 \text{ k}$$
$$\downarrow 3.75 \text{ k}$$
$$0 \leftarrow 9.38 \text{ k}$$

Force in the lower right-hand bolt (equal to force in the upper right-hand bolt) is determined as follows:

$$r = \sqrt{(6.88)^2 + (9.38)^2} = \underline{11.63 \text{ k}} \qquad \blacksquare$$

Should the eccentric load be inclined, it can be broken down into vertical and horizontal components and the moment of each about the center of gravity of the bolt group determined. Various design formulas can be developed which will enable the engineer to directly design eccentric connections, but in all probability the process of assuming a certain number and arrangement of bolts, checking stresses, and redesigning is probably just as satisfactory.

The trouble with this inaccurate but very conservative method of analysis is that we in effect are assuming that there is a linear relation between loads and deformations in the fasteners and further that their yield stress is not exceeded when the ultimate load on the connection is reached. Various experiments have shown that these assumptions are incorrect.

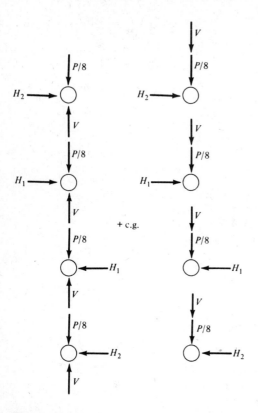

Figure 12-6

Reduced Eccentricity Method

The elastic analysis method just described appreciably overestimates the moment forces applied to the connectors. As a result, quite a few proposals have been made through the years which make use of an effective eccentricity, thus in effect taking into account the slip resistance on the faying or contact surfaces. One set of reduced eccentricity values which were fairly common at one time follow:

1. With one gage line of fasteners and where n is the number of fasteners in the line,

$$e_{\text{effective}} = e_{\text{actual}} - \frac{1 + 2n}{4}$$

2. With two or more gage lines of fasteners symmetrically placed and where n is the number of fasteners in each line.

$$e_{\text{effective}} = e_{\text{actual}} - \frac{1 + n}{2}$$

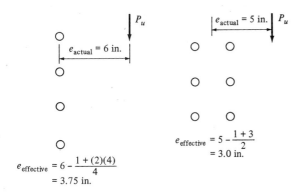

Figure 12-7

The reduced eccentricity values for two fastener arrangements are shown in Fig. 12-7.

To analyze a particular connection with the reduced eccentricity method, the value of $e_{effective}$ is computed as described above and used to compute the eccentric moment. Then the elastic procedure just described is used for the remainder of the calculations. For some connections this reduced eccentricity method is thought to provide safety factors that are too small.

Ultimate Strength Method

The elastic and reduced eccentricity methods for analyzing eccentrically loaded fastener groups are both based upon the assumption that the behavior of the fasteners is elastic. A much more realistic method of analysis is the ultimate strength method, which is described in the next few paragraphs. The values given in the tables of Part 4 of the ASD Manual for eccentrically loaded fastener groups were computed using this method.

If one of the outermost bolts or rivets in an eccentrically loaded connection begins to slip or yield, the connection will not fail. Instead the magnitude of the eccentric load may be increased, the inner bolts will resist more load, and failure will not occur until all of the bolts slip or yield.

The eccentric load tends to cause both a relative rotation and translation of the connected material. In effect this is equivalent to pure rotation of the connection about a single point called the *instantaneous center of rotation*. An eccentrically loaded bolted connection is shown in Fig. 12-8, and the instantaneous center is represented by point O. It is located a distance r_0 from the center of gravity of the bolt group.

The deformations of these bolts are assumed to vary in proportion to their distances from the instantaneous center. The ultimate shear force which one of them can resist is not equal to the pure shear force which a bolt can resist. Rather it is dependent upon the load-deformation relationship in the bolt. Stud-

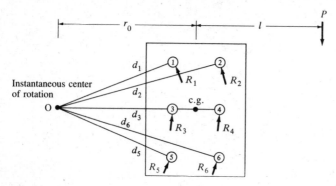

Figure 12-8

ies by Crawford and Kulak[4] have shown that this force may be closely esti-mated with the following expression.

$$R = R_{ult}(1 - e^{-10\Delta})^{0.55}$$

In this formula R_{ult} is the ultimate shear load for a single fastener equaling an experimentally obtained 74 k for a $\frac{3}{4}$-in. A325 bolt, e is the base of the natural logarithm (2.718), and Δ is equal to the total deformation of a bolt experimentally determined as equal to 0.34 in. The coefficients 10.0 and 0.55 were also experimentally obtained. Fig. 12-9 illustrates this load-deformation relationship.

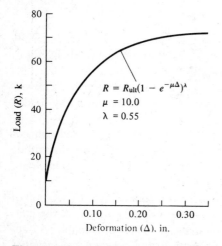

Figure 12-9 Ultimate shear force R in a single bolt at any given deformation.

[4] S. F. Crawford and G. L. Kulak, "Eccentrically Loaded Bolt Connections." *Journal of Structural Division*, ASCE, 97, ST3 (March 1971), pp. 765–783.

This expression clearly shows that the ultimate shear load taken by a particular bolt in an eccentrically loaded connection is affected by its deformation. Thus, the load applied to a particular bolt is dependent upon its position in the connection with respect to the instantaneous center of rotation.

The resisting forces of the bolts of the connection of Fig. 12-8 are represented by R_1, R_2, R_3, and so on. Each of these forces is assumed to act in a direction perpendicular to a line drawn from point 0 to the center of the bolt in question. For this symmetrical connection the instantaneous center of rotation will fall somewhere on a horizontal line through the center of gravity of the bolt group. This is the case because the sum of the horizontal components of the R forces must be zero as also must be the sum of the moments of the horizontal components about point O. The position of point O on the horizontal line may be found by a tedious trial-and-error procedure to be described here.

With reference to Fig. 12-8 the moment of the eccentric load about point O must be equal to the summation of the moments of the R resisting forces about the same point. If we knew the location of the instantaneous center, we could compute R values for the bolts with the Crawford-Kulak formula and determine the load P_u as follows:

$$P_u(r_0 + l) = \Sigma Rd$$

$$P_u = \frac{\Sigma Rd}{r_0 + l}$$

To determine the allowable load P that such a connection can resist according to the ASD Specification, we replace R_{ult} in the Crawford-Kulak formula with the allowable load on one bolt. For instance, if we have $\frac{7}{8}$-in. A325 bolts (threads excluded) in single shear bearing on a sufficient thickness so bearing does not control, R_u will equal $(0.6)(30) = 18$ k.

The location of the instantaneous center is not known, however. Its position is estimated, the R values determined, and P calculated as described. It will be noted that P must be equal to the summation of the vertical components of the R resisting forces (ΣR_v). If the value is computed and equals the P computed by the formula above, we have the correct location for the instantaneous center. If not we try another location, and so on.

In Example 12-2 the author demonstrates the lengthy trial-and-error calculations necessary to locate the instantaneous center of rotation for a symmetrical connection consisting of four bolts. In addition the allowable design strength P of the connection is determined.

To solve such a problem it is very convenient to set the calculations up in a table similar to the one used in the solution to follow. In the table shown the h and v values given are the horizontal and vertical components of the d distances from point O to the center of gravities of the bolts. The bolt which is located at the greatest distance from point O is assumed to have a Δ value of 0.34 in. The Δ values for the other bolts are assumed to be proportionate to their distances from point O. The Δ values so determined are used in the R formula.

A set of tables entitled "Eccentric Loads on Fastener Groups" is presented in Part 4 of the ASD Manual. The values in these tables were determined by the procedure described here. A large percentage of the practical cases that the designer will encounter are included in the tables. It will be noted that the Manual presents a procedure for handling inclined loads as well as vertical ones. Should some other situation not covered be faced, the designer may very well decide to use the more conservative elastic procedure previously described.

■ Example 12-2

The bearing-type $\frac{7}{8}$-in. A325 bolts of the connection of Fig. 12-10 have an allowable strength $= (0.6)(30) = 18$ k. Locate the instantaneous center of rotation of the connection, using the trial-and-error procedure, and determine the value of P.

Solution.   By trial and error.
     Try $r_0 = 3$ in., reference being made to Fig. 12-11

BOLT NO.	h (in.)	v (in.)	d (in.)	Δ (in.)	R (k)	R_v (k)	Rd (in.-k)
1	1.5	3	3.3541	0.211	16.76	7.50	56.21
2	4.5	3	5.4083	0.34	17.67	14.70	95.56
3	1.5	3	3.3541	0.211	16.76	7.50	56.21
4	4.5	3	5.4083	0.34	17.67	14.70	95.56
						$\Sigma = 44.40$	$\Sigma = 303.54$

$$P = \frac{\Sigma Rd}{r_0 + l} = \frac{303.54}{3 + 5} = 37.94 \text{ k} \quad \text{not} = 44.40 \text{ k} \qquad \text{NG}$$

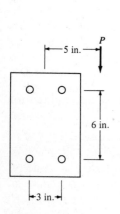

Figure 12-10

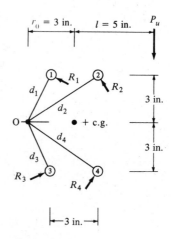

Figure 12-11

After several trials assume $r_0 = 2.40$ in.

BOLT NO.	h (in.)	v (in.)	d (in.)	Δ (in.)	R (k)	R_v (k)	R_d (in.-k)
1	0.90	3	3.1321	0.216	16.82	4.83	52.68
2	3.90	3	4.9204	0.34	17.67	14.01	86.94
3	0.90	3	3.1321	0.216	16.82	4.83	52.68
4	3.90	3	4.9204	0.34	17.67	14.01	86.94
						$\Sigma = 37.68$	$\Sigma = 279.24$

$$P = \frac{279.24}{2.4 + 5} = 37.74 \text{ k almost} = 37.68 \text{ k} \qquad \text{OK}$$

$$P = \underline{\underline{37.7 \text{ k}}} \qquad \blacksquare$$

Although the development of this method of analysis was actually based on bearing-type connections where slip may occur, both theory and load tests have shown that it may conservatively be applied to slip-critical connections.[5]

The method may be expanded to include inclined loads and unsymmetrical bolt arrangements, but the trial-and-error calculations are extraordinarily long for such situations.

Examples 12-3 and 12-4 which follow provide illustrations of the use of the tables of the ASD Manual for both analysis and design of eccentrically loaded bolted connections.

■ Example 12-3
Repeat Example 12-2 using the tables entitled "Eccentric Loads on Fastener Groups" in Part 4 of the ASD Manual.

Solution.

$P = C \times r_v$

$C = 2.10$ from Table XII with $b = 6$ in., $n = 2$, and $l = 5$ in.

$P = (2.10)(0.6 \times 30) = \underline{\underline{37.8 \text{ k}}}$ \qquad OK \quad ■

■ Example 12-4
Determine the number of $\frac{7}{8}$-in. A325 bolts in standard-size holes required for the connection shown in Fig. 12-12. Use A36 steel and assume the connection is to be a bearing type with threads excluded from the shear plane. Further assume that the bolts are in single shear and bearing on $\frac{1}{2}$ in. Use the ASD Manual tables.

---

[5] G. L. Kulak, "Eccentrically Loaded Slip-Resistant Connections," *Engineering Journal*, AISC, 12, no. 2 (2d quarter, 1975), pp. 52–55.

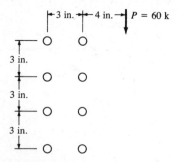

Figure 12-12

Solution.   From Table XII of Part 4 of the Manual, $l = 5.5$ in.
Bolts in single shear and bearing on $\frac{1}{2}$ in.

$$r_v = \text{allowable shear load per fastener}$$
$$= (0.6)(30) = 18 \text{ k} \leftarrow$$
$$r_v = \text{allowable bearing load per fastener}$$
$$= (\tfrac{7}{8})(\tfrac{1}{2})(1.2 \times 58) = 30.4 \text{ k}$$
$$C = \frac{P}{r_v} = \frac{60}{18} = 3.33$$

<u>By interpolation in the table we need 4 bolts in each line</u>

Note: If the situation faced by the designer does not fit the tables given in the ASD Manual, it is recommended that he or she use the conservative elastic procedure to handle the analysis or design problem.   ∎

12-2 BOLTS SUBJECTED TO SHEAR AND TENSION

The bolts used for a good many structural steel connections are subjected to a combination of shear and tension. One obvious case is shown in Fig. 12-13 where a diagonal brace is attached to a column. The 76.03-k component shown in the figure is trying to shear the bolts off at the face of the column, while the 152.05-k component is tending to pull off their heads.

Tests on bearing-type bolts subject to combined shear and tension show that their ultimate strengths can be represented with an elliptical interaction curve as shown in Fig. 12-14. A safety factor has been divided into the values to obtain the curve shown. Particularly note the values F_t and F_v. F_t is the limiting tensile stress if there is no shear, and F_v is the limiting shearing stress if there is no externally applied tension.

The three straight dashed lines shown in Fig. 12-14 very closely represent the test-result interaction curve. Equations for these lines are given in Table 12-1 (Table J3.3 in ASD Specification).

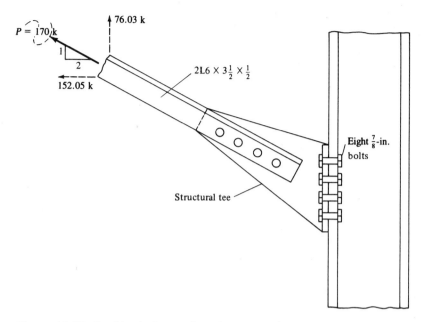

Figure 12-13 Combined shear and tension connection.

Example 12-5 illustrates the calculations involved in checking a high-strength bearing-type bolted connection for shear and tension. (There is another stress situation which may affect this connection. It is called *prying action*, but as it is not discussed until Section 12-4, it is not considered in this example.)

■ Example 12-5

The tension member in Fig. 12-13 is connected to the column shown with eight $\frac{7}{8}$-in. A325 high-strength bolts in a bearing-type connection with the

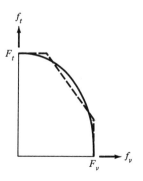

Figure 12-14 Bolts in a bearing-type connection subject to combined shear and tension.

TABLE 12-1 ALLOWABLE TENSION STRESS F_t FOR FASTENERS IN BEARING-TYPE CONNECTIONS

DESCRIPTION OF FASTENERS	THREADS INCLUDED IN SHEAR PLANE	THREADS EXCLUDED FROM SHEAR PLANE
A307 bolts	\multicolumn{2}{c}{$26 - 1.8f_v \leq 20$}	
A325 bolts	$\sqrt{(44)^2 - 4.39f_v^2}$	$\sqrt{(44)^2 - 2.15f_v^2}$
A490 bolts	$\sqrt{(54)^2 - 3.75f_v^2}$	$\sqrt{(54)^2 - 1.82f_v^2}$
Threaded parts, A449 bolts over $1\frac{1}{2}$-in. diameter	$0.43F_u - 1.8f_v \leq 0.33F_u$	$0.43F_u - 1.4f_v \leq 0.33F_u$
A502, gr. 1, rivets	$30 - 1.3f_v \leq 23$	
A502, gr. 2, rivets	$38 - 1.3f_v \leq 29$	

Source: American Institute of Steel Construction, *Manual of Steel Construction Allowable Stress Design,* 9th ed. (Chicago: AISC, 1989), Table J3.3, p. 5-74. Reprinted with the permission of AISC.

threads excluded from the shear plane and standard-size holes. Is this a sufficient number of bolts to resist the applied load according to the ASD Specification?

Solution.

$$\text{Shearing stress } f_v = \frac{76.03}{(8)(0.6)} = 15.84 \text{ ksi}$$

$$\text{Tension stress } f_t = \frac{152.05}{(8)(0.6)} = 31.68 \text{ ksi}$$

$$\text{Allowable tension stress } F_t = \sqrt{(44)^2 - 2.15(15.84)^2}$$

$$37.37 \text{ ksi} > 31.68 \text{ ksi} \qquad\qquad \text{OK}$$

<u>Connection is satisfactory</u> ∎

When an axial tension force is applied to a slip-critical connection, the clamping force will be reduced and the allowable shear strength must be decreased in some proportion to the loss in clamping or prestress. This is accomplished in the ASD Specification (J3.6) by requiring that the allowable slip-critical stress given in Table 11-6 (ASD Table J3.2) be multiplied by the reduction factor $(1 - f_t A_b/T_b)$ in which $f_t A_b$ is the computed total tensile force in one bolt and T_b is the minimum pre-tension load for one bolt in a slip-critical connection as given in Table 11.1 (ASD Table J3.7).

450 Lexington Ave., New York City. (Courtesy of Owen Steel Company, Inc.)

For bolts subject to combined shear and tension which are used in slip-critical connections with standard holes the allowable shear stresses are to be determined as follows:

$$\text{For A325 bolts} \leq \left(1 - \frac{f_t A_b}{T_b}\right) 17.0$$

$$\text{For A490 bolts} \leq \left(1 - \frac{f_t A_b}{T_b}\right) 21.0$$

■ Example 12-6

Rework Example 12-5 with P reduced to 80 k and a slip-critical connection.

Solution.

$$P_v = \left(\frac{1}{\sqrt{5}}\right)(80) = 35.78 \text{ k}$$

$$P_h = \left(\frac{2}{\sqrt{5}}\right)(80) = 71.55 \text{ k}$$

$$f_v = \frac{35.78}{(8)(0.6)} = 7.45 \text{ ksi}$$

$$f_t = \frac{71.55}{(8)(0.6)} = 14.91 \text{ ksi}$$

Allowable shear strength $= \left(1 - \dfrac{f_t A_b}{T_b}\right)17$

$$= \left(1 - \dfrac{14.91 \times 0.6}{39}\right)17 = 13.10 \text{ ksi} < 14.91 \text{ ksi} \quad \text{NG}$$

Connection is unsatisfactory if slip-critical ■

12-3 TENSION LOADS ON BOLTED JOINTS

Bolted and riveted connections subjected to pure tensile loads have been avoided as much as possible in the past by designers. The use of tensile connections was probably "forced on" them more often for wind-bracing systems in tall buildings than for any other situation. There are some other locations where they have been used, however, such as hanger connections for bridges, flange connections for piping systems, and so on. Figure 12-15 shows a hanger connection with an applied tensile load.

Hot-driven rivets and fully tensioned high-strength bolts are not free to shorten, with the result that large tensile forces are produced in them during their installation. These initial tensions are actually close to the yield points. There has always been considerable reluctance among designers to apply tensile loads to connectors of this type for fear that the external loads might easily increase their already present tensile stresses and cause them to fail. The truth of the matter, however, is that when external tensile loads are applied to connections of this type the connectors will probably experience little if any change in stress.

Hot-driven rivets which have cooled and shrunk or fully tensioned high-strength bolts actually prestress the joints in which they are used against tensile loads. (To follow this discussion the student may like to think of a prestressed concrete beam which has external compressive loads applied at each end.) The tensile stresses in the connectors squeeze together the members being con-

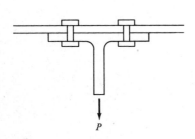

Figure 12-15 Hanger connection.

nected. If a tensile load is applied to this connection at the contact surface, it cannot exert any additional load on the bolts or rivets until the members are pulled apart and additional strains put on the bolts or rivets. The members cannot be pulled apart until a load is applied which is larger than the total tension in the connectors of the connection. This statement means that the joint is prestressed against tensile forces by the amount of stress initially put in the shanks of the connectors.

Another way of saying this is that, if a tensile load P is applied at the contact surface, it tends to reduce the thickness of the plates somewhat, but at the same time the contact pressure between the plates will be correspondingly reduced and the plates will tend to expand by the same amount. The theoretical result then is no change in plate thickness and no change in connector tension. This situation continues until P equals the connector tension. At this time an increase in P will result in separation of the plates, and thereafter the tension in the connector will equal P.

Should the load be applied to the outer surfaces, there will be some immediate strain increase in the connector. This increase will be accompanied by an expansion of the plates even though the load does not exceed the prestress, but the increase will be very slight because the load will go to the plate and connectors roughly in proportion to their stiffness. As the plate is stiffer, it will receive most of the load. An expression can be developed for the elongation of the bolt based on the bolt area and the assumed contact area between the plates. Depending on the contact area assumed, it will be found that, unless P is greater than the bolt tension, its stress increase will be in the range of 10 percent. Should the load exceed the prestress, the bolt stress will rise appreciably.

The preceding rather lengthy discussion is approximate but should explain why an ordinary tensile load applied to a riveted or bolted joint will not change the stress situation very much.

The allowable tensile design strength of bolts, rivets, and threaded parts is given by the expression to follow, which is independent of any initial tightening force:

$$P = F_t A_g$$

Table 11-6 gives values of F_t for the different kinds of connectors, with the values of rivets and threaded parts being quite conservative.

In this expression A_g is the nominal body area of a rivet or the unthreaded portion of a bolt or its threaded part not including upset rods. An upset rod has its ends made larger than the regular rod, and the threads are placed in the enlarged section so that the area at the root of the thread is larger than that of the regular rod. An upset rod is shown in Fig. 12-16. The use of upset rods is not usually economical and should be avoided unless a large order is being made.

If an upset rod is used, the allowable tensile strength of the threaded portion is set equal to $0.33F_u$ times the cross-sectional area at its major thread diameter. This value must be larger than $0.60F_y$ times the nominal body area of the rod before upsetting.

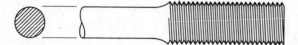

Figure 12-16 A round upset rod.

Example 12-7 illustrates the determination of the strength of a tension connection.

■ Example 12-7

Determine the allowable tensile strength of the bolts of the hanger connection of Fig. 12-15 if eight $\frac{7}{8}$-in. A490 high-strength bolts are used.

Solution.

$$P = (8)(0.6)(54) = \underline{\underline{259.2 \text{ k}}}$$

The load applied to a tensile connection shall be the sum of the external loads plus any tension forces that result from prying action, as described in the next section.   ■

12-4 PRYING ACTION

A further consideration in tensile connections is the possibility of prying action. A tensile connection, shown in Fig. 12-17(a), is subjected to prying action as illustrated in part (b). Should the flanges of the connection be quite thick and stiff or have stiffener plates as shown in Fig. 12-17(c), the prying action will probably be negligible. If they are thin and flexible and have no stiffeners, prying action can be very severe and must not be neglected.

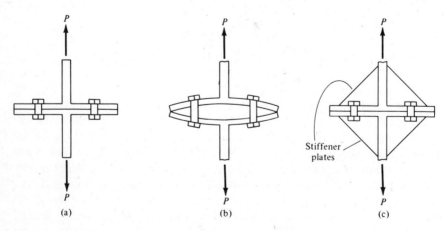

Figure 12-17 Tensile connection subject to prying action.

It is usually desirable to limit the number of rows of rivets or bolts in a tensile connection because a large percentage of the load is carried by the inner rows of multirow connections even at ultimate load. The tensile connection shown in Fig. 12-18 illustrates this point; the prying action will throw a large part of the load to the inner connectors, particularly if the plates are thin and flexible. For connections subjected to pure tensile loads, estimates should be made of possible prying action and its magnitude.

The additional force in the bolts resulting from prying action should be added to the tensile force resulting directly from the applied forces. The actual determination of prying forces is quite complex, and research on the subject is still being conducted. Several empirical formulas that approximate test results have been developed. Among these are the ASD expressions included in this section. The reader should realize that we don't know very much about prying action and our formulas keep changing almost yearly.

Only fully tensioned bolts should be used for connections for which the applied loads subject the bolts to axial tension. This is true whether or not the connections are classified as slip-critical, whether or not they are subject to fatigue loads, and whether or not there is prying action. If snug-tight bolts are used for any of these situations, the tensile loads will immediately start increasing bolt tensions.

Hanger and other tension connections should be so designed as to prevent significant deformations. The most important item in such designs is the use of rigid flanges. Rigidity is more important than bending resistance. To achieve this goal the distance *b* shown in Fig. 12-19 should be made as small as possible, with a minimum value equal to the space required to use a wrench for tightening the bolts. Information concerning wrench clearance dimensions is presented in the tables entitled "Threaded Fasteners Assembling Clearances" on pages 4-137 to 4-139 of the ASD Manual. On pages 4-89 to 4-95 of the Manual, a detailed procedure is presented for designing hanger connections and computing prying forces.

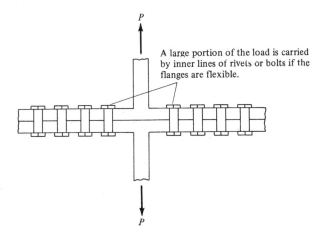

P

A large portion of the load is carried by inner lines of rivets or bolts if the flanges are flexible.

P

Figure 12-18

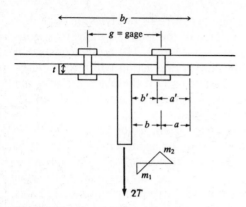

Figure 12-19 Hanger connection.

For reasons of space limitations only one numerical example is presented here. Four detailed examples for designing hangers and computing prying action are included in the ASD Manual.

To determine the prying force in a hanger connection the ASD Manual presents a long and tedious set of empirical equations. Reference should be made to Fig. 12-19 for several of the terms used in these expressions. The nomenclature to follow was taken from page 4-90 in the Manual.

T = tension force (kips) applied to each bolt *not including prying action* (This force is really fictitious unless the tensile load exceeds the prestress due to tensioning the bolts.)

Q = prying force (kips) per bolt at design load

B = allowable tension (kips) per bolt

t_c = flange or angle thickness (inches) required to develop B in bolts with no prying action

$$t_c = \sqrt{\frac{8Bb'}{\rho F_y}}$$

F_y = yield stress (ksi) of flange

ρ = length (inches) of connection, parallel to stem or leg, tributary to each bolt

a = distance (inches) from bolt centerline to the edge of tee flange or angle leg but not more than 1.25 b

d = bolt diameter (inches)

d' = width (inches) of bolt hole parallel to tee stem or angle leg

g = gage (inches)

b = distance (inches) from bolt centerline to the web (This must be sufficient for wrench clearance as provided in the ASD Manual.)

$$b = \frac{g}{2} - \frac{t_w}{2}$$

$$b' = b - \frac{d}{2} \quad \text{(inches)}$$

$$a' = a + \frac{d}{2} \quad \text{(inches)}$$

$$p = \frac{b'}{a'}$$

$\alpha = 0 \le a \le 1.0$, ratio of moment at bolt line to moment at stem line

$$\alpha = \frac{M_2}{\delta M_1}$$

$\alpha' =$ value of α for which required thickness (t_{reqd}) is a minimum, or allowable applied tension per bolt (T_{all}) is a maximum

$\delta =$ ratio of the net area at the bolt line to the gross area at the face of the stem

$$\delta = 1 - \frac{d'}{p}$$

To analyze a particular connection for prying action the ASD Manual (pages 4-91 and 4-92) suggests the following steps:

1. The applied tension per bolt (T) is compared to the allowable tension per bolt (B). Obviously if T is larger than B, we must use more or stronger bolts.
2. Compute the following:

$$\alpha' = \frac{1}{\delta(1 + p)}\left[\left(\frac{t_c}{t}\right)^2 - 1\right]$$

If $\alpha' > 1.0$, $T_{\text{all}} = B\left(\frac{t}{t_c}\right)^2(1 + \delta)$

If $0 \le \alpha' \le 1.0$, $T_{\text{all}} = B\left(\frac{t}{t_c}\right)^2(1 + \delta\alpha')$

If $\alpha' = 0$, $T_{\text{all}} = B$

3. If $T_{\text{all}} < T$, select a section with a thicker flange and repeat step 2.
4. If we need prying force, substitute into the following expressions:

$$\alpha = \frac{1}{\delta}\left[\frac{T/B}{(t/t_c)^2} - 1\right]; \text{ if } \alpha < 0 \text{ set } \alpha = 0$$

$$Q = B\delta\alpha p\left(\frac{t}{t_c}\right)^2$$

A similar procedure for the design of connections with possible prying action is presented in the Manual, as well as several numerical examples for both analysis and design. Example 12-8 presents the analysis of a structural tee hanger with prying action, using the current recommended ASD procedure.

After you wade through all of these definitions and confusing equations, you may very well ask yourself, "What have I learned?" From the perspective of these empirical equations, which are constantly being changed by the various steel specifications, the answer is, "Not very much." However, by recognizing prying action and knowing what to do about it (thicker flanges or stiffeners or more bolts or better detailing), your steel knowledge will have increased significantly.

■ Example 12-8

A 10-in.-long WT8 × 22.5 ($t_f = 0.565$ in., $t_w = 0.345$ in., $b_f = 7.035$ in.) is connected to a W36 × 150 as shown in Fig. 12-20 with four $\frac{7}{8}$-in. A 325 high-strength bolts. If A36 steel is used, are the bolts satisfactory? Include the effect of prying action.

Solution.

$$B = (0.60)(44) = 26.4 \text{ k}$$

$$T = \frac{65}{4} = 16.25 < 26.4 \text{ k} \qquad\qquad \text{OK}$$

Determining the values of b, a, b', ρ, d', δ, and P for use in subsequent expressions,

$$b = 2.0 - \frac{t_w}{2} = 2 - \frac{0.345}{2} = 1.827 \text{ in.} > 1\tfrac{3}{8} \text{ in.}$$

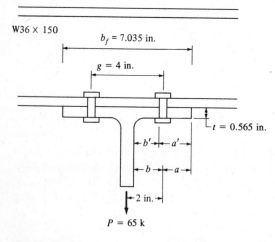

W36 × 150

$b_f = 7.035$ in.

$g = 4$ in.

$t = 0.565$ in.

b' a'

b a

2 in.

$P = 65$ k

Figure 12-20

which is required for wrench clearance (page 4-139 ASD Manual)

$$a = \frac{b_f}{2} - 2.0 = \frac{7.035}{2} - 2.0 = 1.517 \text{ in.}$$

Note that $1.25b = (1.25)(1.827) = 2.284 \text{ in.} > 1.517 \text{ in.}$

$$\therefore \text{ Use } a = 1.517 \text{ in.}$$

$$b' = b - \frac{d}{2} = 1.827 - \frac{0.875}{2} = 1.389 \text{ in.}$$

$$\rho = \frac{10}{2} = 5 \text{ in.}$$

$$d' = \frac{15}{16} \text{ in.} = 0.937 \text{ in.}$$

$$\delta = 1 - \frac{d'}{\rho} = 1 - \frac{0.937}{5} = 0.813 \text{ in.}$$

$$p = \frac{b'}{a'} = \frac{1.389}{1.954} = 0.711 \text{ in.}$$

$$t_c = \sqrt{\frac{(8)(26.4)(1.389)}{(5)(36)}} = 1.277 \text{ in.}$$

$$\alpha' = \frac{1}{0.813(1 + 0.711)}\left[\left(\frac{1.277}{0.565}\right)^2 - 1\right] = 2.953 \text{ in.}$$

As $\alpha' > 1.0$:

$$T_{\text{all}} = (26.4)\left(\frac{0.565}{1.277}\right)^2(1 + 0.813)$$

$$= 9.37 \text{ k} < 16.25 \text{ k} \qquad\qquad \text{NG}$$

Bolts are unsatisfactory ■

12-5 HISTORICAL NOTES ON RIVETS

For many years rivets were the accepted method for connecting the members of steel structures. Today, however, they no longer provide the most economical connections. They are still occasionally used for fasteners, but their use has declined to such a degree that most steel fabricators in the United States have discontinued riveting altogether. It is desirable, however, for the designer to be familiar with rivets even though he or she will seldom if ever design riveted structures. He or she may have to analyze an existing riveted structure for new loads or for an expansion of the structure. The purpose of these sections is to present a very brief introduction to the analysis and design of rivets.

Handling hot rivets, Chicago, Ill. (Courtesy of IR Construction Products Company.)

The rivets used in construction work were usually made of a soft grade of steel which would not become brittle when heated and hammered with a riveting gun to form the head. The usual rivet consisted of a cylindrical shank of steel with a rounded head on one end. It was heated in the field to a cherry-red color (approximately 1800°F), inserted in the hole, and a head formed on the other end, usually with a portable rivet gun operated by compressed air. The rivet gun, which had a depression in its head to give the rivet head the proper shape, applied a rapid succession of blows to the rivet.

For riveting done in the shop the rivets were most often heated to a light cherry-red color and driven with a pressure-type riveter. This type of riveter, usually called a bull riveter, squeezed the rivet with a pressure of perhaps as high as 50 to 80 tons (445 to 712 kN) and drove the rivet with one stroke. Because of this great pressure the rivet in its soft state was forced to fill the hole very satisfactorily. This type of riveting was much to be preferred over that done with the pneumatic hammer, but no greater nominal strengths were allowed by riveting specifications. The bull riveters were built for much faster operation than were the portable hand riveters, but the latter riveters were needed for places that are not easily accessible (i.e., field erection).

As the rivet cooled it shrank, or contracted and squeezed together the parts being connected. The squeezing effect caused, by friction, considerable transfer of stress between the parts being connected. The amount of friction was not dependable, however, and the specifications did not permit its inclusion in the allowable strength of a connection. Rivets shrink diametrically as well as lengthwise and actually become somewhat smaller than the holes which they are assumed to fill. (Permissible strengths for rivets are given in terms of the nominal cross-sectional areas of the rivets before driving.)

Some shop rivets were driven cold with tremendous pressures. Obviously the cold-driving process worked better for the smaller-size rivets, $\frac{3}{4}$ in. in diameter or less, although larger ones have been successfully used. Cold-driven rivets fill the holes better, eliminate the cost of heating, and are stronger because the steel is cold-worked. There is a reduction of clamping force, however, since the rivets do not shrink after driving.

12-6 TYPES OF RIVETS

The rivets used in ordinary construction work were $\frac{3}{4}$ in. and $\frac{7}{8}$ in. in diameter, but they could be obtained in standard sizes from $\frac{1}{2}$ in. to $1\frac{1}{2}$ in. in $\frac{1}{8}$-in. increments. (The smaller sizes were used for small roof trusses, signs, small towers, etc., while the larger sizes were used for very large bridges or towers and very tall buildings.) The use of more than one or two sizes of rivets or bolts on a single job is usually undesirable because it is expensive and inconvenient to punch different-size holes in a member in the shop, and the installation of different-size rivets or bolts in the field may be confusing. Some cases arise where it is absolutely necessary to have different sizes, as where smaller rivets or bolts are needed for keeping the proper edge distance in certain sections, but these situations should be avoided if possible.

Rivet heads were usually round in shape, called button heads; but if clearance requirements dictated, the head was flattened or even countersunk and chipped flush. These situations are shown in Fig. 12-21.

The countersunk and chipped-flush rivets do not have sufficient bearing areas to develop full strength and should be discounted 50 percent in design (ASD Specification J3.3). A rivet with a flattened head was to be preferred to a countersunk rivet, but if a smooth surface was required, the countersunk and chipped-flush rivet were necessary. This latter type of rivet was appreciably more expensive than the button head in addition to being weaker, and it was not used unless absolutely necessary.

There are three ASTM classifications for rivets for structural steel applications as described in the following paragraphs.

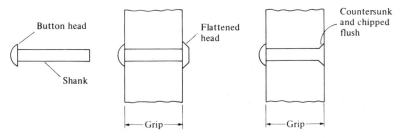

Figure 12-21 Types of rivets.

ASTM Specification A502, Grade 1

A502, grade 1, rivets were used for most structural work. They had a low-carbon content of about 0.80 percent, were weaker than the ordinary structural carbon steel, and had a higher ductility. The fact that these rivets were easier to drive than the higher-strength rivets was the main reason that A502, grade 1, rivets were most often used, regardless of the strength of the steel used in the structural members.

ASTM Specification A502, Grade 2

A502, grade 2, carbon-manganese rivets have higher strengths than the grade 1 rivets and were developed for the higher-strength steels. Their higher strength permits the designer to use fewer rivets in a connection and thus smaller gusset plates.

ASTM Specification A502, Grade 3

A502, grade 3, rivets have the same nominal strengths as the grade 2 rivets, but they have much higher resistance to atmospheric corrosion, equal to approximately four times that of carbon steel without copper.

12-7 ALLOWABLE STRENGTH OF RIVETED CONNECTIONS—RIVETS IN SHEAR

The factors determining the allowable strength of a rivet are its grade, its diameter, and the thickness and arrangement of the pieces being connected. The actual distribution of stress around a rivet hole is difficult to determine, if it can be determined at all, and to simplify the calculations it is assumed to vary uniformly over a rectangular area equal to the diameter of the rivet times the thickness of the plate.

The strength of a rivet in bearing equals the design bearing unit stress of the rivet times the diameter of the shank of the rivet times the thickness of the member that bears on the rivet. The strength of a rivet in single shear is the allowable shearing stress times the cross-sectional area of the shank of the rivet. Should a rivet be in double shear, its allowable shearing strength is considered to be twice its single shear value.

The allowable tension and shearing stresses for rivets and A307 bolts are given in Table 11-6. These values are for static loads only. Notice that the allowable shear stress for A307 bolts is not affected if the bolt threads are in the shear plane.

Examples 12-9 and 12-10 illustrate the calculations necessary to determine the allowable design strengths of existing connections or to design riveted connections. Little comment is made here concerning A307 bolts: all the calculations for these fasteners are made exactly as they are for rivets except that the allowable shear stresses given by the ASD Specification are different. Only one brief example with these common bolts (12-11) is included.

■ Example 12-9

Determine the allowable design strength P of the bearing-type connection shown in Fig. 12-22. A36 steel and A502, grade 1, rivets are used in the connection, and it is assumed that standard-size holes are used and that edge distances and center-to-center distances are sufficient for the $1.2F_u$ allowable bearing stress to be used.

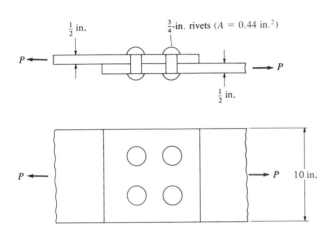

Figure 12-22

Solution. Allowable tensile force applied to plates:

$$A_g = (\tfrac{1}{2})(10) = 5.00 \text{ in.}^2$$
$$A_n = [(\tfrac{1}{2})(10) - (2)(\tfrac{7}{8})(\tfrac{1}{2})] = 4.125 \text{ in}^2 = A_e$$
$$P = 0.60F_y A_g = (0.60)(36)(5.00) = 108 \text{ k}$$
$$P = 0.50F_u A_e = (0.50)(58)(4.125) = 119.6 \text{ k}$$

Rivets in single shear and bearing on $\tfrac{1}{2}$ in.:

Allowable shearing strength of rivets $= (0.44)(17.5)(4) = 30.8 \text{ k} \leftarrow$
Allowable bearing strength of rivets $= (\tfrac{3}{4})(\tfrac{1}{2})(1.2)(58)(4) = 104.4 \text{ k}$
$$P = 30.8 \text{ k}$$ ■

■ Example 12-10

How many $\tfrac{7}{8}$-in. A502, grade 1, rivets are required for the connection shown in Fig. 12-23 if the plates are A36, standard-size holes are used, and the edge distances and center-to-center distances are sufficient for the $1.2F_u$ allowable bearing stress to be used?

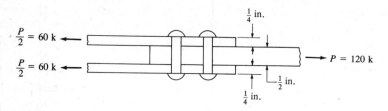

Figure 12-23

Solution. Rivets in double shear and bearing on $\frac{1}{2}$ in.:

Allowable shear strength of 1 rivet = $(2 \times 0.6)(17.5) = 21$ k ←
Allowable bearing strength of 1 rivet = $(\frac{7}{8})(\frac{1}{2})(1.2 \times 58) = 30.4$ k

$$\text{Number of rivets required} = \frac{120}{21} = 5.71$$

Use six $\frac{7}{8}$-in. rivets

■ **Example 12-11**
Repeat Example 12-10 using $\frac{7}{8}$-in. A 307 bolts.

Solution. Bolts in double shear and bearing on $\frac{1}{2}$ in.:

Allowable shear strength of 1 bolt = $(2)(0.6)(10) = 12$ k ←
Allowable bearing strength of 1 bolt = $(\frac{7}{8})(\frac{1}{2})(1.2 \times 58) = 30.4$ k

$$\text{Number of bolts required} = \frac{120}{12} = 10$$

Use ten $\frac{7}{8}$-in. A307 bolts

PROBLEMS

For each of the problems the following information is to be used unless otherwise indicated: (a) A36 steel; (b) standard-size holes; (c) edge distances and center-to-center distances sufficient to use $F_p = 1.2F_u$; (d) threads of bolts excluded from shear plane.
    12-1 to 12-7   Determine the resultant load on the most stressed bolt in the eccentrically loaded connections shown, using the elastic method.

12-1 (*Ans.* 10.77 k)

12-2 $P = 40$ k

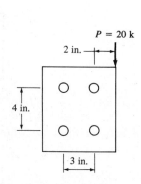

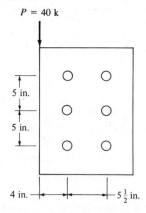

12-3 (*Ans.* 16.16 k)   **12-4**

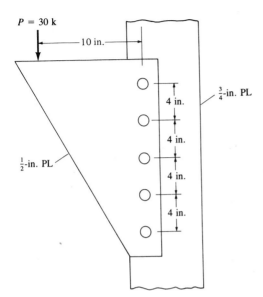

P = 30 k

10 in.

4 in.

4 in.

4 in.

4 in.

$\frac{3}{4}$-in. PL

$\frac{1}{2}$-in. PL

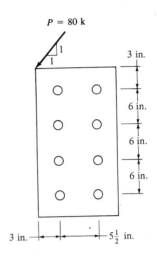

P = 80 k

3 in.

6 in.

6 in.

6 in.

3 in.

$5\frac{1}{2}$ in.

12-5 (*Ans.* 19.31 k)   **12-6**

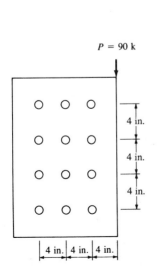

P = 90 k

4 in.

4 in.

4 in.

4 in. 4 in. 4 in.

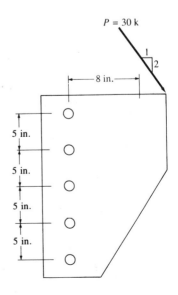

P = 30 k

1
2

8 in.

5 in.

5 in.

5 in.

5 in.

12-7 (*Ans.* 27.00 k)

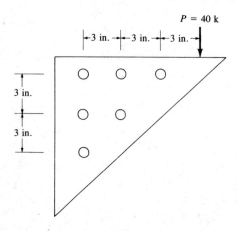

12-8 Repeat Prob. 12-2 using the reduced eccentricity method.

12-9 Using the elastic method, determine the allowable design strength of the bearing-type connection shown. The bolts are $\frac{7}{8}$-in. A325 and are in single shear and bearing on $\frac{1}{2}$ in. (*Ans.* 59.74 k)

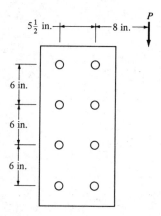

12-10 Using the elastic method, determine the allowable design strength P for the slip-critical connection shown. The $\frac{3}{4}$-in. A325 bolts are in "double shear."

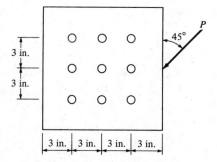

12-11 Repeat Prob. 12-9 using the tables entitled "Eccentric Loads on Fastener Groups" in Part 4 of the ASD Manual. (*Ans.* 71.73 k)

12-12 Repeat Prob. 12-10 using the tables entitled "Eccentric Loads on Fastener Groups" in Part 4 of the ASD Manual.

12-13 Is the bearing-type connection shown sufficient to resist the 100-k load which passes through the center of gravity of the bolt group? (*Ans.* Yes: $F_t = 36.27$ ksi $> f_t = 22.63$ ksi)

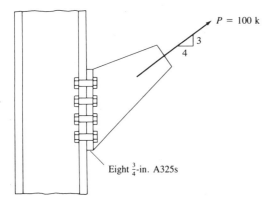

12-14 Repeat Prob. 12-13 using slip-critical bolts and $P = 75$ k.

12-15 If the load shown passes through the center of gravity of the bolt group, how large can it be if a bearing-type connection is used? (*Ans.* 301.1 k)

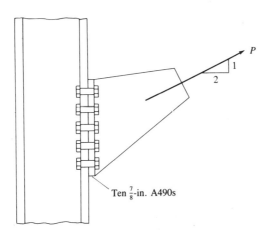

12-16 Repeat Prob. 12-15 using A325 bolts.

12-17 Determine the number of $\frac{3}{4}$-in. A325 bolts required in the angles and in the flange of the W shape shown if a bearing-type connection is used. (*Ans.* 10 bolts)

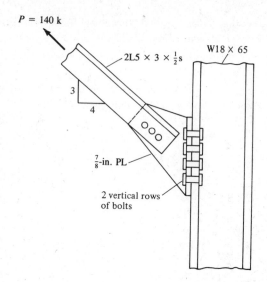

12-18 Are the bolts shown in this 16-in.-long hanger satisfactory to resist direct tension and prying? Eight $\frac{7}{8}$-in. A325 bolts are spaced 4 in. on center longitudinally.

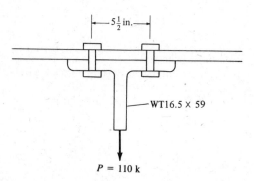

12-19 Determine the allowable design strength P of the rivets in the connection shown if $\frac{7}{8}$-in. A502, grade 1, rivets are used. (*Ans.* 126.3 k)

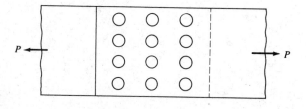

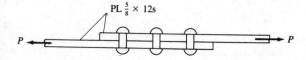

12-20 The truss tension member shown consists of a single-angle $5 \times 3 \times \frac{5}{16}$ and is connected to a $\frac{1}{2}$-in. gusset plate with five $\frac{7}{8}$-in. A502, grade 1, rivets. Determine the allowable P of the rivets only.

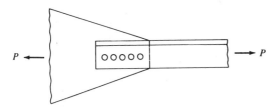

12-21 How many $\frac{3}{4}$-in. A502, grade 1, rivets are needed to carry the load shown? (*Ans.* 12.94, say 14)

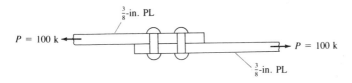

12-22 Repeat Prob. 12-21 using A307 bolts.

12-23 How many A502, grade 1, rivets with 1-in. diameters need to be used for the butt joint shown? (*Ans.* 5.82, say 6)

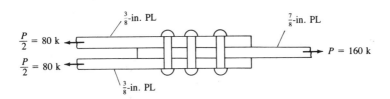

12-24 How many $\frac{7}{8}$-in. A307 bolts are required for the connection shown?

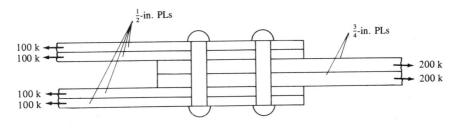

12-25 For the connection shown, $P = 400$ k; determine the number of $\frac{7}{8}$-in. A502, grade 2, rivets required. (*Ans.* 15.12, say 16)

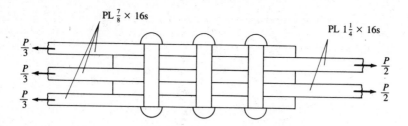

12-26 Determine the allowable design strength P for the connection shown if 1-in. A502, grade 2, rivets are used. Include the strength of the plates.

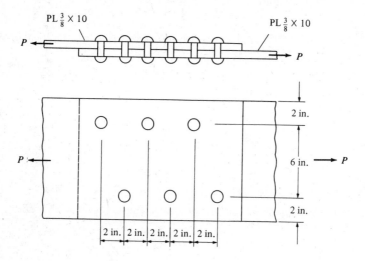

12-27 For the beam shown, what is the required spacing of $\frac{3}{4}$-in. A 307 bolts if $V = 100$ k? (*Ans.* 3.26 in., say $3\frac{1}{4}$ in.)

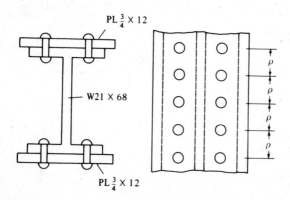

12-28 Is the connection shown sufficient to resist the 70-k load which passes through the center of gravity of the rivet group?

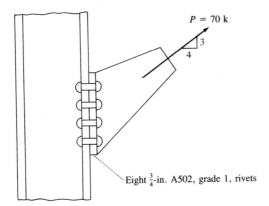

Eight $\frac{3}{4}$-in. A502, grade 1, rivets

12-29 If the load shown passes through the center of gravity of the rivet group, how large can it be? (*Ans.* 122.3 k)

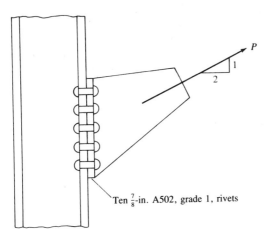

Ten $\frac{7}{8}$-in. A502, grade 1, rivets

13

Welded Connections

13-1 GENERAL

Welding is a process in which metallic parts are connected by heating their surfaces to a plastic or fluid state and allowing the parts to flow together and join (with or without the addition of other molten metal). It is impossible to determine when welding originated, but it was several thousand years ago. Metalworking, including welding, was quite an art in ancient Greece at least three thousand years ago, but welding had undoubtedly been performed for many centuries before those days. Ancient welding was probably a forging process in which the metals were heated to a certain temperature (not to the melting stage) and hammered together.

Although modern welding has been available for a good many years, it has only come into its own in the last few decades for the building and bridge phases of structural engineering. The adoption of structural welding was quite slow for several decades because many engineers thought that welding had two great disadvantages—(a) welds had reduced fatigue strength as compared with riveted and bolted connections and (b) it was impossible to ensure a high quality of welding without unreasonably extensive and costly inspection.

These negative feelings persisted for many years, although tests seemed to indicate that neither reason was valid. Regardless of the validity of these fears, they were widely held and undoubtedly slowed down the use of welding, particularly for highway bridges and to an even greater extent for railroad bridges.

Fabrication of plate girders for Connecticut expressway
bridge. (Courtesy of Lincoln Electric Company.)

Today most engineers agree that welded joints have considerable fatigue
strength. They will also admit that the rules governing the qualification of
welders, the better techniques applied, and the excellent workmanship requirements of the (AWS) American Welding Society specifications make the inspection of welding a much less difficult problem. Consequently welding is today
permitted for almost all structural work other than for some bridges.

On the subject of fear of welding it is interesting to consider welded ships.
Ships are subjected to severe impactive loadings which are difficult to predict,
yet naval architects use all-welded ships with great success. A similar discussion can be made for airplanes and aeronautical engineers. The slowest adoption of structural welding was for railroad bridges. These bridges are undoubtedly subjected to heavier live loads than highway bridges, larger vibrations,
and more stress reversals; but are their stress situations as serious and as
difficult to predict as those for ships and planes?

13-2 ADVANTAGES OF WELDING

Today it is possible to make use of the many advantages which welding offers,
since the fatigue and inspection fears have been largely eliminated. Several of
the many welding advantages are discussed in the following paragraphs.

1. To most persons the first advantage is in the area of economy, because
the use of welding permits large savings in pounds of steel used. Welded structures allow the elimination of a large percentage of the gusset and splice plates

necessary for riveted or bolted structures as well as the elimination of rivet or bolt heads. In some bridge trusses it may be possible to save up to 15 percent or more of the steel weight by using welding. Welding also requires appreciably less labor than does riveting, because one welder can replace the standard four-person riveting crew.

2. Welding has a much wider range of application than does riveting or bolting. Consider a steel pipe column and the difficulties of connecting it to other steel members by bolting. A bolted connection may be virtually impossible, but a welded connection will present no difficulties whatsoever. The student can visualize many similar situations where welding has a decided advantage.

3. Welded structures are more rigid because the members are often welded directly to each other. The connections for bolted structures are often made through connection angles or plates which deform due to load transfer, making the entire structure more flexible. On the other hand, greater rigidity can be a disadvantage where simple end connections with little moment resistance are desired. For such cases designers must be careful as to the type of joint they specify.

4. The process of fusing pieces together gives the most truly continuous structures. It results in one-piece construction, and because welded joints are as strong as or stronger than the base metal no restrictions have to be placed on the joints. This continuity advantage has permitted the erection of countless slender and graceful statically indeterminate steel frames throughout the world. Some of the more outspoken proponents of welding have referred to riveted and bolted structures, with their heavy plates and large number of rivets or bolts, as looking like tanks or armored cars when compared with the clean, smooth lines of welded structures. For a graphic illustration of this advantage the student should compare the moment-resisting connections of Fig. 14-4.

5. It is easier to make changes in design and to correct errors during erection (and less expensive) if welding is used. A closely related advantage has certainly been illustrated in military engagements during the past few wars by the quick welding repairs made to military equipment under battle conditions.

6. Another item that is often important is the relative silence of welding. Imagine the importance of this fact when working near hospitals or schools or when making additions to existing buildings. Anyone with close to normal hearing who has attempted to work in an office within several hundred feet of a bolted or riveted job will agree this is an advantage.

7. Fewer pieces are used, and as a result time is saved in detailing, fabrication, and field erection.

13-3 TYPES OF WELDING

Although both gas and arc welding are available, almost all structural welding is arc welding. Sir Humphry Davy discovered in 1801 how to create an electric arc by bringing close together two terminals of an electric circuit of relatively high voltage. Although he is generally given credit for the development of modern welding, a good many years elapsed after his discovery before welding was actually performed with the electric arc. (His work was of the greatest importance to the modern structural world, but many people say his greatest discovery was not the electric arc but rather a laboratory assistant whose name was Michael Faraday.) Several Europeans formed welds of one type or another in the 1880s with the electric arc, while in the United States the first patent for arc welding was given to Charles Coffin of Detroit in 1889.[1]

The figures in this chapter show the necessity of supplying additional metal to the joints being welded to give satisfactory connections. In electric-arc welding the metallic rod, which is used as the electrode, melts off into the joint as it is being made. When gas welding is used, it is necessary to introduce a metal rod known as a *filler* or *welding rod*.

In gas welding a mixture of oxygen and some suitable type of gas is burned at the tip of a torch or blowpipe held in the welder's hand or by an automatic machine. The gas used in structural welding is usually acetylene, and the process is called oxyacetylene welding. The flame produced can be used for flame cutting of metals as well as for welding. Gas welding is rather easy to learn, and the equipment used is rather inexpensive. It is, however, a slow process as compared with other means of welding, and is normally used for repair and maintenance work and not for the fabrication and erection of large steel structures.

In arc welding an electric arc is formed between the pieces being welded and an electrode held in the operator's hand with some type of holder or by an automatic machine. The arc is a continuous spark which upon contact brings the electrode and the pieces being welded to the melting point. The resistance of the air or gas between the electrode and the pieces being welded changes the electrical energy into heat. A temperature of somewhere between 6,000 and 10,000°F is produced in the arc. At the end of the electrode small droplets or globules of the molten metal are formed and are forced by the arc across to the pieces being connected, penetrating the molten metal to become a part of the weld. The amount of penetration can be controlled by the amount of current consumed. Since the molten droplets of the electrodes are actually propelled to the weld, arc welding can be successfully used for overhead work.

A pool of molten steel can hold a fairly large amount of gases in solution and if not protected from the surrounding air will chemically combine with oxygen and nitrogen. After cooling, the welds will be relatively porous due to the little pockets formed by the gases. Such welds are relatively brittle and have much less resistance to corrosion. A weld can be shielded by using an electrode coated with certain mineral compounds. The electric arc causes the coating to melt and creates an inert gas or vapor around the area being welded. The vapor acts as a shield around the molten metal and keeps it from coming freely in contact with the surrounding air. It also deposits in the molten metal

---

[1] *Procedure Handbook of Arc Welding Design and Practice*, 11th ed. (Lincoln Electric Company, 1957), Part 1.

a slag that has less density than the base metal and comes to the surface to protect the weld from the air while the weld cools. After cooling, the slag can easily be removed by peening and wire brushing (such removal being absolutely necessary before painting or application of another weld layer). Fig. 13-1 shows the elements of the shielded arc welding process. Shielded metal arc welding is frequently abbreviated here with the letters SMAW.

The type of welding electrode used is very important as it decidedly affects the weld properties such as strength, ductility, and corrosion resistance. Quite a number of different types of electrodes are manufactured, the type to be used for a certain job being dependent upon the type of metal being welded, the amount of material which needs to be added, the position of the work, and so on. The electrodes fall into two general classes—the *lightly coated electrodes* and the *heavily coated electrodes*.

The heavily coated electrodes are normally used in structural welding because the melting of their coatings produces very satisfactory vapor shields around the work as well as slag in the weld. The resulting welds are stronger, more resistant to corrosion, and more ductile than are those produced with lightly coated electrodes. When the lightly coated electrodes are used, no attempt is made to prevent oxidation and no slag is formed. The electrodes are lightly coated with some arc-stabilizing chemical such as lime.

Another type of welding is the submerged (or hidden) arc welding (SAW). In this method the arc is covered with a mound of granular fusible material and thus hidden from view. A bare metal electrode is fed from a reel and melted and deposited as filler material. SAW welds are quickly and efficiently made and are of high quality, exhibiting high impact strength and corrosion resistance and good ductility. Furthermore they provide deeper penetration, with the result that the area effective in resisting loads is larger. A large per-

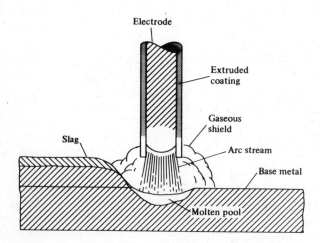

Figure 13-1 Elements of the shielded metal arc welding process (SMAW). From *Procedure Handbook of Arc Welding Design and Practice,* 11th ed. (Lincoln Electric Company, 1957).

centage of the welding done for bridge structures is SAW. If a single electrode is used, the size of the weld obtained with a single pass is limited. Multiple electrodes may be used, however, with no limit on weld size. Positions for SAW welds should be flat or horizontal.

13-4 WELDING INSPECTION

Three steps must be taken to ensure good welding for a particular job: (a) establishing good welding procedures, (b) use of prequalified welders, and (c) employment of competent inspectors in shop and field.

When the procedures established by the AWS and AISC for good welding are followed and when welders are used who have previously been required to prove their ability, good results will probably be obtained. However, to make absolutely sure, well-qualified inspectors are needed.

Good welding procedure involves the selection of proper electrodes, current, and voltage; the properties of base metal and filler; and the position of welding, to name only a few factors. The usual practice for large jobs is to employ welders who have certificates showing their qualifications. In addition, it

Lincoln ML-3 Squirtwelder mounded on a self-propelled trackless trailer deposits this $\frac{1}{4}$-in. web-to-flange weld at 28 in./min. (Courtesy of Lincoln Electric Company.)

is not a bad practice to have each person make an identifying mark on each weld so that those persons frequently doing poor work can be located. This practice is likely to improve the general quality of work performed by all welders on the job.

Visual Inspection

Another factor that will cause welders to perform better work is just the presence of an inspector who they feel knows good welding when he or she sees it. A good inspector should have done welding and spent much time observing the work of good welders. From this experience he or she should be able to know if a welder is obtaining satisfactory fusion and penetration. The inspector should be able to recognize good welds by their shape, size, and general appearance. For instance, the metal in a good weld should approximate its original color after it has cooled. If it has been overheated, it may have a rusty and reddish-looking color. An inspector can use various scales and gages to check the sizes and shapes of welds.

Visual inspection by a qualified person will probably give a good indication of the quality of welds but is not a perfect source of information as to the subsurface condition of the weld. There are several methods for determining the internal soundness of a weld. These include the use of penetrating dyes and magnetic particles, ultrasonic testing, and radiographic procedures. These methods can be used to detect internal defects such as porosity, weld penetration, and presence of slag.

Liquid Penetrants

Various types of dyes can be spread over weld surfaces, and they will penetrate into surface cracks of the weld. After the dye has penetrated into the crack, the excess surface material is wiped off and a powdery developer is used to draw the dye out of the cracks. The outlines of the cracks can then be seen with the dye. Several variations of this method are used to improve the visibility of the defects, such as the use of fluorescent dyes. After the dye is drawn from the cracks, the cracks are made to stand out brightly by examination under black light.[2]

Magnetic Particles

In this method the weld being inspected is magnetized electrically. Cracks that are at or near the surface of the weld cause north and south poles to form on each side of the cracks. Dry powdered iron filings or a liquid suspension of

---

[2] James Hughes, "It's Superinspector," *Steelways* 25, no. 4 (New York: AISI, September/October 1969), pp. 19–21.

particles is placed on the weld. The patterns of these particles form when many of them cling to the cracks show the locations of cracks and indicate their size and shape. A disadvantage of this method is that, if multilayer welds are used, the method has to be applied to each layer.

Ultrasonic Testing

In recent years the steel industry has applied ultrasonics to the manufacture of steel. Although the equipment is expensive, the method is quite useful in welding inspection as well. Sound waves are sent through the material being tested and are reflected from the opposite side of the material. These reflections are shown on a cathode ray tube. Defects in the weld will affect the time of the sound transmission. The operator can read the picture on the tube and then locate flaws and learn how severe they are.

Radiographic Procedures

The more expensive radiographic methods can be used to check occasional welds in important structures. From these tests it is possible to make good estimates of the percentage of bad welds in a structure. The use of portable X-ray machines where access is not a problem and the use of radium or radioactive cobalt for making pictures are excellent but expensive methods of testing welds. These methods are satifactory for groove welds (such as for the welding of important stainless steel piping at chemical and nuclear projects) but are not satisfactory for fillet welds because the pictures are difficult to interpret. Fillet and groove welds are described in the next section. A further disadvantage of these methods is the radioactive danger. Careful procedures have to be used to protect the technicians as well as nearby workers. On the average construction job this danger probably requires night inspection of welds when only a few workers are near the inspection area. (Normally a very large job would be required before the use of the extremely expensive radioactive materials could be justified.)

A properly welded connection can always be made much stronger (perhaps $1\frac{1}{2}$ or 2 times) than the parts being connected. As a result, the actual strength is much higher than is required by the specifications. The reasons for this extra strength are that the electrode wire is made from premium steel, the metal is melted electrically (as is done in the manufacture of high-quality steels), and the cooling rate is quite rapid. As a result, it is probably rare for a welder to make a weld of less strength than required by the design.

13-5 CLASSIFICATION OF WELDS

There are three separate classifications of welds, based upon the types of welds made, positions of welds, and types of joints.

Type of Weld

The two main types of welds are *fillet welds* and *groove welds*. In addition there are plug and slot welds which are not as common in structural work. These four types of welds are shown in Fig. 13-2.

The fillet welds will be shown to be weaker than groove welds; however, most structural connections (about 80 percent) are made with fillet welds. Any person who has had experience in steel structures will understand why fillet welds are more common than are groove welds. Groove welds (which are welds made in grooves between the members to be joined) are used when the members to be connected are lined up in the same plane. To use them in every situation would mean that the members would have to fit almost perfectly, and unfortunately the average steel structure does not fit together in that manner. Many students have seen steelworkers pulling and ramming steel members to get them in position. When members are allowed to lap over each other, larger tolerances are allowable in erection, and fillet welds are used. Nevertheless groove welds are quite common for many connections such as column splices, butting of beam flanges to columns, and so on, and they make up about 15 percent of structural welding.

Groove welds may be *complete-penetration* welds, which extend for the full thickness of the part being connected, or *partial-penetration* welds, which extend for only part of the member thickness. Table 13-1 shows that allowable stresses will be smaller for certain types of partial-penetration welds than for complete-penetration ones.

A plug weld is a circular weld passing through one member to another and joining the two together. A slot weld is a weld formed in a slot or elongated hole which joins one member to the other member through the slot. The slot may be partly or fully filled with weld material. These two expensive types of welds may occasionally be used when members lap over each other and the desired length of fillet welds cannot be obtained. They may also be used to stitch together parts of a member, as in the fastening of cover plates to a built-up member.

A plug or slot weld is not generally considered to be suitable for transferring tensile forces perpendicular to the faying surface. The reason is that there

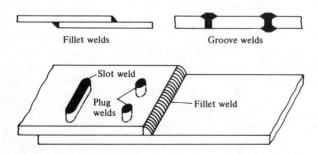

Figure 13-2 Weld types.

TABLE 13-1 ALLOWABLE STRESS ON WELDS

TYPE OF WELD AND STRESS[a]	ALLOWABLE STRESS	REQUIRED WELD STRENGTH LEVEL[b, c]
COMPLETE-PENETRATION GROOVE WELDS		
Tension normal to effective area	Same as base metal	"Matching" weld metal shall be used.
Compression normal to effective area	Same as base metal	Weld metal with a strength level equal to or less than "matching" weld metal is permitted.
Tension or compression parallel to axis of weld	Same as base metal	
Shear on effective area	0.30 × nominal tensile strength of weld metal (ksi)	
PARTIAL-PENETRATION GROOVE WELDS[d]		
Compression normal to effective area	Same as base metal	Weld metal with a strength level equal to or less than "matching" weld metal is permitted.
Tension or compression parallel to axis of weld[e]	Same as base metal	
Shear parallel to axis of weld	0.30 × nominal tensile strength of weld metal (ksi)	
Tension normal to effective area	0.30 × nominal tensile strength of weld metal (ksi), except tensile stress on base metal shall not exceed 0.60 × yield stress of base-metal	
FILLET WELDS		
Shear on effective area	0.30 × nominal tensile strength of weld metal (ksi)	Weld metal with a strength level equal to or less than "matching" weld metal is permitted.
Tension or compression parallel to axis of weld[e]	Same as base metal	
PLUG AND SLOT WELDS		
Shear parallel to faying surfaces (on effective area)	0.30 × nominal tensile strength of weld metal (ksi)	Weld metal with a strength level equal to or less than "matching" weld metal is permitted.

TABLE 13-1 (continued)

Source: American Institute of Steel Construction, *Manual of Steel Construction Allowable Stress Design,* 9th ed. (Chicago: AISC, 1989), Table J2.5, p. 5-70. Reprinted with the permission of AISC.

Note: The design of connected material is governed by Chapters D through G of ASD specification. Also see Commentary Sect. J2.4.

[a] For definition of effective area, see Sect. J2.

[b] For "matching" weld metal, see Table 4.1.1, AWS D1.1.

[c] Weld metal one strength level stronger than "matching" weld metal will be permitted.

[d] See Sect. J2.1b for a limitation on use of partial-penetration groove welded joints.

[e] Fillet welds and partial-penetration groove welds joining the component elements of built-up members, such as flange-to-web connections, may be designed without regard to the tensile or compressive stress in these elements parallel to the axis of the welds.

is not usually much penetration of the weld into the member behind the plug or slot—and yet resistance to tension is provided primarily by penetration.

Structural designers accept plug and slot welds as being satisfactory for stitching the different parts of a member together, but many designers are not happy to use these welds for the transmission of shear forces. The penetration of the welds from the slots or plugs into the other members is questionable, and in addition there can be critical voids down in the welds, which cannot be detected with the usual inspection procedures.

Position

Welds are referred to as being flat, horizontal, vertical, and overhead. They are listed in the preceding sentence in order of their economy, with the flat welds being the most economical and the overhead welds being the most expensive. A fairly good welder can do a very satisfactory job with a flat weld, but it takes the very best welder to do a good job with an overhead weld. Although the flat welds can often be made with an automatic machine, much structural welding is done by hand. The assistance of gravity is not necessary for the forming of good welds, but it does speed up the process. The globules of the molten electrodes can be forced into the overhead welds against gravity and good welds will result, but they are slow and expensive to make, so it is desirable to avoid them whenever possible. These types of welds are shown in Fig. 13-3.

Type of Joint

Welds can be further classified according to the type of joint used: butt, lap, tee, edge, corner, and so on. These joint types are shown in Fig. 13-4.

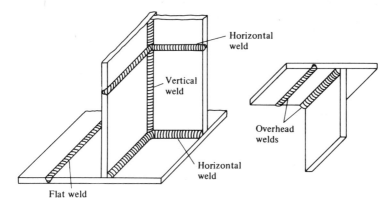

Figure 13-3

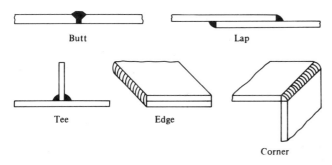

Figure 13-4

13-6 WELDING SYMBOLS

Fig. 13-5 presents the welding symbols developed by the AWS. With this excellent shorthand system a great deal of information can be presented in a small space on engineering plans and drawings. These symbols remove the necessity of drawing in the welds and making long descriptive notes. It is certainly desirable for steel designers and drafters to use this standard system. Should most of the welds on a drawing be of the same size, a note to that effect can be given and the symbols omitted, except for the off-size welds.

The purpose of this figure is not to show every possible type of symbol but rather to give a general idea of the appearance of welding symbols and the information which they can show. The reader can refer to the detailed information published by the AWS and reprinted in many handbooks (including the ASD Manual). At first glance the information presented in Fig. 13-5 is probably quite confusing. For this reason a few very common symbols for fillet welds are presented in Fig. 13-6, with an explanation of each.

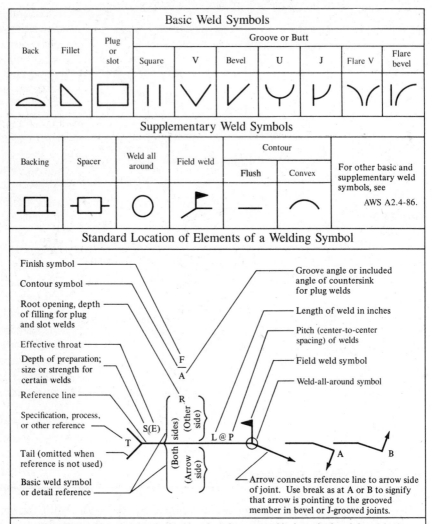

Basic Weld Symbols

Back	Fillet	Plug or slot	Groove or Butt						
			Square	V	Bevel	U	J	Flare V	Flare bevel

Supplementary Weld Symbols

Backing	Spacer	Weld all around	Field weld	Contour		For other basic and supplementary weld symbols, see AWS A2.4-86.
				Flush	Convex	

Standard Location of Elements of a Welding Symbol

Finish symbol

Contour symbol

Root opening, depth of filling for plug and slot welds

Effective throat

Depth of preparation; size or strength for certain welds

Reference line

Specification, process, or other reference

Tail (omitted when reference is not used)

Basic weld symbol or detail reference

Groove angle or included angle of countersink for plug welds

Length of weld in inches

Pitch (center-to-center spacing) of welds

Field weld symbol

Weld-all-around symbol

Arrow connects reference line to arrow side of joint. Use break as at A or B to signify that arrow is pointing to the grooved member in bevel or J-grooved joints.

Note: Size, weld symbol, length of weld, and spacing must read in that order from left to right along the reference line. Neither orientation of reference line nor location of the arrow alters this rule.

The perpendicular leg of ⊿, ⟍, ⊢, ⟋ weld symbols must be at left.

Arrow and Other Side welds are of the same size unless otherwise shown. Dimensions of fillet welds must be shown on both the Arrow Side and the Other Side symbol.

Flag of field-weld symbol shall be placed above and at right angle to reference line of junction with the arrow.

Symbols apply between abrupt changes in direction of welding unless governed by the All Around symbol or otherwise dimensioned.

These symbols do not explicitly provide for the case that frequently occurs in structural work, where duplicate material (such as stiffeners) occurs on the far side of a web or gusset plate. The fabricating industry has adopted this convention: that when the billing of the detail material discloses the existence of a member on the far side as well as on the near side, the welding shown for the near side shall be duplicated on the far side.

Figure 13-5 Welded joints—standard symbols. From American Institute of Steel Construction, *Manual of Steel Construction Allowable Stress Design,* 9th ed. (Chicago: AISC, 1989), p. 4-155. Reprinted with the permission of AISC.

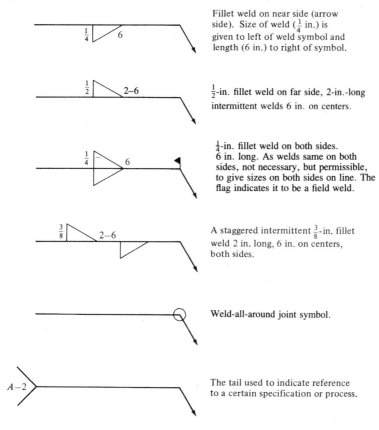

Fillet weld on near side (arrow side). Size of weld ($\frac{1}{4}$ in.) is given to left of weld symbol and length (6 in.) to right of symbol.

$\frac{1}{2}$-in. fillet weld on far side, 2-in.-long intermittent welds 6 in. on centers.

$\frac{1}{4}$-in. fillet weld on both sides. 6 in. long. As welds same on both sides, not necessary, but permissible, to give sizes on both sides on line. The flag indicates it to be a field weld.

A staggered intermittent $\frac{3}{8}$-in. fillet weld 2 in. long, 6 in. on centers, both sides.

Weld-all-around joint symbol.

The tail used to indicate reference to a certain specification or process.

Figure 13-6 Samples of welding symbols.

13-7 GROOVE WELDS

When complete-penetration groove welds are subjected to axial tension or axial compression, the weld stress is assumed to equal the load divided by the net area of the weld. Three types of groove welds are shown in Fig. 13-7. The square groove joint (a) is used to connect relatively thin material up to roughly $\frac{5}{16}$ in. (8 mm) thick. As the material becomes thicker, it is necessary to use the single-vee groove welds (b) and the double-vee groove welds (c). For these two welds the members are beveled before welding to permit full penetration of the weld.

The groove welds shown in Fig. 13-7 are said to have *reinforcement*. Reinforcement is added weld metal that causes the throat dimension to be greater than the thickness of the welded material. Because of reinforcement, groove welds may be referred to as 100, 125, 150 percent, and so on, welds according to the amount of extra thickness at the weld. There are two major reasons for having reinforcement: (a) reinforcement gives a little extra strength because the extra metal takes care of pits and other irregularities, and (b) the welder can easily make the weld a little thicker than the welded material. It would be

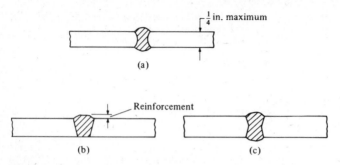

Figure 13-7

a difficult if not impossible task to make a perfectly smooth weld with no places that were thinner or thicker than the material welded.

Reinforcement undoubtedly makes groove welds stronger and better when they are to be subjected to static loads. When the connection is to be subjected to repeated and vibrating loads, however, reinforcement is not as satisfactory because stress concentrations seem to develop in the reinforcement and contribute to earlier failure. For such cases a common practice is to provide reinforcement and grind it off flush with the material being connected. This is often required in highway bridge specifications.

Fig. 13-8 shows some of the edge preparations that may be necessary for groove welds. In part (a) a bevel with a feathered edge is shown. When feathered edges are used there is a problem with burn-through. This may be lessened if a *land* is used such as shown in part (b) of the figure or a backup strip as shown in part (c). The backup strip is often a $\frac{1}{4}$-in. copper plate. Weld metal does not stick to copper, and copper also has a very high conductivity which is useful in carrying away excess heat and reducing distortion. Sometimes steel backup strips are used, but they will become a part of the weld and are left in place. A land should not be used with a backup strip because there is a high possibility that a gas pocket might be formed, preventing full penetration.

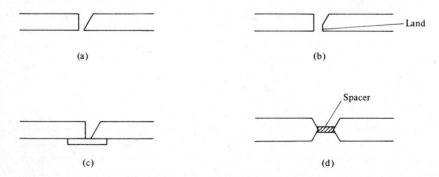

Figure 13-8 Edge preparation for groove welds. (a) Bevel with feathered edge. (b) Bevel with a land. (c) Bevel with a backup plate. (d) Double bevel with a spacer.

When double bevels are used as shown in Fig. 13-8(d), spacers are sometimes provided to prevent burn-through. The spacers are removed after one side is welded.

From the standpoints of strength, resistance to impact and stress repetition, and amount of filler metal required, groove welds are much to be preferred to fillet welds. From other standpoints, however, they are not so attractive, and the vast majority of structural welding is fillet welding. Groove welds have higher residual stresses, and the preparations (such as the scarfing and veeing) of the edges of members for groove welds are expensive, but the major disadvantages probably lie in the problems involved in getting the pieces to fit together in the field. The advantages of fillet welds in this respect were described in Section 13-5. For these reasons field butt joints are not used very often, except on small jobs and where members may be fabricated a little long and cut in the field to the lengths necessary for precise fitting.

13-8 FILLET WELDS

Tests have shown that fillet welds are stronger in tension and compression than they are in shear, so the controlling fillet-weld stresses given by the various specifications are shearing stresses. When practical, it is desirable to try to arrange welded connections so they will be subjected to shearing stresses only and not to a combination of shear and tension or shear and compression.

Fillet welds when tested to failure seem to fail by shear at angles of about 45° through the throat. Their strength is therefore assumed to equal the allowable shearing stress times the theoretical throat area of the weld. The theoretical throats of several fillet welds are shown in Fig. 13-9. The throat area equals the theoretical throat distance times the length of the weld. In this figure the root of the weld is the point where the faces of the original metal pieces intersect, and the theoretical throat of the weld is the shortest distance from the root of the weld to its diagrammatic face.

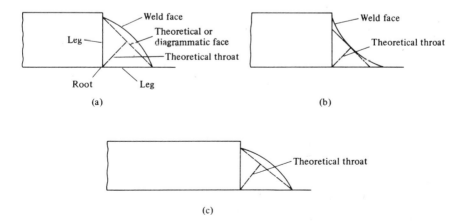

Figure 13-9 (a) Convex surface. (b) Concave surface. (c) Unequal-leg fillet weld.

For the 45° or equal-leg fillet the throat dimension is 0.707 times the leg of the weld; the dimension has a different value for fillet welds with unequal legs. The desirable fillet weld has a flat or slightly convex surface, although the convexity of the weld does not add to its calculated strength. At first glance the concave surface would appear to give the ideal fillet-weld shape because stresses could apparently flow smoothly and evenly around the corner with little stress concentration. Years of experience, however, have shown that single-pass fillet welds of a concave shape have a greater tendency to crack upon cooling, and this factor has proved to be of greater importance than their smoother stress distribution.

When a concave weld shrinks, the surface is placed in tension tending to cause cracks, whereas when the surface of a convex weld shrinks, it does not place the outer surface in tension. Rather, as the face shortens it is placed in compression. See Fig. 13-10.

Another item of importance pertaining to the shape of fillet welds is the angle of the weld with the pieces being welded. The desirable value of this angle is in the vicinity of 45°. For 45° fillet welds the leg sizes are equal, and such welds are referred to by the leg sizes (as a $\frac{1}{4}$-in. fillet weld). Should the leg sizes be different (not a 45° weld), both leg sizes are given in describing the weld, (as a $\frac{3}{8}$- by $\frac{1}{2}$-in. fillet weld).

The automatic SAW method provides a deeper penetration than does the usual shielded arc welding process. As a result the ASD permits the designer to use a larger throat area for welds made by this process. In Section J2.2a the ASD Specification states that the effective throat thickness for SAW fillet welds with legs $\frac{3}{8}$ in. or less may equal the leg sizes. For larger leg sizes the effective throat thicknesses are considered to equal the theoretical throat thicknesses plus 0.11 in.

13-9 ALLOWABLE STRENGTH OF WELDS

For this discussion reference is made to Fig. 13-11. As previously indicated, the stress in a weld is considered to equal the load P divided by the effective throat area of the weld. This method of determining the strength of fillet welds is used regardless of the direction of load. Tests have shown that transverse

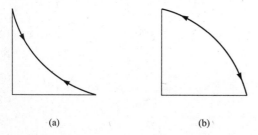

(a)   (b)

Figure 13-10 (a) Concave weld surface put in tension by shrinking. (b) Convex weld surface put in compression by shrinking.

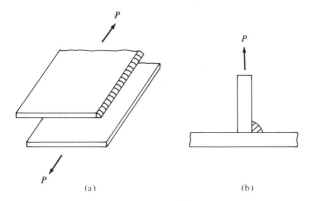

Figure 13-11 (a) Longitudinal fillet weld. (b) Transverse fillet weld.

fillets are without question about a third stronger than are longitudinal fillets, but this fact is not recognized by most specifications in order to simplify design calculations. One reason why transverse fillet welds are stronger is because they are more uniformly stressed for their entire length, while the longitudinal fillets are stressed unevenly due to varying deformations along the length of the weld. Another reason for their greater strength is given by tests which show failure occurs at an angle other than 45°, giving them a larger effective throat area.

13-10 ASD REQUIREMENTS

When welds are made, the electrode material should have properties of the base metal. If the properties are comparable, the weld metal is referred to as the "matching" base metal. Table 4.1.1 of AWS Specification D1.1 gives details concerning "matching" weld metals.

Table 13-1 gives allowable stresses for various types of welds. Should a complete-penetration groove weld be used, the allowable stress in the weld is the same as in the base material. If the connection is in compression normal to the effective area of the weld metal, a weld metal with a strength less than the "matching" weld metal may be used. If in tension, a "matching" weld metal must be used.

For fillet welds the allowable stresses for shear on the effective area of the welds are equal to 0.30 times the nominal tensile strength of the weld metal, but the stress in the base material may not be greater than $0.60F_y$ for tension. Values are also given in Table 13-1 for partial-penetration groove welds and for plug and slot welds.

The filler metal electrodes for shielded arc welding are listed as E60XX, E70XX, and so on. In this classification the letter E represents an electrode, while the first set of digits (which may be 60, 70, 80, 90, 100, or 110) indicates the nominal or minimum ultimate tensile strength of the weld in ksi. The remaining two digits designate the type of coating to be employed with a certain electrode.

The strength is the important matter to the designer, and as a result the last two digits are usually listed as XX. Thus we will say E70XX or just E70, which represents an electrode with an ultimate tensile strength of 70 ksi. The electrodes must match the base metal. Only two electrodes need to be considered for the usual grades of steel:[3] E70XX electrodes for steels with F_y from 36 to 60 ksi and E80XX electrodes for steels with $F_y = 65$ ksi.

In addition to the allowable stresses given in Table 13-1 there are several other ASD provisions applying to welding. Among the more important are the following:

1. The minimum length of a fillet weld may not be less than four times the nominal leg size of the weld. Should its length be less than this value, the weld size considered effective must be reduced to one-quarter of the weld length.

2. The maximum size of a fillet weld along edges of material less than $\frac{1}{4}$ in. thick equals the material thickness. For thicker material the weld may not be larger than the material thickness less $\frac{1}{16}$ in., unless the weld is designated on the drawings to be specially built out to give a full throat thickness.

3. The minimum-size fillet welds are given in Table 13-2 and vary from $\frac{1}{8}$ in. for $\frac{1}{4}$ in. or less thickness of material, up to $\frac{5}{16}$ in. for material over $\frac{3}{4}$ in. in thickness. The smallest practical weld size is about $\frac{1}{8}$ in., and the most economical size is probably about $\frac{5}{16}$ in. The $\frac{5}{16}$-in. weld is about the largest size that can be made in one pass with the SMAW process and $\frac{1}{2}$ in. with the SAW process.

These minimum sizes are not based on strength considerations but rather on the fact that thick materials have a quenching or rapid cooling effect on small welds. If this happens, the result is often a loss in weld ductility. In addition the thicker material tends to restrain the weld material from shrinking as it cools, with the result that weld cracking can be a problem.

TABLE 13-2 MINIMUM SIZE OF FILLET WELDS

MATERIAL THICKNESS OF THICKER PART JOINED (IN.)	MINIMUM SIZE OF FILLET WELD[a] (IN.)
To $\frac{1}{4}$ inclusive	$\frac{1}{8}$
Over $\frac{1}{4}$ to $\frac{1}{2}$	$\frac{3}{16}$
Over $\frac{1}{2}$ to $\frac{3}{4}$	$\frac{1}{4}$
Over $\frac{3}{4}$	$\frac{5}{16}$

Source: American Institute of Steel Construction, *Manual of Steel Construction Allowable Stress Design,* 9th ed. (Chicago: AISC, 1989), Table J2.4, p. 5-67. Reprinted with the permission of AISC.

[a] Leg dimension of fillet welds. Single-pass welds must be used.

---

[3] W. T. Segui, *Fundamentals of Structural Steel Design* (Boston: PWS-Kent, 1989), p. 276.

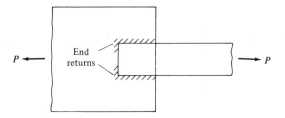

Figure 13-12

You will particularly note that the minimum sizes given in Table 13-2 are dependent on the thicker of the two parts being joined. Regardless of the value given in the table, it does not have to exceed the thickness of the thinner part. It may be larger, however, if so required by the calculated strength.

4. When practical, end returns (sometimes called boxing) should be made for fillet welds as shown in Fig. 13-12. The length of returns should not be less than twice the nominal size of the weld. When they are not used, it is considered good practice by some designers to subtract two times the weld size from the effective weld length. End returns are useful in reducing the high stress concentrations which occur at the ends of welds, particularly for connections where there is considerable vibration and eccentricity of load. The ASD Specification (J2.2a) says that the effective length of fillet welds includes the length of the end returns used.

5. When longitudinal fillet welds are used for the connection of plates or bars, their length may not be less than the perpendicular distance between them because of shear lag, discussed in Chapter 3. Furthermore, the distance between fillet welds may not be greater than 8 in. for end connections unless the member is designed on the basis of its effective area in accordance with ASD Specification B3.

6. For lap joints the minimum amount of lap permitted is equal to five times the thickness of the thinner part joined but may not be less than 1 in. The purpose of this minimum lap is to keep the joint from rotating excessively (see Fig. 11.1[a]) when the connected parts are loaded.

13-11 DESIGN OF SIMPLE FILLET WELDS

Examples 13-1 through 13-3 illustrate the calculations necessary to determine the allowable strength of various fillet-welded connections, while Example 13-4 presents the design of such a connection. In these and other problems weld lengths are selected no closer than to the nearest $\frac{1}{4}$ in., because closer work cannot be expected in shop or field.

■ Example 13-1

Determine the allowable shear force that may be applied to a 1-in. length of a $\frac{5}{16}$-in. fillet weld using (a) the shielded metal arc process (SMAW) and (b) the

The all-welded 56-story Toronto Dominion Bank Tower. (Courtesy of Lincoln Electric Company.)

submerged arc process (SAW). Use the ASD Specification and E70 electrodes with a minimum tensile strength of 70 ksi.

Solution.
(a) SMAW process:

Effective throat thickness of weld $= (0.707)(\frac{5}{16}) = 0.221$ in.

Allowable capacity $= (0.221)(0.30 \times 70)$

$$= \underline{\underline{4.64 \text{ k/in.}}}$$

(b) SAW process:

From ASD (J2.2a) the effective throat thickness of weld = $\frac{5}{16}$ in.

$$\text{Allowable capacity} = (\tfrac{5}{16})(0.30 \times 70) = \underline{\underline{6.56 \text{ k/in.}}}$$ ∎

Fillet welds may not be designed using an allowable stress that is greater than permitted on the adjacent members being connected. For instance, the allowable shear stress on the effective area of fillet welds is 0.30 times the tensile strength of the electrode but may not exceed the allowable stress in the base material ($0.60F_y$ in tension). It should be remembered from the discussion in Section 13-8 that fillet welds are assumed in design to transmit loads by shear on the effective area regardless of the direction of the loads or of the position of the welds at the connections.

Examples 13-2 and 13-3 illustrate the calculations necessary to determine the allowable loads that can be applied to plates connected with SMAW and SAW fillet welds. In each of these examples the allowable strength per inch of the welds controls and is multiplied by the total length of the welds to give the total allowable capacity of the connections.

■ Example 13-2

What is the allowable capacity of the connection shown in Fig. 13-13 if A36 steel, the ASD Specification, and E70 electrodes are used? The $\frac{7}{16}$-in. fillet welds shown were made by the SMAW process.

Solution.

Effective throat thickness = t_e = $(0.707)(\tfrac{7}{16})$ = 0.309 in.

Allowable capacity of weld/in. = $(0.309)(0.30 \times 70)$ = 6.489 k/in. ←

Allowable tensile capacity of weld = $(20)(6.489)$ = 129.8 k

Allowable tensile capacity of PL = $(\tfrac{3}{4} \times 8)(0.60 \times 36)$ = 129.6 k ←

$$P = 129.6 \text{ k}$$ ∎

■ Example 13-3

Repeat Example 13-2 using SAW welds.

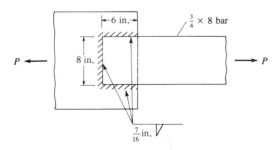

Figure 13-13

Solution.

Effective throat thickness = t_e = $(0.707)(\frac{7}{16})$ + 0.11 = 0.419 in.

Allowable shear capacity of weld/in. = $(0.419)(0.30 \times 70)$ = 8.799 k/in. ←

Allowable shear capacity of weld = $(20)(8.799)$ = 176 k

Allowable tensile capacity of PL = $(\frac{3}{4} \times 8)(0.60 \times 36)$ = 129.6 k←

$$P = 129.6 \text{ k}$$ ■

■ Example 13-4

Using A36 steel, the ASD Specification, and E70 electrodes, design SMAW fillet welds to resist a full capacity load on the $\frac{3}{8}$ × 6-in. member shown in Fig. 13-14.

Solution.

$$P = (\tfrac{3}{8})(6)(22) = 49.5 \text{ k}$$
$$\text{Max weld size} = \tfrac{3}{8} - \tfrac{1}{16} = \tfrac{5}{16} \text{ in.} \qquad \text{(ASD Section J2.2b)}$$
$$\text{Min weld size} = \tfrac{3}{16} \text{ in. (Table 13-2)}$$

$$\underline{\text{Use } \tfrac{5}{16}\text{-in. weld}}$$

Effective thickness = t_e = $(0.707)(\frac{5}{16})$ = 0.221 in.

Allowable capacity of weld/in. = $(0.221)(0.30 \times 70)$ = 4.64 k/in. ←

$$\text{Length required} = \frac{49.5}{4.64} = 10.67 \text{ in.}$$

Use end return not less than $2 \times \tfrac{5}{16} = \tfrac{10}{16}$ (say 1 in.)

$$10.67 - 2 = 8.67 \text{ in. or } 4\tfrac{1}{2} \text{ in. each side}$$

However, <u>use 6 − 1 = 5-in. welds each side as required by ASD Section</u> <u>J2.2b plus end returns.</u> ■

On some occasions the lengths available for the usual longitudinal fillet welds are not sufficient for the load to be resisted. For the situation shown in

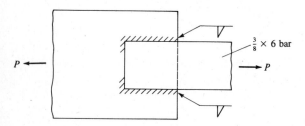

$\frac{3}{8}$ × 6 bar

Figure 13-14

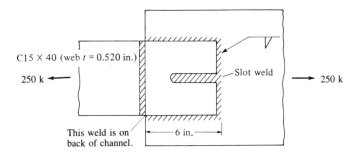

Figure 13-15

Fig. 13-15 it may be possible to develop sufficient strength by welding along the back of the channel at the edge of the plate if sufficient space is available.

Another possibility is the use of slot welds as illustrated in Example 13-5. There are several ASD requirements pertaining to slot welds which need to be mentioned here. This specification says that the width of a slot may not be less than the member thickness plus $\frac{5}{16}$ in. (rounded off to the next greater odd $\frac{1}{16}$ in., since structural punches are made in odd-sixteenths diameters) nor may it be greater than $2\frac{1}{4}$ times the weld thickness. For members up to $\frac{5}{8}$ in. thick the weld thickness must equal the plate thickness, and for members greater than $\frac{5}{8}$ in. thick the weld thickness may not be less than one-half the member thickness or $\frac{5}{8}$ in. The maximum length permitted for slot welds is ten times the weld thickness. The limitations given in specifications for the maximum sizes of plug or slot welds are caused by the detrimental shrinkage which occurs around these types of welds when they exceed certain sizes. Should holes or slots larger than those specified be used, it is desirable to use fillet welds around the borders of the holes or slots rather than using a slot or plug weld. Slot and plug welds are normally used in conjunction with fillet welds in lap joints. Sometimes plug welds are used to fill in the holes temporarily used for erection bolts for beam and column connections. They may or may not be included in the calculated strength of these joints.

The strength of a plug or slot weld is equal to its allowable stress times its nominal area in the shearing plane. This area is equal to the area of contact at the base of the plug or slot. The length of a slot weld can then be determined by

$$L = \frac{\text{load}}{(\text{width})(\text{allowable stress})}$$

Example 13-5 illustrates the design of the welds necessary to connect a channel to a plate. The calculations show that the ordinary side and end fillet welds do not provide sufficient strength. It is decided to use a slot weld to provide the remaining load resistance needed.

■ Example 13-5
Design SMAW fillet welds to connect a C15 × 40 to the plate shown in Fig. 13-15. The load to be resisted is 250 k, and E70 electrodes are to be used. As

shown in the figure the channel may lap over the plate by only 6 in. due to space limitations. See ASD Specification J2.3. A weld may not be placed on the back of the channel.

Solution. Due to limited space use

$$\text{Max weld size} = \text{web } t - \tfrac{1}{16} = \tfrac{1}{2} - \tfrac{1}{16} = \tfrac{7}{16} \text{ in.}$$
$$\text{Effective throat thickness} = t_e = (0.707)(\tfrac{7}{16}) = 0.309 \text{ in.}$$
$$\text{Capacity of weld/in.} = (0.309)(0.30 \times 70) = 6.489 \text{ k/in.} \leftarrow$$
$$\text{Length required} = \frac{250}{6.489} = 38.53 \text{ in.} > 27 \text{ in. available}$$

$$\therefore \text{ Use a slot weld.}$$

$$\text{Min width of slot} = 0.520 + \tfrac{5}{16} = \tfrac{13}{16} \text{ in.}$$
$$\text{Max width} = 2\tfrac{1}{4} \times \text{ weld thickness}$$
$$= (2\tfrac{1}{4})(\text{web } t \text{ of channel}) = (2\tfrac{1}{4})(\tfrac{1}{2})$$
$$= 1\tfrac{1}{8} \text{ say } \tfrac{17}{16} \text{ (to odd } \tfrac{1}{16})$$
$$\underline{\text{Use } \tfrac{15}{16} \text{ in.}}$$

$$\text{Capacity of } \tfrac{7}{16}\text{-in. fillet weld} = (6.489)(6 + 6 + 15 - \tfrac{15}{16}) = 169.1 \text{ k}$$
$$\text{Load to be resisted by slot weld} = 250 - 169.1 = 80.9 \text{ k}$$

$$\text{Length required for slot weld} = \frac{80.9}{(\tfrac{15}{16})(0.30 \times 70)} = 4.11 \text{ in.} \qquad \underline{(\text{say } 4\tfrac{1}{2} \text{ in.})}$$

$$\text{Max length permitted by ASD} = (10)(\tfrac{1}{2})$$
$$= 5.00 \text{ in.} > 4\tfrac{1}{2} \text{ in.} \qquad \text{OK}$$

$$\underline{\text{Use } \tfrac{15}{16} \times 4\tfrac{1}{2}\text{-in. slot weld}}$$

Alternate Solution.
Should space be available on the back of the channel next to the plate, a $\tfrac{7}{16}$-in. fillet weld would carry $(15)(6.489) = 97.3 \text{ k} > 80.9 \text{ k}$. This would be a much more economical solution. ∎

13-12 DESIGN OF FILLET WELDS FOR TRUSS MEMBERS

Should the members of a welded truss consist of single angles, double angles, or similar shapes and be subjected to static axial loads only, the ASD (J1.9) permits their connections to be designed by the same procedures described in the preceding section. The designers can select the weld size, calculate the to- tal length of the weld required, and place the welds around the member ends as they see fit.[4] (It would not be logical, of course, to place the weld all on one

---

[4] *Welding Journal* (January 1942), pp. 44–45.

side of a member such as for the angle of Fig. 13-16 because of the rotation possibility.) Example 13-6 illustrates the simple calculations involved in designing the welds for the ends of a truss member.

■ Example 13-6

Using the ASD Specification, A36 steel, and E70 electrodes, design side and end fillet SMAW welds for the full capacity of a 6 × 4 × $\frac{1}{2}$-in. angle tension member with the long leg connected. Assume static load.

Solution.

$$\text{Tensile capacity of } L = (4.75)(22) = 104.5 \text{ k}$$
$$\text{Max weld size} = \tfrac{1}{2} - \tfrac{1}{16} = \tfrac{7}{16} \text{ in.} \qquad \text{(ASD J2.2b)}$$
$$\text{Min weld size} = \tfrac{3}{16} \text{ in.} \qquad \text{(from Table 13-2)}$$

Use $\tfrac{5}{16}$-in. weld as maximum size which can be made in one pass

$$\text{Effective thickness} = t_e = (0.707)(\tfrac{5}{16}) = 0.221 \text{ in.}$$
$$\text{Capacity of weld/in.} = (0.221)(0.30 \times 70) = 4.64 \text{ k/in.} \leftarrow$$

$$\text{Length required} = \frac{104.5}{4.64} = 22.52 \text{ in.} \qquad \text{(say 23 in.)}$$

Place welds as shown in Fig. 13-16 ■

The student should carefully note that the centroid of the welds and the centroid of the statically loaded angle do not coincide in the connection selected in Example 13-6 and shown in Fig. 13-16. Should a welded connection

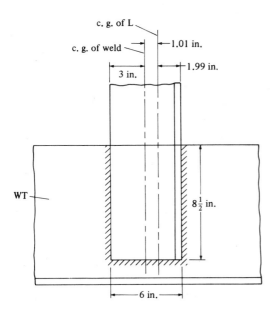

Figure 13-16

be subjected to repeated stresses (such as those occurring in a bridge member), it is considered desirable to place the welds so that their centroid will coincide with the centroid of the member (or the resulting torsion must be accounted for in design). If the member being connected is symmetrical, the welds will be placed symmetrically, but if the member is not symmetrical, the welds will not be symmetrical.

The force in an angle, such as the one shown in Fig. 13-17, is assumed to act along its center of gravity. If the center of gravity of weld resistance is to coincide with the angle force, it must be asymmetrically placed, or in this figure L_1 must be longer than L_2. (When angles are connected by rivets or bolts, there is usually an appreciable amount of eccentricity, but in a welded joint eccentricity can be fairly well eliminated.) The information necessary to handle this type of weld design can be easily expressed in equation form, but only the theory behind the equations is presented here.

For the angle shown in Fig. 13-17 the force acting along line L_2 (designated here as P_2) can be determined by taking moments about L_1. The member force and the weld resistance are to coincide, and the moments of the two about any point must be zero. If moments are taken about L_1, the force P_1 (which acts along line L_1) will be eliminated from the equation, and P_2 can be determined. In a similar manner P_1 can be determined by taking moments along L_2 or by $\Sigma V = 0$. Example 13-7 illustrates the design of fillet welds of this type. A similar problem is handled in Example 13-8 except an end fillet weld is included. The center of gravity and resistance of the end weld are known and can be easily included in the moment equations.

There are other possible solutions for the design of the welds for the angle considered in these two examples. Although the $\frac{7}{16}$-in. weld is the largest one permitted at the rounded edges of the $\frac{1}{2}$-in. angle and at its end, a larger weld could be used on the other side next to the outstanding leg. From a practical point of view, however, the welds should be the same size because different-size welds slow the welder down in that he or she has to change electrodes to make different sizes.

According to our fatigue discussion in Chapter 4, if the estimated number of loading cycles during the structure's estimated life exceeds 20,000, it will be necessary to study the stress range of the connection at service loads. This

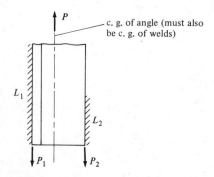

Figure 13-17

study may very well result in larger connections as required by Appendix K of the ASD Specification.

■ Example 13-7

Using the ASD Specification, A36 steel, E70 electrodes, and the SMAW process, design side fillet welds for the full capacity of the $5 \times 3 \times \frac{1}{2}$-in. angle tension member shown in Fig. 13-18. Assume the member is subjected to repeated stress variations, making any connection eccentricity undesirable.

Solution.

$$\text{Tensile capacity of } L$$
$$P = (22)(3.75) = 82.50 \text{ k} \leftarrow$$
$$\text{assuming } U = 0.85$$
$$P = (0.50 \times 58)(3.75) = 92.44 \text{ k}$$
$$\text{Max weld size} = \frac{1}{2} - \frac{1}{16} = \frac{7}{16} \text{ in.}$$
$$\underline{\text{Use } \tfrac{5}{16}\text{-in. weld}}$$

Effective throat thickness $= t_e = (0.707)(\frac{5}{16}) = 0.221$ in.

Capacity of weld/in. $= (0.221)(0.30 \times 70) = 4.641$ k/in. $\leftarrow$

$$\text{Total weld length required} = \frac{82.50}{4.641} = 17.78 \text{ in.}$$

Taking moments about L_1 (see Fig. 13-18) to determine force P_2

$$(82.50)(1.75) - 5.00P_2 = 0$$
$$P_2 = 28.88 \text{ k}$$
$$P_1 = P - P_2 = 82.50 - 28.88 = 53.62 \text{ k}$$
$$L_1 = \frac{53.62}{4.641} = 11.55 \text{ in.} \qquad \underline{(\text{say } 11\tfrac{1}{2} \text{ in.})}$$
$$L_2 = \frac{28.88}{4.641} = 6.22 \text{ in.} \qquad \underline{(\text{say } 6\tfrac{1}{2} \text{ in.})}$$

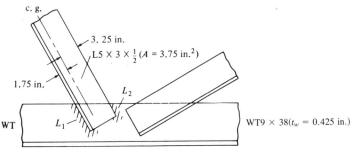

Figure 13-18

Use end returns $2 \times \frac{5}{16} = \frac{10}{16}$ (say 1 in.). The end return lengths may be subtracted from the side weld lengths, making them $\underline{\underline{11 \text{ in.}}}$ and $\underline{\underline{5\frac{1}{2} \text{ in.}}}$, respectively.

Check block shearing strength, assuming the dimensions shown in Fig. 13-19.

$$T_{bs} = 0.30F_u A_v + 0.50F_u A_t$$
$$= (0.30)(58)(11 + 5\tfrac{1}{2})(0.425) + (0.50)(58)(5 \times 0.425)$$
$$= 183.6 \text{ k} > 82.50 \text{ k} \qquad\qquad\qquad\quad \text{OK} \quad \blacksquare$$

■ Example 13-8

Rework Example 13-8 using fillet welds along the sides and end of the angle.

Solution.   Assuming $\frac{5}{16}$-in. weld (capacity = 4.641 k/in. from Example 13-7):

$$\text{Strength of end weld} = (5.00)(4.641) = 23.20 \text{ k}$$

Taking moments about L_1 to determine force P_2:

$$(82.50)(1.75) - (2.50)(23.20) - 5.00P_2 = 0$$
$$P_2 = 17.28 \text{ k}$$
$$P_1 = 82.5 - 17.28 - 23.20 = 42.02 \text{ k}$$
$$L_1 = \frac{42.02}{4.641} = 9.05 \text{ in.} \quad \underline{\underline{\text{(say 9 in.)}}}$$
$$L_2 = \frac{17.28}{4.641} = 3.71 \text{ in.} \quad \underline{\underline{\text{(say 4 in.)}}} \qquad \blacksquare$$

It is rather convenient for design purposes to know the allowable strength of a $\frac{1}{16}$-in. fillet weld 1 in. long. Though this size is below the minimum permissible size given in Table 13-2, the calculated strength of such a weld is useful for determining weld sizes for calculated forces. For a 1-in.-long SMAW weld and an E70 electrode we have

$$(0.707)(\tfrac{1}{16})(1.0)(0.30)(70) = 0.928 \text{ k/in.}$$

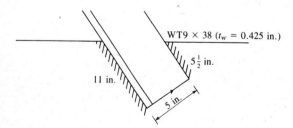

WT9 × 38 $(t_w = 0.425 \text{ in.})$

$5\frac{1}{2}$ in.

11 in.

5 in.

Figure 13-19

If we are designing a fillet weld to carry a force of 4.5 k/in., the required weld size will be $4.5/0.928 = 4.85$ sixteenths of an inch, say $\frac{5}{16}$ in.

13-13 SHEAR AND TORSION

Fillet welds are frequently loaded with eccentrically applied loads, with the result that the welds are subjected to either shear and torsion or to shear and bending. Fig. 13-20 presents the difference between the two situations. Shear and torsion, shown in part (a) of the figure, is the subject of this section while shear and bending, shown in part (b), is the subject of Section 13-14.

As is the case for eccentrically loaded bolt groups (Section 12-1), the ASD Specification provides the permissible design strength of welds but does

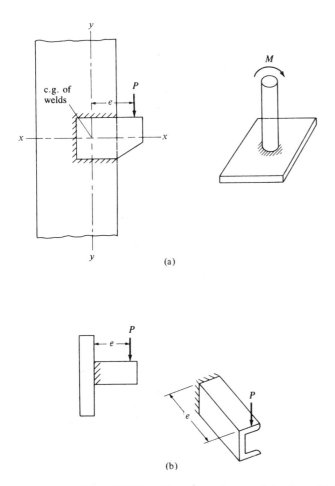

(a)

(b)

Figure 13-20 (a) Welds subjected to shear and torsion. (b) Welds subjected to shear and bending.

not specify a method of analysis for eccentrically loaded welds. The method to be used is left up to the designer.

Elastic Method

In the very conservative elastic method friction or slip resistance between the connected parts is neglected and the connected parts are assumed to be perfectly rigid.

For this discussion the welded bracket of Fig. 13-20(a) is considered. The pieces being connected are assumed to be completely rigid, as they were in bolted connections. The effect of this assumption is that all deformation occurs in the weld. The weld is subjected to a combination of shear and torsion as was the eccentrically loaded bolt group considered in Section 12-1. The force caused by torsion can be computed from the familiar expression

$$f = \frac{Td}{J}$$

In this expression T is the torsion, d is the distance from the center of gravity of the weld to the point being considered, and J is the polar moment of inertia of the weld. It is usually more convenient to break the force down into its vertical and horizontal components. In the expressions to follow, h and v are the horizontal and vertical components of the distance d. (These formulas are almost identical to those used for determining stresses in bolt groups subject to torsion.)

$$f_h = \frac{Tv}{J} \qquad f_v = \frac{Th}{J}$$

These components are combined with the usual direct shearing stress, which is assumed to equal the reaction divided by the total length of the welds. For design of a weld subject to shear and torsion it is convenient to assume a 1-in. weld and compute the stresses on a weld of that size. Should the assumed weld be overstressed, a larger weld is required; if understressed, a smaller one is desirable.

Although the calculations will in all probability show the weld to be overstressed or understressed, the math does not have to be repeated because a ratio can be set up to give the weld size for which the load would produce a stress exactly equal to the design stress. The student should note that the use of a 1-in. weld simplifies the units because 1 in. of length of weld is 1 in.^{2} of weld and the computed stresses can be said to be either kips per square inch or kips per inch of length. Should the calculations be based on some size other than a 1-in. weld, the student will have to be very careful in keeping the units straight, particularly in obtaining the final weld size. To further simplify the calculations the welds are assumed to be located at the edges along which the fillet welds are placed rather than at the centers of their effective throats. As the throat dimensions are rather small, this assumption changes the results very little. Example 13-9 illustrates the calculations involved in determining the weld size required for a connection subjected to a combination of shear and torsion.

■ Example 13-9

For the bracket shown in Fig. 13-21(a), determine the fillet-weld size required if E70 electrodes, the ASD Specification, and the SMAW process are used.

Solution. Assuming a 1-in. weld as shown in Fig. 13-21(b):

$$A = 18 \text{ in.}^2$$

$$\bar{x} = \frac{(4)(2)(2)}{18} = 0.89 \text{ in.}$$

$$I_x = (\tfrac{1}{2})(1)(10)^3 + (2)(4)(5)^2 = 283.3 \text{ in.}^4$$
$$I_y = (2)(\tfrac{1}{3})(0.89^3 + 3.11^3) + (10)(0.89)^2 = 28.5 \text{ in.}^4$$
$$J = 283.3 + 28.5 = 311.8 \text{ in.}^4$$

Most stressed portions of weld are greatest distance from weld center of gravity (*A* and *B* in Fig. 13-21[b]).

$$f_h = \frac{Tv}{J} = \frac{(15 \times 11.11)(5)}{311.8} = 2.67 \text{ k/in.}^2$$

$$f_v = \frac{Th}{J} = \frac{(15 \times 11.11)(3.11)}{311.8} = 1.66 \text{ k/in.}^2$$

$$f_s = f_{\text{shear}} = \frac{15}{18} = 0.83 \text{ k/in.}^2$$

$$f_r = f_{\text{resultant}} = \sqrt{(1.66 + 0.83)^2 + (2.67)^2} = 3.65 \text{ k/in.}^2$$

Allowable capacity of a 1-in. fillet weld (E70 electrode)
$$= (0.707)(1.00)(0.30 \times 70) = 14.85 \text{ k/in.}$$

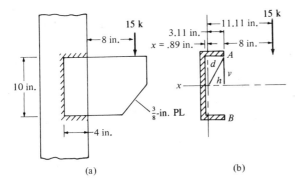

(a)   (b)

Figure 13-21

$$\text{Weld size required} = \frac{3.65}{14.85} = 0.246 \text{ in.} \qquad (\text{say } \tfrac{1}{4} \text{ in.})$$

$$\text{Use } \tfrac{1}{4}\text{-in. fillet weld} \qquad \blacksquare$$

Ultimate Strength Method

An ultimate strength analysis of eccentrically loaded welded connections is much more realistic than is the very conservative elastic procedure just described. For the discussion to follow, the eccentrically loaded fillet weld of Fig. 13-22 is considered. As for eccentrically loaded bolted connections, the load tends to cause a relative rotation and translation between the parts connected by the weld.

Even if the eccentric load is of such a magnitude as to cause the most stressed part of the weld to yield, the entire connection will not fail. The load may be increased, the less stressed fibers will begin to resist more of the load, and failure will not occur until all the weld fibers yield. The weld will tend to rotate about its instantaneous center of rotation. The location of this point (which is indicated by the letter O in the figure) is dependent upon the location of the eccentric load, the geometry of the weld, and the deformations of the different elements of the weld.

If the eccentric load P_u is vertical and if the weld is symmetrical about a horizontal axis through its center of gravity, the instantaneous center will fall somewhere on the horizontal x axis. Each differential element of the weld will provide a resisting force R. As shown in Fig. 13-22 each of these resisting forces is assumed to act perpendicular to a ray drawn from the instantaneous center to the center of gravity of the weld element in question. The sum of the moments of the resisting forces of all the elements of the weld about point O must be equal and opposite to the moment of the eccentric load about the same point.

$$P_u(e' + e) = \Sigma(Rds)(d)$$

$$P_u = \frac{\Sigma(Rds)(d)}{e' + e}$$

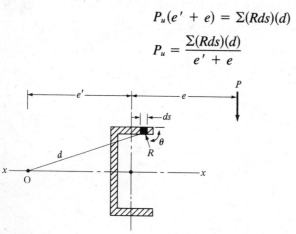

Figure 13-22

Studies have been made to determine the maximum shear forces which eccentrically loaded weld elements can withstand.[5,6] The results which depend on the load-deformation relationship of the weld elements may be represented either with curves or in formula fashion. The ductility of the entire weld is governed by the maximum deformation of the weld element which first reaches its limit (it is probably the element the greatest distance from the weld's instantaneous center). In the expression to follow, R is the ultimate shear force for a single weld element subjected to eccentric shear, R_{ult} is the ultimate pure shear force for a single weld element, Δ is the deformation of a weld element, and μ and λ are regression coefficients.

$$R = R_{ult}(1 - e^{-\mu\Delta})^\lambda$$

We can assume a location of the instantaneous center, determine R values for the different elements of the weld, and compute the value of P_u from the equation given. If this value of P_u does not equal the sum of the vertical components of the R values, we will need to assume another location for point O, and so on.

To determine the allowable loads given in the tables entitled "Eccentric Loads on Weld Groups" in Part 4 of the ASD Manual, the value of R for each element is

$$R = (0.707)(\text{leg size})(0.30)(\text{nominal tensile strength}$$
$$\text{of weld metal})(\text{length of element})$$

A numerical example of this tedious trial-and-error procedure is not presented, but two examples using the tables are included. In these tables the allowable eccentric load P applied to a particular welded connection can be determined from the following expression, in which C is a coefficient taken from the ASD tables mentioned above, C_1 is a coefficient depending on the electrode number and given in Table 13-3, D is the weld size in sixteenths of an

TABLE 13-3 C_1 COEFFICIENTS

ELECTRODE	E60	E70	E80	E90	E100	E110
F_v (ksi)	18.0	21.0	24.0	27.0	30.0	33.0
C_1	0.857	1.0	1.14	1.29	1.43	1.57

Source: American Institute of Steel Construction, *Manual of Steel Construction Allowable Stress Design,* 9th ed. (Chicago: AISC, 1989), p. 4-72. Reprinted with the permission of AISC.

---

[5] L. J. Butler, S. Pal, and G. L. Kulak, "Eccentrically Loaded Weld Connections," *Journal of Structural Division,* 98, ST5, (May 1972) pp. 989–1005.

[6] G. L. Kulak and P. A. Timmler, "Tests on Eccentrically Loaded Fillet Welds," Dept. of Civil Engineering, University of Alberta, Edmonton, December 1984.

inch, and l is the length of the weld in question, which is identified with each of the Manual tables.

$$P = CC_1Dl$$

The Manual tables were prepared for vertical loads, but they present a procedure for adjusting the table values for eccentric loads that are inclined from the vertical.

Should a particular eccentrically loaded connection not be covered by the tables, it is suggested that the elastic method be used. Examples 13-10 and 13-11 illustrate the use of the ASD tables.

■ Example 13-10

Using the "Eccentric Loads on Weld Groups" tables of Part 4 of the ASD Manual, determine the allowable load P which the $\frac{5}{16}$-in. weld of Fig. 13-23 can support. Assume E70 electrodes are used.

Solution.

$$k = \frac{4}{12} = 0.333$$

$$xl = \frac{(2)(4)(2)}{20} = 0.80 \text{ in.}$$

$$x = \frac{0.80}{12} = 0.0667$$

x can also be determined by entering Table XXIII in the Manual with $k = 0.333$ and determining the value by interpolation at the bottom of the table.

$$al = 8 + 4 - 0.80 = 11.20 \text{ in.}$$

$$a = \frac{11.20}{12} = 0.933$$

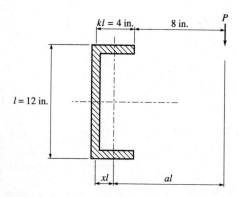

Figure 13-23

C from Table XXIII by interpolation $= 0.510$
$C_1 = 1.0$ for E70 electrodes (Table 13-3)

$$D = 5$$
$$P = CC_1Dl = (0.510)(1.0)(5)(12) = \underline{\underline{30.6 \text{ k}}}$$ ∎

■ Example 13-11

Determine the weld size for the $\frac{7}{16}$-in. weld (E70 electrodes) shown in Fig. 13-24, using the "Eccentric Loads on Weld Groups" tables of Part 4 of the ASD Manual.

Solution.

$$k = \frac{6}{16} = 0.375$$

$$x = \frac{3}{16} = 0.1875$$

$$al = 6 + 3 = 9 \text{ in.}$$

$$a = \frac{9}{16} = 0.5625$$

C from Table XXI in Manual by interpolation $= 1.283$
$C_1 = 1.0$ for E70 electrodes (see Table 13-3)

$$D = \frac{P}{CC_1l} = \frac{100}{(1.283)(1.0)(16)} = 4.87 \text{ sixteenths of an in.}$$

$$\text{Use } \tfrac{5}{16} \text{ in.}$$ ∎

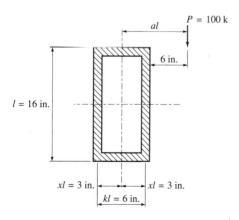

Figure 13-24

13-14 SHEAR AND BENDING

The welds shown in Fig. 13-20(b) and in Fig. 13-25 are subjected to a combination of shear and bending.

For short welds of this type the usual practice is to consider a uniform variation of shearing stress. If, however, the bending stress is assumed to be given by the flexure formula, the shear does not vary uniformly for vertical welds but as a parabola with a maximum value $1\frac{1}{2}$ times the average value. These stress and shear variations are shown in Fig. 13-26.

The student should carefully note that the maximum shearing stresses and the maximum bending stresses occur at different locations. Therefore, it is probably not necessary to combine the two stresses at any one point. If the weld is capable of withstanding the worst shear and the worst moment individually, it is probably satisfactory. In Example 13-12, however, a welded connection subjected to shear and bending is designed by the usual practice of assuming a uniform shear distribution in the weld and combining the value vectorially with the maximum bending stress.

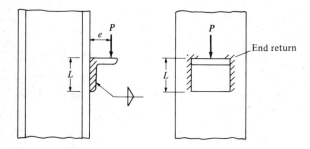

Figure 13-25

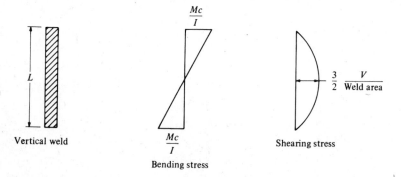

Figure 13-26 Stress and shear variations in a vertical weld subject to shear and bending.

■ **Example 13-12**

Using E70 electrodes, the SMAW process, and the ASD Specification, determine the weld size required for the connection of Fig. 13-25 if $P = 30$ k, $e = 2\frac{1}{2}$ in., and $L = 8$ in. Assume that the member thicknesses do not control weld size.

Solution.

$$f_s = \frac{30}{(2)(8)} = 1.875 \text{ k/in.}$$

$$f = \frac{(30 \times 2.5)(4)}{(\frac{1}{12})(1)(8)^3(2)} = 3.51 \text{ k/in.}$$

$$f_r = \sqrt{(1.875)^2 + (3.51)^2} = 3.98 \text{ k/in.}$$

$$\text{Weld size required} = \frac{3.98}{(0.707)(1)(0.30 \times 70)} = 0.268 \qquad (\text{say } \tfrac{5}{16} \text{ in.}) \quad ■$$

The subject of shear and bending is a very practical one, as it is the situation commonly faced in moment-resisting connections. This topic is continued at length in Section 13-15.

13-15 DESIGN OF MOMENT-RESISTING CONNECTIONS

It is not the purpose of the author to describe all the different arrangements of bolted and welded moment-resisting connections. He instead attempts to present the basic theory of transferring shear and moment from the beam to another member and provides one simple numerical example. If this information is clearly understood, the reader should be able to satisfactorily design moment-resisting connections whatever the arrangements of the bolts and/or welds.

For continuous structures the connections are designed to resist the full calculated moments. Fig. 13-27(a) shows a common type of moment-resisting connection. In the connection shown the tensile force at the top of the beam is transferred by fillet welds to the top plate and by groove welds from the plate to the column. For easier welding the top plate may be tapered as shown in Fig. 13-27(b). The student may have noticed such tapered plates used for facilitating welding in other situations.

Sometimes the beam flanges are groove-welded flush with the column on one end and connected to the beam on the other end with the type of connection just described. This practice is illustrated in Example 13-13.

■ **Example 13-13**

Design moment-resistant connections of the type shown in Fig. 13-28 for the ends of a W18 × 46. The beam, which consists of A36 steel, has end reac-

Top flanges of beams connected by strap plates over a girder, while lower flanges are butt-welded to the girder web, Ainsley Building, Miami, Fla. (Courtesy of Lincoln Electric Company.)

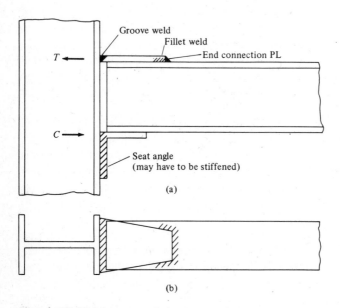

Figure 13-27 Moment-resisting connection.

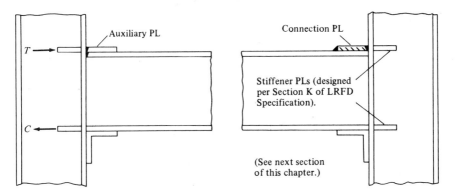

Figure 13-28

tions of 42 k and end moments of 165 ft-k. SMAW welds and E70 electrodes are to be used. It is assumed that a $6 \times 4 \times \frac{3}{4}$-in. seat L has previously been selected.

Solution.   Shear connection:
    Try $\frac{1}{4}$-in. fillet welds on seat L

$$\text{Depth required} = \frac{42}{(2)(0.707)(\frac{1}{4})(0.30 \times 70)}$$

$$= 5.66 \text{ in.}$$

Use 6 in. each side

Moment connections on flush end:
    Assuming a bevel butt weld for full flange width:

$$T = \text{force to be carried}$$

$$= \text{moment divided by center-to-center distance of flanges}$$

$$= \frac{12 \times 165}{18.06 - 0.605} = 113.4 \text{ k}$$

Strength of butt weld in tension for full width of flange = $(6.060)(0.605)(24) = 88.0$ k. Tension to be resisted by auxiliary PL = $113.4 - 88.0 = 25.4$ k. Assuming a $\frac{3}{8}$-in.-thick PL, its width will equal

$$\frac{25.4}{\frac{3}{8} \times 24} = 2.82 \text{ in.} \qquad \text{(say 3 in.)}$$

Assuming a $\frac{3}{16}$-in. fillet weld on auxiliary PL as shown in Fig. 13-29:

$$\text{Length of fillet weld} = \frac{25.4}{(\frac{3}{16})(0.707)(0.30 \times 70)} - 3$$

$$= 6.12 \text{ in.} \qquad \text{(say 4 in. each side)}$$

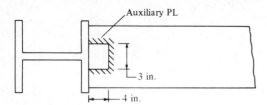

Figure 13-29

Allow 1 in. for bevel groove weld.

$$\text{Use auxiliary PL}\tfrac{3}{8} \times 3 \times 4$$

Notes:

1. On page 4–106 in the ASD Manual it is stated that based on research at the University of California and Lehigh University the full plastic moment capacity of a beam can be developed with a full penetration groove weld connecting the girder flange to the column. *Based on this information the plate and fillet weld designed for each of the flanges above are unnecessary.*

2. Should wind moment be involved allowable stresses may be increased by one-third to resist wind forces, says ASD Specification A5.2.

Moment connection at nonflush end:
  Assuming T and C a distance apart equal to the beam depth:

$$T = \frac{12 \times 165}{18.06} = 109.6 \text{ k}$$

Assuming top connection has a width something less than the width of the beam flange (say 5 in.), its thickness can be found as follows:

$$t = \frac{109.6}{(5)(24)} = 0.913 \text{ in.} \qquad (\text{say } 1.0 \text{ in.})$$

Assume $\tfrac{3}{8}$-in. fillet welds of PL:

$$\text{Weld length required} = \frac{109.6}{(3/8)(0.707)(0.30 \times 70)} = 19.69 \text{ in.}$$

$$\text{See Fig. 13-30}$$

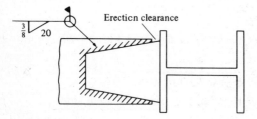

Figure 13-30

PROBLEMS

13-1 A $\frac{5}{16}$-in. fillet-weld SMAW process is used to connect the members shown. Determine the design load that can be applied to this connection including the strength of the member according to the ASD Specification, using A36 steel and E70 electrodes. (*Ans.* 64.8 k)

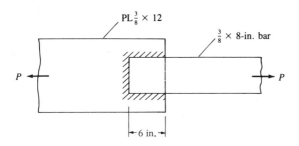

13-2 Rework Prob. 13-1 using the SAW process.

13-3 Rework Prob. 13-1 using A572, grade 65, steel and E80 electrodes. (*Ans.* 106.1 k)

13-4 Design maximum-size fillet welds to develop the full strength of the A36 bar shown. Use E70 electrodes and the SMAW process.

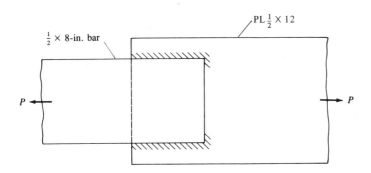

13-5 Rework Prob. 13-4 using the SAW process. (*Ans.* Use $\frac{7}{16}$-in. welds 8 in. long each side as required by ASD Specification J2.2b. Includes end returns.)

13-6 Rework Prob. 13-4 using side welds and a vertical end weld on the $\frac{1}{2} \times 8$ bar. Also use A572, grade 65, steel and E80 electrodes.

13-7 Rework Prob. 13-4 using side welds and welds at the end of the PL $\frac{1}{2} \times 8$ at the connection. (*Ans.* Use $\frac{7}{16}$-in. welds 8 in. long on end and 3 in. each side)

13-8 The PL$\frac{5}{8} \times 8$ shown consists of A36 steel and is to be connected to a gusset plate with $\frac{5}{16}$-in. SMAW fillet welds. Determine the length L required to develop the full strength of the bar if E70 electrodes are used.

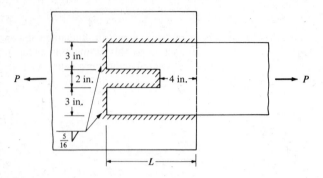

13-9 Design maximum-size side SMAW fillet welds to develop the full tensile strength of an L6 × 4 × $\frac{1}{2}$ × 4 using E70 electrodes and A36 steel. The member is connected on the 6-in. leg and is subject to alternating loads. (*Ans.* Use $5\frac{1}{2}$- and 11-in. side welds)

13-10 Rework Prob. 13-9 using side welds and a weld at the end of the angle.

13-11 Rework Prob. 13-10 using A588 steel and E80 electrodes. (*Ans.* Use $3\frac{1}{2}$- and 10-in. side welds)

13-12 One leg of an 8 × 8 × $\frac{3}{4}$ angle is to be connected with side welds and an end weld to a plate behind to develop a load P of 174 k (A36). Balance the SMAW fillet welds around the center of gravity of the angle, use maximum weld size, and assume E70 electrodes.

13-13 It is desired to design $\frac{5}{16}$-in. SMAW fillet welds necessary to connect a C10 × 30 made from A36 steel to a $\frac{3}{8}$-in. gusset plate. End, side, and slot welds may be used to develop the full tensile capacity of the channel. No welding is permitted on the back of the channel. Use E70 electrodes. It is assumed that due to space limitations the channel can lap over the gusset plate by a maximum of 10 in. (*Ans.* 10-in. side welds and 10-in. end weld and $\frac{17}{16}$ × 3 in. slot weld)

13-14 Rework Prob. 13-13 using A572, grade 60, steel and E70 electrodes and $\frac{5}{16}$-in. fillet welds.

13-15 Using the elastic method, determine the maximum force per inch to be resisted by the fillet weld shown. (*Ans.* 8.83 k/in.)

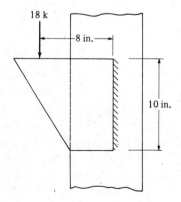

13-16 Using the elastic method, determine the maximum force to be resisted per inch by the fillet weld shown.

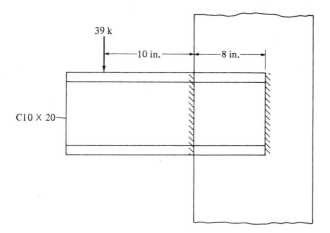

13-17 Using the elastic method, rework Prob. 13-16 using welds on the top and bottom of the channel in addition to those shown in the figure. (*Ans.* 4.36 k/in.)

13-18 Using the elastic method, determine the maximum force per inch to be resisted by the fillet welds shown. Also determine the required weld thickness using A36 steel, E70 electrode, and SMAW welds.

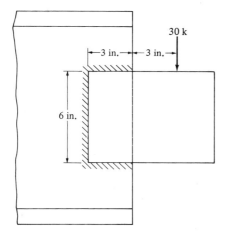

13-19 Determine the maximum eccentric load P that can be applied to the connection shown if $\frac{1}{4}$-in. SMAW fillet welds are used. Assume plate thickness $= \frac{1}{2}$ in., E70 electrodes, and A36 steel.   **(a)** Use elastic method.   **(b)** Use ASD tables and ultimate strength method. (*Ans.* [a] 11.0 k [b] 14.66 k)

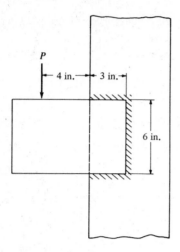

13-20 Rework Prob. 13-19 using $\frac{3}{8}$-in. fillet welds and a vertical weld 10 in. high.

13-21 Determine the SMAW fillet-weld size required for the connection of Prob. 13-15 with the load increased to 30 k and the height of the weld is 12 in. Use A36 steel and E70 electrodes. **(a)** Use elastic method. **(b)** Use ASD tables and ultimate strength method. (*Ans.* [a] 0.694 in., say $\frac{3}{4}$ in. [b] $\frac{8.14}{16}$ in., say $\frac{9}{16}$ in.)

13-22 Using E70 electrodes, the SMAW process, and A36 steel, determine the fillet weld size required for the bracket shown. **(a)** Use elastic method. **(b)** Use ASD tables and ultimate strength method.

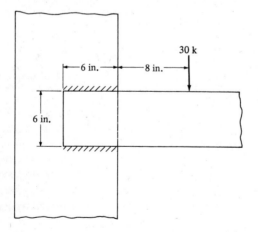

13-23 Rework Prob. 13-22 with the load increased from 30 to 50k and the weld lengths increased from 6 to 8 in. (*Ans.* [a] 1.06 in., say $1\frac{1}{16}$ in. [b] 12.86/16 in., say $\frac{13}{16}$ in.)

13-24 Determine the SMAW fillet-weld size required for the connection shown. Use A36 steel and E70 electrodes. **(a)** Use elastic method. **(b)** Use tables and ultimate strength method.

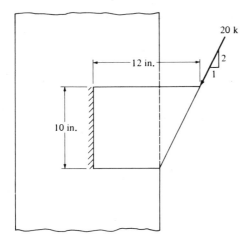

13-25 Determine the SMAW fillet-weld size required for the connection shown. Use A36 steel and E70 electrodes. What angle thickness should be used?   **(a)** Use elastic method.   **(b)** Use ASD tables and ultimate strength method. (*Ans.* [a] 0.271 in., say $\frac{5}{16}$-in. weld and $\frac{3}{8}$-in. angle [b] 3.69/16 in., say $\frac{1}{4}$-in. weld and $\frac{5}{16}$-in. angle)

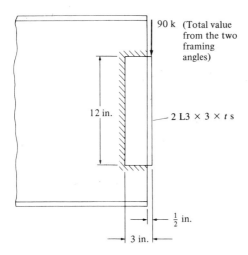

13-26 Assuming the SMAW process is to be used, determine the fillet-weld size required for the connection shown. Use A36 steel, E70 electrodes, and the elastic method.

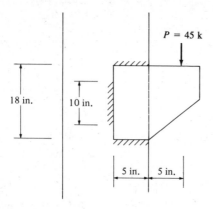

13-27 Determine the fillet-weld size required for the connection shown. Use A36 steel, E70 electrodes, and the SMAW process. **(a)** Use elastic method. **(b)** Use ASD tables and ultimate strength method. (*Ans.* (a) 1.06 in. Say $\frac{17}{16}$ in., (b) 0.78 in. Say $\frac{13}{16}$ in.)

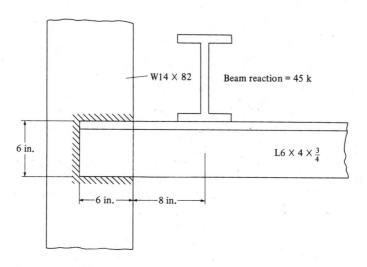

13-28 Determine the value of the load P that can be applied to the connection shown, if $\frac{3}{8}$-in. fillet welds are used. Use A36 steel, E70 electrodes, SAW, and the elastic method.

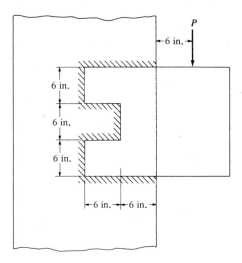

13-29 Determine the length of $\frac{1}{4}$-in. SMAW fillet welds 12 in. on center required to connect the cover plates for the section shown, at a point where the external shear is 95 k. Use A36 steel and E70 electrodes. (*Ans.* 4.81 in., say 5 in.)

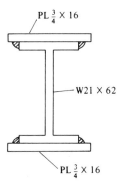

PL $\frac{3}{4}$ X 16

W21 X 62

PL $\frac{3}{4}$ X 16

13-30 The welded plate girder shown has an external shear of 720 k at a particular section. Determine the fillet-weld size required to fasten the plates to the web if the SMAW process is used. Use A36 steel and E70 electrodes.

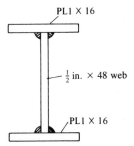

PL1 X 16

$\frac{1}{2}$ in. X 48 web

PL1 X 16

13-31 Design SMAW welded moment-resisting connections for the ends of a W24 × 76 to resist an R of 80 k and an M of 150 ft-k. Use A36 steel and E70 electrodes. Assume the column flange is 16 in. wide. The moment is to be resisted by full-penetration groove welds in the flanges, and the shear is to be resisted by welded clip angles along the web. (*Ans.* One *answer* is $\frac{1}{4}$-in. fillet weld 11 in. long each side of web and 5 in. groove weld in flanges.)

13-32 The beam shown is assumed to be attached at its ends with moment-resisting connections. Select the beam, assuming full lateral support, A36 steel, SMAW, and E70 electrodes. Use a connection of the type used in Prob. 13-31.

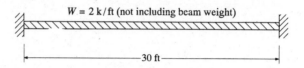

$W = 2$ k/ft (not including beam weight)

—30 ft—

13-33 The two W16 × 40 beams shown are to be made continuous across the supporting W24 × 84 girder. If the end reactions are each 60 k and their end moments are 50 ft-k, design the connections using E70 electrodes, SMAW welds, and A36 steel. (*One ans.* $\frac{5}{16}$-in. welds $6\frac{1}{2}$ in. long on seat angles. Full penetration groove welds on flanges of W16 × 40 beams.)

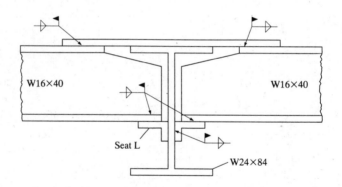

W16×40   W16×40

Seat L

W24×84

14

Building Connections

This chapter is concerned with the actual beam-to-beam and beam-to-column connections commonly used in steel buildings. Under present-day steel specifications four types of fasteners are permitted for these connections: welds, unfinished bolts, high-strength bolts, and rivets.

Selection of the type of fastener or fasteners for a particular structure usually involves many factors, including requirements of local building codes, relative economy, preference of designer, availability of good welders or riveters, loading conditions (as static or fatigue loadings), preference of fabricator, and equipment available. It is impossible to list a definite set of rules from which the best type of fastener can be selected for any given structure. One can only give a few general statements which may be helpful in making a decision.

1. Unfinished bolts are often economical for light structures subject to small static loads and for secondary members (such as purlins, girts, bracing, etc.) in larger structures.
2. Field bolting is very rapid and involves less-skilled labor than welding or riveting. The purchase price of high-strength bolts, however, is rather high.
3. If a structure is later to be disassembled, welding is probably ruled out, leaving the job open to bolts.

4. For fatigue loadings fully tensioned high-strength bolts and welds are very good.
5. Welding requires the smallest amounts of steel, probably provides the most attractive-looking joints, and also has the widest range of application to different types of connections.
6. When continuous and rigid fully moment-resisting joints are desired, welding will probably be selected.
7. Welding is almost universally accepted as being satisfactory for shop-work. For fieldwork it is very popular in some areas of the United States, while in a few others it is stymied by the fear that field inspection is rather questionable.
8. Rivets can be rapidly installed in the shop with heavy riveters but nevertheless are very seldom if ever used.
9. For fieldwork, rivets are almost extinct.

A most interesting article entitled "Choosing the Best Structural Fastener" by Henry J. Stetina appeared in the November 1963 issue of *Civil Engineering*. It would be well worth the student's time to read this article.

A valuable reference on the relative economy of various types of beam-to-beam and beam-to-column connections is given in a handbook published by the United States Steel Corporation (now USX Corporation) in October 1963, entitled *Building Design Data*.

14-2 TYPES OF BEAM CONNECTIONS

According to their rotational characteristics under load, connections can be classed as being rigid, simple, or semirigid. A connection that does not rotate at all or has complete moment resistance is a *rigid connection,* while a connection that is completely flexible and free to rotate and has no moment resistance is a *simple connection.* A *semirigid connection* has resistance falling somewhere between the simple and rigid types.

The ASD Specification (A2.2) classifies these three types of connections: Type 1 (rigid), Type 2 (simple), and Type 3 (semirigid).

From a practical standpoint, since there are no connections that are completely rigid or completely flexible, it is a common practice to classify them on a percentage of moment developed to complete rigidity or complete moment resistance. (A measure of the rotational characteristics of a particular connection cannot be practically obtained by a theoretical method, and it is necessary to run tests and plot curves of the relationships between moments and rotations for each type of connection.) A rough rule is simple connections have 0–20 percent rigidity, semirigid 20–90 percent, and rigid above 90 percent. A typical moment-rotation curve for these types of connections is shown in Fig. 14-1.

Each of these three general types of connections is briefly discussed in this section with little mention of the specific types of connectors used. The remainder of the chapter is concerned with detailed designs of these connections using specific types of fasteners. In this discussion the author probably overem-

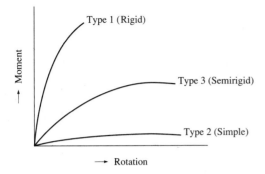

Figure 14-1 Typical end moment rotation curves.

phasizes the semirigid and rigid type connections, because a very large percentage of the building designs with which the average designer works will have simple connections. A few descriptive comments are given in the following paragraphs concerning each of these three types of connections.

Simple connections are quite flexible and are assumed to allow the beam ends to be substantially free to rotate downward under load as true simple beams should. Although simple connections do have some moment resistance (or resistance to end rotation), it is assumed to be negligible, and they are assumed to be able to resist shear only. Several types of simple connections are shown in Fig. 14-2. More detailed descriptions of each of these connections and their assumed behavior under load are given in later sections of this chapter. In this figure most of the connections are made entirely with the same type of fastener; in actual practice two types of fasteners are often used for the same connection. For example, a very common practice is to shop-weld the web angles to the beam web and field-bolt them to the column or girder.

Semirigid connections have appreciable resistance to end rotation, thus developing appreciable end moments. In design practice it is quite common for the designer to assume all connections are either simple or rigid with no consideration given to those situations in between, thereby simplifying the analysis. Should such an assumption be made for a true semirigid connection, an opportunity for appreciable moment reduction may be missed. To understand this possibility the reader is referred to the moment diagrams shown in Fig. 14-3 for a group of uniformly loaded beams supported with connections having different percentages of rigidity. This figure shows that the maximum moments in a beam vary greatly with different types of end connections. For example, the maximum moment in the semirigid connection of Fig. 14-3(d) is only 50 percent of the maximum moment in the simply supported beam of part (a) and only 75 percent of the maximum moment in the rigidly supported beam of part (b).

Actual semirigid connections are used fairly often, but usually no advantage is taken of their moment-reducing possibilities in the calculations. Perhaps one factor that keeps the design profession from taking advantage of them more often is the ASD Specification (A2), which states that consideration of a

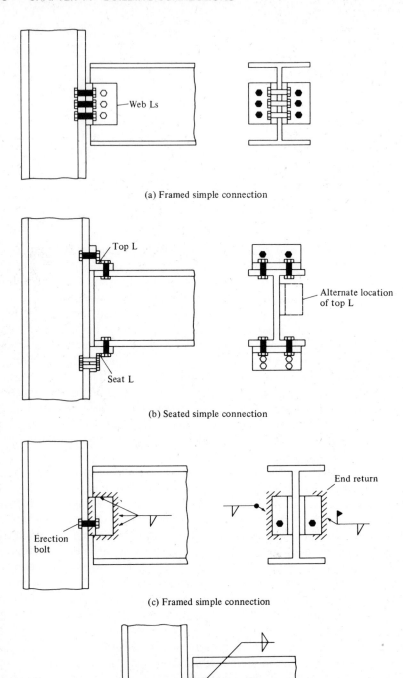

(a) Framed simple connection

(b) Seated simple connection

(c) Framed simple connection

(d) Single-plate shear connection

Figure 14-2 Some simple connections.

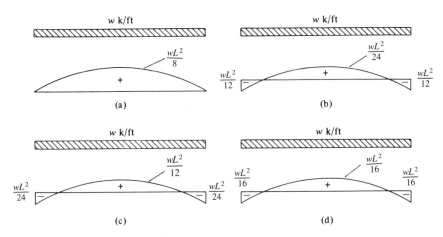

Figure 14-3 (a) Simple connections (0 percent). (b) Rigid connections (100 percent) (c) Semirigid connections (50 percent). (d) Semirigid connections (75 percent).

connection as being semirigid is permitted only upon presentation of evidence that it is capable of providing a certain percentage of the end restraint furnished by a completely rigid connection.

A second deterring factor is the need for a method of analysis which falls in between analysis for simple beams and analysis for statically indeterminate

A semirigid beam-to-column connection, Ainsley Building, Miami, Fla. (Courtesy of Lincoln Electric Company.)

structures with completely rigid joints. The student can see that analysis of a building by moment distribution could be drastically affected if the end connections were assumed to have varying percentages of moment restraint. This subject is discussed in detail by Bruce Johnston and Edward H. Mount in a paper entitled "Analysis of Building Frames with Semi-Rigid Connections."[1] The student is also referred to Chapter 8 of *Advanced Design in Structural Steel* by John E. Lothers (Prentice Hall, 1960) for an excellent discussion of this subject. Several types of semirigid connections are shown in Fig. 14-4.

Rigid connections theoretically allow no rotation at the beam ends and thus transfer 100 percent of the moment of a fixed end. Connections of this type may be used for tall buildings in which wind resistance is developed by providing continuity between the members of the building frame. Several rigid con-

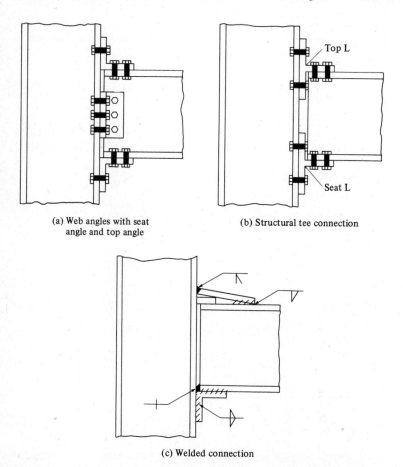

(a) Web angles with seat
angle and top angle

(b) Structural tee connection

(c) Welded connection

Figure 14-4 Some semirigid connections. These may be rigid if column web stiffeners are used.

[1] *Transactions ASCE* 107 (1942), p. 993.

nections which provide almost 100 percent restraint are shown in Fig. 14-5. It will be noticed in the figure that column web stiffeners may be required for some of these connections to provide sufficient resistance to rotation. The design of these stiffeners is discussed in Section 14-11.

The moment connection shown in part (d) is rather popular with steel fabricators, and the end-plate connection of part (e) has also been frequently used in recent years.[2]

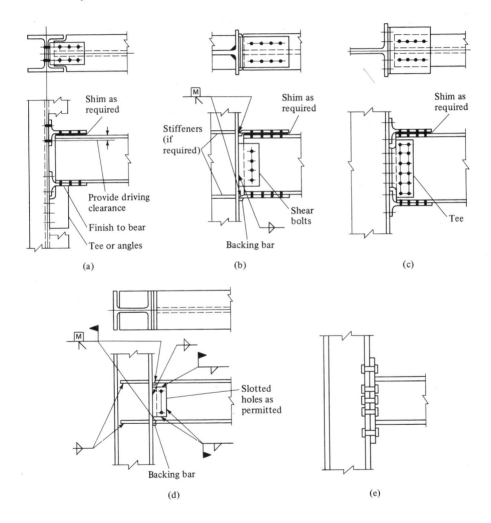

(a)   (b)   (c)

(d)   (e)

Figure 14-5 Rigid or moment-resisting connections. From American Institute of Steel Construction, *Manual of Steel Construction Allowable Stress Design*, 9th ed, (Chicago: AISC, 1989), p. 4-127. Reprinted with the permission of AISC.

[2] J. D. Griffiths, "End-Plate Moment Connections—Their Use and Misuse," *Engineering Journal*, AISC, 21, no. 1 (1st quarter, 1984), pp. 32–34.

14-3 STANDARD BOLTED BEAM CONNECTIONS

Several types of standard bolted connections are shown in Fig. 14-6. These connections are usually designed to resist shear only, as testing has proved this practice to be quite satisfactory. Part (a) shows a connection between beams with the so-called *framed connection*. This type of connection consists of a pair of flexible web angles usually shop-connected to the web of the supported beam and field-connected to the supporting beam or column. When two beams are being connected, it is usually necessary to keep their top flanges at the same elevation, with the result that the top flange of one will have to be cut back (called *coping*) as shown in Fig. 14-6(b). For such connections we must check block shear as discussed in Section 3-7.

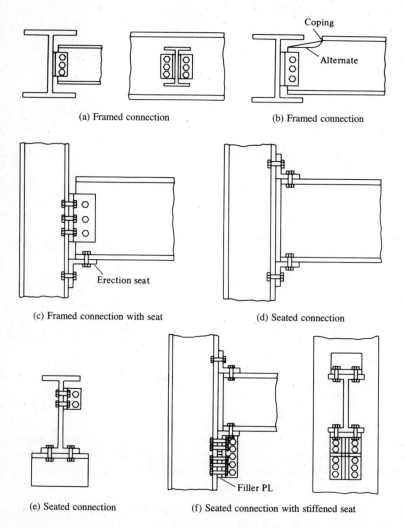

(a) Framed connection   (b) Framed connection

(c) Framed connection with seat   (d) Seated connection

(e) Seated connection   (f) Seated connection with stiffened seat

Figure 14-6 Framed and seated simple connections.

Simple connections of beams to columns can be either framed or seated. In Fig. 14-6(c) a framed connection is shown in which two web angles are connected to the beam web in the shop, after which bolts are placed through the angles and column in the field. It is often convenient to have an angle, called an *erection seat,* to support the beam during erection. Such an angle is shown in the figure.

The seated connection has an angle under the beam similar to the erection seat just mentioned, which is shop-connected to the column. In addition, there is another angle, usually on top of the beam, which is field-connected to the beam and column. A seated connection of this type is shown in Fig. 14-6(d). Should space limitation prove a problem above the beam, the top angle may be placed in the optional location shown in part (e). The top angle, at either of the locations mentioned, is very helpful in keeping the top flange of the beam from being accidentally twisted out of place during construction.

The amount of load which can be supported by the types of connections shown in parts (c), (d), and (e) of Fig. 14-6 is severely limited by the flexibility or bending strength of the horizontal legs of the seat angles. For heavier loads it is necessary to use stiffened seats such as the one shown in part (f).

The majority of these connections are selected by referring to standard tables. The ASD Manual has excellent tables for selecting bolted or welded beam connections of the types shown in Fig. 14-6. After a rolled-beam section has been selected, it is quite convenient to refer to these tables and select one of the standard connections, which will be suitable for the vast majority of cases.

In order to make these standard connections have as little moment resistance as possible, the angles used in making up the connections are usually light and flexible. To qualify as simple end supports, the ends of the beams should be as free as possible to rotate downward. Fig. 14-7 shows the manner in which framed and seated end connections will theoretically deform as the ends of the beams rotate downward. The designer does not want to do anything that will hamper these deformations.

For the rotations shown in Fig. 14-7 to occur there must be some deformation of the angles. As a matter of fact, if end slopes of the magnitudes which are computed for simple ends are to occur, the angles will actually bend enough to be stressed beyond their yield points. If this situation occurs, they will be permanently bent, and the connections will quite closely approach true simple ends. The student should now see why it is desirable to use rather thin angles and large gages for the bolt spacing if flexible simple end connections are the goal of the designer.

These connections do have some resistance to moment. When the ends of the beam begin to rotate downward, the rotation is certainly resisted to some extent by the tension in the top bolts, even if the angles are quite thin and flexible. Neglecting the moment resistance of these connections results in conservative selection of beam sizes. If moments of any significance are to be resisted, more rigid-type joints need to be provided than are available with the framed and seated connections.

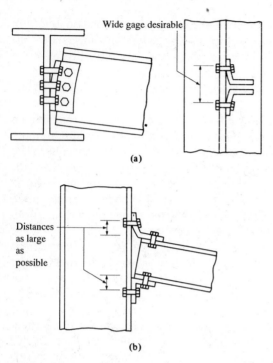

Figure 14-7 (a) Bending of framed-beam connection. (b) Bending of seated-beam connection.

14-4 ASD MANUAL STANDARD CONNECTION TABLES

In Part 4 of the ASD Manual a series of tables is presented with which the designer may select several different types of standard connections. There are tables for bolted or welded two-angle framed connections, seated beam connections, stiffened seated beam connections, eccentrically loaded connections, single-angle framed connections, and others.

In Sections 14-5 through 14-10 a few standard connections are selected using the tables in the Manual. The author hopes that these will be sufficient to introduce the reader to the ASD tables and enable him or her to make designs using the other tables with little difficulty. There are on the market today various computer programs with which these connections can be designed.

14-5 DESIGNS OF STANDARD BOLTED FRAMED CONNECTIONS

For small and low-rise buildings (and that means most buildings) simple framed connections of the types shown in Fig. 14-6(a) and (b) are usually used to connect beams to girders or to columns. The angles used are rather thin

($\frac{5}{8}$ in. is the arbitrary maximum thickness used in the ASD Manual) so they will have the necessary flexibility shown in Fig. 14-7. The angles will develop some small moments (supposedly not more than 20 percent of full fixed-end conditions), but they are neglected in design.

It will be noted in the examples to follow that, to calculate net areas effective in resisting shear and tension, the ASD Manual uses holes with diameters equal to the *nominal fastener diameters plus $\frac{1}{16}$ in.* You will remember that in ASD Section B2 (which was applicable to net area calculations for tension members) hole diameters were specified as being nominal hole diameters plus $\frac{1}{8}$ in.

The framing angles extend out from the beam web by $\frac{1}{2}$ in. as shown in Fig. 14-8. This protrusion, which is often referred to as the *setback,* is quite useful in fitting members together during steel erection.

Several standard bolted framed connections for simple beams are designed in this section with the tables provided in Part 4 of the ASD Manual. In these tables the following abbreviations for different bolt conditions are used:

1. A325-SC and A490-SC (slip-critical connections)
2. A325-N and A490-N (bearing-type connections with threads included in the shear planes)
3. A325-X and A490-X (bearing-type connections with threads excluded from shear planes)
4. A307 (common bolts)

To determine the design capacity of bolted framed beam connections, it is necessary to check the shearing and bearing capacities of the bolts as well as the shearing and bearing capacities of the framing or connection angles. To make these checks the following ASD tables are used: I-D (shearing strength of bolts), I-E (bearing strength of connections), I-F (bearing strength of bolts with different edge distances), I-G (block shearing strengths), and II-A and II-B (bolt shearing strengths for bearing-type and slip-critical connections).

It can be seen in Tables II-A and II-B that the standard connection angle lengths (L values in the tables) vary from $5\frac{1}{2}$ to $29\frac{1}{2}$ in. If the bolts are staggered, however, the lengths (L' values in the tables) vary from 7 to 31 in. The vertical spacing of fasteners used in these tables is 3 in.; a $1\frac{1}{4}$-in. end distance

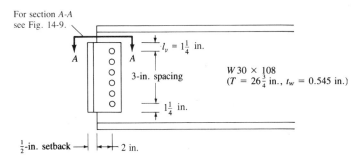

For section A-A
see Fig. 14-9.

$l_v = 1\frac{1}{4}$ in.

3-in. spacing

$1\frac{1}{4}$ in.

$W\,30 \times 108$
($T = 26\frac{3}{4}$ in., $t_w = 0.545$ in.)

$\frac{1}{2}$-in. setback

2 in.

Figure 14-8

is used for the angles as permitted by ASD Table J3.5. This latter distance will be adjusted as required by ASD Specification J3.4 when oversized or slotted bolts are used. The vertical spacing may also be varied just so long as it is within the parameters of the ASD Specification for center-to-center bolt spacings.

It is thought that the minimum depth of framing angles should be at least equal to one-half of the distance between the web toes of the beam fillets (called the T distances and given in the properties tables of Part 1 of the Manual). This minimum depth is used so as to provide sufficient stability during steel erection.

Example 14-1 presents the design of standard framing angles for a simply supported beam using bearing-type bolts in standard-size holes. In this example the design strengths of the bolts and the angles are taken from the appropriate tables. The author also shows how the table values can be checked with the procedures presented in Chapters 11 and 12.

■ Example 14-1

Select a framed simple end connection for the W30 × 108 ($t_w = 0.545$ in.) shown in Fig. 14-8 if the end reaction R equals 110 k. the beam consists of A36 steel, and $\frac{3}{4}$-in. A325-N bolts are to be used in standard-size $\frac{13}{16}$-in. holes.

Solution.

Table II-A Shear Strength of Bolts. From this table for $\frac{3}{4}$-in. A325-N bolts and a reaction of 110 k, a six-row connection with an angle thickness of $\frac{5}{16}$ in. and a length $L = 17\frac{1}{2}$ in. is selected. (This length seems satisfactory in comparison with the T of this shape of $26\frac{3}{4}$ in.) These bolts will carry 111 k in shear.

Checking the preceding value: the six bolts are in double shear and will resist $(6)(2)(0.44)(21.0) = 110.9$ k.

Table II-C Shear Capacity of Connection Angles. From this table the design shear capacity of the $\frac{5}{16}$-in. angles for a six-row connection of $\frac{3}{4}$-in. A325-N bolts is seen to equal 137 k.

This value is based on the shearing strength vertically through the holes for two angles as determined from the formula to follow, in which d is the bolt diameter. To compute shear capacity we use $d + \frac{1}{16}$ in. rather than $d + \frac{1}{8}$ in. as for tension members. The allowable shear stress for these end connections is specified as $0.30 F_u$ by the ASD Specification (J4).

$$R = 2t[(L \text{ or } L') - n(d + \tfrac{1}{16})]0.3F_u$$
$$R = (2 \times \tfrac{5}{16})[(17.5) - (6)(\tfrac{3}{4} + \tfrac{1}{16})](0.3)(58) = 137.3 \text{ k}$$

Table I-E Bearing Capacity of Bolts. For one $\frac{3}{4}$-in. A325-N bolt bearing on 1 in., the table gives a value of 52.2 k. As we have six bolts bearing on 0.545 in., our total bearing capacity is

$$(6)(52.2)(0.545) = 170.7 \text{ k} > 11.0 \text{ k}.$$

This value can be checked using an allowable bearing stress equal to $1.2F_u$ as specified in ASD Specification J3.7:

$$(6)(1.2 \times 58)(\tfrac{3}{4})(0.545) = 170.7 \text{ k}$$

Table I-F Bearing Capacity of Angles.   The top bolt with a $1\frac{1}{4}$-in. edge distance from the top (as per ASD Table I-F) can resist 36.3 k for the two angles if it is bearing on 1 in.; however, it is bearing on $2 \times \frac{5}{16}$ in. of angle thickness.

$$\text{Top bolt carries } (1)(36.3)(2 \times \tfrac{5}{16}) = \ \ 22.7 \text{ k}$$
$$\text{Other 5 bolts carry } (5)(52.2)(2 \times \tfrac{5}{16}) = \underline{163.1 \text{ k}}$$
$$\text{Total} = 185.8 \text{ k} > 110 \text{ k}$$

This value can be checked with the allowable bearing stresses of ASD Equations J3-3 and J3-1 as follows, where L_e is the edge distance for the top bolt:

Allowable bearing stress on end bolt $= F_P = L_e F_u/2d \le 1.2\, F_c$

(ASD Equation J3-3)

$$\text{Allowable total bearing} = \text{allowable bearing on edge bolt}$$
$$+ \text{ allowable bearing on other bolts}$$
$$\text{Allowable total bearing} = (F_P \text{ from ASD Equation J3-3 or J3-2})(2t)(d)$$
$$+ (1.2F_u)(2t)(d)(\text{number of other bolts})$$

$$= \left(\frac{1.25 \times 58}{2 \times \tfrac{3}{4}}\right)(2)(\tfrac{5}{16})(\tfrac{3}{4})$$

$$+ (1.2)(58)(2)(\tfrac{5}{16})(\tfrac{3}{4})(5) = 185.8 \text{ k}$$

To select the lengths of the angle legs it is necessary to study the dimensions given in Fig. 14-9, which is a view along section A-A in Fig. 14-8. On the lower right part of the figure are shown the minimum clearances needed

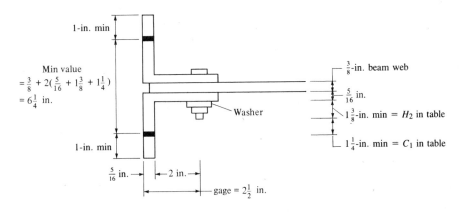

Min value
$= \tfrac{3}{8} + 2(\tfrac{5}{16} + 1\tfrac{3}{8} + 1\tfrac{1}{4})$
$= 6\tfrac{1}{4}$ in.

1-in. min

$\tfrac{5}{16}$ in.   2 in.

gage $= 2\tfrac{1}{2}$ in.

1-in. min

Washer

$\tfrac{3}{8}$-in. beam web

$\tfrac{5}{16}$ in.

$1\tfrac{3}{8}$-in. min $= H_2$ in table

$1\tfrac{1}{4}$-in. min $= C_1$ in table

Figure 14-9

for insertion and tightening of the bolts. These are the H_2 and C_1 distances shown and are obtained for $\frac{3}{4}$-in. bolts from the table "Assembling Clearances" in Part 4 of the ASD Manual.

For the legs bolted to the beam web a $2\frac{1}{2}$-in. gage is used. Using a minimum edge distance of 1 in. here, we will make these angle legs $3\frac{1}{2}$ in. For the outstanding legs the minimum gage is

$$\tfrac{5}{16} + 1\tfrac{3}{8} + 1\tfrac{1}{4} = 2\tfrac{15}{16} \text{ in.} \qquad \text{(say 3 in.)}$$

Including a 1-in. edge distance we will make this a 4-in. leg.

$$\underline{\text{Use } 2 \text{ L4} \times 3\tfrac{1}{2} \times \tfrac{5}{16} \times 1 \text{ ft } 5\tfrac{1}{2} \text{ in.}} \qquad \blacksquare$$

The most common framing angles used by steel fabricators are the $4 \times 3\frac{1}{2} \times \frac{5}{16}$ and $4 \times 3\frac{1}{2} \times \frac{3}{8}$ with the 4-in. legs outstanding. (The $3\frac{1}{2} \times 3 \times \frac{5}{16}$ and $3\frac{1}{2} \times 3 \times \frac{3}{8}$ angles are preferred by some.) Many fabricators use short slots in the outstanding legs to facilitate steel erection. In addition, they try to keep the same gage lines in the columns as much as possible.

The design of another bolted framed connection is presented in Example 14-2. The problem is very close to that of Example 14-1 except that the top flange is coped 2 in.

■ Example 14-2

Select a framed beam connection for the W27 × 94 shown in Fig. 14-10. This simply supported beam consists of A36 steel and has an end reaction R equal to 110 k. The connection is to be made with $\frac{7}{8}$-in. A325-N bolts used in standard $\frac{15}{16}$-in. holes.

Solution.

Table II-A Shear Strength of Bolts. From this table we try a five-row connection with a $\frac{3}{8}$-in. angle of length $14\frac{1}{2}$ in. (OK with respect to 24-in. T) and a capacity of 126 k.

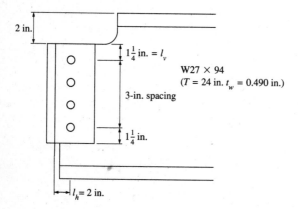

2 in.

$1\frac{1}{4}$ in. $= l_v$

W27 × 94
($T = 24$ in. $t_w = 0.490$ in.)

3-in. spacing

$1\frac{1}{4}$ in.

$l_h = 2$ in.

Figure 14-10

Table II-C Shear Capacity of Connection Angles.   From this table the design shear capacity of the trial connection is 128 k > 110 k.

Table I-E Bearing Capacity of Bolts.   With the usual $1\frac{1}{4}$-in. edge distance to cope and a 3-in. bolt spacing, the allowable bearing capacity is

$$(60.9)(5)(0.490) = 149.2 \text{ k} > 110 \text{ k}$$

Table I-F Bearing Capacity of Angles.   The bearing capacity for the top bolt with a $1\frac{1}{4}$-in. edge distance can resist $(1)(36.3)(2 \times \frac{3}{8})$, and for the other four bolts equals $(4)(60.9)(2 \times \frac{3}{8})$, Total = 209.9 k.

Table I-G Block Shear Capacity.   Entering the table with $l_v = 1\frac{1}{4}$ in. and $l_h = 2$ in. we find, $C_1 = 1.38$, and C_2 for $\frac{7}{8}$-in. bolt and $n = 5$ is 2.10.

$$\text{Allowable reaction } R_{bs} = (C_1 + C_2)F_u t$$
$$= (1.38 + 2.10)(58)(0.490)$$
$$= 98.90 \text{ k} < 110 \text{ k NG}$$

Try six-row connection ($L = 17\frac{1}{2}$ in.)

$$C_1 = 1.38, C_2 = 2.72$$
$$R_{bs} = (1.38 + 2.72)(58)(0.490) = 116.5 \text{ k} > 110 \text{ k} \qquad \text{OK}$$

Use 2 L5 $\times 3\frac{1}{2} \times \frac{3}{8} \times$ 1 ft $5\frac{1}{2}$ in.

Length of legs selected as described in solution for Example 14-1.   ■

14-6 DESIGNS OF STANDARD WELDED FRAMED CONNECTIONS

Table III of Part 4 of the ASD Manual presents the information needed for designing welded framing angles for beams. This table is normally used where the angles are attached to the beams in the shop and then field-bolted to other members. Should the framing angles be welded to both members, the weld values provided in Table IV of the Manual are considered more appropriate.

The weld used to connect the angles to the beam is called Weld A as shown in Fig. 14-11. If a weld is used to connect the beam to another beam or column, that weld is called Weld B as shown in Fig. 14-12.

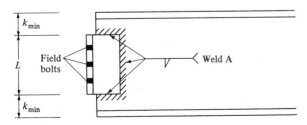

Figure 14-11 Weld connecting angle to beam.

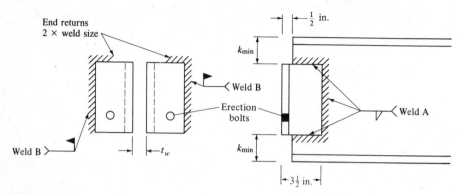

Figure 14-12

For the usual situations $4 \times 3\frac{1}{2}$-in. angles are used with the $3\frac{1}{2}$-in. legs connected to the beam webs. The 4-in. outstanding legs will usually accommodate the standard gages for the bolts going into the other members. The angle thickness selected equals the weld size plus $\frac{1}{16}$ in. or the minimum value given in Table II-A for the bolts. The angle lengths are the same as those used for the nonstaggered bolt cases (that is, $5\frac{1}{2}$ through $29\frac{1}{2}$ in.).

The design strengths of the welds to the beam webs (Weld A) given in Table III of the Manual were computed taking into account eccentricities using the instantaneous center method which was briefly discussed in Chapter 13. To select a connection of this type, the designer picks a weld size from Table III and then goes to Table II-A to determine the number of bolts required for connection to the other member. This procedure is illustrated in Example 14-3.

■ Example 14-3

Design a framed beam connection to be welded (SMAW) to a W30 × 90 beam ($t_w = 0.470$ in., $T = 26\frac{3}{4}$ in.) and then bolted to another member. The beam reaction R is 125 k, the steel is A36, the weld is E70, and the bolts are $\frac{3}{4}$-in. A325-N.

Solution.

Table III Weld A Design Strength. From this table one of several possibilities is a $\frac{1}{4}$-in. weld $20\frac{1}{2}$ in. long with a design strength of 158 k. The depth or length is compatible with the T value of $26\frac{3}{4}$ in. For a $\frac{1}{4}$-in. weld the minimum angle thickness is $\frac{5}{16}$ in. as per ASD Specification J2.2, and minimum beam web thickness is 0.51 in., which is greater than the 0.470 in. furnished. The weld capacity is reduced proportionately to the following:

$$\frac{0.470}{0.51}(158) = 145.6 \text{ k} > 125 \text{ k} \qquad \text{OK}$$

Table II-A Shear Strength of Bolts. The angle length selected is $20\frac{1}{2}$ in., and this corresponds to a seven-row bolted connection in Table II-A. From this latter table we find that seven rows of $\frac{3}{4}$-in. A325-N bolts will support 130 k and an angle $\frac{5}{16}$ in. thick is required.

$$\text{Use 2 L4} \times 3\tfrac{1}{2} \times \tfrac{5}{16} \times 1 \text{ ft } 8\tfrac{1}{2} \text{ in.}$$ ■

Table IV in the Manual (Part 4) provides the information necessary for designing all-welded standard framed connections, that is, with welds A and B as shown in Fig. 14-12. The design strengths for welds A (as in Table III) have taken into account the eccentricity of the load, but the values given for welds B do not.

Erection bolts are often necessary for erecting these beams. They are usually placed near the bottom of the angles so they will not appreciably reduce the flexibility of the angles. In some situations they may be necessary at the top of these angles to stabilize the joint during erection. Such bolts may be removed at a later date if it is felt that they provide too much restraint against rotation.

Example 14-4 illustrates the design of a framed beam connection connected with welds A and B.

■ **Example 14-4**
Select a standard framed connection with welds A and B for a W30 × 90 ($t_w = 0.470$ in., $T = 26\tfrac{3}{4}$ in.) with a reaction equal to 138 k. The welds are E70 and SMAW.

Solution.
  Table IV Welds A and B Design Strengths. For weld A one possibility is a $\frac{1}{4}$-in. weld 20 in. long (compatible with T) which will support 151 k. A $\frac{5}{16}$-in. weld B for this case will support 152 k. The minimum web thickness is 0.51 in. As a result the weld capacity is reduced proportionately as follows:

$$\frac{0.470}{0.51}(151) = 139.2 \text{ k} > 135 \text{ k} \qquad \text{OK}$$

$$\text{Use 2 L4} \times 3 \times \tfrac{3}{8} \times 1 \text{ ft } 8 \text{ in.}$$ ■

14-7 SINGLE-PLATE SHEAR CONNECTIONS

A rather economical type of flexible connection which is being used more and more frequently and which is presented in the ASD Manual is the single-plate shear connection illustrated in Figure 14-2(d). The bolt holes are prepunched in the plate and the web of the beam. Then the plate is shop-welded to the supporting beam or column, and lastly the beam is bolted to the plate in the field. Steel erectors like this kind of connection because of its simplicity. They are particularly pleased with it when there is a beam connected to each side of a girder as shown in Fig. 14-13(a). All they have to do is bolt the beam webs to the single plate on each side of the girder. Should web clip angles be used for

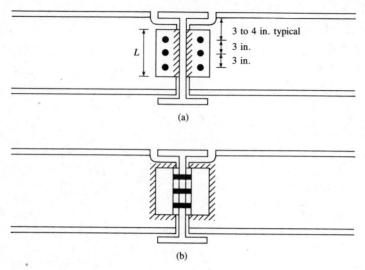

(a)

(b)

Figure 14-13 (a) Single-plate framing connection. (b) Simple connection with web angles.

such a connection, the bolts will have to pass through the angles on each side of the girder as well as the girder web as shown in Fig. 14-13(b). This is a little more difficult field operation.

With the single-plate connection the reaction or shear load is assumed to be distributed equally among the bolts passing through the beam web. It is also assumed that relatively free rotation occurs between the end of the member and the supporting beam or column. Because of these assumptions this type of connection is often referred to as a *shear tab* connection. Various studies and tests have shown that these connections can develop some end moments depending on the number and size of the bolts and their arrangement, the thicknesses of the plate and beam web, the span-depth ratio of the beam, the type of loading, and the flexibility of the supporting element.

Fig. 14-14 shows a single-plate shear connection, with a set of assumed dimensions. Table X of Part 4 of the ASD Manual provides the allowable reactions for various single-plate shear connections. The values in this table are for A36 steel and fillet welds made with E70XX electrodes. The plate thicknesses are $t \leq d_b/2 + 1/16$ in., where d_b is the bolt diameter.

The values given for these connections are based on work at the University of California at Berkeley.[3] They are satisfactory for composite and noncomposite beams, standard or slotted holes, fully tightened or snug-tight bolts, and beams of all steel grades. They may, however, be overly conservative for snug-

[3] A. Astaneh-Asl, K. M. McMullin, and S. M. Call, "Design of Single-Plate Framing Connections," Report UCB/SEMM, University of California-Berkeley (July 1988).

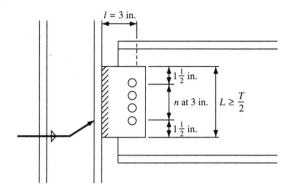

Figure 14-14 Single-plate shear connection.

tight bolts in slotted holes. To design a single-plate shear connection the following steps are suggested in the ASD Manual:

1. Select a trial plate size from Table X in ASD Manual.
2. Compute plate capacity in yielding (kips).

$$R_o = (0.4)(36)Lt$$

3. Calculate fillet-weld size D in sixteenths of an inch, R_o/LC, where C is obtained from ASD Table XI with $al = 3$ in. or $n \times 1$ in., whichever is larger, and $k = 0$. Research has shown that the fillet-weld size does not need to be greater than $0.75t$.
4. Calculate bolt capacity, $R_b = Cr_v$, where C is obtained from Table XI assuming $l = 3$ in. and where r_v is the capacity (kips) of one bolt in single shear.
5. Determine the net shear fracture capacity of the plate (kips).

$$R_{ns} = 0.3F_u A_{net} = (0.3)(58)[L - n(d_b + \tfrac{1}{16})]t$$

6. Check the plate's bearing capacity with the following expression, in which n is the number of bolts:

$$P_b \le 1.2F_u td_b n$$

7. Check block shear in the beam if it is coped.

■ Example 14-5
Design a single-plate shear connection of the type shown in Fig. 14-14 for a 70-k reaction for a W30 × 99 ($t_w = 0.520$ in., $T = 26\frac{3}{4}$ in.). Use A36 steel, $\frac{7}{8}$-in. A325-N snug-tight bolts, and E70XX electrodes.

Solution.   From Table X try a $\frac{5}{16} \times 21$-in. shear plate with seven bolts and $\frac{1}{4}$-in. fillet welds with a capacity of 76.4 k. Notice that the L of 21 in. is $>T/2$ of $26.75/2 = 13.375$ in. and that the thickness of the plate $= \frac{5}{16}$ in. $< d_b/2 + \frac{1}{16}$ in. $= \frac{7}{8}/2 + \frac{1}{16} = 0.50$ in.

Checking our selection:
Plate yielding

$$R_o = (0.4)(36)(21)(\tfrac{5}{16}) = 94.5 \text{ k} > 70 \text{ k} \qquad \text{OK}$$

Weld size: Referring to Table XIX in Part 4 of the Manual and assuming al = the larger of 3 in. or $n \times 1$ in. = 7 in. and $k = 0$

$$a = \tfrac{7}{21} = 0.333$$

$C = 1.073$ by interpolation

$$D = \frac{70}{21 \times 1.073} = 3.11 \text{ sixteenths of an inch}$$

But research says need not be $> 0.75t$.

$$(0.75)\left(\frac{5}{16}\right) = 0.234 \text{ in.} < \frac{1}{4} \text{ in.} \qquad \text{OK}$$

Bolt capacity: C from Table XI = 6.07 with $n = 7$ and $l = 3$ in.

$$R_b = (6.07)(0.6 \times 21.0) = 76.5 \text{ k} > 70 \text{ k} \qquad \text{OK}$$

Shear fracture

$$P_{\text{ns}} = (0.3)(58)[21 - 7(\tfrac{7}{8} + \tfrac{1}{16})]\tfrac{5}{16} = 78.5 \text{ k} > 70 \text{ k} \qquad \text{OK}$$

Bearing

$$R_b = (1.2)(58)(\tfrac{5}{16})(\tfrac{7}{8})(7) = 133.2 \text{ k} > 70 \text{ k} \qquad \text{OK}$$

Use $\tfrac{5}{16} \times 21$-in. shear plate, 7 bolts, $\tfrac{1}{4}$-in. welds ■

14-8 END-PLATE SHEAR CONNECTIONS

Another type of framing connection is the *end-plate shear connection*. It consists of a plate set in a vertical position welded flush against the end of a beam. Bolts connect the plate to the column or other member. To use this type of connection it is necessary to carefully control the length of the beam and the squaring of its ends so that the end plates will be vertical. Camber must also be considered in its effect on the position of the end plate. Table IX of Part 4 of the Manual is devoted to these connections. After a little practice in erecting members with end-plate shear connections, fabricators seem to like to use them. There is still trouble in getting the dimensions just right, and they are not as commonly used as the single-plate connectors.

14-9 DESIGNS OF WELDED SEATED BEAM CONNECTIONS

Another type of fairly flexible beam connection is obtained by using a beam seat such as the one shown in Fig. 14-15. Beam seats are obviously of advantage to the worker performing the steel erection. The connections for these an-

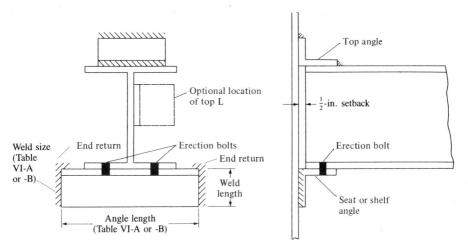

Figure 14-15 Seated beam connection.

gles may be bolts or welds, but only the welded type is considered here because of lack of space. For such a situation the seat angle would usually be shop-welded to the column and field-welded to the beam. When welds are used, the seat angles, which are also called *shelf angles,* may be punched for erection bolts as shown in the figure. These holes can be slotted if desired to permit easy alignment of the members.

A seated connection may be used only when a top angle is used, as shown in Fig. 14-15. This angle, which provides lateral support for the beam, may be placed on top of the beam, or at the optional location shown on the side of the beam. As the top angle is not usually assumed to resist any of the load, its size is probably selected by judgment. Fairly flexible angles are used which will bend away from the column or girder to which it is connected when the beam tends to rotate downward under load. This desired situation was illustrated in Fig. 14-7(b). A common top-angle size is $4 \times 4 \times \frac{1}{4}$ in.

As will be seen in the ASD tables, unstiffened seated beam connections of practical sizes can support only fairly light loads up to as high as 63.6 k if 1-in.-thick seat angles of A36 steel are used. For loads of this size two vertical end welds on the seat are sufficient. The top angle is welded only on its toes so that, as the beam tends to rotate, this thin flexible angle will be free to pull away from the column.

The design loads shown in Tables VI-A and VI-B were calculated on the basis of a $\frac{3}{4}$-in. setback rather than the nominal $\frac{1}{2}$-in. value used for web-framing angles. This value is used to provide for possible mill underrun in the beam length. The values given in Table VI-A are based on beams and seat angles both made from A36 steel. The values in Table VI-B are based on A36 seat angles and $F_y = 50$ ksi beams. The numbers in both of these tables represent the maximum loads which can be placed out on the horizontal outstanding legs of the seat angles.

Table VI-C shows the design strengths of the seat-angle welds. These values were determined from the elastic (or vector) method which was described in Section 13-13. Example 14-6 illustrates the selection of a welded unstiffened beam seat and top angles for a beam.

■ Example 14-6

Design a welded unstiffened seated beam connection to support a beam reaction of 40 k. The beam is to be a W24 × 55 ($t_w = \frac{3}{8}$ in.). It is assumed that there is sufficient length along the beam web to install an 8-in.-long seat angle. Use A36 steel and E70 SMAW fillet welds.

Solution.

   Table VI-A.   For a $\frac{3}{8}$-in. beam web thickness it is necessary to go to an 8-in. outstanding seat angle leg to the value 47.6 k. For this value a 1-in. thick seat angle is required as shown in the table.

   Table VI-C.   It is possible to support the 40-k load with a 5 × $3\frac{1}{2}$ in. angle and a $\frac{5}{8}$-in. weld, or a 6 × 4 angle with a $\frac{1}{2}$-in. weld, or a 7 × 4 angle with a $\frac{3}{8}$-in. weld. However, with an 8 × 4 angle only a one-pass weld of $\frac{5}{16}$ in. is needed.

<u>Use seat L8 × 4 × 1 × 0 ft 8 in. with a top L4 × 4 × $\frac{1}{4}$</u>   ■

14-10 STIFFENED SEATED BEAM CONNECTIONS

When beams are supported by seated connections and when the factored reactions become fairly large (say above 60 k depending on the steel grade), it is necessary to stiffen the seats. These larger reactions cause moments in the outstanding or horizontal legs of the seat angles which cannot be supported by standard-thickness angles unless they are stiffened in some manner. Typical stiffened seated connections are shown in Fig. 14-6(f) and in Fig. 14-16.

   Stiffened seats may be bolted or welded. Bolted seats may be stiffened with a pair of angles as shown in Fig. 14-16(a). Structural tee stiffeners either bolted or welded may be used. A welded one is shown in part (b). Also commonly used are welded two-plate stiffeners such as the one shown in part (c).

   Sample designs of both bolted and welded stiffened seat connections are presented in Part 4 of the ASD Manual. These designs make use of Tables VII and VIII in the Manual.

14-11 COLUMN WEB STIFFENERS

If a column to which a beam is being connected bends appreciably at the connection, the moment resistance of the connection will be reduced no matter how good the connection may be. Furthermore, if the top connection plate in pulling away from the column tends to bend the column flange as shown in Fig. 14-17(a), the middle part of the weld may be greatly overstressed (like the prying action we discussed in Chapter 12 for bolts).

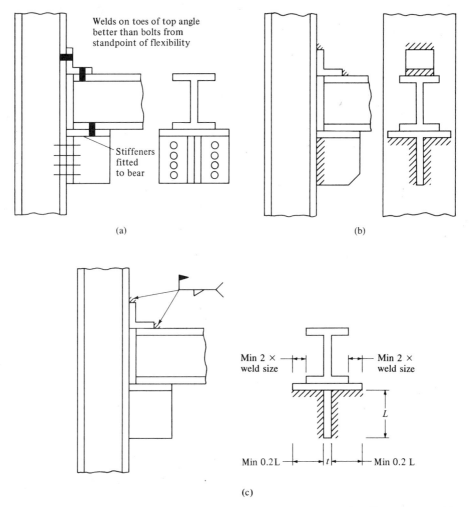

Figure 14-16 (a) Bolted stiffened seat with angles. (b) Welded stiffened seat with structural tee. (c) Welded two-plate stiffened seat.

When there is a danger of the column flange bending as described here, we must make sure that the desired moment resistance of the connection is provided. This may be done by either using a heavier column with stiffer flanges or by introducing column web stiffener plates as shown in Fig. 14-17(b). It is almost always desirable to use a heavier column because column web stiffener plates are quite expensive and a nuisance to use.

Column web stiffener plates are somewhat objectionable to architects as they find it convenient to run pipes and conduits inside their columns, but this objection can easily be overcome. If the connection is to only one column flange, the stiffener does not have to run for more than half the column depth as shown in Fig. 14-17(b). If connections are made to both column flanges, the column stiffener plates may be cut back to allow the passage of pipes, conduits, and so on, as shown in part (c) of the figure.

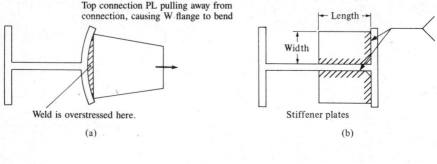

(a)

(b)

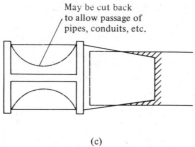

(c)

Figure 14-17

For this discussion the force applied from the beam flange to the column is referred to as P_{bf}. Should its value be larger than any one of the following resisting forces, it will be necessary to provide column web stiffeners, says ASD Specification K1. In the expressions to follow d_c is the depth of the column web between the web toes of the fillets, and t_f is the thickness of the beam flange or connection plate from which the concentrated force is being applied.

Local Flange Buckling

If our column flange thickness t_f is less than the value given by Equation K1-1, it will be necessary to provide column stiffeners. If stiffeners are needed for one flange, they need only extend for one-half of the column depth. If stiffeners are needed for both flanges, they must extend from flange to flange. Stiffeners are needed if

$$t_f < 0.4\sqrt{\frac{P_{bf}}{F_{yc}}}$$ (ASD Equation K1-1)

In Part 3 of the Manual this equation is written in the following form, where P_{fb} is the maximum column web resisting force at the beam compression flange. If P_{bf} is larger than P_{fb}, we need stiffeners.

$$P_{bf} > \frac{F_{yc}t_f^2}{0.16} = P_{fb}$$

Values of P_{fb} are given in the column tables of the Manual for W and S shapes with F_y values of 36 and 50 ksi. Should the force be located within the middle 15 percent of the column flange, no stiffeners are needed according to ASD Section K1.2.

Compression Buckling of Web

It is necessary to use one or two stiffeners opposite the compression flange when d_c (the column web depth clear of fillets) is greater than the value given by the expression to follow, in which t_{wc} is the column web thickness:

$$d_c > \frac{4100t_{wc}{}^3\sqrt{F_{yc}}}{P_{bf}} \qquad \text{(ASD Equation K1-8)}$$

This expression can be written in a more convenient form for use in the column tables. If the applied beam flange force P_{bf} is equal to or less than P_{wb} (the maximum column web resisting force at the beam compression flange), stiffeners are not required. Values of P_{wb} are given in the column tables.

$$P_{bf} \leq P_{wb} = \frac{4100t_{wc}{}^3\sqrt{F_{yc}}}{d_c}$$

Column Web Yielding

When flanges of beams or moment connection plates are welded to a column flange, a pair of column web stiffeners will be needed when the calculated value of A_{st} is positive.

$$A_{st} = \frac{P_{bf} - F_{yc}t_{wc}(t_b + 5k)}{F_{yst}} \qquad \text{(ASD Equation K1-9)}$$

In this expression P_{bf} is the force delivered to the flange, F_{yc} and F_{yst} are the yield stresses of the column web and stiffeners, respectively, and t_b is the thickness of the flange or moment connection plate providing the force. This expression has been rewritten in the Manual as follows, where no stiffeners are required if P_{bf} is less than the computed value shown:

$$P_{bf} = [F_{yc}t_{wc}]t_b + [F_{yc}t_w 5k] = P_{wi}t_b + P_{wo}$$

The values of P_{wi} and P_{wo} are computed for each of the W and S shapes listed in the column section of the Manual.

The following general rules are provided in Section K1 of the ASD Specification for the design of stiffeners:

1. The width of the stiffener plus one-half of the column web thickness should not be less than one-third the width of the beam flange or of the moment connection plate which applies the concentrated force.
2. The thickness of the stiffeners t_s may not be less than one-half the flange thickness (t_b in Fig. 14-18) or the plate delivering the force.

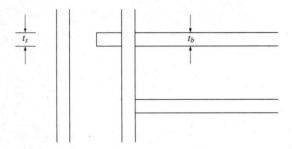

Figure 14-18

3. If there is a moment connection applied to only one flange of the column, the length of the stiffener plate does not have to exceed one-half the column depth.
4. The stiffener plate should be welded to the column web with a sufficient strength to carry the force caused by the unbalanced moment on the opposite sides of the column.

For the column given in Example 14-7 it is necessary to use column web stiffeners or to select a larger column. These two alternatives are considered in the solution.

■ Example 14-7

It is assumed that a W12 × 87 column consists of A36 steel and is subjected to 185-k C and T forces transferred by Type 1 (rigid) connections through the flanges of the W14 × 68 beam. It will be found that this column is not satisfactory to resist these forces. (a) Select a larger W12 column section which will be satisfactory. (b) Using a W12 × 87 column, design column web stiffeners including the stiffener connections, using E70 SMAW welds.

Solution.

Beam is a W14 × 68 ($b_f = 10.035$ in., $t_f = t_b = 0.720$ in.)
Column is a W12 × 87 ($d = 12.53$ in., $t_w = 0.515$ in., $t_f = 0.810$ in.)

Check to see if forces transferred to column are too large. Take values for P_{fb}, P_{wb}, P_{wi}, and P_{wo} from the ASD column tables for a W12 × 87.

$$P_{fb} = 148 \text{ k} < 185 \text{ k} \qquad\qquad \text{NG}$$
$$P_{wb} = 354 \text{ k} > 185 \text{ k} \qquad\qquad \text{OK}$$
$$P_{bf} = P_{wi}t_b + P_{wo} = (19)(0.810) + 139 = 154.4 \text{ k} < 185 \text{ k} \qquad \text{NG}$$

∴ Either a larger column or column web stiffeners are required.
Selecting a larger column:

By examination a W12 × 96 will not do, because P_{fb} in the table is <185 k.

Try a W12 × 106

$$P_{fb} = 221 \text{ k} > 185 \text{ k} \qquad\qquad \text{OK}$$
$$P_{wb} = 588 \text{ k} > 185 \text{ k} \qquad\qquad \text{OK}$$
$$P_{bf} = (22)(0.810) + 185 = 202.8 \text{ k} > 185 \text{ k} \qquad \text{OK}$$

Use W12 × 106

Design of web stiffeners using a W12 × 87 column and the suggested ASD rules presented before this example:

$$\text{Required stiffener area} = \frac{185 - 148}{36} = 1.03 \text{ in.}^2$$

$$\text{Minimum width} = \left(\frac{1}{3}\right)(10.035) - \frac{0.515}{2} = 3.09 \text{ in.}$$

$$\text{Minimum thickness} = \frac{t_b}{2} = \frac{0.720}{2} = 0.36 \text{ in.}$$

$$\text{Required thickness} = \frac{1.03}{3.09} = 0.33 \text{ in.} \qquad (\text{Use } \tfrac{3}{8} \text{ in.})$$

$$\text{Required width} = \frac{1.03}{0.375} = 2.75 \text{ in.}$$

(say 4 in. as a practical dimension)

$$\text{Minimum length} = \frac{d}{2} - t_f = \frac{12.53}{2} - 0.810 = 5.46 \text{ in.} \qquad (\text{say 6 in.})$$

Use 2 PL$\frac{3}{8}$ × 4 × 0 ft 6 in. (see Fig. 14-19)

Minimum weld size = $\frac{1}{4}$ in. based on t_w of column (ASD Table J2.4)

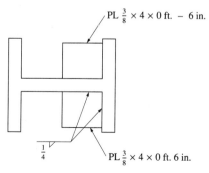

PL $\frac{3}{8}$ × 4 × 0 ft. − 6 in.

PL $\frac{3}{8}$ × 4 × 0 ft. 6 in.

Figure 14-19

Required weld length for both plates

$$= \frac{185 - 148}{(\frac{1}{4})(0.707)(0.3 \times 70)} = 9.96 \text{ in.} \qquad \underline{\text{(say 6 in. each plate)}} \qquad \blacksquare$$

14-12 CONNECTION DESIGN AIDS—HANDBOOKS AND COMPUTER PROGRAMS

To simplify the use of the ASD Manual tables for selecting simple connections, the AISC has published a small book entitled *Allowable Stress Designs of Simple Shear Connections.*[4] (A similar book is available for LRFD.) Included in the book are framed beam, seated beam, end-plate, single-plate, and single angle connections.

Several computer programs are available for the design and checking of connections. They are economical to buy, save a great deal of time, and reduce the possibility of numerical mistakes. Anyone who has struggled through connection designs, particularly the selection of dimensions, will appreciate these programs.

One very satisfactory program is CONXPRT. It was developed by Structural Engineers Inc. of Radford, VA and is marketed by the AISC. It is applicable to double-angle, shear end-plate, and single-plate simple framing connections. Not only is a program such as CONXPRT useful for designs, but it also is very good for checking existing designs. Just imagine that you are working for a consulting engineer who receives the detailed drawings for a large steel building from the steel fabricator and asks you to check the adequacy of all the connections. If you check them with the handbook and a pocket calculator you are facing a monumental job—but with a computer program the work can be handled quickly and accurately.

PROBLEMS

14-1 Determine the maximum end reaction that can be transferred through the web angle connection shown. The steel is A36, and the bolts are $\frac{3}{4}$-in. A325-N and are used in standard-size holes. (*Ans. R* = 74.2 k)

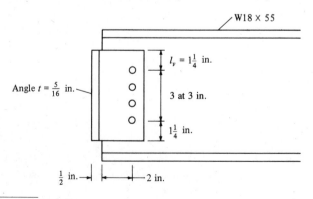

---

[4] American Institute of Steel Construction (1990, Chicago) 119 pages.

14-2 Repeat Prob. 14-1 using $\frac{7}{8}$-in. A325-N bolts.

14-3 Repeat Prob. 14-1 using 1-in. A324-X bolts. (*Ans. R* = 78.8 k)

14-4 Using the ASD Manual, select a pair of bolted standard web angles for a W36 × 150 beam with a 250-k reaction. The bolts are to be $\frac{7}{8}$-in. A325-N in standard-size holes arranged as shown. The steel is A36.

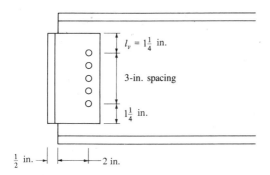

14-5 Repeat Prob. 14-4 using 1-in. A325-N bolts. (*Ans.* 2L6 × 4 × $\frac{5}{8}$ × 1 ft 11$\frac{1}{2}$ in., 8-row connection)

14-6 Repeat Prob. 14-1 using $\frac{3}{4}$-in. A325 S.C. bolts.

14-7 Design a framed beam connection for a W27 × 84 to support a beam reaction of 150 k. The bolts are to be 1-in. A325-X in standard-size holes, and A36 steel is to be used. The edge distances and bolt spacings are the same as those shown for Prob. 14-4. (*Ans.* 2 L6 × 4 × $\frac{1}{2}$ × 1 ft 2$\frac{1}{2}$ in., 5-row connection)

14-8 Repeat Prob. 14-1 using 1-in. A325-N bolts in $1\frac{1}{16}$ × $1\frac{5}{16}$-in. short slots with long axes perpendicular to the transmitted force. Angle *t* is $\frac{5}{8}$ in.

14-9 Design a framed beam connection for a W18 × 50 to support a beam reaction of 75 k. The beam's top flange is to be coped for a 2-in. depth, and $\frac{7}{8}$-in. A325-X bolts in standard-size holes are to be used as shown. Use A36 steel. (*Ans.* 2 L5 × 3$\frac{1}{2}$ × $\frac{5}{8}$ × 1 ft 2$\frac{1}{2}$ in., 5-row connection)

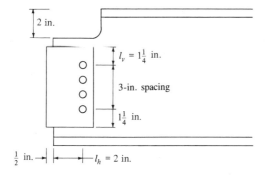

14-10 Repeat Prob. 14-7 with a beam reaction of 190 k and using A242 steel (F_y = 50 ksi and F_u = 70 ksi).

14-11 Select a framed beam connection for a W33 × 130 beam (A36 steel) with a beam reaction of 140 k. The web angles are to be welded with E70 electrodes (weld A in the Manual) and are to be field-connected to the girder with $\frac{7}{8}$-in. A325-N bolts. (*One ans.* 2 L4 $\times$ $3\frac{1}{2}$ $\times$ $\frac{3}{8}$ $\times$ 1 ft $8\frac{1}{2}$ in., $\frac{1}{4}$-in. weld A, 7 rows bolts)

14-12 Repeat Prob. 14-11 using SMAW shop and field welds (welds A and B in the Manual).

14-13 Select a framed beam connection from the ASD Manual for a W30 $\times$ 124 consisting of A36 steel (F_y = 36 ksi), using SMAW E70 shop and field welds. The beam reaction is 185 k. (*Ans.* 2 L4 $\times$ 3 $\times$ $\frac{1}{2}$ $\times$ 1 ft 10 in., $\frac{5}{16}$-in. weld A, $\frac{3}{8}$-in. weld B)

14-14 Repeat Prob. 14-13 for R = 135 k.

14-15 Select a seated beam connection bolted with $\frac{3}{4}$-in. A325-N bolts in standard-size holes for the following data: Beam is W16 $\times$ 67 consisting of A36 steel, R is 42 k, and the column gage is $5\frac{1}{2}$ in. (*Ans.* L8 $\times$ 4 $\times$ 1 with 3 rows bolts)

14-16 Repeat Prob. 14-15 using $\frac{7}{8}$-in. A325-X bolts to attach the beam flange to the seat and using E70 SMAW fillet welds for the other connections.

14-17 Repeat Prob. 14-15 using E70 SMAW fillet welds for all connections and with sufficient room for an 8-in.-long seat angle. (*Ans.* L8 $\times$ 4 $\times$ 1 $\times$ 0 ft 8 in., $\frac{5}{16}$-in weld horizontally and vertically)

14-18 Design a single-plate shear connection for a reaction of 60 k from a W27 $\times$ 94 consisting of A36 steel. Welds are to be made with E70XX electrodes, and bolts are to be $\frac{3}{4}$-in. A490-N, snug-tight, and with a 3-in. spacing.

14-19 Repeat Prob. 14-18 with the reaction 75 k, the bolts $\frac{7}{8}$ in., and the beam coped $1\frac{1}{2}$ in. (*Ans.* PL$\frac{3}{8}$ $\times$ $4\frac{1}{4}$ $\times$ 1 ft 6 in. with 6 bolts and $\frac{5}{16}$-in. welds)

15

Design of Steel Buildings

15-1 INTRODUCTION TO LOW-RISE BUILDINGS

The material in this section and the next pertains to the design of low-rise steel buildings up to several stories in height, while Sections 15-3 to 15-10 provide information concerning common types of building floors and Section 15-11 presents common types of roof construction. Sections 15-12 and 15-13 are concerned with walls, partitions, and fireproofing. The sections thereafter present general information relating to high-rise or multistory buildings.

The low-rise buildings considered include apartment houses, office buildings, warehouses, schools, and institutional buildings which are not very tall with respect to their least lateral dimensions. The separating factor between the buildings discussed here and those mentioned in the multistory sections is the matter of wind forces and not the actual height of the building in question. The usual rule of thumb is that if the height of the building is not greater than twice its least lateral dimension, provision for wind forces is unnecessary. For buildings of these dimensions the walls and partitions probably provide sufficient resistance to wind forces except in unusual circumstances. Following such a rule of thumb, however, is a dangerous engineering practice. Each structure should be considered on the basis of its dimensions, location, surrounding structures, and so on. Then a decision can be made as to whether a definite system of lateral bracing needs to be used.

15-2 TYPES OF STEEL FRAMES USED FOR BUILDINGS

Steel buildings are usually classified as being in one of four groups according to their type of construction. These types are bearing-wall construction, skeleton construction, long-span construction, and combination steel and concrete framing. More than one of these construction types can be used in the same building. Each of these types is briefly discussed in the following paragraphs.

Bearing-Wall Construction

Bearing-wall construction is the most common type of single-story light commercial construction. The ends of beams or joists or light trusses are supported by the walls that transfer the loads to the foundation. The old practice was to rapidly thicken load-bearing walls as buildings became taller. For instance, the wall on the top floor of a building might be one or two bricks thick while the lower walls might be increased by one brick thickness for each story as we come down the building. As a result this type of construction was usually thought to have an upper economical limit of about two or three stories. A great deal of research has been conducted for load-bearing construction in re-

Coliseum in Spokane, Wash. (Courtesy of Bethlehem Steel Corporation.)

cent decades, and it has been discovered that thin load-bearing walls may be quite economical for many buildings up to 10 or 20 or even more stories.

The average engineer is not very well versed in the subject of bearing-wall construction, with the result that he or she may often specify complete steel or reinforced-concrete frames where bearing-wall construction might have been just as satisfactory and more economical. Bearing-wall construction is not very resistant to seismic loadings and does have an erection disadvantage for buildings of more than one story. For such cases, it is necessary to place the steel floor beams and trusses floor by floor as the masons complete their work below, thus requiring alternation of the masons and ironworkers.

Bearing plates are usually necessary under the ends of the beams or light trusses which are supported by the masonry walls, because of the relatively low bearing strength of the masonry. Although theoretically the beam flanges may on many occasions provide sufficient bearing without bearing plates, the plates are almost always used—particularly where the members are so large and heavy that they must be set by a steel erector. The plates are usually shipped loose and set in the walls by the masons. Setting them in their correct positions and at the correct elevation is a critical part of the construction. Should they not be properly set, there will be some delay in correcting their positions. If a steel erector is used he or she will probably have to make an extra trip to the job.

When the ends of a beam are enclosed in a masonry wall, some type of wall anchor is desirable to prevent the beam from moving longitudinally with respect to the wall. The usual anchors consist of bent steel bars passing through beam webs. These so called *government anchors* are shown in Fig. 15-1. Occasionally clip angles attached to the web are used instead of government anchors. These are shown in Fig. 15-1 (b). Should longitudinal loads of considerable magnitude be anticipated, regular vertical anchor bolts may be used at the beam ends.

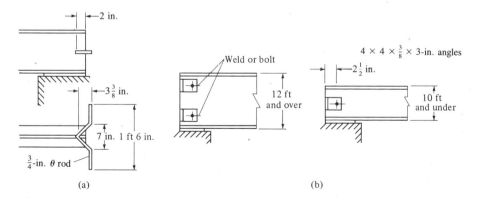

Figure 15-1 (a) Government anchor. (b) Angle wall anchors. From American Institute of Steel Construction, *Manual of Steel Construction Allowable Stress Design,* 9th ed. (Chicago: AISC, 1989), p. 4-136. Reprinted with the permission of AISC.

For small commercial and industrial buildings bearing-wall construction is quite economical when the clear spans are not greater than roughly 35 or 40 ft. If the clear spans are much greater, it becomes necessary to thicken the walls and use pilasters to ensure stability. For these cases it may often be more economical to use intermediate columns if permissible.

Skeleton Construction

In skeleton construction the loads are transmitted to the foundations by a framework of steel beams and columns. The floor slabs, partitions, exterior walls, and so on, are all supported by the frame. This type of framing, which can be erected to tremendous heights, is often referred to as beam-and-column construction.

In beam-and-column construction the frame usually consists of columns spaced 20, 25, or 30 ft apart with beams and girders framed into them from both directions at each floor level. One very common method of arranging the members is shown in Fig. 15-2. The girders are placed in the longer direction between the columns, while the beams are framed between the girders in the short direction. With various types of floor construction other arrangements of beams and girders may be used.

For skeleton framing the walls are supported by the steel frame and are generally referred to as *nonbearing* or *curtain walls*. The beams supporting the exterior walls are called *spandrel beams*. These beams, illustrated in Fig. 15-3, can usually be placed so that they will serve as the lintels for the windows.

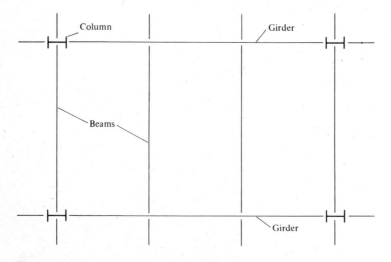

Figure 15-2 Beam-and-column construction.

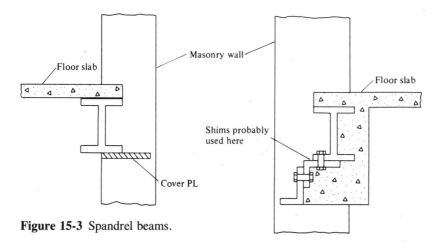

Figure 15-3 Spandrel beams.

Long-Span Steel Structures

When it becomes necessary to use very large spans between columns as for field houses, auditoriums, theaters, hangars, or hotel ballrooms, the usual skeleton construction may not be sufficient. Should the ordinary rolled W sections be insufficient, it may be necessary to use cover-plated beams, plate girders, box girders, large trusses, arches, rigid frames, and the like. When depth is limited, cover-plated beams, plate girders, or box girders may be called upon to do the job. Should depth not be so critical, trusses may be satisfactory. For very large spans arches and rigid frames are often used. These various types of structures are referred to as *long-span structures*. Fig. 15-4 shows a few of these types of structures.

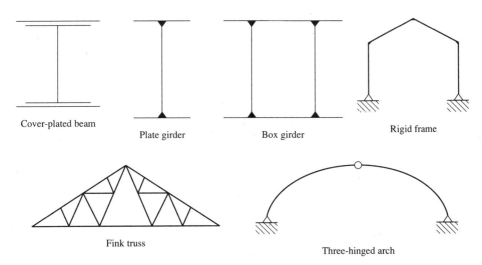

Figure 15-4 Long-span structures.

Combination Steel and Concrete Framing

A tremendous percentage of the buildings erected today make use of a combination of reinforced concrete and structural steel. If reinforced-concrete columns were used in very tall buildings, they would be rather large on the lower floors and take up considerable space. Steel column shapes surrounded by and bonded to reinforced concrete may be used and are referred to as combination or composite columns.[1]

15-3 COMMON TYPES OF FLOOR CONSTRUCTION

Concrete floor slabs of one type or another are used almost universally for steel-frame buildings. They are strong and have excellent fire ratings and good acoustic ratings. On the other hand, appreciable time and expense are required to provide the formwork necessary for most slabs. Concrete floors are heavy, they must include some type of reinforcing bars or mesh, and there may be a problem involved in making them watertight. Among the many types of concrete floors used today for steel-frame buildings are the following:

1. Concrete slabs supported with open-web steel joists (Section 15-4).
2. One-way and two-way reinforced-concrete slabs supported on steel beams (Section 15-5).
3. Concrete slab and steel beam composite floors (Section 15-6).
4. Concrete-pan floors (Section 15-7).
5. Steel-decking floors (Section 15-8).
6. Flat slab floors (Section 15-9).
7. Precast concrete slab floors (Section 15-10).

Among the several factors to be considered in selecting the type of floor system to be used for a particular building are loads to be supported; fire rating desired; sound and heat transmission; dead weight of floor; ceiling situation below (to be flat or have beams exposed); facility of floor for locating conduits, pipes, wiring, and so on; appearance; maintenance required; time required to construct; and depth available for floor.

The student can obtain a great deal of information about these and other construction practices by referring to various engineering magazines and catalogs, particularly *Sweet's Catalog File* published by McGraw-Hill Information Systems Company of New York. The author cannot make too strong a recommendation for the student to examine these books to see the tremendous amount of data available there which may be of use in his or her engineering practice. The sections to follow present brief descriptions of the floor systems mentioned in this section along with some discussions of their advantages and chief uses.

---

[1] J. C. McCormac, *Structural Steel Design LRFD Method* (New York: Harper & Row, 1989), pp. 440–456.

15-4 CONCRETE SLABS ON OPEN-WEB STEEL JOISTS

Perhaps the most common type of floor slab in use for small steel-frame buildings is the slab supported by open-web steel joists. The joists are small parallel chord trusses whose members are often made from bars (hence the common name *bar joist*) or small angles or other rolled shapes. Steel forms or decks are usually attached to the joists by welding or self-drilling, self-tapping screws, then concrete slabs are poured on top. This is one of the lightest types of concrete floors and also one of the most economical. A sketch of an open-web joist floor is shown in Fig. 15-5.

Open-web joists are particularly well suited to building floors with relatively light loads and for structures where there is not too much vibration. They have been used a good deal for fairly tall buildings, but generally speaking they are better suited for the shorter buildings. They are very satisfactory for supporting floor and roof slabs for schools, apartment houses, hotels, office buildings, restaurant buildings, and similar low-level buildings.

Open-web joists must be braced laterally to keep them from twisting or buckling and also to keep the floors from being too springy. Lateral support is provided by *bridging,* which consists of continuous horizontal rods fastened to the top and bottom chords of the joists or of diagonal cross bracing. Bridging is desirably used at spaces not exceeding 7 ft on centers.

Open-web joists are easy to handle and are quickly erected. If desired, a ceiling can be attached to the bottom of the joists or suspended therefrom. The open spaces in the webs are admirably suited for placing conduits, ducts, wiring, piping, and so on. The joists should be either welded to supporting steel beams or well anchored in the masonry walls. When concrete slabs are placed on top of the joists, they are usually from 2 to $2\frac{1}{2}$ in. thick. Nearly all of the many concrete slabs cast in place or precast on the market today can be used successfully on top of open-web joists.

These joists are selected from manufacturers' catalogs. The student can learn a great deal about open-web joists by studying various manufacturers' literature. An exceptionally useful reference of this type is *Sweet's Catalog File.*

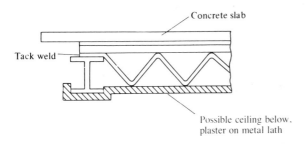

Figure 15-5 Open-web joists.

The joists are designated as J-Series when the material used has a minimum yield point of 36 ksi (248 MPa), and as H-Series when the chord material consists of steel with a minimum yield point of 50 ksi (345 MPa). There are also long-span joists L-J Series and the deep long-span joists (DLJ Series with $F_y = 36$ ksi) intended for roof construction. The DLH Series are deep long-span joists with minimum yield points of 50 ksi. In the tables the number preceding the joist number is the nominal depth of the joist in inches, the letter or letters indicate the joist series, and the number to the right designates the chord section.

The tables provided in the manufacturers' catalogs give the safe uniform loads in pounds per foot which the joists can support. Also provided are the maximum end reaction in pounds which a joist can support from the standpoint of shear and the maximum uniform live load in pounds per foot which it can support without having a deflection larger than $\frac{1}{360}$ of the span.

15-5 ONE-WAY AND TWO-WAY REINFORCED-CONCRETE SLABS

One-Way Slabs

A very large number of concrete floor slabs in old office and industrial buildings consisted of one-way slabs about 4 in. thick supported by steel beams 6 to 8 ft on centers. These floors were often referred to as concrete arch floors because at one time brick or tile floors were constructed in approximately the same shape, that is, in the shape of arches with flat tops.

A one-way slab is shown in Fig. 15-6. The slab spans in the short direction shown by the arrows in the figure. One-way slabs are usually used when the long direction is two or more times the short direction. In such cases the short span is so much stiffer than the long span that almost all of the load is carried by the short span. The short direction is the main direction of bending and will be the direction of the main reinforcing bars in the concrete, but temperature and shrinkage steel is needed in the other direction.

A typical cross section of a one-way slab floor with supporting steel beams is shown in Fig. 15-7. When steel beams or joists are used to support reinforced-concrete floors, it may be necessary to encase them in concrete or other materials to provide the required fire rating. Such a situation is shown in the figure.

It may be necessary to leave steel lath protruding from the bottom flanges or soffits of the beam for the purposes of attaching plastered ceilings. Should such ceilings be required to cover the beam stems, this floor system will lose a great deal of its economy.

One-way slabs have an advantage when it comes to formwork in that the forms can be supported entirely by the steel beams with no vertical shoring needed. They have a disadvantage in that they are much heavier than most of

Short-span joists in the Univac Center of Sperry-Rand Corporation, Blue Bell, Pa. (Courtesy of Bethlehem Steel Corporation.)

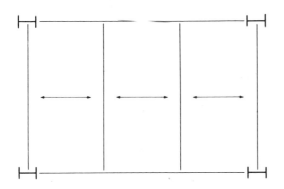

Figure 15-6 One-way slab.

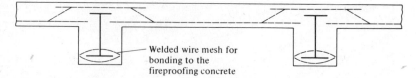

Figure 15-7 Cross section of one-way slab floor.

the newer lightweight floor systems. The result is that they are not used as often as formerly for lightly loaded floors; but when a floor to support heavy loads is desired, or a rigid floor or a very durable floor, the one-way slab may be a very desirable selection.

Two-Way Concrete Slabs

The two-way concrete slab is used when the slabs are square or nearly so with supporting beams under all four edges. The main reinforcing runs in both directions. Other characteristics are similar to those of the one-way slab.

15-6 COMPOSITE FLOORS

Composite floors have steel beams (rolled sections, cover-plated beams, or built-up members) bonded together with concrete slabs in such a manner that the two act as a unit in resisting the total loads which the beam sections would otherwise have to resist alone. The steel beams can be smaller when composite floors are used, because the slabs act as part of the beams.

A particular advantage of composite floors is that they utilize concrete's high compressive strength by keeping all or nearly all of the concrete in compression, and at the same time stress a larger percentage of the steel in tension than is normally the case with steel-frame structures. The result is less steel tonnage in the structure. A further advantage of composite floors is an appreciable reduction in total floor thickness, which is particularly important in taller buildings.

Two types of composite floor systems are shown in Fig. 15-8. The steel beam can be completely encased in the concrete and the horizontal shear transferred by friction and bond (plus some shear reinforcement if necessary). This type of composite floor is usually uneconomical. A second possibility is shown in part (b), where the steel beam is bonded to the concrete slab with some type of shear connectors. Various types of shear connectors have been used during the past few decades, including spiral bars, channels, angles, studs, and so on, but economic considerations usually lead to the use of round studs welded to the top flanges of the beams in place of the other types mentioned. Typical studs are $\frac{1}{2}$ to $\frac{3}{4}$ in. in diameter and 2 to 4 in. in length.

Cover plates may occasionally be welded to the bottom flanges of rolled steel sections. The student can see that with the slab acting as a part of the

The 104-ft joists for the Bethlehem Catholic High School, Bethlehem, Pa. (Courtesy of Bethlehem Steel Corporation.)

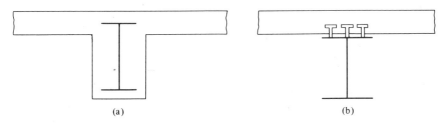

Figure 15-8 Composite floors. (a) Steel beam encased in concrete. (b) Steel beam bonded to concrete slab with shear connectors.

beam there is quite a large area available on the compressive side of the beam. By adding plates to the tensile flange a little better balance is obtained. Chapter 16 is devoted entirely to the design of composite sections.

15-7 CONCRETE-PAN FLOORS

There are several types of pan floors which are constructed by placing concrete in removable pan molds. (Some special light corrugated pans are also available which can be left in place.) Rows of the pans are arranged on wooden floor forms, and the concrete is placed over the top of them, producing a floor cross

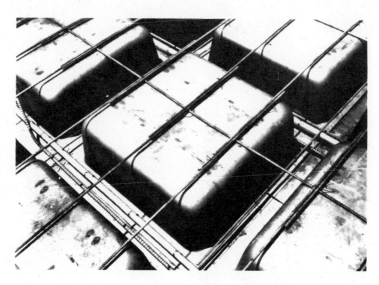

One-piece metal dome pans. (Courtesy of Gateway Erectors, Inc.)

Temple Plaza parking facility, Salt Lake City, Utah. (Courtesy of Ceco Steel Products Corporation.)

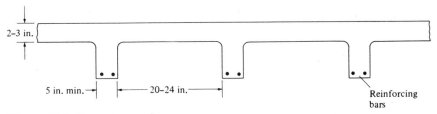

Figure 15-9 Concrete-pan floor.

section such as the one shown in Fig. 15-9. Joists are formed between the pans, giving a tee-beam-type floor.

These floors, which are suitable for fairly heavy loads, are appreciably lighter than the one-way and two-way concrete slab floors. They require a good deal of formwork, including appreciable shoring underneath the stems. Labor is thus higher than for many floors, but saving due to weight reduction and reuse of standard-size pans may make them competitive. If suspended ceilings are required, pan floors will have a decided economic disadvantage.

Two-way construction is available—that is, with ribs or stems running in both directions. Pans with closed ends are used, and the result is a waffle-type floor. This type of floor is usually used when the floor panels are square or nearly so. Two-way construction can be obtained for reasonably economical prices and makes a very attractive ceiling below with fairly good acoustical properties.

15-8 STEEL-DECKING FLOORS

Typical cross sections of steel-decking floors are shown in Fig. 15-10; several other variations are available. *Today formed steel decking with a concrete topping is by far the most common type of floor system used for office and apart-*

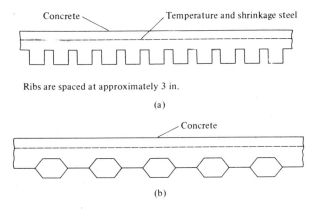

Figure 15-10 Steel-decking floors.

ment buildings. It is also popular for hotels and other buildings where the loads are not very large.

A particular advantage of steel-decking floors is that the decking immediately forms a working platform. The light steel sheets are quite strong and can span up to 20 ft or more. Due to the considerable strength of the decking the concrete does not have to be particularly strong. This permits the use of lightweight concrete as thin as 2 or $2\frac{1}{2}$ in.

The cells in the decking are convenient for conduits, pipes, and wiring. The steel is usually galvanized, and if exposed underneath can be left as it comes from the manufacturer or painted as desired. Should fire resistance be necessary, a suspended ceiling with metal lath and plaster may be used. The same is necessary if a flat ceiling is required below for the types of decking shown in Fig. 15-10, but decking is available with a flat soffit.

15-9 FLAT SLABS

Formerly, flat slab floors were limited to reinforced-concrete buildings, but today it is possible to use them in steel-frame buildings. A flat slab is reinforced in two or more directions and transfers its loads to the supporting columns without the use of beams and girders protruding below. The supporting concrete beams and girders are made so wide that they are the same depth as the slab.

Flat slabs are of great value when the panels are approximately square, when more headroom is desired than is provided with the normal beam and girder floors, when heavy loads are anticipated, and when it is desired to place the windows as near to the tops of the walls as possible. Another advantage is the flat ceiling produced for the floor below. Although the large amounts of reinforcing steel required cause additional costs, the simple formwork cuts expenses decidedly. The significance of simple formwork will be understood when it is realized that over one-half of the cost of the average poured reinforced-concrete floor slab is in the formwork.

For some reinforced-concrete frame buildings with flat slab floors, it is necessary to flare out the tops of the columns, forming column capitals, and perhaps thicken the slab around the column with the so-called drop panels. These items, which are shown in Fig. 15-11, may be necessary to prevent shear failures in the slab around the column.

It is possible today in steel-frame buildings to use short steel cantilever beams connected to the steel columns and embedded in the slabs. These beams serve the purposes of the flared columns and drop panels in ordinary flat slab construction. This arrangement is often called a *steel grillage* or *column head.* The flat slab is not a very satisfactory type of floor system for the usual tall building where lateral forces (wind or earthquake) are appreciable, because protruding beams and girders are desirable to serve as part of the lateral bracing system.

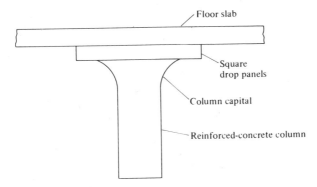

Figure 15-11 A flat slab floor for a reinforced-concrete building.

15-10 PRECAST CONCRETE FLOORS

Precast concrete sections are more commonly associated with roofs than they are with floor slabs, but their use for floors is increasing. They are quickly erected and reduce the need for formwork. Lightweight aggregates are most often used in the concrete, making the sections light and easy to handle. Some of the aggregates used make the slabs nailable and easily cut and fitted on the job. For floor slabs with their fairly heavy loads, the aggregates should be of a quality that will not greatly reduce the strength of the resulting concrete.

101 Hudson St., Jersey City, N.J. (Courtesy of Owen Steel Company, Inc.)

The reader is again referred to *Sweet's Catalog File* at this point. In these catalogs a great amount of information is available on the various types of precast floor slabs on the market today. A few of the common types available are listed below, and a cross section of each type mentioned is presented in Fig. 15-12. Due to slight variations in the upper surfaces of precast sections it is necessary to use a mortar topping of 1 to 2 in. before asphalt tile or other floor coverings can be installed.

1. *Precast concrete planks* are roughly 2 to 3 in. thick and 12 to 24 in. wide and are placed on joists up to 5 or 6 ft on center. They usually have tongue-and-groove edges.
2. *Hollow-cored slabs* are sections roughly 6 to 8 in. deep and 12 to 18 in. wide and have hollow, perhaps circular cores formed in a longitudinal direction, thus reducing their weights by approximately 50 percent (as compared with solid slabs of the same dimensions). These sections, which can be used for spans of from roughly 10 to 25 ft, may be prestressed.
3. *Prefabricated concrete block systems* are slabs made by tying together precast concrete blocks with steel rods (may also be prestressed). These

Interns' living quarters, Good Samaritan Hospital, Dayton, Ohio. (Courtesy of Flexi-core Company, Inc.)

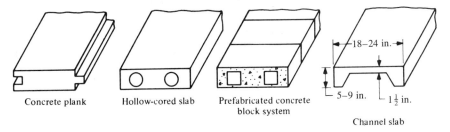

Concrete plank   Hollow-cored slab   Prefabricated concrete
block system

Channel slab

Figure 15-12 Precast concrete roof and floor slabs.

very seldom used floors vary from 4 to 8 in. deep and can be used for spans of roughly 8 to 32 ft.

4. *Channel slabs* are used for spans of approximately 8 to 24 ft. Rough dimensions of these slabs are given in Fig. 15-12.

15-11 TYPES OF ROOF CONSTRUCTION

The types of roof construction commonly used for steel-frame buildings include concrete slabs on open-web joists, steel-decking roofs, and various types of precast concrete slabs. For industrial buildings the roof system in predominant use today is one which involves the use of a cold-rolled steel deck. In addition, the other types mentioned in the preceding sections on floor types are occasionally used, but they usually are unable to compete economically. Among the factors to be considered in selecting the specific types of roof construction are strength, weight, span, insulation, acoustics, appearance below, and type of roof covering to be used.

The major differences between floor-slab selection and roof-slab selection probably occur in the considerations of strength and insulation. The loads applied to roofs are generally much smaller than those applied to floor slabs, thus permitting the use of many types of lightweight aggregate concretes which may be appreciably weaker. Roof slabs should have good insulation properties, or they will have to have insulation materials placed on them and covered by the roofing. Among the many types of lightweight aggregates used are wood fibers, zonolite, foams, sawdust, gypsum, and expanded shale. Although some of these materials decidedly reduce concrete strengths, they provide very light roof decks with excellent insulating properties.

Precast slabs made with these aggregates are light, are quickly erected, have good insulating properties, and can usually be sawed and nailed. For poured concrete slabs on open-web joists, several lightweight aggregates work very well (zonolite, foams, gypsum, etc.), and the resulting concrete can easily be pumped up to the roofs, thereby facilitating construction. By replacing the aggregates with certain foams, concrete can be made so light it will float in water. Needless to say, the strength of the resulting concrete is quite low.

Steel decking with similar thin slabs of lightweight and insulating concrete placed on top make very good and economical roof decks. A competitive variation consists of steel decking with rigid insulation board placed on top, followed by the regular roofing material. The other concrete-slab types are hard pressed to compete economically with these types for lightly loaded roofs. Other types of poured concrete decks require much more labor.

15-12 EXTERIOR WALLS AND INTERIOR PARTITIONS

Exterior Walls

The purposes of exterior walls are to provide resistance to atmospheric conditions including insulation against heat and cold, satisfactory sound-absorption and light-refraction characteristics, sufficient strength, and fire ratings. They should look good and be reasonably economical.

For many years exterior walls were constructed of some type of masonry, glass, or corrugated sheeting. In recent decades, however, the number of satisfactory materials for exterior walls has increased tremendously. Available and commonly used today are precast concrete panels, insulated metal sheeting, and many other prefabricated units. Of increasing popularity are the light prefabricated sandwich panels which consist of three layers. The exterior surface is made from aluminum, stainless steel, ceramics, plastics, and others. The center of the panel consists of some type of insulating material such as fiberglass or fiberboard, while the interior surface is made from metal, plaster, masonry, or some other attractive material.

Interior Partitions

The main purpose of interior partitions is to divide the inside space of a building into rooms. Partitions are selected for appearance, fire rating, weight, or acoustical properties. Partitions are bearing or nonbearing.

Bearing partitions support gravity loads in addition to their own weights and are thus permanently fixed in position. They can be constructed from wood or steel studs or from masonry units and faced with plywood, plaster, wallboard, or other material.

Nonbearing partitions do not support any loads in addition to their own weights and thus may be fixed or movable. Selection of the type of material to be used for a partition is based on the answers to the following questions: Is the partition to be fixed or movable? Is it to be transparent or opaque? Is it to extend all the way to the ceiling? Is it to be used to conceal piping and electrical conduits? Are there fire-rating and acoustical requirements? The more common types are made of metal, masonry, or concrete. For design of the floor slabs in a building that has movable partitions, some allowance should be given to the fact that the partitions may be moved. Probably the most usual practice is to increase the floor-design live load by 15 or 20 psf over the entire floor area.

15-13 FIREPROOFING OF STRUCTURAL STEEL

Although structural steel members are incombustible, their strength is tremendously reduced at temperatures normally reached in fires when the other materials of a building burn. Many disastrous fires have occurred in empty buildings where the only fuel was the buildings themselves. Steel is an excellent heat conductor, and nonfireproofed steel members may transmit enough heat from one burning compartment of a building to ignite materials with which they are in contact in adjoining sections of the building.

Steels (particularly those with rather high carbon contents) may actually increase a little in strength as they are heated from room temperature up to approximately 600°F. As temperatures are raised into the 800° to 1000°F range, steel strengths are drastically reduced, and at 1200°F they have little strength left.

Typical ratios of yield stresses at high temperatures to yield stresses at room temperatures are approximately 0.77 at 800°F, 0.63 at 1000°F, and 0.37 at 1200°F.[2]

The fire resistance of structural steel members can be greatly increased by applying to them fire protective covers such as concrete, gypsum, mineral fiber sprays, special paints, and others. The thickness and kind of fireproofing used depends on the type of structure, the degree of fire hazard, and economics.

In the past, concrete was commonly used for fire protection. Although concrete is not a particularly good insulation material, it is very satisfactory when applied in thicknesses of $1\frac{1}{2}$ to 2 in. or more, due to its mass. Furthermore the water in concrete (16 to 20 percent when fully hydrated) improves its fireproofing qualities appreciably. The boiling off of the water from the concrete requires a great deal of heat. As the water or steam escapes, it greatly reduces the concrete temperature. This is identical to the behavior of steel steam boilers, where the steel temperature is held to maximum values of only a few hundred degrees Fahrenheit due to the escape of the steam. It is true, however, that in very intense fires the boiling off of the water may cause severe cracking and spalling of the concrete.

Though concrete is an everyday construction material and though in mass it is quite a satisfactory fireproofing material, its installation cost is extremely high and its weight is large. As a result, for most steel construction, spray-on fireproofing materials have almost completely replaced concrete.

The spray-on materials usually consist of either mineral fibers or cementious fireproofing materials. The mineral fibers formerly used were asbestos. Due to the health hazards associated with this material, its use has been discontinued, and other fibers are now used by manufacturers. The cementious

---

[2] American Institute of Steel Construction, *Manual of Steel Construction Allowable Stress Design*, 9th ed. (Chicago: AISC, 1989), pp. 6-3 to 6-4.

fireproofing materials are composed of gypsum, perlite, vermiculite, and others. Sometimes, when plastered ceilings are required in a building, it is possible to hang the ceilings and light-gage furring channels by wires from the floor systems above and use the plaster as the fireproofing.

The cost of fireproofing structural steel buildings is high and hurts steel in its economic competition with other materials. As a result, a great deal of research is being conducted by the steel industry on imaginative new fireproofing methods. Among these ideas is the coating of steel members with expansive and insulative paints. When heated to certain temperatures, these paints will char, foam, and expand, forming an insulative shield around the members.

Other techniques involve the isolation of some steel members outside of the building, where they will not be subject to damaging fire exposure, and the circulation of liquid coolants inside box- or tube-shaped members in the building. It is probable that major advances will be made with these and other fireproofing ideas in the near future.[3]

15-14 INTRODUCTION TO HIGH-RISE BUILDINGS

Space is not taken in this introductory text to discuss the design of multistory buildings in detail. The material of the next few sections gives the student a general idea of the problems involved in the design of such buildings and does not present elaborate design examples.

Office buildings, hotels, apartment houses, and other buildings of many stories are quite common in the United States, and the trend is toward an even larger number of tall buildings in the future. Available land for building in our heavily populated cities is becoming scarcer and scarcer, and costs are becoming higher and higher. Tall buildings require a smaller amount of this expensive land to provide required floor space. Other factors contributing to the increased number of multistory buildings are new and better materials and construction techniques.

There are, on the other hand, several factors that may limit the heights to which buildings will be erected.

1. Certain city building codes prescribe maximum heights for buildings.
2. Foundation conditions may not be satisfactory for supporting buildings of many stories.
3. Floor space may not be rentable above a certain height. Someone will always be available to rent the top floor or two of a 250-story building, but floor numbers 100 through 248 may not be so easy to rent.
4. There are several cost factors that tend to increase with taller buildings. Among these factors are elevators, plumbing, heating and air-conditioning, glazing, exterior walls, and wiring.

---

[3] W. A. Rains, "A New Era in Fire Protective Coatings for Steel," *Civil Engineering* (New York: ASCE, September 1976), pp. 80–83.

Whether a multistory building is used for an office building, a hospital, a school, an apartment house, or whatever, the problems of design are generally the same. The construction is usually of the skeleton type in which the loads are transmitted to the foundation by a framework of steel beams and columns. The floor slabs, partitions, and exterior walls are all supported by the frame. This type of framing, which can be erected to tremendous heights, may also be referred to as beam-and-column construction. Bearing-wall construction is not often used for buildings of more than a few stories, although it has on occasion been used for buildings up to 20 or 25 stories. The columns of a skeleton frame are usually spaced 20, 25, or 30 ft on centers, with beams and girders framing into them from both directions (see Fig. 15-2). On some of the floors, however, it may be necessary to have much larger open areas between columns for dining rooms, ballrooms, and so on. For such cases very large beams (perhaps plate girders) may be needed to support column loads for many floors above.

It is usually necessary in multistory buildings to fireproof the members of the frame with concrete, gypsum, or some other material. The exterior walls are usually constructed with concrete or masonry units, although an increasing number of modern buildings are being erected with large areas of glass in the exterior walls.

For these tall and heavy buildings the usual spread footings may not be sufficient to support the loads. If the bearing strength of the soil is high, steel grillage footings may be sufficient, but for poor soil conditions pile or pier foundations may be necessary.

For multistory buildings the beam-to-column systems are said to be super-imposed on top of each other story by story or tier by tier. The column sec-

Broadview Apartments, Baltimore, Md. (Courtesy of Lincoln Electric Company.)

tions can be fabricated for one, two, or more floors, with the two-story lengths probably being the most common. Theoretically, column sizes can be changed at each floor level, but the costs of the splices involved usually more than cancel any savings in column weights. Columns of three or more stories in height are difficult to erect. The two-story heights work out very well most of the time.

15-15 DISCUSSION OF LATERAL FORCES

For tall buildings, lateral forces must be considered as well as gravity forces. In past decades the usual practice was to consider wind forces in a building-frame design when the height was two or more times the least lateral dimension. The usual framing of buildings with a smaller height-width ratio has sufficient stiffness to resist forces of large magnitude without any special design provisions. Perhaps the common 2-1 ratio is a little high, however, for some modern buildings with their large areas of glass in the exterior walls.

High wind pressures on the sides of tall buildings produce overturning moments. These moments are probably resisted without difficulty by the axial strengths of the columns, but the horizontal shears produced on each level may be of such magnitude as to require the use of special bracing or moment-resisting connections.

Unless they are fractured, the floors and walls provide a large part of the lateral stiffness of tall buildings. Although the amount of such resistance may be several times that provided by the lateral bracing, it is difficult to estimate and may not be reliable. Today so many modern buildings have light movable interior partitions, glass exterior walls, and lightweight floors that only the steel frame should be assumed to provide the required lateral stiffness.

A building must be not only sufficiently braced laterally to prevent failure but also prevented from deflecting so much as to injure its various parts. Another item of importance is the provision of sufficient bracing to give the occupants a feeling of safety. They might not have this feeling in tall buildings which have a great deal of lateral movement in times of high winds. There have actually been tales of occupants of the upper floors of tall buildings complaining of seasickness on very windy days.

The horizontal deflection of a multistory building due to wind or seismic loading is called *drift*. It is represented by Δ in Fig. 15-13. Drift is measured by the *drift index*, Δ/h, where h is the height or distance to the ground.

The usual practice in the design of multistory steel buildings is to provide a structure with sufficient lateral stiffness to keep the drift index between approximately 0.0015 and 0.0030 radians for the worst storms that occur in a period of approximately 10 years. These so-called 10-year storms might have wind in the range of about 90 mph, depending on location and weather records. It is felt that when the index does not exceed these values, the users of the building will be reasonably comfortable.

In addition multistory buildings should be designed to withstand safely 50-year storms. For such cases the drift index will be larger than the range men-

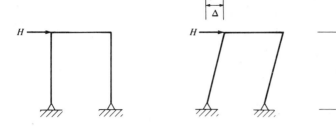

Figure 15-13

tioned, with the result that the occupants will suffer some discomfort. The 1450-ft World Trade Center in New York City deflects or sways up to about 3 ft in 10-year storms (drift index = 0.0021), while in hurricane winds it will sway up to about 7 ft (drift index = 0.0048).

Many areas of the world, including the western part of the United States, fall in earthquake territory, and in those areas it is necessary to consider seismic forces in design for tall or low buildings. During an earthquake there is an acceleration of the ground surface. This acceleration can be broken down into vertical and horizontal components. Usually the vertical component of the acceleration is assumed to be negligible, but the horizontal component can be severe.

Most buildings can be designed with little extra expense to withstand the forces caused during an earthquake of fairly severe intensity. On the other hand, earthquakes during recent years have clearly shown that the average building which is not designed for earthquake forces can be destroyed by earthquakes that are not particularly severe. The usual practice is to design buildings for additional lateral loads (representing the estimate of the earthquake forces) which are equal to some percentage (5 to 10) of the weight of the building and its contents.

Many people look upon the seismic loads to be used in design as being merely percentage increases of the wind loads. This is not altogether correct, however, as seismic loads are different in their action and are not proportional to the exposed area but to the building weight above the level in question.

The lateral forces require the use of more steel even though load factors are reduced for wind and earthquake forces. This additional steel will be used for bracing or moment-resisting connections as described in the next section.

15-16 TYPES OF LATERAL BRACING

A steel building frame with no lateral bracing is shown in Fig. 15-14(a). Should the beams and columns shown be connected together with the standard ("simple beam") connections, the frame would have little resistance to the lateral forces shown. Assuming the joints to act as frictionless pins, the frame would be laterally deflected as shown in Fig. 15-14(b).

Chase Manhattan Bank Building, New York City. (Courtesy of Bethlehem Steel Corporation.)

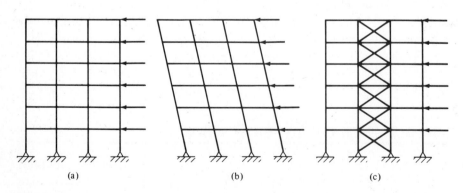

(a)　　　　　(b)　　　　　(c)

Figure 15-14

To resist these lateral deflections, the simplest method from a theoretical standpoint is the insertion of full diagonal bracing as shown in Fig. 15-14(c). From a practical standpoint, however, the student can easily see that in the average building full diagonal bracing would often be in the way of doors, windows, and other wall openings. Furthermore, many buildings have movable interior partitions, and the presence of interior cross bracing would greatly reduce this flexibility. Usually diagonal bracing is convenient in solid walls in and around elevator shafts, stairwells, and other walls in which few or no openings are planned.

Another very common method of providing resistance to lateral forces involves the use of moment-resisting connections as illustrated in Fig. 15-15. This *bracket-type bracing* may be used economically to provide lateral bracing for lower-rise buildings. As buildings become taller, however, bracket-type bracing is not very economical nor is it very satisfactory in limiting lateral deflections.

Some ways in which we can transmit lateral forces to the ground for buildings in the 20- to 60-story range are shown in Fig. 15-16. The X bracing system of part (a) works very well except that it might get in the way. Furthermore, as compared with the systems shown in Fig. 15-16(b) and (c), the floor beams have longer spans and will have to be larger.

The K bracing system (b) provides more freedom for the placing of openings than does the full X bracing. K braces are supported at midspan, with the result that the floor beams will have smaller moments. The K bracing system (like the knee bracing system) uses less material than does X bracing. If we need more room than is available with the K bracing system, we can go to the full-story knee bracing system shown in Fig. 15-16(c).[4]

In the usual building the floor system (beams and slabs) is assumed to be rigid in the horizontal plane, and the lateral loads are assumed to be concen-

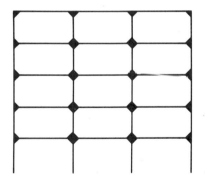

Figure 15-15 Bracket-type bracing.

[4]E. H. Gaylord, Jr., and C. N. Gaylord, *Structural Engineering Handbook*, 2d ed. (New York: McGraw-Hill, 1979), pp. 19-77 to 19-111.

(a) (b) (c)

Figure 15-16 (a) X brace. (b) K brace. (c) Full-story knee brace.

trated at the floor levels. Floor slabs and girders acting together provide considerable resistance to lateral forces. Investigation of steel buildings that have withstood high wind forces have shown that the floor slabs distribute the lateral forces so that all of the columns on a particular floor have equal deflections. When rigid floors are present, they spread the lateral shears to the columns or walls in the building. When lateral forces are particularly large, as in very tall buildings or where seismic forces are being considered, certain specially designed walls may be used to resist large parts of the lateral forces. These walls are called *shear walls*.

It is not necessary to brace every panel in a building. Usually the bracing can be placed in the outside walls with less interference than in the inside walls where movable partitions may be desired. Probably the bracing of the outside panels alone is not enough, and some interior panels may need to be braced. It is assumed that the floors and beams are sufficiently rigid to transfer the lateral forces to the braced panels. Three possible arrangements of braced panels are shown in Fig. 15-17. A symmetrical arrangement is probably desirable to prevent uneven lateral deflection in the building and thus torsion.

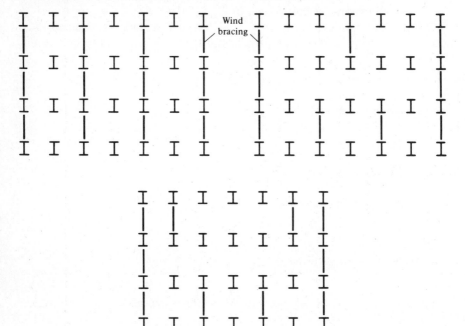

Figure 15-17

·Bracing around elevator shafts is usually permissible, while for other locations it will often interfere with windows, doors, movable partitions, glass exterior walls, open spaces, and so on. Should bracing be used around an elevator shaft as shown in Fig. 15-18(a) and should calculations show that the drift index is too large, it may be possible to use a *hat truss* on the top floor as shown in part (b). Such a truss will substantially reduce drift. If a hat truss is not feasible because of interference with other items, one or more *belt trusses* may be possible as shown in Fig. 15-18(c). A belt truss will appreciably reduce drift, although not as much as a hat truss.

The bracing systems described so far are not efficient for buildings above approximately 60 stories. These taller buildings have very large lateral wind and perhaps seismic forces applied hundreds of feet above the ground. The designer needs to develop a system that will resist these loads without failure and in such a manner that lateral deflections are not so great as to frighten the occupants. The bracing methods used for these buildings are usually based on a tubular frame concept.

With the tubular system a vertical tubular cantilever frame is created much like the tube of Fig. 15-19. The tube consists of the building columns and girders in both the longitudinal and transverse directions of the building. The idea is to create a tube that will act like a continuous chimney or stack.

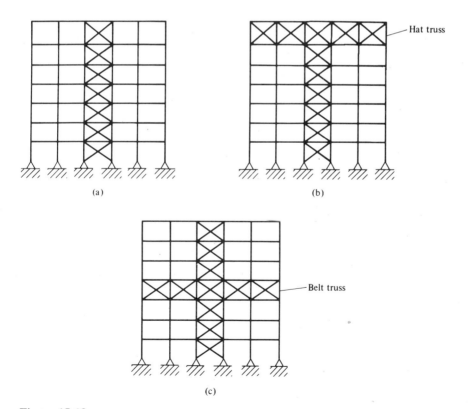

(a)

(b)

(c)

Figure 15-18

Figure 15-19 Solid-wall tube.

To build the tube the exterior columns are spaced close together, from 3 or 4 ft up to 10 or 12 ft on center. They are connected with spandrel beams at the floor levels as shown in Fig. 15-20(a).

Further improvements of the tubular system can be made by cross bracing the frame with X bracing over many stories as illustrated in Fig. 15-20(b). This latter system is very stiff and efficient and is very helpful in distributing gravity loads rather uniformly over the exterior columns.

Another variation on this system is the tube-within-a-tube system. Interior columns and girders are used to create additional tubes. In tall buildings it is

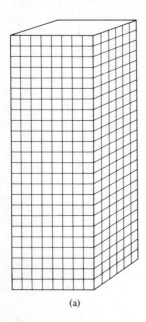

(a)

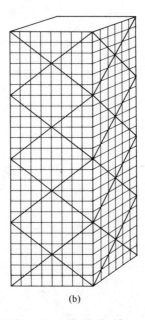

(b)

Figure 15-20 (a) Tubular frame. (b) Cross-braced tubular frame.

common to group elevator shafts, utility shafts, and stairs, and these systems can include shear walls and braced bents.[5]

15-17 ANALYSIS OF BUILDINGS WITH DIAGONAL WIND BRACING FOR LATERAL FORCES

Full diagonal cross bracing has been described as being an economical type of wind bracing for tall buildings. Where this type of bracing cannot practically be used due to interference with windows, doors, or movable partitions, moment-resisting brackets are often used. Should, however, very tall narrow buildings (with height–least-width ratios of 5.0 or greater) be constructed, lateral deflections may become a problem with moment-resisting joints. The joints can satisfactorily resist the moments, but the deflections may be excessive.

Should maximum wind deflections be kept under 0.002 times the building height, there is little chance of injury to the building, says ASCE Subcommittee 31.[6] The deflection is to be computed neglecting any resistance supplied by floors and walls. To keep lateral deflections within this range, it is necessary to use deep knee bracing, K bracing, or full diagonal cross bracing when the height–least-width ratio is about 5.0; and for greater values of the ratio, full diagonal cross bracing[7] or some other method such as the tubular frame is needed.

The student may often see bracing used in buildings for which he or she would think wind stresses were negligible. Such bracing stiffens up a building appreciably and serves the useful purpose of plumbing the steel frame during erection. Before bracing is installed, the members of a steel building frame may be twisted in all sorts of directions. Connecting the diagonal bracing should pull members into their proper positions.

When diagonal cross bracing is used, it is desirable to introduce initial tension into the diagonals. This prestressing will make the building frame tight and reduce its lateral deflection. Furthermore, the members can support compressive stress due to their pretensioning. Since the members can resist compression, the horizontal shear to be resisted will be assumed to split equally between the two diagonals. For buildings with several bay widths, equal shear distribution is usually assumed for each bay.

Should the diagonals not be initially tightened, rather stiff sections should be used so they will be able to resist appreciable compressive forces. The direct axial forces in the girders and columns can be found for each joint from the shear forces assumed in the diagonals. Usually the girder axial forces so

[5] R. N. White, P. Gergely, and R. G. Sexsmith, *Structural Engineering,* vol 3 (New York: Wiley 1974), pp. 537–546.

[6] "Wind Bracing in Steel Buildings," *Transactions ASCE* **105** (1940), pp. 1713–1739.

[7] L. E. Grinter, *Theory of Modern Steel Structures* (New York: Macmillan, 1962), p. 326.

Transfer truss, 150 Federal Street, Boston, Mass. (Courtesy of Owen Steel Company, Inc.)

computed are too small to consider, but the values for columns can be quite important. The taller the building becomes, the more critical are the column axial forces caused by lateral forces.

For types of bracing other than cross bracing, similar assumptions can be made for analysis. The student is referred to pages 333–339 of *Theory of Modern Steel Structures* by L. E. Grinter (New York: Macmillan, 1962) for discussion of this subject.

15-18 MOMENT-RESISTING JOINTS

For a large percentage of buildings under eight to ten stories, the beams and girders are connected to each other and to the columns with simple end-framed connections of the types described in Chapter 14. As buildings become taller, it is absolutely necessary to use a definite wind-bracing system or moment-resisting joints. Moment-resisting joints may also be used in lower buildings where it is desired to take advantage of continuity and the consequent smaller beam sizes and depths and shallower floor construction. Moment-resisting brackets may also be necessary in some locations for loads that are applied eccentrically to columns.

Several types of moment-resisting connections which may be used as wind bracing are shown in Fig. 15-21. The design of connections of these types was also presented in Chapters 13 and 14. The average design company through the years will probably develop a file of moment-resisting connections from their previous designs. When they have a wind moment of such and such a value, they merely refer to their file and select one of their former designs which will provide the required moment resistance.

In the following paragraphs a few comments are made about each of the types of connections shown in Fig. 15-21. The letter preceding each of these paragraphs corresponds to part of the figure.

(a) The top-angle and seat-angle connection shown, sometimes called an *angle bracket,* provides the minimum acceptable type of wind connection. Even if web angles, shown by a dashed line, are added, the moment resistance is not appreciably increased. This type of wind connection is not satisfactory for very tall buildings.

(b) The *split-beam brackets* have been frequently used as wind connections in multistory buildings. When large girder reactions have to be transferred by this excellent type of connection, the seats may have to be stiffened as shown by the dashed line.

(c) This heavy bulky obsolete connection was used to resist very large moments. Although it can reduce girder moments decidedly, its high fabrication cost surely cancels the saving in girder weight.

(d) and (e) These are the two most commonly used moment-resisting connections today. It is difficult to say that one is more economical than the other.

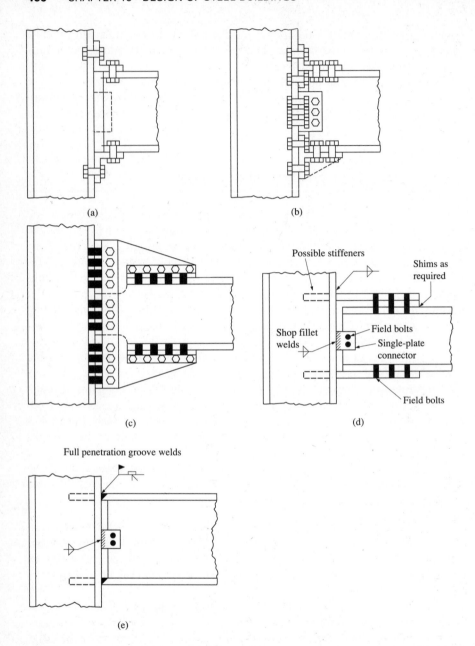

Figure 15-21 (a) Angle bracket. (b) Split-beam connection. (c) Heavy bracket. (d) Shop-welded, field-bolted moment connection. (e) Field-welded, field-bolted moment connection.

Some fabricators like one and some the other. Single-plate (or shear tab) connectors are shown in the figures, but framing angles may be used instead. Column stiffeners (shown dotted in the figures) are sometimes required by ASD Specification K1 for local flange bending, local web yielding, compression buckling of the web, and so on. Stiffeners are a nuisance and cost an appreciable amount of money. As a result we will, when the specification shows they are needed, try to avoid them by increasing column sizes by one or two series.

A very satisfactory variation of these last two connections involves an end plate as shown on page 4-116 of the ASD Manual. This type of connection may, however, be used only for static load situations.

15-19 ANALYSIS OF BUILDINGS WITH MOMENT-RESISTING JOINTS FOR LATERAL LOADS

Approximate methods have been very popular for many years for analyzing multistory buildings for lateral forces. Among the several reasons for this popularity have been the following:

1. Large building frames are statically indeterminate to a very high degree, and their analysis by an "exact" method using a pocket calculator is a lengthy and difficult problem.
2. The resistance to lateral forces supplied by the walls and floors in a tall building is difficult to estimate accurately, and the results of analysis by an "exact" method are therefore not very precise.
3. Before the members of a statically indeterminate structure can be designed their forces have to be determined, but these forces are dependent upon their sizes. Application of an approximate analysis should yield forces from which very good estimates of member sizes can be made for preliminary design.

The analysis of multistory buildings is a very difficult problem, and past practice has been to use one of the several approximate methods of analysis available. Today, however, with the availability of electronic computers it is feasible to make "exact" analyses in appreciably less time than required to make the approximate analyses (without the use of computers). The more accurate values obtained permit the use of smaller members. Computer usage saves money in analysis time and in the use of smaller members.

In the pages that follow a brief review of the portal and cantilever approximate methods is presented. No consideration is given in either of these methods to the elastic properties of the members. These omissions can be very serious in unsymmetrical frames and in very tall buildings. For example, consider the changes in member sizes in a very tall building. In such a building there will probably not be a great deal of change in beam sizes from the top floor to the bottom floor. For the same loadings and spans the changed sizes would be due to the large wind moments in the lower floors. The change, however, in column sizes from top to bottom would be tremendous. The result is that the relative sizes of columns and beams on the top floors are entirely different from their relative sizes on the lower floors. When this fact is not considered, it causes large errors in the analysis.

If the height of a building is roughly five or more times its least lateral dimension, it is generally felt that a more precise method of analysis should be

The 20-story United Founders Life Tower in Oklahoma City, Okla. (Courtesy of Lincoln Electric Company.)

used than the methods described here. There are several excellent approximate methods which make use of the elastic properties and which give values closely approaching the results of the "exact" methods. These include the Factor method, the Witmer method of K percentages, and the Spurr method. Should an exact method be desired, the slope-deflection and moment-distribution methods are available. If the slope-deflection procedure is used, the designer will have the problem of solving a large number of simultaneous equations. The problem is not so serious, however, if a digital computer is used.

The Portal Method

The most common approximate method of analyzing tall building frames for lateral loads is the portal method. The chief advantage of this method, which is said to be satisfactory for most buildings up to 25 stories,[8] is its simplicity. A detailed description of this method, including the basis for the assumptions made, can be found in most structural analysis textbooks. Only a brief listing of the steps involved in its application is given here.

1. The horizontal shears on each level are arbitrarily distributed between the columns. One commonly used procedure is to assume the shear divides between the columns in the ratio of one part to exterior columns and two parts to interior columns. Another common distribution (and the one used in the illustrative problem to follow) is to assume the shear V taken by each column is in proportion to the floor area it supports.

2. The moment M in each column is assumed to equal the column shear times half the column height (thus assuming a point of contraflexure at middepth).

3. The girder moments M are determined by joints, by noting that the sum of the girder moments at any joint equals the sum of the column moments at that joint. These calculations are easily made by starting at the upper left joint and working joint by joint across to the right; after which the next level is handled left to right, and so on.

4. The shear V in each girder is assumed to equal its moment divided by half the girder length.

5. Finally, the column axial forces S are determined by summing up the beam shears and other column axial forces at each joint. These calculations are again handled conveniently by working from left to right and from the top floor down.

Fig. 15-22 shows a building frame which is to be analyzed by the portal method. The frames are assumed to be placed 20 ft 0 in. on centers and the exterior walls are assumed to be subjected to a wind pressure of 20 lb per vertical square foot. From these data the horizontal loads shown at each floor level are calculated.

---

[8] "Wind Bracing in Steel Buildings," *Transactions ASCE* **105** (1940), p. 1723.

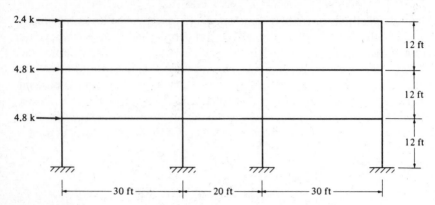

Figure 15-22

This frame is analyzed in Fig. 15-23 by the portal method. The arrows shown on the figure give the direction of the girder shears and the column axial forces. The student can visualize the stress condition of the frame if he or she assumes the frame is tending to be pushed over from left to right by the wind, stretching the left exterior columns and compressing the right exterior columns.

The Cantilever Method

Another simple method of analyzing building frames for lateral forces is the cantilever method. This method is said to be a little more desirable for high narrow buildings than the portal method and may be used satisfactorily for buildings with heights not in excess of 25 to 35 stories.[9] It is not as popular as the portal method.

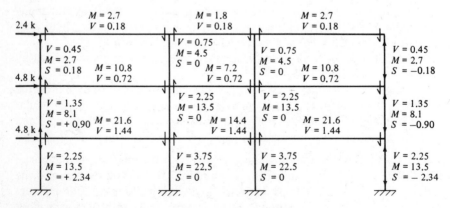

Figure 15-23 Frame analysis by the portal method.

---

[9] Ibid.

Instead of initially assuming the horizontal shears to be divided between the columns on each level in some proportion, the first assumption in the cantilever method pertains to the column axial forces. The axial force in a particular column is assumed to vary in direct proportion to its distance from the center of gravity of the group of columns on that level. With the wind blowing from left to right, the columns to the left of the center of gravity will be in tension while those to the right will be in compression. The steps involved in analyzing a building frame for lateral forces by the cantilever method are as follows:

1. To obtain the column axial forces, moments are taken above an asumed plane of contraflexure through the middepth of the columns on each level. As an illustration of this procedure, moments are taken here to determine the axial forces in the columns on the top level of the building frame of Fig. 15-22. A free body is drawn in Fig. 15-24 of the part of the building above the assumed plane of contraflexure on the top level. The value of S is determined:

$$\Sigma M_x = 0$$
$$(2.4)(6) + (30)(S) - (50)(S) - (80)(4S) = 0$$
$$S = 0.0424$$

2. The girder shears are determined joint by joint from the column axial forces.
3. The girder moments are determined by multiplying the girder shears by the half-girder lengths.
4. The column moments are found joint by joint from the girder moments.
5. The column shears are obtained by dividing the column moments by the half-column heights.

The analysis of the frame of Fig. 15-22 by the cantilever method is illustrated in Fig. 15-25. Arrows representing the directions of the column axial forces and the girder shears are again shown.

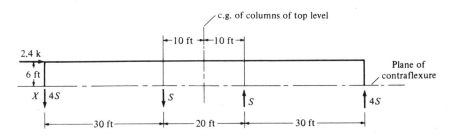

Figure 15-24

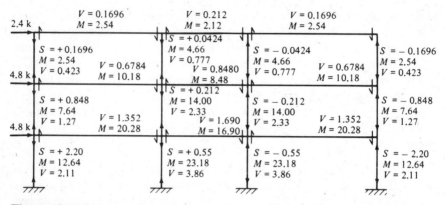

Figure 15-25 Frame analysis by the cantilever method.

15-20 ANALYSIS OF BUILDINGS FOR GRAVITY LOADS

Simple Framing

If simple framing is used, the design of the girders is quite simple because the shears and moments in each girder can be determined by statics. The gravity loads applied to the columns are relatively easy to estimate, but the column

"Topping out" of Blue Cross–Blue Shield Building in Jacksonville, Fla. (Courtesy of Owen Steel Company, Inc.) The Christmas tree is an old north European custom used to ward off evil spirits. It is also used today to show that the steel frame was erected with no lost-time accidents to personnel.

moments may be a little more difficult. If the girder reactions on each side of the interior column of Fig. 15-26 are equal, theoretically no moment will be produced in the column at that level. This situation is probably not realistic, because it is highly possible for the live load to be applied on one side of the column and not on the other (or at least be unequal in magnitude). The results will be column moments. If the reactions are unequal, the moment produced in the column will equal the difference between the reactions times the distances to the center of gravity of the column.

Exterior columns may often have moments due to spandrel beams opposing the moments caused by the floor loads on the inside of the column. Nevertheless, gravity loads will generally cause the exterior columns to have larger moments than the interior columns.

To estimate the moment applied to a column above a certain floor, it is probably reasonable to assume that the unbalanced moment at that level splits evenly between the column above and below. In fact, such an assumption is often on the conservative side, as the column below may be larger than the one above.

Rigid Framing

For buildings with moment-resisting joints it is a little more difficult to estimate the girder moments and make preliminary designs. If the ends of each girder are assumed to be completely fixed, the moments for uniform loads are as shown in Fig. 15-27(a). Conditions of complete fixity are probably not realized, with the result that the end moments are smaller than shown. There is a corresponding increase in the positive centerline moments approaching the simple moment ($wL^2/8$) shown in Fig. 15-27(b). Probably a moment diagram somewhere in between the two extremes is more realistic. Such a moment diagram is represented by the dotted line of Fig. 15-27(a). A reasonable procedure is to assume a moment in the range of $wL^2/10$, where L is the clear span.

When the beams and girders of a building frame are rigidly connected to each other, a continuous frame is the result. From a theoretical standpoint an

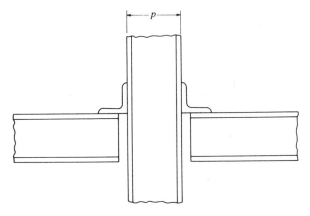

Figure 15-26 Simple framing of interior column.

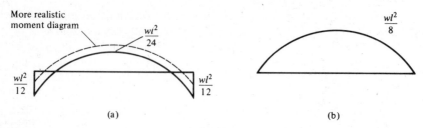

Figure 15-27 (a) Fixed-end beam. (b) Simple beam.

accurate analysis of such a structure cannot be made unless the entire frame is handled as a unit.

Before this subject is pursued further, it should be realized that in the upper floors of tall buildings the wind moments will be small, and the framed and seated connections described in Chapter 14 will provide sufficient moment resistance. For this reason the beams and girders of the upper floors may very well be designed on the basis of simple beam moments, while those of the lower floors may be designed as continuous members with moment-resistant connections because of the larger wind moments.

For the design of beams and girders it is necessary to consider two loading conditions: $(DL + LL)$ and $(DL + LL + WL)\frac{3}{4}$. In this latter expression WL represents lateral forces, whether due to wind or wind and earthquake, and the $\frac{3}{4}$ factor represents the one-third increase in allowable stresses when lateral forces are involved. The reasoning for this increase in allowable values permitted by the ASD specification (A5.2) is somewhat dubious.[10] For the upper floors the wind moments will not increase the girder sizes because of the allowable-stress increase, but on lower floors girder size will begin to increase.

From a strictly theoretical viewpoint there are several live-load conditions which need to be considered to obtain maximum shears and moments at various points in a continuous structure. For the building frame shown in Fig. 15-28 it is desired to place live loads to cause maximum positive moment in span AB. A qualitative influence line for positive moment at the centerline of this span is shown in part (a). This influence line shows that, to obtain maximum positive moment at the centerline of span AB, the live loads should be placed as shown in Fig. 15-28(b).

To obtain maximum negative moment at point B or maximum positive moment in span BC, other loading situations need to be considered. With the availability of computers, more of this detailed analysis is being done every day. The student, however, can see that unless the designer uses a computer, he or she probably will not have sufficient time to go through all of these theoretical situations. Furthermore, it is doubtful if the accuracy of our analysis methods would justify all of the work anyway. It is, however, often feasible to take out two stories of the building at a time as a free body and analyze that

---

[10] D. S. Ellifritt, "The Mysterious $\frac{1}{3}$ Stress Increase," *Engineering Journal* 14, No. 4, October 1977 (New York: AISC), p. 138–140.

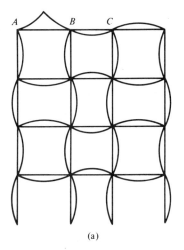

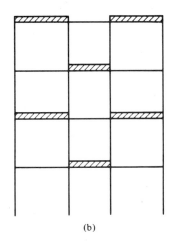

(a)   (b)

Figure 15-28

part by one of the "exact" methods such as the successive correction method of moment distribution.

Before such an "exact" analysis can be made it is necessary to make an estimate of the member sizes, probably based on the results of analysis by one of the approximate methods.

15-21 DESIGN OF MEMBERS

The girders can be designed for the two load combinations mentioned in Section 15-20. For the top several floors the second condition will not control design, but farther down in the building it will control the sizes for the reasons described in the next paragraph.

Each of the members of a building frame must be designed for all of the dead loads which it supports, but it may be possible to design some members for lesser loads than their full theoretical live-load values. For example, it seems unlikely in a building frame of several stories that the absolutely maximum-design live load will occur on every floor at the same time. The lower columns in a building frame are designed for all of the dead loads above but probably for a percentage of live load appreciably less than 100 percent. Some specifications require that the beams supporting floor slabs be designed for full dead and live loads but permit the main girders to be designed under certain conditions for reduced live loads. (This reduction is based on the improbability that a very large area of a floor would be loaded to its full live-load value at any one time.) The ASCE Code provides some very commonly used reduction expressions.[11]

[11] *American Society of Civil Engineers Minimum Design Loads for Buildings and Other Structures,* ASCE 7–88 (New York: ASCE), p. 6.

Should simple framing be used, the girders will be proportioned for simple beam moments plus the moments caused by the lateral loads. For continuous framing the girders will be proportioned for $wL^2/10$ (for uniform loads) plus the moments caused by the lateral loads. An interesting comparison of designs by the two methods is shown in pages 717–719 of Beedle et al., *Structural Steel Design* (New York: Ronald, 1964). It may be necessary to draw the moment diagram for the two cases (gravity loads and lateral loads) and add them together to obtain the maximum positive moment out in the span. The value so computed may very well control the girder size in some of the members.

A major part of the design of a multistory building is involved in setting up a column schedule which shows the loads to be supported by the various columns story by story. The gravity forces can be estimated very well, while the shears, axial forces, and moments caused by lateral forces can be roughly approximated for the preliminary design from some approximate method (portal, cantilever, factor, or other).

If both axes of the columns are free to sway, the column effective lengths theoretically must be calculated for both axes as described in Chapter 5. If diagonal bracing is used in one direction, thus preventing sidesway, the K value will be less than 1.0 in that direction. When frames are braced in the narrow direction, it is reasonable to use $K = 1.0$ for both axes.

After the sizes of the girders and columns are tentatively selected for a two-story height, an "exact" analysis can be performed and the members redesigned. The two stories taken out for analysis are often referred to as a *tier*. This process can be continued tier by tier down through the building. Many tall buildings have been designed and are performing satisfactorily in which this last step (the two-story "exact" analysis) was omitted.

16

Composite Construction

16-1 INTRODUCTION

When a concrete slab is supported by steel beams and there is no provision for shear transfer between the two, the result is a noncomposite section. In noncomposite construction the load undoubtedly causes the slab to deflect along with the beam, with the result that some of the load is carried by the slab. Unless there is a great deal of bond between the two, however (as would be the case where the steel beam is completely encased in the concrete, or where a system of mechanical shear connectors is provided), the load carried by the slab is small and may be neglected.

For many years steel beams and reinforced-concrete slabs were used together with no consideration being made for any composite effect. In recent decades, however, it has been shown that a great strengthening effect can be obtained by tying the two together to act as a unit in resisting loads. Steel beams and concrete slabs joined together compositely can often support a one-third, one-half, or even greater increase in load than could the steel beams alone in noncomposite action.

Composite construction for highway bridges was given the green light by the adoption of the 1944 AASHTO Specifications which approved the method. Since about 1950 the use of composite bridge floors has rapidly increased until today they are commonplace all over the United States. In these bridges the longitudinal shears are transferred from the stringers to the reinforced-concrete slab or deck with shear connectors (to be described in Section 16-7), causing the slab or deck to assist in carrying the bending moments. This type of section is shown in Fig. 16-1(a).

The first approval for composite building floors was given by the 1952 AISC Specification, and today they are very common. These floors may either be encased in concrete (very rare due to expense) as shown in Fig. 16-1(b) or be nonencased with shear connectors as shown in (c). If the steel sections are encased in concrete, the shear transfer is made by bond and friction between the beam and the concrete and by the shearing strength of the concrete along the dotted lines shown in Fig. 16-1(b).

Today formed steel deck (illustrated in Fig. 16-2) is used for almost all composite building floors. The initial examples in this chapter, however, pertain to the calculations for composite sections where formed steel deck is not used. Sections that make use of formed steel deck are described later in the chapter.

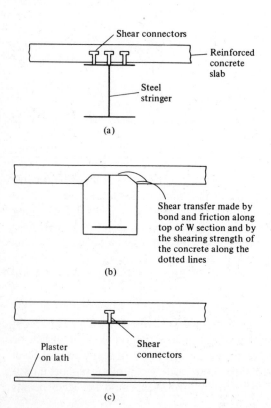

Figure 16-1 (a) Composite bridge floor with shear connectors. (b) Encased section for building floors. (c) Building floors with shear connectors.

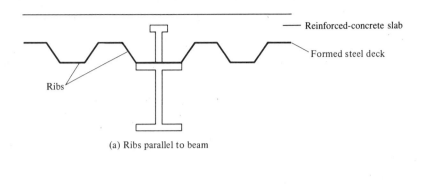

(a) Ribs parallel to beam

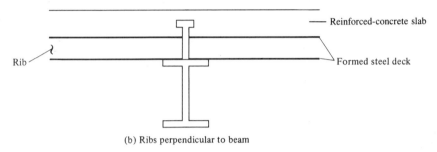

(b) Ribs perpendicular to beam

Figure 16-2 Composite sections using formed steel deck.

16-2 ADVANTAGES OF COMPOSITE CONSTRUCTION

The floor slab in composite construction acts not only as a slab for resisting the live loads but also as an integral part of the beam. It actually serves as a large cover plate for the upper flange of the steel beam, appreciably increasing the beam's strength.

A particular advantage of composite floors is that they make use of concrete's high compressive strength by putting a large part of the slab in compression. At the same time a larger percentage of the steel is kept in tension (also advantageous) than is normally the case in steel-frame structures. The result is less steel tonnage required for the same loads and spans (or longer spans for the same sections). Composite sections have greater stiffness than noncomposite sections, and they have smaller deflections—perhaps only 20 to 30 percent as large. Furthermore, tests have shown that the ability of a composite structure to take overload is decidedly greater than for a noncomposite structure.

A further advantage of composite construction is the possibility of having smaller overall floor depths—a fact of particular importance for tall buildings. Smaller floor depths permit reduced building heights with the consequent advantages of lower costs for walls, plumbing, wiring, ducts, elevators, and foundations. Another important advantage available with reduced beam depths

Bridge on John F. Kennedy Expressway, Chicago, Ill. (Courtesy of Robert Mc-Cullough, Chicago, Ill.)

is a saving in fireproofing costs for the beams. A disadvantage for composite construction is the cost of furnishing and installing the shear connectors. This extra cost will usually exceed the cost reductions mentioned when spans are short and lightly loaded.

16-3 DISCUSSION OF SHORING

After the steel beams are erected, the concrete slab is placed on them. The formwork, wet concrete, and other construction loads must, therefore, be supported by the beams or by temporary shoring. Should no shoring be used, the steel beams must support all of these loads as well as their own weights. Most specifications say that after the concrete has gained 75 percent of its 28-day strength, the section has become composite and all loads applied thereafter may be considered to be supported by the composite section. When shoring is used, it supports the wet concrete and the other construction loads. It does not really support the weight of the steel beams unless they are given an initial upward deflection (probably impractical). When the shoring is removed (after the concrete gains at least 75 percent of its 28-day strength), the weight of the slab is transferred to the composite section, not just to the steel beams. The student can see that if shoring is used it will be possible to use lighter and thus cheaper steel beams. The question then arises, Will the saving in steel cost be greater than the extra cost of shoring? Probably the answer is No. The usual decision is to use heavier steel beams and do without shoring for several reasons.

1. Apart from reasons of economy, the use of shoring is a tricky operation, particularly where settlement of the shoring is possible, as is often the case in bridge construction.

2. Both theory and load tests show that the ultimate strengths of composite sections of the same sizes are the same whether shoring is used or not. If lighter steel beams are selected for a particular span because shoring is used, the result is a smaller ultimate strength.

3. Another disadvantage of shoring is that after the concrete hardens and the shoring is removed, the slab will participate in composite action in supporting the dead loads. The slab will be placed in compression by these long-term loads and will have substantial creep and shrinkage parallel to the beams. The result will be a great decrease in the stress in the slab with a corresponding increase in the steel stresses. The probable consequence is that most of the dead load will be supported by the steel beams anyway, and composite action will apply only to the live loads as though shoring had not been used.

16-4 EFFECTIVE FLANGE WIDTHS

There is a problem involved in estimating how much of the slab acts as part of the beam. Should the beams be rather closely spaced, the bending stress in the slab will be fairly uniformly distributed across the compression zone. If the distances between beams are large, however, bending stresses will vary quite a bit and nonlinearly across the flange. The farther a particular part of the slab or flange is away from the steel beam, the smaller will be its bending stress. Specifications attempt to handle this problem by replacing the actual slab with a narrower or effective slab which has a constant stress. This equivalent slab is deemed to support the same total compression as is supported by the actual slab. The effective width of the slab b_e is shown in Fig. 16-3.

The portion of the slab or flange which can be considered to participate in the composite beam action is controlled by the specifications. ASD Specification I1 states that the effective width of the concrete slab on each side of the beam centerline is to be taken as equal to the least of the values to follow. This same set of rules applies whether the slab exists on one or both sides of the beam.

1. One-eighth of the span of the beam measured center to center of supports for both simple and continuous spans.

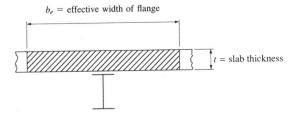

b_e = effective width of flange

t = slab thickness

Figure 16-3

2. One-half of the distance from the beam centerline to the centerline of the adjacent beam.
3. The distance from the beam centerline to the edge of the slab.

The AASHTO requirements for determining effective flange widths are somewhat different. The maximum total flange width may not exceed one-fourth of the beam span, 12 times the least thickness of the slab, or the distance center to center of the beams. Should the slab exist on only one side of the beam, its effective width may not exceed one-twelfth of the beam span, six times the slab thickness, or one-half of the distance from the centerline of the beam to the centerline of the adjacent beam.

16-5 STRESS CALCULATIONS FOR NONENCASED COMPOSITE SECTIONS

For stress calculations the properties of a composite section are computed with the transformed area method. In this method the cross-sectional area of one of the two materials has to be replaced or transformed into an equivalent area of the other. For composite design, it is customary to replace the concrete with an equivalent area of steel, whereas the reverse procedure is used in the working stress design method for reinforced-concrete design.

In the transformed area procedure the concrete and steel are assumed to be bonded tightly together so that their strains will be the same at equal distances from the neutral axis. The unit stress in either material can then be said to equal its strain times its modulus of elasticity (ϵE_c for the concrete or ϵE_s for the steel). The unit stress in the steel is then $\epsilon E_s / \epsilon E_c = E_s / E_c$ times as great as the corresponding unit stress in the concrete. The E_s / E_c ratio is referred to as the modular ratio n; therefore, n in.^{2} of concrete are required to resist the same total stress as 1 in.^{2} of steel; and the cross-sectional area of the slab (A_c) is replaced with a transformed or equivalent area of steel equal to A_c/n and called A_{ctr} herein.

The American Concrete Institute (ACI) Building Code suggests that the following expression may be used for calculating the modulus of elasticity of concrete weighing from 90 to 155 lb/ft^3:

$$E_c = w^{1.5} 33\sqrt{f_c'}$$

In this expression w is the weight of the concrete in pounds per cubic foot, and f_c' is its specified compression strength in pounds per square inch.

From the preceding information we can see that the modulus of elasticity of the concrete and thus the modular ratio are functions of both the strength and the unit weight of the concrete. For the purpose of computing flexural stresses, however, the ASD Specification (I2) states that the equivalent or transformed area of the concrete for either normal-weight concrete or lightweight concrete is to be determined by dividing its area by the value of n for normal-weight concrete of the strength specified. For deflection computations, however, this same specification states that the transformed properties of the concrete must be determined using an n that is based upon both strength and weight.

When composite sections with appropriate shear connectors are constructed without shoring, the ASD Specification (I2.2) states that steel stresses are to be computed before composite action is developed, assuming all of the moment is resisted by the steel sections. The stress so obtained may not exceed the values previously considered for steel beams and specified in ASD Specification F1.1. The value of F_b used could possibly be reduced for lateral-torsional buckling, but $0.66F_y$ is usually used, considering the lateral bracing supplied by the formwork or the formed steel deck. In the flexure formulas that follow, M_D is the moment due to the weight of the steel beam and the wet concrete, and M_L is the moment due to the loads applied after concrete hardening.

$$f_s = \frac{M_D}{S_s} \leq F_b \quad \text{before concrete hardens}$$

As shown above, the steel section alone is assumed to resist all loads applied before the concrete hardens. The composite section is assumed to resist all loads applied thereafter, and the total steel stress is determined with the expression to follow, in which $S_{tr\,bot}$ is the section modulus of the transformed composite section referred to the tension flange.

$$f_s = \frac{M_D}{S_s} + \frac{M_L}{S_{tr\,bot}} \leq 0.9F_y \quad \text{after concrete hardens}$$

The total of the precomposite and composite steel stresses may not not exceed $0.90F_y$. The requirements of the ASD Specification for composite sections are based on ultimate load considerations even though they are written in terms of working stresses. As a result, actual stresses under service loads are higher than our formulas seem to indicate, and the steel stress is limited to the 90 percent value (ASD Section I2.2).

Supposedly, the limiting of the steel stress calculated as described makes sure that the steel beam remains in an elastic condition and thus prevents permanent deformation under severe loads and has no effect on the ultimate moment capacity of the composite section. This same ASD section provides requirements that limit the permissible stresses in the compression flanges of the steel sections during construction loading.

It will be noted that the steel sections for such members are exempt from the compact section requirements and that there is no limit on the unbraced length of the compression flanges.

The transformed section modulus referred to the extreme fiber of the concrete in compression ($S_{tr\,top}$) is used to calculate the stress in the concrete for the loads applied after hardening.

$$f_c = \frac{M_L}{nS_{tr\,top}} \leq 0.45f_c' \quad \text{after concrete hardens}$$

The ASD Commentary (I2) states that, to avoid overly conservative proportions between the concrete slab and the steel beam, the concrete stress is to be calculated on the basis of the composite section. The stress so obtained is limited to the usual maximum ACI working stress value of $0.45 f_c'$.

Example 16-1 presents the calculation of the stresses in a composite floor with no shoring.

■ Example 16-1

Determine the stresses for the composite beam section shown in Fig. 16-4, according to the ASD Specification. Adequate shear connectors are assumed to be present, and no shoring is used. Assume simple spans and the following data:

$LL = 100$ psf

Partition weight $= 18$ psf

4-in. concrete slab weight $= 50$ psf

$f_c' = 3000$ psi

$f_c =$ allowable concrete compression stress $= 0.45 f_c' = 1350$ psi

$n = 9$

A36 steel

Solution.   Calculation of moments:

Loads applied before concrete hardens (75 percent of 28-day strength)

$$\text{Slab} = (8)(50) = 400 \text{ lb/ft}$$
$$\text{Beam} = \underline{31} \text{ lb/ft}$$
$$\text{Total} = 431 \text{ lb/ft}$$

$$M_D = \frac{(0.431)(27)^2}{8} = 39.3 \text{ ft-k}$$

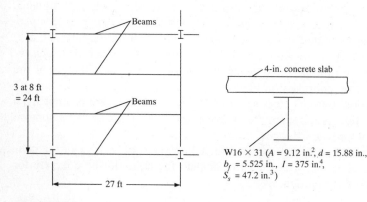

Beams

4-in. concrete slab

3 at 8 ft
= 24 ft

Beams

27 ft

W16 × 31 ($A = 9.12$ in.², $d = 15.88$ in.,
$b_f = 5.525$ in., $I = 375$ in.⁴,
$S_s = 47.2$ in.³)

Figure 16-4

Loads applied after concrete hardens

$$\text{Partitions} = (8)(18) = 144 \text{ lb/ft}$$
$$LL = (8)(100) = \underline{800} \text{ lb/ft}$$
$$\text{Total} = 944 \text{ lb/ft}$$

$$M_L = \frac{(0.944)(27)^2}{8} = 86.0 \text{ ft-k}$$

Effective width of flange:

$$b = (2)(\tfrac{1}{8})(12 \times 27) = 81 \text{ in.} \leftarrow$$
$$b = (2)(\tfrac{1}{2})(8 \times 12) = 96 \text{ in.}$$

Properties of composite section (see Fig. 16-5):

$$A = 9.12 + (4)(9.00) = 45.12 \text{ in.}^2$$

$$y_b = \frac{(9.12)(7.94) + (4)(9.00)(17.88)}{45.12} = 15.87 \text{ in.}$$

$$I = 375 + (9.12)(7.93)^2 + (\tfrac{1}{12})(9.00)(4)^3 + (4)(9.00)(2.01)^2 = 1142 \text{ in.}^4$$

$$S_{tr \, bot} = \frac{1142}{15.87} = \underline{\underline{72 \text{ in.}^3}}$$

$$S_{tr \, top} = \frac{1142}{4.01} = \underline{\underline{284.8 \text{ in.}^3}}$$

Review of stresses:
Before concrete hardens

$$f_s = \frac{M_D}{S_s} = \frac{(12)(39.3)}{47.2} = 9.99 \text{ ksi} < 0.66F_y = 24 \text{ ksi} \qquad \text{OK}$$

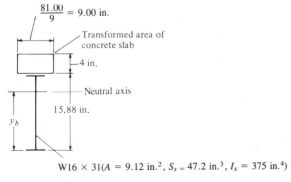

$$\frac{81.00}{9} = 9.00 \text{ in.}$$

Transformed area of concrete slab

4 in.

Neutral axis

15.88 in.

y_b

W16 × 31($A = 9.12$ in.^{2}, $S_s = 47.2$ in.^{3}, $I_x = 375$ in.^{4})

Figure 16-5

After concrete hardens

$$f_s = \frac{M_D}{S_s} + \frac{M_L}{S_{tr\,bot}} = 9.99 + \frac{(12)(86.0)}{72} = 24.32 \text{ ksi} < 0.90F_y = 32.4 \text{ ksi} \quad \text{OK}$$

$$f_c = \frac{M_L}{nS_{tr\,top}} = \frac{(12)(86.0)}{(9)(284.8)} = 0.402 \text{ ksi} < 0.45f_c' = 1.35 \text{ ksi} \qquad \text{OK} \qquad \blacksquare$$

The concrete compressive stresses are usually not significant for the average composite section. Should 50-ksi steel and narrow concrete flanges (or both) be used, however, concrete stresses may be critical. Concrete stresses are higher with shored construction than with unshored construction. On the other hand the use of partially composite construction (which is described in Section 16-11 of this chapter) provides some reduction in concrete stresses.

Should the neutral axis fall in the slab, the concrete below may not be considered to contribute to the moment of inertia (except for deflection calculations), as the concrete is assumed to be cracked and to have no tensile strength. It is true that concrete has little strength in tension—but it does have some. As a result some people feel that all of the concrete slab depth should be counted in the composite section properties regardless of the location of the neutral axis. The tensile stress at which the concrete is usually assumed to crack (the *modulus of rupture*) is specified by the ACI to equal approximately $7.5\sqrt{f_c'}$. For the concrete in the preceding example, this is $7.5\sqrt{3000} = 411$ psi. (You should realize that once the concrete cracks in tension its tensile strength is truly negligible.) If we count all of the slab in the computations for the concrete and steel stresses, we will usually get about the same answers. *The ASD Specification, however, does not permit this calculation.*

When there are reinforcing bars parallel to the steel section and within the effective slab width, they may be used in determining the properties of composite sections if appropriate shear connectors are furnished in accordance with ASD Section I4.

16-6 COMPOSITE SECTION PROPERTIES FROM ASD MANUAL

In the composite section tables of Part 2 of the ASD Manual, approximate moment of inertia and section moduli values ($\bar{I}_{tr}$ and $\bar{S}_{tr}$) are provided for many commonly used composite sections. Fig. 16-6 shows the nomenclature used for these tables.

In the Manual the moment of inertia values for the transformed area of the concrete have been simplified by using $\frac{1}{12}(A_{ctr})(t)^2$ with $t = 1$. The result is that the I and S values obtained are a little small and are referred to as $\bar{I}_{tr}$ and $\bar{S}_{tr}$. Though this puts us on the conservative side, the reader may desire to use

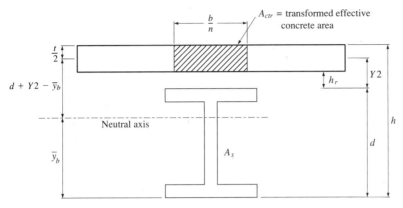

Figure 16-6 Terms used in composite section tables.

the correct theoretical values. These may easily be obtained with simple adjustments as shown in the following expressions:

$$I_{tr} = \bar{I}_{tr} + (\tfrac{1}{12})(A_{ctr})(t^2 - 1)$$

$$S_{tr} = (I_{tr})\left(\frac{\bar{S}_{tr}}{\bar{I}_{tr}}\right)$$

$$y_b = \frac{S_{tr}}{I_{tr}}$$

Remember that if the neutral axis falls in the concrete slab, only the concrete in compression above the neutral axis may be used for t and A_{ctr}. We can easily see if the neutral axis is located in the slab, because we will be in the shaded portion of the tables. Example 16-2 shows the determination of the section properties of the composite section of Example 16-1 using the ASD Manual.

■ Example 16-2
Using the ASD Manual, determine the values of I_{tr}, S_{tr}, and y_b for the composite section considered in Example 16-1.

Solution.

$A_{ctr} = $ *transformed effective concrete area* $= (\tfrac{81}{9})(4) = 36$ in.^{2}

$Y2 = $ distance from top of steel section to center of slab $= \tfrac{4}{2} = 2$ in.

From tables by interpolation:

$$\bar{I}_{tr} = 1096 \text{ in.}^4$$
$$\bar{S}_{tr} = 69.15 \text{ in.}^3$$

Correcting these values:

$$I_{tr} = 1096 + (\tfrac{1}{12})(36)(4^2 - 1) = \underline{\underline{1141 \text{ in.}^4}}$$

$$S_{tr} = 1141\left(\frac{69.15}{1096}\right) = \underline{\underline{72 \text{ in.}^3}}$$

$$y_b = \frac{I_{tr}}{S_{tr}} = \frac{1141}{72} = \underline{\underline{15.85 \text{ in.}}}$$

These values check very well with those determined in Example 16-1.   ■

Should the neutral axis be located in the slab, we can determine the value of the reduced effective concrete thickness t by using a trial-and-error procedure or the following quadratic expression. After t is determined, we can go to the Manual tables and find $\bar{I}_{tr}$ and $\bar{S}_{tr}$.

$$t = A_s \left\{ \frac{-1 + \sqrt{1 + \dfrac{2\left(\dfrac{b}{n}\right)}{A_s}\left(\dfrac{d}{2} + h_r + t_0\right)}}{\left(\dfrac{b}{n}\right)} \right\} \le t_0$$

16-7 SHEAR TRANSFER

The concrete slabs may rest directly on top of the steel beams or the beams may be completely encased in concrete for fireproofing purposes. The longitudinal shear can be transferred between the two by natural bond when the beams are encased. When not encased, mechanical connectors must transfer the load. Fireproofing is not necessary for bridges, and the slab is placed on top of the steel beams. Bridges are subject to heavy impactive loads, and the bond between the beams and the deck, which is easily broken, is considered negligible. For this reason shear connectors are designed to resist all of the shear between bridge slabs and beams.

Various types of shear connectors have been tried, including spiral bars, channels, zees, angles, and studs. Several of these types of connectors are shown in Fig. 16-7. Economic considerations have usually led to the use of round studs welded to the top flanges of the beams. These studs are available in diameters from $\frac{1}{2}$ to 1 in. and in lengths from 2 to 8 in., but the most commonly used sizes are $\frac{3}{4}$ or $\frac{7}{8}$ in. and 2 and 4 in. in length. They actually consist of rounded steel bars welded on one end to the steel beams. The other end is upset to prevent vertical separation of the slab from the beam. These studs can be quickly attached to the steel beams with stud-welding guns by semiskilled workers.[1]

---

[1] Ivan M. Viest, R. S. Fountain, and R. C. Singleton, *Composite Construction in Steel and Concrete* (New York: McGraw-Hill, 1958), 50–51.

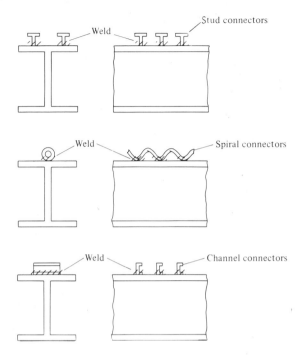

Figure 16-7 Shear connectors.

Shop installation of shear connectors is initially more economical, but there is a growing tendency to use field installation. There are two major reasons for this trend: the connectors may easily be damaged during transportation and setting of the beams, and they serve as a hindrance to the workers walking along the top flanges during the early phases of construction. Furthermore, most composite sections make use of formed steel deck, and for them it is absolutely necessary to install the connectors in the field.

For composite sections it is permissible to use normal-weight stone concrete (made with aggregates conforming to ASTM C33) or lightweight concrete weighing not less than 90 lb/ft^3 (made with rotary kiln-produced aggregates conforming to ASTM C330).

The shear values permitted by the ASD Specification for buildings for various types of connectors are presented in Table 16-1 for normal-weight concrete. The values given for q are the ultimate strengths of the connectors divided by a factor of safety of 2.5 to obtain working load values. The same allowable loads apply for studs longer than the ones listed in the table.

When lightweight concretes are used, it is necessary to reduce the allowable connector shear loads. This is done by multiplying the allowable shear loads given in Table 16-1 by the appropriate coefficient from Table 16-2.

The ASD Specification requires that shear connectors have at least 1 in. of concrete cover in all directions except for those installed in the ribs of formed steel decks. In addition, unless they are located directly over the web, their di-

TABLE 16-1 ALLOWABLE HORIZONTAL SHEAR LOAD FOR ONE CONNECTOR (q), KIPS

	SPECIFIED COMPRESSIVE STRENGTH OF CONCRETE (f'_c), KSI		
CONNECTOR[a]	3.0	3.5	≥4.0
½" dia. × 2" hooked or headed stud	5.1	5.5	5.9
⅝" dia. × 2½" hooked or headed stud	8.0	8.6	9.2
¾" dia. × 3" hooked or headed stud	11.5	12.5	13.3
⅞" dia. × 3½" hooked or headed stud	15.6	16.8	18.0
Channel C3 × 4.1	$4.3w$[b]	$4.7w$[b]	$5.0w$[b]
Channel C4 × 5.4	$4.6w$[b]	$5.0w$[b]	$5.3w$[b]
Channel C5 × 6.7	$4.9w$[b]	$5.3w$[b]	$5.6w$[b]

Source: American Institute of Steel Construction, *Manual of Steel Construction Allowable Stress Design,* 9th ed. (Chicago: AISC, 1989), Table I4.1.

Note: Applicable only to concrete made with ASTM C33 aggregates.

[a] The allowable horizontal loads tabluated are also permitted for studs longer than shown.

[b] w = length of channel, inches.

TABLE 16-2 COEFFICIENTS FOR USE WITH CONCRETE MADE WITH C330 AGGREGATES

SPECIFIED COMPRESSIVE STRENGTH OF CONCRETE (f'_c)	AIR DRY UNIT WEIGHT OF CONCRETE, PCF						
	90	95	100	105	110	115	120
≤4.0 ksi	0.73	0.76	0.78	0.81	0.83	0.86	0.88
≥5.0ksi	0.82	0.85	0.87	0.91	0.93	0.96	0.99

Source: American Institute of Steel Construction, *Manual of Steel Construction Allowable Stress Design,* 9th ed. (Chicago: AISC, 1989), Table I4.2.

ameter shall not be greater than 2.5 times the flange thickness to which they are welded. Tests have shown if this rule is not met, connectors will tend to tear out before their full shear-resisting capacity is reached.

Shear connectors must be capable of resisting both horizontal and vertical movement because there is a tendency for the slab and beam to separate vertically as well as to slip horizontally. Because of the tendency to slip vertically, the longitudinal center-to-center spacing of connectors is limited by the ASD Specification to a maximum value equal to eight times the slab thickness. (The maximum is 36 in. when formed steel deck is used.) The upset heads of the studs help to prevent vertical separation. The minimum permissible center-to-center spacing is six connector diameters along the longitudinal axis and four diameters in the transverse direction.

Careful attention should be given to the ASD method of determining the horizontal shear to be taken by the connectors. Rather than basing designs on the shear computed by the VQ/I formula, shear is estimated at ultimate load conditions. When a composite beam is being tested, failure will probably occur with a crushing of the concrete. At that time it seems reasonable to assume that the concrete and steel have both reached a plastic stress condition.

For this discussion reference is made to Fig. 16-8. Should the neutral axis fall in the slab, the maximum horizontal shear (or horizontal force on the plane between the concrete and the steel) is said to be $A_s F_y$; and if the neutral axis is in the steel section, the maximum horizontal shear is said to be $0.85 f_c' A_c$. (For the student unfamiliar with the strength design theory for reinforced concrete, the average stress at failure on the compression side of a concrete beam is usually assumed to be $0.85 f_c'$.)

From this information, expressions for V_h (the shear to be taken by the connectors) are written below. These values are divided by 2 to estimate conditions at working loads. The ASD Specification says that the two expressions for V_h are to be solved and the smaller value used. V_h is the total shear to be resisted between the point of maximum positive moment in a simple span and the end of the beam. (For continuous beams, it would be between the point of maximum positive moment and the point of contraflexure.) The number of shear connectors required on each side of the point of maximum positive moment can be determined by dividing V_h by q, the strength of one connector of the size and type being used. As previously described, the ASD permits a uniform spacing of connectors.

$$V_h = \frac{0.85 f_c' A_c}{2} \qquad \text{(ASD Equation 14-1)}$$

$$V_h = \frac{A_s F_y}{2} \qquad \text{(ASD Equation 14-2)}$$

$$N_1 = \text{Number of connectors} = \frac{V_h}{q}$$

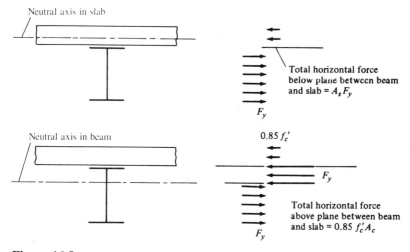

Figure 16-8

When longitudinal reinforcing steel (with area A_s' and specified minimum yield stress F_{yr}) is located within the effective width of the concrete flange and when it has been included in the calculated properties of the composite section, the first of the V_h expressions should be written as follows:

$$V_h = \frac{0.85 f_c' A_c}{2} + \frac{A_s' F_{yr}}{2}$$

Shear connectors are relatively flexible, with the result that there is some slippage between the steel girders and the concrete slabs. Since slip does occur, there is a redistribution of load between the connectors similar to that discussed for bolted connections in Section 11-13. Because of this redistribution of unequal connector loads, the ASD Specification permits, for the usual building with its design based on uniform loads, a uniform spacing of connectors between the points of maximum positive moment and the points of zero moment.

The ASD Commentary (C-I4) says that usually shear connectors can be spaced uniformly between points of maximum and zero moments. There are certain loading situations, however, where the connectors need to be placed closer together over part of the distance. As an illustration, a uniformly loaded beam with two equal and symmetrically placed concentrated loads is considered. For such a case the moment at the concentrated loads may be almost as large as the moment at midspan. Thus there should be almost as many shear connectors from each end of the beam to the adjacent concentrated load as the

Channel-section shear connectors, Grand Rapids, Mich. (Courtesy of Lincoln Electric Company.)

number needed from the beam end to its centerline. To handle this situation the ASD Specification states that the number of connectors between the concentrated load and the nearest point of zero moment shall not be less than determined by the following formula:

$$N_2 = \frac{N_1(M\beta/M_{\max} - 1)}{\beta - 1} \qquad \text{(ASD Equation I4-5)}$$

where

M = moment (less than the maximum moment) at a concentrated load point

N_1 = number of connectors required between point of maximum moment and point of zero moment, determined by the relationship V_h/q or V_h'/q, as applicable (V_h' is defined later in this section)

$\beta = \dfrac{S_{tr}}{S_s}$ or $\dfrac{S_{eff}}{S_s}$ as applicable

For continuous beams, connectors may be spaced uniformly between the points of maximum negative moment and the points of zero moment, says ASD Specification I4.

The resisting moment of a steel beam and concrete slab increases directly with the number of shear connectors used, up to a maximum value when the required number of connectors for full composite action is used. At times it may not be feasible to use shear connectors; at other times it may be feasible or necessary to use only a part of the total number required, thus providing only *partial composite action*. For this reason the ASD Specification (I2) permits the use of full composite action and of partial composite action. A detailed discussion of this topic is presented in Section 16-11.

16-8 CONTINUOUS SPANS

The ASD Specification (Chapter I) and the AASHTO Specification (Article 10.38.4.2) permit the use of continuous composite design. If shear connectors are used as required by the applicable specification in the negative moment regions, the reinforcing bars which are parallel to the steel girders and within the calculated effective width of the slab can be used in calculating the properties of the composite section. The bars must be properly anchored in the positive moment region.

The composite section properties are calculated by the elastic theory, and any concrete in tension is neglected. The total cross-sectional area of a section in the negative moment region equals the area of the steel girder plus the area of the reinforcing bars plus the transformed area of any concrete which is in compression, as in encased construction. The centroid of the section and its moment of inertia can be calculated by the usual methods.

For continuous composite sections with reinforcing bars that are considered part of the composite section in negative moment regions, the total horizontal shear that the shear connectors between interior supports and each adjacent point of contraflexure must resist is to equal the following, in which A_{sr} is the total longitudinal reinforcing at the interior support within the effective flange width, and F_{yr} is the specified minimum yield stress of the reinforcing bars:

$$V_h = \frac{F_{yr}A_{sr}}{2} \qquad \text{(ASD Equation I4-3)}$$

The shear connectors in the negative moment area may be spaced uniformly between the point of maximum negative moment and the points of zero moment.

16-9 DEFLECTIONS

Deflections for composite beams may be calculated by the same methods used for other types of beams. The student must be careful to compute deflections for the various types of loads separately. For example, there are dead loads applied to the steel section alone (if no shoring is used), dead loads applied to the composite section, and live loads applied to the composite section.

The long-term creep effect in the concrete in compression causes deflections to increase with time. These increases, however, are usually not considered significant for the average composite beam. If the designer feels that it is a matter of importance, he or she should compute long-term deflections using values of $2n$ for the modular ratio when calculating the composite section properties to be used for deflections.

Should lightweight concrete be used, the actual modulus of elasticity of that concrete E_c (which may be rather small) should be used in calculating I_{tr} for deflection computations. For stress calculations we use E_c for normal-weight concrete.

Generally speaking, shear deflections are neglected, although on occasion they can be quite large.[2] The steel beams can be cambered for all or some portion of deflections. It may be feasible in some situations to make a floor slab a little thicker in the middle than on the edges to compensate for deflections.

The designer may want to control vibrations in composite floors subject to pedestrian traffic or other moving loads. This may be the case where we have large open floor areas with no damping furnished by partitions, as in shopping malls. For such cases dynamic analyses should be made.[3]

---

[2] L. S. Beedle et al., *Structural Steel Design* (New York: Ronald Press, 1964), p. 452.

[3] Thomas M. Murray, "Design to Prevent Floor Vibrations," *Engineering Journal,* AISC, 12, no. 3 (3d quarter, 1975). pp. 82–87.

16-10 PROPORTIONING COMPOSITE SECTIONS

Composite construction is of particular advantage economically when loads are heavy, spans are long, and beams are spaced at fairly large intervals. For steel-frame buildings, composite construction is economical for spans varying roughly from 25 to 50 ft, with particular advantage in the longer spans. For bridges, simple spans have been economically constructed up to approximately 120 ft and continuous spans 50 or 60 ft longer. Composite bridges are generally economical for simple spans greater than about 40 ft and for continuous spans greater than about 60 ft.

For building design calculations the spans are often considered to be simply supported, but the steel beams do not generally have perfectly simple ends. The result of this situation is that some negative moment may occur at the beam ends with possible cracking of the slab above. To prevent or minimize cracking, some extra steel can be placed in the top of the slab, extending 2 or 3 ft out into the slab. The amount of steel added is in addition to the temperature and shrinkage requirements specified by ACI.

Occasionally cover plates are used on the bottom of the steel sections. The student can see that, with the slab acting as part of the beam, there is a very large compressive area available, and the addition of cover plates to the tension flange will provide a little better balance. Such plates will clearly increase the strength of composite sections. They may also be used to obtain the same strengths with shallower overall sections (though deflections will be larger). In general, however, cover plates are quite expensive, and they are not often used. As most of their cost is in fabrication, heavy cover plates do not cost proportionately more than thin ones; it is logical from an economical standpoint, therefore, to use larger cover plates or none at all. Cover plates are not usually economical when lightweight concrete is used, because the high modular ratios of such concrete greatly reduce the effective concrete areas.

In tall buildings where headroom is a problem, it is desirable to use the smallest overall floor thicknesses. For buildings, minimum depth-span ratios of approximately $1/24$ are recommended if the loads are fairly static and $1/20$ if the loads are of such a nature as to cause appreciable vibration. The thicknesses of the floor slabs are known (from the concrete design), and the depths of the steel beams can be fairly well estimated from these ratios. For bridges the AASHTO gives a desirable minimum ratio of $1/25$ for total depth to span and $1/30$ for steel beam depth to span. Should shallower sections be used, there is the usual requirement that the resulting deflections may not exceed those developed had the recommended ratios been followed.

Perhaps the greatest difficulty in composite design has been the selection of economical beam sizes, as a lengthy trial-and-error process has been involved. In the past few years, however, much data has been published, which appreciably abbreviates the problem. For example, the ASD Manual includes a set of tables for selecting sizes. These tables, developed for building floors which are given for 36- and 50-ksi steels and for 4- to 12-in. slabs, are used for the design of a composite section in Example 16-3.

■ Example 16-3

Design a composite section using A36 steel and the ASD Specification for the situation shown in Fig. 16-9. No shoring is to be used, the beams are not to be coped, and simple spans are assumed. Determine dead-load and live-load deflections. The following additional data is supplied:

$LL = 120$ psf

Ceiling weight $= 10$ psf

Partition weight $= 15$ psf

Concrete weight $= 150$ lbs/ft^3

$f_c' = 3$ ksi

$f_c = 1.35$ ksi

$n = 9$

Use Manual tables to select a section and then review design with the transformed area method using data from the Manual. Compute deflections and design shear connectors.

Solution.

Calculation of moments and shears:

Construction loads

$$\text{Slab} = (\tfrac{4}{12})(150)(9) = 450 \text{ lb/ft}$$
$$\text{Assumed beam weight} = \underline{35} \text{ lb/ft}$$
$$\text{Total} = 485 \text{ lb/ft}$$

$$M_b = \frac{(0.485)(27)^2}{8} = 44.2 \text{ ft-k}$$

Loads applied after concrete hardens

$$\text{Ceiling} = (9)(10) = 90 \text{ lb/ft}$$
$$\text{Partitions} = (9)(15) = 135 \text{ lb/ft}$$
$$LL = (9)(120) = \underline{1080} \text{ lb/ft}$$
$$\text{Total} = 1305 \text{ lb/ft}$$

$$M_L = \frac{(1.305)(27)^2}{8} = 118.9 \text{ ft-k}$$

Maximum moment

$$M_{max} = M_D + M_L = 44.2 + 118.9 = 163.1 \text{ ft-k}$$

Maximum shear

$$V_{max} = (\tfrac{27}{2})(0.485 + 1.305) = 24.2 \text{ k}$$

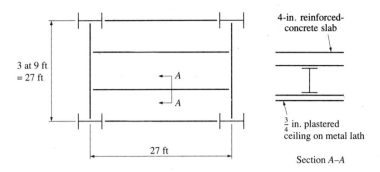

Figure 16-9

Effective width of slab:

$$b = (\tfrac{1}{4})(27.0) = 6.75 \text{ ft} = 81 \text{ in.} \leftarrow$$
$$b = 9 \text{ ft} = 108 \text{ in.}$$

Required section modulus:

$$S_{tr} \text{ for } M_{\max} = \frac{(12)(163.1)}{24} = 81.6 \text{ in.}^3$$

Assuming full lateral bracing for compression flange

$$S_s \text{ for } M_D = \frac{(12)(44.2)}{24} = 22.1 \text{ in.}^3$$

Select section from tables ($S_{tr} = 81.6 \text{ in.}^3$, $Y2 = 2 \text{ in.}$, $A_{ctr} = (\tfrac{81}{9})(4) = 36 \text{ in.}^2$):

Try a W18 × 35 ($A_s = 10.3 \text{ in.}^2$, $d = 17.70 \text{ in.}$, $t_w = 0.300 \text{ in.}$, $t_f = 0.425 \text{ in.}$, $I_s = 510 \text{ in.}^4$, $S_s = 57.6 \text{ in.}^3$)

$\bar{S}_{tr} = 84 \text{ in.}^3$ with $Y2 = 2 \text{ in.}$ and $A_{ctr} = 30 \text{ in.}^2$

Pick $\bar{I}_{tr}$ and $\bar{S}_{tr}$ from the Manual and correct:

By interpolation

$$\bar{I}_{tr} = 1452.8 \text{ in.}^4$$
$$\bar{S}_{tr} = 84.6 \text{ in.}^3$$

Then with corrections

$$I_{tr} = 1452.8 + (\tfrac{1}{12})(36)(4^2 - 1) = 1497.8 \text{ in.}^3$$

$$S_{tr} = 1497.8\left(\frac{84.6}{1452.8}\right) = 87.2 \text{ in.}^3$$

$$y_b = \frac{1497.8}{87.2} = 17.18 \text{ in.}$$

$$y_{top} = 4 + 17.70 - 17.18 = 4.52 \text{ in.}$$

$$S_{tr\ top} = \frac{1497.8}{4.52} = 331.4 \text{ in.}^3$$

Compute stresses:
  Before concrete hardens

$$f_s = \frac{(12)(44.2)}{57.6} = 9.21 \text{ ksi} < 0.66\ F_y = 24 \text{ ksi} \qquad\qquad \text{OK}$$

After concrete hardens

$$f_s = 9.21 + \frac{(12)(118.9)}{87.2} = 25.57 \text{ ksi} < 0.9\ F_y = 32.4 \text{ ksi} \qquad \text{OK}$$

$$f_c = \frac{(12)(118.9)}{(9)(331.4)} = 0.478 \text{ ksi} < 0.45\ f_c' = 1.35 \text{ ksi} \qquad\qquad \text{OK}$$

$$f_v = \frac{24.2}{(17.70)(0.300)} = 4.54 \text{ ksi} < 0.4F_y = 14.4 \text{ ksi} \qquad\qquad \text{OK}$$

Compute deflections:
  For construction loads before concrete hardens

$$\Delta_{DL} = \frac{(5)\left(\dfrac{485}{12}\right)(12 \times 27)^4}{(384)(29 \times 10^6)(510)} = 0.39 \text{ in.}$$

For loads applied after concrete hardens

$$\Delta_{LL} = \frac{(5)\left(\dfrac{1305}{12}\right)(12 \times 27)^4}{(384)(29 \times 10^6)(1497.8)} = 0.36 \text{ in.} < \left(\frac{1}{360}\right)(12 \times 27) = 0.9 \text{ in.}$$

$$\text{OK}$$

Shear connectors for full composite action:
  Try $\frac{3}{4}$-in. diameter $\times$ 3-in. studs
  Maximum stud diameter per ASD Specification I4

$$2.5t_f = 2.5 \times 0.425 = 1.06 \text{ in.} > 0.75 \text{ in.} \qquad\qquad \text{OK}$$

Total horizontal shear (tables give $V_h = 185$ k or we can compute)

$$V_h = \frac{0.85 f_c' A_c}{2} = \frac{(0.85)(3)(81 \times 4)}{2} = 413.1 \text{ k}$$

$$V_h = \frac{A_s F_y}{2} = \frac{(10.3)(36)}{2} = 185.4 \text{ k} \leftarrow$$

$q = 11.5$ k/connector from Table 16-1

$$\text{Connectors required} = N = \frac{V_n}{q} = \frac{185.4}{11.5}$$

$$= 16.12 \text{ each side of the point of maximum moment.}$$

Use 33 $\frac{3}{4}$-in.-diameter $\times$ 3-in. studs spaced uniformly, 16 on each side of centerline and 1 at centerline with a W18 $\times$ 35 ■

16-11 PARTIALLY COMPOSITE BEAMS

For this discussion we assume a steel section that when made composite with the concrete slab will have an allowable resisting moment of 200 ft-k. When we select a section from the Manual it has (when made composite with the slab) an allowable resisting moment of 250 ft-k. If we now provide shear connectors for full composite action, the section will have an allowable resisting moment of 250 ft-k. But we need only 200 ft-k.

It seems very logical to decide to provide only enough connectors to develop an allowable strength of 200 ft-k. In this way we can reduce the number of connectors and save some money (perhaps a good bit if this section is repeated many times in the structure). The resulting section is a *partially composite section,* one that does not have a sufficient number of connectors to develop the section to its greatest resisting moment. (If some years later we were to decide to expand this building and increase its loading we might very well wish we had made the system fully composite.)

The shear V_h required for full composite action is reduced to V_h' by the expression to follow. The value so obtained is used to calculate the number of shear connectors required for partial composite action.

$$V_h' = V_h \left(\frac{S_{\text{reqd}} - S_s}{S_{tr} - S_s} \right)^2$$

It is usually felt that the total allowable strength of the shear connectors used in a particular beam V_h' should not be less than 25 percent of the allowable shear strength required for full composite action $(0.25V_h)$. This minimum allowable shear strength is considered to be necessary to prevent excessive slipping and loss of stiffness.

Section I4 of the ASD Specification states that the reduced horizontal shear V_h' used to calculate the effective section modulus S_{eff} of the partially composite section is to be taken as q times the number of connectors to be used between the point of maximum moment and the nearest point of zero moment.

$$V_h' = qN \geq 0.25V_h$$

This value may not be less than $0.25V_n$. Should V'_h/V_h be as low as 0.5 to 0.75, the resulting reduction of connectors is thought to be particularly worthwhile from an economical standpoint. A 10 to 30 percent reduction in the number of shear connectors does not usually reduce S_{tr} by more than 5 or 10 percent.

The live-load deflections occurring after composite action develops will be a little larger in a partially composite section than in a section with a sufficient number of connectors to develop full composite action. The effective moment of inertia for partial composite action may be computed with the expression

$$I_{eff} = I_s + \sqrt{\frac{V'_h}{V_h}}(I_{tr} - I_s) \qquad \text{(ASD Equation 14-4)}$$

The Δ_{LL} deflection for a partially composite section will equal I_{tr}/I_{eff} times Δ_{LL} for a fully composite section.

In Example 16-4 the section selected in Example 16-3 is redesigned as a partially composite section, and Δ_{LL} is recalculated.

■ Example 16-4

Redesign the shear connectors of Example 16-3 so they are sufficient only for the maximum estimated moment to be applied to the section, and recalculate the live-load deflection.

Solution.

Connectors:

S_{eff} = required S_{tr} = 81.6 in.^{3} from Prob. 16-3 solution

$$V'_h = V_h\left(\frac{S_{eff} - S_s}{S_{tr} - S_s}\right)^2 = 185.4\left(\frac{81.6 - 57.6}{87.2 - 57.6}\right)^2 = 121.9 \text{ k}$$

$\frac{1}{4}$ full $V_h = (\frac{1}{4})(185.4) = 46.35 \text{ k} < 121.9\text{k}$ ⟶ OK

$$N_1 = \frac{V'_h}{q} = \frac{121.9}{11.5} = 10.6 \text{ each side of centerline}$$

Use 22 $\frac{3}{4}$-in.-diameter × 3$\frac{1}{2}$-in. studs

Deflection:

$V'_h = 11.5 \times 11 = 126.5 \text{ k}$

$$I_{eff} = I_s + \sqrt{\frac{V'_h}{V_h}}(I_{tr} - I_s)$$

$$= 510 + \sqrt{\frac{126.5}{185.4}}(1497.8 - 510) = 1326 \text{ in.}^4$$

$$\Delta_{LL} = \frac{1497.8}{1326} \times 0.36 = \underline{0.407 \text{ in.}} < \left(\frac{1}{360}\right)(12 \times 27) = 0.9 \text{ in.} \quad \text{OK} \quad ■$$

16-12 COMPOSITE BEAMS WITH FORMED STEEL DECK

As previously mentioned, a very large proportion of composite slabs are constructed with formed steel decks. The theory for handling these sections is the same as for slabs without formed steel decks. There are some special requirements for using this type of composite section, however, and these are given in Section I5 of the ASD Specification. Some of these are listed below and illustrated in Fig. 16-10.

1. Rib heights are limited to a maximum value of 3 in.
2. The average width of the concrete ribs or haunches may not be less than 2 in. but for calculations may not be taken as more than the minimum clear width near the top of the steel deck.
3. Shear connectors may not have diameters larger than $\frac{3}{4}$ in., and they must extend at least $1\frac{1}{2}$ in. above the top of the steel deck.
4. The concrete slab above the steel deck must have a thickness of at least 2 in.
5. If the ribs are perpendicular to the steel beam, the concrete below the top of the steel deck is to be neglected in calculating the transformed section properties and the area A_c used for calculating V_h. Should the ribs be parallel to the steel beams, this concrete may be counted in making the calculations.

When shear connectors are placed in the ribs of steel decks, the ASD Specification (I5) requires that their allowable capacity determined from Table 16-1 be reduced and gives expressions for doing this. For instance, if the ribs are running perpendicular to the steel beam, the reduction factor to be multiplied by q is given in ASD Equation I5-1. A different reduction factor is provided for composite sections where the ribs are parallel to the beams.

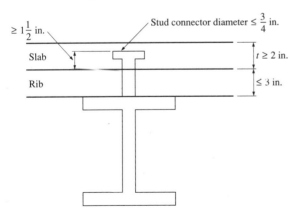

Figure 16-10 Some requirements for composite sections with formed steel deck.

$$\left(\frac{0.85}{\sqrt{N_r}}\right)\left(\frac{w_r}{h_r}\right)\left(\frac{H_s}{h_r} - 1.0\right) \le 1.0 \qquad \text{(ASD Equation 15-1)}$$

where

h_r = nominal rib height (inches)

H_s = length of the connector after welding (inches). For calculating the reduction factor it may not be longer than $h_r + 3$ in.

N_r = number of stud connectors in one rib. It may not be equal to more than 3 even if more connectors are installed.

w_r = average width of the concrete ribs (inches).

The studs used for composite sections with steel decks may be welded directly through the deck or through holes prepunched in the deck or cut in place. The most common practice is to weld them through the deck, but special procedures prescribed by the manufacturer are necessary if the deck thickness is larger than 16 gage for single thickness or 18 gage each sheet for double thickness, or if a galvanized coating heavier than 1.25 oz/ft² is used.

Example 16-5 presents the design of a composite beam with formed steel deck with the ribs perpendicular to the beam.

■ Example 16-5

Repeat Example 16-3 using formed steel deck (ribs perpendicular to the steel beam). The rib height is assumed to be 2 in., and the concrete slab is 2 in. thick. Assume w_r is $2\frac{1}{2}$ in. Fig. 16-11 shows the steel deck and its dimensions.

Solution.

Moments and shears:

Construction loads

$$
\begin{aligned}
\text{Slab + beam weight from Example 16-3} &= 485 \text{ lb/ft} \\
\text{Assumed deck wt} = (7)(9) &= \underline{63 \text{ lb/ft}} \\
\text{Total} &= 548 \text{ lb/ft}
\end{aligned}
$$

$$M_D = \frac{(0.548)(27)^2}{8} = 49.9 \text{ ft-k}$$

Loads applied after concrete hardens

$$M_L = 118.9 \text{ ft-k from Example 16-3}$$

Maximum moment

$$M_{max} = M_D + M_L = 49.9 + 118.9 = 168.8 \text{ ft-k}$$

Maximum shear

$$V_{max} = (\tfrac{27}{2})(0.548 + 1.305) = 25.0 \text{ k}$$

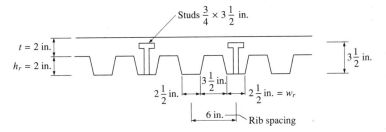

Figure 16-11 Sketch of formed steel deck.

Effective width of slab = 81 in. as before
Required section modulus:

$$S_{tr} \text{ for } M_{max} = \frac{(12)(168.8)}{24} = 84.4 \text{ in.}^3$$

$$S_s \text{ for } M_D = \frac{(12)(49.9)}{24} = 25 \text{ in.}^3$$

Select section from tables ($S_{tr} = 84.4$ in.^{3}, $Y2 = 3$ in., and $A_{ctr} = [\frac{81}{9}][2] = 18$ in.^{2}:

Try a W18 × 35 ($A_s = 10.3$ in.^{2}, $t_w = 0.300$ in., $t_f = 0.425$ in., $d = 17.70$ in., $I_s = 510$ in.^{4}, $S_s = 57.6$ in.^{3}, max $V_h = 185$ k

Determine $\bar{I}_{tr}$ and $\bar{S}_{tr}$ from the Manual and correct:
  By interpolation

$$\bar{I}_{tr} = 1427.2 \text{ in.}^4$$

$$\bar{S}_{tr} = 87.2 \text{ in.}^3$$

$$I_{tr} = 1427.2 + (\tfrac{1}{12})(18)(2^2 - 1) = 1431.7 \text{ in.}^4$$

$$S_{tr} = 1431.7\left(\frac{87.2}{1427.2}\right) = 87.5 \text{ in.}^3$$

$$y_b = \frac{1431.7}{87.5} = 16.36 \text{ in.}$$

$$y_{top} = 17.70 + 4 - 16.36 = 5.34 \text{ in.}$$

$$S_{tr\,top} = \frac{1431.7}{53.4} = 268.1 \text{ in.}^3$$

Check stresses and deflections as in Example 16-3:
  Results not shown, but they are satisfactory.
Shear connectors for full composite action:
  Using $\frac{3}{4}$-in. diameter × $3\frac{1}{2}$-in. studs
  Maximum stud diameter per ASD Specification I4 is

$$2.5 \times 0.425 = 1.06 \text{ in.} > \tfrac{3}{4} \text{ in.} \qquad \text{OK}$$

$$V_h \text{ from tables} = 185 \text{ k}$$

$$q = 11.5 \text{ k from Table 16-1}$$

Computing stud reduction factor assuming $N_r = 1$, $H_s = 3.5$ in., and given $h_r = 2$ in., $w_r = 2\frac{1}{2}$ in.

$$\text{Reduction factor} = \left(\frac{0.85}{\sqrt{1}}\right)\left(\frac{2.5}{2}\right)\left(\frac{3.5}{2} - 1.0\right) = 0.797$$

Reduced $q = (0.797)(11.5) = 9.2$ k

$$N_{reqd} = \frac{185}{9.2} = 20.1 \quad \text{(say 41 studs)}$$

<u>20 studs on each side of centerline</u>

Design studs for partial composite action:

$$V_h' = 185\left(\frac{84.4 - 57.6}{87.5 - 57.6}\right)^2 = 148.6 \text{ k} > (\tfrac{1}{4})(185) = 46.25 \text{ k} \qquad \text{OK}$$

$$N = \frac{148.6}{9.2} = 16.1$$

<u>16 studs each side of centerline</u>

We have to put the studs in the deck ribs, which are spaced 6 in. on centers. So we put them in at 6 in. or 12 in., as shown in Fig. 16-12. The spacing may not be more than 8 times the slab $t = (8)(2) = 16$ in., says ASD Specification I4.

Check deflection for partial composite action:

Not shown, but it is satisfactory. ∎

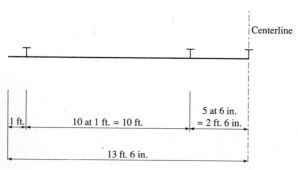

Figure 16-12 Stud spacing for Example 16-5.

16-13 ENCASED SECTIONS

For fireproofing purposes the steel beams in building floors may be completely encased in concrete. Under certain conditions the horizontal shears between the slabs and beams can be considered to be transferred by natural bond and friction between the two. The ASD Specification (I1) says that for this transfer to be permissible the encasing concrete must be placed integrally with the slab concrete and must cover the steel by at least 2 in. on the sides and bottom (or soffit). It is further required that the top of the steel section be at least $1\frac{1}{2}$ in. below the top of the slab and 2 in. above the bottom of the slab. Finally, the encasing concrete must have adequate mesh or other reinforcing for its full depth and across the soffit of the beam. The exact amount, which is not specified by the ASD Specification, can be very nominal in size (from "chicken wire size" on up). Encased section requirements are illustrated in Fig. 16-13.

The ASD Specification states that steel sections which are to be encased in concrete and are not to be shored must be able to resist the moment M_B due to the weight of the steel beam and the wet concrete.

$$f_s = \frac{M_B}{S_s} \leq 0.66F_y \qquad \text{before hardening}$$

Bending stresses after the concrete has hardened are to be computed with the composite section properties. *Notice that the moment M_D due to load applied after the concrete hardens plus the moment M_L due to the weight of the section are both assumed to be applied to the composite section.* The composite section must be able to resist these moments without f_s exceeding $0.66F_y$ or f_c exceeding $0.45f_c'$.

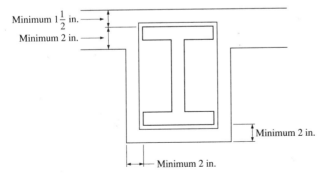

Minimum $1\frac{1}{2}$ in.

Minimum 2 in.

Minimum 2 in.

Minimum 2 in.

Figure 16-13 Encased section requirements.

$$f_s = \frac{M_D + M_L}{S_{tr\,bot}} \le 0.66F_y$$

$$f_c = \frac{M_D + M_L}{nS_{tr\,top}} \le 0.45f_c'$$

$\left.\right\}$ after hardening

An alternative method is permitted which says that the steel beam alone must be able to resist all moments without a steel stress exceeding $0.76F_y$. This value might be used where a reduction in calculations is desired. The higher allowable stress recognizes the fact that composite action permits the section to support more load than the steel beam could alone. Should this latter method be used, it is unnecessary to compute the properties of the composite section.

Example 16-6 illustrates the calculations involved in reviewing the design of an encased section.

■ Example 16-6

Compute the bending stresses before and after hardening for the nonshored section of Fig. 16-14, using the following data:

A36 steel

$f'_c = 3$ ksi

$n = 9$

Effective flange width = 60 in.

$M_D = 40$ ft-k

$M_L = 120$ ft-k

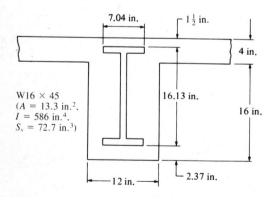

W16 × 45
(A = 13.3 in.²,
I = 586 in.⁴,
S_x = 72.7 in.³)

7.04 in. 1½ in. 4 in. 16.13 in. 16 in. 12 in. 2.37 in.

Figure 16-14

Solution. Calculated properties of composite section:
Neglecting concrete area below flange

$$A = 13.3 + \frac{(4)(60)}{9} = 13.3 + 26.67 = 39.97 \text{ in.}^2$$

$$y_b = \frac{(13.3)(10.44) + (26.67)(18)}{39.97} = 15.48 \text{ in.}$$

$$I = 586 + (13.3)(5.045)^2 + (\tfrac{1}{12})(\tfrac{60}{9})(4)^3 + (26.67)(2.52)^2 = 1129 \text{ in.}^4$$

$$S_{bot} = \frac{1129}{15.48 - 2.37} = 86.12 \text{ in.}^3$$

$$S_{top} = \frac{1129}{4.52} = 249.8 \text{ in.}^3$$

Before concrete hardens

$$f_s = \frac{(12)(40)}{72.7} = \underline{\underline{6.60 \text{ ksi}}} < 24 \text{ ksi} \qquad\qquad \text{OK}$$

After concrete hardens

$$f_s = \frac{(12)(40 + 120)}{86.12} = \underline{\underline{22.29 \text{ ksi}}} < 24 \text{ ksi} \qquad\qquad \text{OK}$$

$$f_c = \frac{(12)(40 + 120)}{(9)(249.8)} = \underline{\underline{0.854 \text{ ksi}}} < 0.45 f_c' = 1.35 \text{ ksi} \qquad \text{OK} \blacksquare$$

For buildings continuous composite construction with encased sections is permissible. For continuous construction the positive moments are handled exactly as has been illustrated by the preceding examples. For negative moments, however, the transformed section is taken as shown in Fig. 16-15. The crosshatched area represents the concrete in compression and all concrete on the tensile side of the neutral axis (that is, above the axis) is neglected. Example 16-7 illustrates the review of a composite section for negative moment.

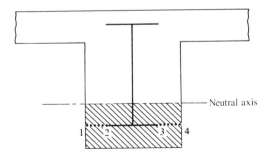

Neutral axis

Figure 16-15

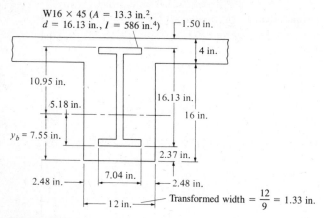

W16 × 45 (A = 13.3 in.^{2},
d = 16.13 in., I = 586 in.^{4})

1.50 in.

4 in.

10.95 in.

5.18 in.

16.13 in.

16 in.

y_b = 7.55 in.

2.37 in.

2.48 in.

7.04 in.

2.48 in.

12 in.

Transformed width = $\dfrac{12}{9}$ = 1.33 in.

Figure 16-16

■ Example 16-7

For the composite section shown in Fig. 16-16 and previously considered in Example 16-6, determine the bending stresses for negative moments of M_D = 50 ft-k and M_L = 65 ft-k, assuming no shoring is used. Other data are the same as for Example 16-6.

Solution.   Properties of section:

$$(1.33y_b)\left(\frac{y_b}{2}\right) = (13.3)(10.44 - y_b)$$

where 10.44 is the distance from soffit to center of gravity of W section.

$$y_b = 7.55 \text{ in.}$$
$$I_{tr} = 586 + (13.3)(2.89)^2 + (\tfrac{1}{3})(1.33)(7.55)^3 = 888 \text{ in.}^4$$

Review of stresses:
   Before concrete hardens

$$f_s = \frac{(12)(50)(8.07)}{586} = \underline{\underline{8.26 \text{ ksi}}} < 24.0 \text{ ksi} \qquad\qquad \text{OK}$$

After concrete hardens

$$f_c = \frac{(12)(115)(7.55)}{(9)(888)} = \underline{\underline{1.30 \text{ ksi}}} < 1.35 \text{ ksi} \qquad\qquad \text{OK}$$

$$f_s = \frac{(12)(115)(10.95)}{888} = \underline{\underline{17.0 \text{ ksi}}} < 24.0 \text{ ksi} \qquad\qquad \text{OK} \blacksquare$$

PROBLEMS

16-1 Using the ASD Specification, calculate the bending stresses for the section shown. The section is used for a simple span of 24 ft and is to have a uniform dead load of 40 psf applied after composite action develops and a uniform live load of 150 psf. Assume no shoring and $n = 9$. (*Ans.* $f_s = 6.79$ ksi and 22.23 ksi, $f_c = 0.484$ ksi)

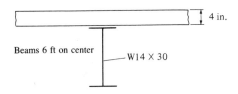

Beams 6 ft on center

W14 × 30

4 in.

16-2 Rework Prob. 16-1 using a W8 × 18 with a $\frac{1}{2}$ × 8-in. cover plate on the tension flange instead of the W14 × 30.

16-3 Repeat Prob. 16-1 using formed steel deck with 2-in. rib heights. Steel deck weighs 7 psf. (*Ans.* $f_s = 7.65$ ksi and 20.78 ksi, $f_c = 0.417$ ksi)

16-4 How many $\frac{7}{8}$-in.-diameter studs $3\frac{1}{2}$-in. long are required for the section of Prob. 16-1 if it is designed to be fully composite? If it is designed to be partially composite? Use A36 steel an 3-ksi concrete.

16-5 Apply the question from Prob. 16-4 to Prob. 16-2. Use $\frac{3}{4}$-in. dia. studs 3 in. long. (*Ans.* 30 if fully composite, 10 if not)

16-6 Compute the deflection for the composite section of Prob. 16-1 before the concrete hardens and then after it hardens under full load.

16-7 Apply the question from Prob. 16-6 to the section of Prob. 16-2. (*Ans.* 0.818 in. before hardening, 0.490 in. more after hardening)

16-8 Apply the question from Prob. 16-6 to the Section of Prob. 16-3.

16-9 Using the ASD Manual and A36 steel, design a nonencased composite section for the simple span beams shown if a 4-in. concrete slab (150 lb/ft³) with $f'_c = 3$ ksi is used. The total live load is 1.4 k/ft.
(a) Select the beams. (*Ans.* W18 × 40)
(b) Determine the number of $\frac{3}{4}$-in.-diameter × $3\frac{1}{2}$-in headed studs required for full composite action. (*Ans.* 37 total studs)
(c) Compute the service live-load deflection. (*Ans.* 0.34 in.)
(d) Check the beam shear stress. (*Ans.* 4.64 ksi)

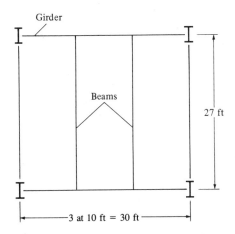

Girder

Beams

27 ft

3 at 10 ft = 30 ft

16-10 Repeat Prob. 16-9 using $F_y = 50$ ksi and $f_c' = 4$ ksi.

16-11 A36 beams 9 ft on center and spanning 40 ft are to be selected to support a 4-in.-deep concrete slab ($f_c' = 3$ ksi) on a 3-in.-deep formed steel deck. The ribs for the steel deck, which are perpendicular to the steel beams, have average widths of $2\frac{1}{2}$ in. The total dead load is to be 0.80 k/ft for the length of the beams, and the service live load is 1.25 k/ft.

   (a) Select the beams. (*Ans.* W24 × 62)

   (b) Determine the number of $\frac{3}{4}$-in.-diameter × $4\frac{1}{2}$-in. headed studs required for full composite action. (*Ans.* 41 each side of centerline)

   (c) Compute the live-load deflection, and check the beam shear. (*Ans.* $\Delta_{LL} = 0.46$ in., $f_v = 4.02$ ksi)

16-12 Repeat Prob. 16-11 using $F_y = 50$ ksi.

16-13 Repeat Prob. 16-11 with spans of 34 ft. (*Ans.* W18 × 40, 33 connectors each side of centerline, $\Delta_{LL} = 0.562$ in., $f_v = 6.18$ ksi)

16-14 Repeat Prob. 16-13 using $F_y = 36$ ksi.

16-15 Select an A36 steel section to support a live load of 100 psf. The beams are to have 36-ft simple spans and are to be spaced 8 ft 6 in. on center. Construction is unshored, concrete weighs 150 lb/ft³, f_c' is 3 ksi. A metal deck with ribs perpendicular to the steel beams is used with a 2-in. concrete slab. The ribs are 2 in. deep and have average widths of $2\frac{1}{2}$ in. Design $\frac{3}{4}$-in.-diameter × $3\frac{1}{2}$-in. headed studs for full composite action, and calculate live-load deflection. (*Ans.* W18 × 40, 23 studs each side of centerline, $\Delta_{LL} = 0.64$ in.)

16-16 Repeat Prob. 16-15 with a live load of 200 psf.

16-17 Using the transformed area method, compute the stresses in the encased section shown if no shoring is used. The section is assumed to be used for a simple span of 30 ft and to have a uniform dead load of 30 psf applied after composite action develops and a uniform live load of 120 psf. Assume $n = 9$, A36 steel, $f_c' = 3$ ksi, and concrete weighing 150 lb/ft³. (*Ans.* $f_s = 10.10$ ksi and 23.84 ksi, $f_c = 1.035$ ksi)

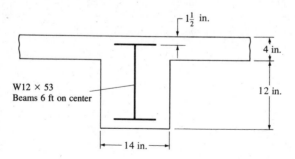

17

Cover-Plated Beams, Built-up Wide-Flange Sections, and Plate Girders

17-1 INTRODUCTION

Should the largest available W section be insufficient to support the anticipated loads for a certain span, several alternatives are possible. Perhaps the most economical solution uses a higher-strength steel W section. If the required strength cannot be obtained in this way, we can use one of the following: (a) two or more regular W sections side by side (an expensive solution), (b) a cover-plated beam, (c) a built-up wide-flange section, (d) a plate girder, or (e) a steel truss.

If two or more rolled beams or channels are used side by side as a flexural member, they must be connected at intervals not greater than 5 ft. Such connections, involving through-bolts and separators, diaphragms, and so on, are described in section F6 of the ASD Specification. Several of the above members can be made composite with concrete slabs. The next two sections of this chapter are concerned with cover-plated beams and built-up wide flanges, while the remainder of the chapter is devoted to plate girders.

Swinging a plate girder into place on Goat Island Bridge across the Niagara River, Niagara Falls, N.Y. (Courtesy of Bethlehem Steel Corporation.)

17-2 COVER-PLATED BEAMS

In addition to being practical for cases where the moments to be resisted are slightly in excess of those which can be supported by the deepest W sections, there are other useful applications for cover-plated beams. On some occasions the total depth may be so limited that the resisting moments of W sections of the specified depth are too small. For instance, the architect may show a certain maximum depth for beams in the drawings for a building. In a bridge, beam depths may be limited by clearance requirements. Cover-plated beams will frequently prove to be a very satisfactory solution for situations like these. Furthermore, there may be economical uses for cover-plated beams where the depth is not limited and where there are standard W sections available to support the loads. A smaller W section than required by the maximum moment can be selected and have cover plates attached to its flanges. These plates can be cut off where the moments are smaller, with resulting saving of steel. Applications of this type are quite common for continuous beams.

Should the depth be fixed and a cover-plated beam seem to be a feasible solution, the usual procedure will be to select a standard section which has a depth leaving room for top and bottom cover plates, then select the cover plate sizes. A cover-plated beam is shown in Fig. 17-1. If the section has been selected, the moment of inertia of the entire section equals the moment of iner-

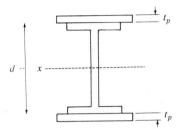

Figure 17-1 Cover-plated beam.

tia of the W section (I_s) plus the moment of inertia of the unknown plate sizes. The moment of inertia of these plates about the x axis (neglecting the minute values about their own axes) equals $A(d/2)^2$ for each plate.

If d represents the distance between the centers of gravity of the flange cover plates, the moment of inertia of the entire section can be expressed as

$$I_{reqd} = I_s + 2A\left(\frac{d}{2}\right)^2$$

It is usually more convenient to work with section modulus values than with moments of inertia. The section modulus required can be written *approximately* as follows (noting that $d/2$ is not exactly the correct distance to the outermost fiber of the plates and that the two section moduli values are not correctly added together, since their c distances to their extreme fibers are not equal):

$$S_{reqd} = S_s + \frac{2A(d/2)^2}{d/2}$$

$$= S_s + Ad$$

From this last expression it is possible to estimate very closely the cover plate area required. The value actually obtained will be very slightly on the small side due to the slightly incorrect distance between the outermost fibers used in the derivation. A cover-plated beam design is illustrated in Example 17-1. The connections between the cover plates and the W section of this example are not designed, as this matter was addressed in earlier chapters (e.g., Example 11-3).

■ Example 17-1

Select a beam limited to a maximum depth of 28.50 in. for the load and span of Fig. 17-2. The beam is assumed to have full lateral support for its compression flange and to have an allowable bending stress of 20 ksi. Use a W27 × 178 with cover plates.

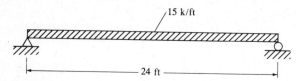

Figure 17-2

Solution.   Assume beam weight = 225 lb/ft:

$$M = \frac{(15.225)(24)^2}{8} = 1096.2 \text{ ft-k}$$

$$S_{\text{reqd}} = \frac{(12)(1096.2)}{20} = 657.7 \text{ in}^2$$

Try W27 × 178 ($d = 27.81$ in., b = 14.085 in., $I = 6990$ in.⁴, $S = 502$ in.³) (see Fig. 17-3)
Assume $\frac{5}{16}$ in. plates

$$S_{\text{reqd}} = S_s + Ad$$
$$657.7 = 502 + (A)(28.12)$$
$$A = 5.54 \text{ in.}^2$$

Try one PL$\frac{5}{16}$ × 20 each flange ($A = 6.25$ in.²)
Compute actual S furnished:

$$S = \frac{6990 + (2)(6.25)(14.06)^2}{14.22} = 665.3 \text{ in.}^3 > 657.7 \text{ in.}^3 \qquad \text{OK}$$

Use W27 × 178 with one PL$\frac{5}{16}$ × 20 each flange   ■

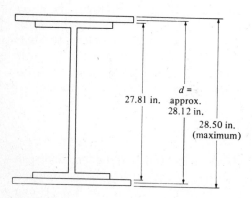

Figure 17-3

17-3 BUILT-UP WIDE-FLANGE SECTIONS

In this section the author initially defines the terms *built-up wide-flange sections* and *plate girders* because the reader may quite understandably find it difficult to distinguish between the two.

In Chapter G of the ASD Specification a clear distinction is made between beams (whether rolled shapes or built-up wide-flange sections) and plate girders. It is stated that plate girders are to be distinguished from beams on the basis of web slenderness. This slenderness is measured by the ratio h/t_w, where h is the clear distance between flanges of a beam or girder and t_w is the web thickness. (See Fig. 17-4.)

To be defined as a beam a section may be a rolled or a built-up shape, but its h/t_w ratio may not be greater than $760/\sqrt{F_b}$, which is equal to 162 for A36 steel assuming $F_b = 22$ ksi. This is the web depth-thickness ratio above which there may be vertical buckling of the web before the flexural stress reaches the steel's yield stress. (The ratio can be increased by using transverse stiffeners.)

With the ASD Specification, overall economy can often be obtained when built-up wide-flange sections are used instead of plate girders. With these sections the webs selected are sufficiently thick to carry shear without buckling and thus do not need web stiffeners. Even though these sections with their unstiffened webs will be heavier than plate girders for the same spans and loads, their overall costs will often be less because of their smaller fabrication costs. In addition, design calculations will be appreciably reduced. Example 17-2 illustrates the design of a built-up wide-flange beam.

■ Example 17-2
Design a 60-in.-deep built-up wide-flange section with no intermediate stiffeners for a 70-ft simple span to support a uniform load of 3 k/ft in addition to its own weight. Assume that end bearing stiffeners are not required by Section K of the ASD Specification and that full lateral bracing is to be provided for the compression flange. Use A36 steel, SMAW fillet welds, and E70 electrodes.

Solution.
Maximum shear and moment:

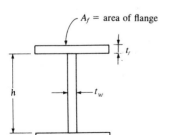

Figure 17-4 A built-up wide-flange or a plate girder?? It is a plate girder if $h/t_w > 760/\sqrt{F_b}$. Otherwise it is a built-up wide-flange beam.

Assume beam wt = 200 lb/ft

$$M = \frac{(3.2)(70)^2}{8} = 1960 \text{ ft-k}$$

$$V = (35)(3.2) = 112 \text{ k}$$

Web size required:

For web to be compact (Table B5.1, ASD Specification)

$$\frac{h}{t_w} \le \frac{640}{\sqrt{F_y}} = \frac{640}{\sqrt{36}} = 107$$

Assume $h = 60 - 2 = 58$ in.

$$\text{Min } t_w = \frac{58}{107} = 0.542 \text{ in.} \qquad \text{(Say } \tfrac{9}{16} \text{ in.)}$$

Try $\tfrac{9}{16} \times 58$ web ($A_w = 32.62$ in.^{2})

$$\frac{h}{t_w} = \frac{58}{\frac{9}{16}} = 103.1 < 107 \qquad\qquad \text{OK}$$

Checking shear stress as required by ASD Section F4

$$f_v = \frac{112}{(\frac{9}{16})(58)} = 3.43 \text{ ksi}$$

$$k_v = 5.34$$

$$C_v = \frac{(45,000)(5.34)}{(36)(103.1)^2} = 0.628 < 0.8$$

$$F_v = \tfrac{36}{2.89}(0.628) = 7.82 \text{ ksi} > 3.43 \text{ ksi}$$

Since $\frac{h}{t_w}$ is <260 (see ASD Specification F5) and since the web shear stress is less than the maximum value permitted by ASD Equation F4-2, stiffeners are not needed. (A further discussion of this topic is presented in Section 17-7.)

Selection of flange plates:

After we have picked a web size, the area needed for the flange plates can be determined by writing an expression for the section modulus of the whole built-up wide-flange section in terms of h, t_w, t_f, and A_f with reference made to Fig. 17-4.

$$S_{reqd} = \frac{(12)(1960)}{24} = 980 \text{ in.}^3$$

$$S_{reqd} = \frac{\frac{1}{12}t_w h^3}{\frac{h}{2} + t_f} + \frac{2A_f \left(\frac{h}{2} + \frac{t_f}{2}\right)^2}{\frac{h}{2} + t_f}$$

Assume $t_f = \dfrac{3}{4}$ in.

$$980 = \frac{\left(\dfrac{1}{12}\right)\left(\dfrac{9}{16}\right)(58)^3}{\dfrac{58}{2}+\dfrac{3}{4}} + \frac{2A_f\left(\dfrac{58}{2}+\dfrac{0.75}{2}\right)^2}{\dfrac{58}{2}+\dfrac{3}{4}}$$

A_f = area of 1 flange = 11.59 in.^{2} (Say $\frac{3}{4} \times 16$)

Use $\frac{9}{16} \times 58$ web and one PL$\frac{3}{4} \times 16$ each flange

17-4 INTRODUCTION TO PLATE GIRDERS

Plate girders are large I-shaped sections built up from plates and rolled sections. They have resisting moments somewhere between those of rolled beams and steel trusses. Several plate girder arrangements are shown in Fig. 17-5. Possible riveted or bolted girders are shown in parts (a) and (b) of the figure, while several welded types are shown in parts (c) through (f). Nearly all plate girders constructed today are welded, although they frequently have bolted field splices. If you will compare the complicated bolted girders of parts (a) and (b) with the simple welded girders of parts (c) through (f), you will immediately understand why welded girders are so commonly used.

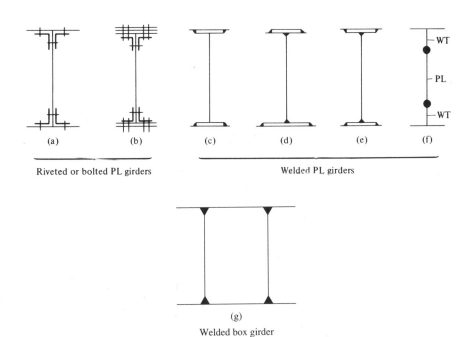

| (a) | (b) | (c) | (d) | (e) | (f) |

Riveted or bolted PL girders   Welded PL girders

(g)

Welded box girder

Figure 17-5 Plate girders.

The welded girder of Fig. 17-5(d) is arranged to reduce overhead welding as compared to the girder of part (c), but in so doing may be creating a little worse corrosion situation if the girder is exposed to the weather. Notice that the girder of part (f) is an ideal hybrid girder with the WTs made with a higher grade of steel than the web plate. A box girder, illustrated in part (g), is occasionally used where moments are large and depths are quite limited. Box girders also have great resistance to torsion and lateral buckling.

Plates and shapes can be arranged to form plate girders of almost any reasonable proportions. This fact may seem to give them a great advantage for all situations, but for the smaller sizes the advantage is usually cancelled by the higher fabrication costs. For example, it is possible to replace a W36 with a plate girder roughly twice as deep which will require considerably less steel and will have much smaller deflections; however, the higher fabrication costs will almost always rule out such a possibility.

Most steel highway bridges built today for spans less than about 80 ft are steel beam bridges. For longer spans the plate girder begins to compete very well economically. Where loads are extremely large as for railroad bridges, plate girders are competitive for spans as low as 45 or 50 ft.

The upper economical limits of bridge plate girder spans depend on several factors, including whether the bridge is simple or continuous, whether a highway or railroad bridge is involved, the largest section that can be shipped in one piece, and so on. Generally speaking plate girders are very economical for railroad bridges for spans from 50 to 130 ft (15 to 40 m) and for highway

Buffalo Bayou Bridge, Houston, Tex., a 270-ft span. (Courtesy of Lincoln Electric Company.)

bridges for spans from 80 to 150 ft (24 to 46 m). However, they are often very competitive for much longer spans, particularly when continuous. In fact they are actually common for 200-ft (61 m) spans and have been used for many spans in excess of 400 ft (122 m).

Though plate girders are most commonly used for bridges, they are sometimes used in buildings where they are called upon to support heavy concentrated loads. Occasionally a large ballroom or dining room with no interfering columns is desired on a lower floor of a multistory building. Such a situation is illustrated in Fig. 17-6. The plate girder shown must support some tremendous column loads for many stories above. The usual building plate girder of this type is simple to analyze because it probably does not have moving loads, although some building girders may be called upon to support traveling cranes.

The usual practical alternative to plate girders in the spans for which they are economical is the truss. In general, plate girders have the following advantages over trusses:

1. The pound price for fabrication of plate girders is lower than for trusses but higher than for rolled-beam sections.
2. Erection is cheaper and faster than for trusses.
3. Due to the compactness of plate girders, vibration and impact are not serious problems.
4. Plate girders require smaller vertical clearance than do trusses.
5. The plate girder has fewer critical points for stresses than do trusses.
6. A bad connection here or there is not as serious as in a truss, where such a situation could spell disaster.
7. There is less danger of injury to plate girders in an accident as compared to trusses. Should a truck run into a bridge plate girder, the girder would probably just bend a little, but a similar accident with a bridge truss member could cause a broken member and perhaps failure.
8. A plate girder is more easily painted than a truss.

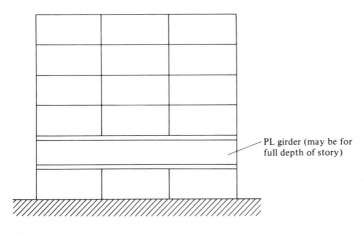

PL girder (may be for full depth of story)

Figure 17-6

On the other hand, plate girders are usually heavier than trusses for the same spans and loads, and they have a further disadvantage in the large number of connections required between webs and flanges.

17-5 MAJOR ITEMS TO BE CONSIDERED IN PLATE GIRDER DESIGN

Several problems in plate girder design are different from those faced in the design of regular rolled sections. These include allowable bending stresses, allowable shear stresses, stiffener designs, and tension field behavior. A few introductory remarks concerning these items are presented in this section, and they are discussed in more detail in the sections that follow.

Allowable Bending Stresses

The allowable bending stress for almost all W sections is $0.66F_y$. For the allowable stresses to equal this value, it is necessary for the flanges and webs to be compact (ASD table B5.1) and for the flanges to be continuously connected to the webs (ASD Specification F1.1). Plate girders do not meet all of these requirements, and it is necessary to reduce their allowable stresses accordingly. Two additional factors may cause further allowable bending stress reductions, the plate girder and hybrid girder factors (see Section 17-6).

Allowable Shear Stresses

Plate girders are usually quite deep, and if the h/t_w requirements are met to use the $0.4F_y$ allowable shear stress value, webs will be very large, heavy, and expensive. As a result plate girder webs are made much thinner than required for compactness. It would thus seem that their allowable shear strengths would be greatly reduced, but, as we will see, this is not necessarily the case.

Intermediate Stiffeners

Compression buckling of webs can be decidedly inhibited and allowable shear stresses kept quite high if vertical intermediate stiffeners are used as shown in Fig. 17-7. Depending on the spacing and stiffness of these stiffeners, the designer will be able to use some or all of the plate girder shear strengths that would be available if the webs were compact. This is true even though h/t_w ratios are two or more times the maximum compact values. A rather large percentage of the time required to design plate girders is devoted to stiffener designs.

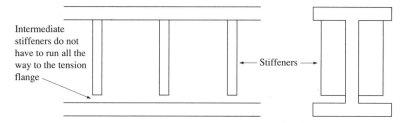

Figure 17-7 Intermediate stiffeners.

Bearing Stiffeners

It is almost always necessary to use bearing stiffeners (designed as columns) at end support reactions and perhaps also at locations where heavy concentrated loads are applied to the top flanges.

Tension Field Action

It has been mentioned that allowable shear stresses in the webs of plate girders are controlled by their h/t_w ratios and by the use or nonuse of intermediate stiffeners. Should the designer also consider the postbuckling behavior of plate girders with intermediate stiffeners (referred to as *tension field action*), he or she will learn that allowable web shear stresses can be appreciably increased (ASD Specification G3). If we take advantage of tension field action, we will have a higher F_v and perhaps a thinner web, but the intermediate stiffeners will be more expensive.

17-6 PROPORTIONS OF PLATE GIRDERS

Depth

The depths of plate girders vary from about one-sixth to one-fifteenth of their spans with average values of one-tenth to one-twelfth, depending on the particular conditions of the job. One condition that may limit the proportions of the girder is the largest size that can be fabricated in the shop and shipped to the job. There may be a transportation problem, such as clearance requirements that limit maximum depths to 10 or 12 ft along the shipping route.

The shallower girders will probably be used when the loads are light, and the deeper ones when very large concentrated loads need to be supported as from the columns in a tall building. If there are no depth restrictions for a particular girder, it will probably pay the designer to make rough designs and corresponding cost estimates to arrive at a depth decision. (Computer solutions will be very helpful here in preparing alternate designs.)

Indiana Harbor Works, East Chicago, Ind. (Courtesy of I.R. Construction Products Co.)

Web Size

After the total girder depth is estimated, the general proportions of the girder can be established from the maximum shear and the maximum moment. As previously described for I-shaped sections in Section 9-1, the web of a beam carries nearly all of the shearing stress; this shearing stress is assumed by the ASD Specification to be uniformly distributed throughout the web. The web depth can be closely estimated by taking the total girder depth and subtracting a reasonable value for the depths of the flanges (roughly 1 to 2 in. each). The web depths are usually selected to the nearest even inch because these plates are not stocked in fractional dimensions.

As a plate girder bends, its curvature creates vertical compression in the web as illustrated in Fig. 17-8. This is due to the downward vertical component of the compression flange bending stress and the upward vertical component of the tension flange bending stress.

The web must have sufficient vertical buckling strength to withstand the squeezing effect shown in Fig. 17-8. The ASD Specification (Gl) handles this problem by providing two maximum permissible web-depth-to-thickness ratios. The ratio to use depends on the spacing of intermediate stiffeners. The purpose of these limiting values is to prevent the vertical buckling of a girder flange into the web before the flexural stress reaches its yield stress F_{yf}.

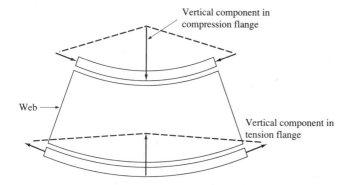

Figure 17-8 Squeezing of plate girder web.

1. If no transverse stiffeners are used or if they are used but are spaced at greater distances on center than $1\frac{1}{2}$ times the distance between flanges, the maximum ratio is

$$\frac{h}{t_w} \leq \frac{14,000}{\sqrt{F_{yf}(F_{yf} + 16.5)}}$$ (ASD Equation G1-1)

2. If transverse stiffeners are used and spaced no more than $1\frac{1}{2}$ times the distance between flanges, the maximum ratio is

$$\frac{h}{t_w} \leq \frac{2,000}{\sqrt{F_{yf}}}$$ (ASD Equation G1-2)

For this discussion it is assumed that stiffener spacing is greater than 1.5 times the distance between the flanges. If we use A36 steel, our web may not have a depth-thickness ratio less than $760/\sqrt{F_b} = 760/\sqrt{22} = 162$ and still be classed as a plate girder, nor may it be larger than $14,000/\sqrt{36(36 + 16.5)} = 322$. Thus for a trial girder section using A36 steel a depth-thickness ratio for the web of somewhere between 162 and 322 is tried. From a corrosion standpoint the usual practice is to use some absolute minimum thickness. For bridge girders, $\frac{3}{8}$ in. is a common minimum, while $\frac{1}{4}$ or $\frac{5}{16}$ in. is probably the minimum values for the more sheltered building girders.

Flange

After the web dimensions are selected, the next step is to select the area of the flange. The goal, of course, is to select a flange of sufficient area which will not be overstressed in bending. For this discussion reference is made to Fig. 17-9. The total bending strength of the plate girder equals the bending strength of the web plus the bending strength of the flanges.

The moment of inertia of the entire plate girder equals the moment of inertia of the web about its own centroidal axis plus the area of each flange times

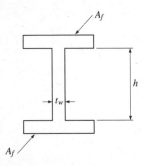

Figure 17-9 Girder cross section.

the distance from its center of gravity to the centroid of the section squared (thus neglecting the very small moment of intertia of the flange about its own centroidal axis). To simplify the expression, the centers of gravity of the flanges are assumed to fall exactly at the top and bottom of the web (a fairly good assumption).

The approximate gross moment of inertia of the plate girder can now be written as follows, where A_f is the area of one flange.

$$I = \frac{t_w h^3}{12} + 2A_f \left(\frac{h}{2}\right)^2$$

It is usually more convenient to work with section modulus values in girder design as in ordinary beams. The gross section modulus of the girder can be approximately written as

$$S = \frac{t_w h^3/12}{h/2} + \frac{2A_f(h/2)^2}{h/2}$$

$$= \frac{t_w h^2}{6} + A_f h$$

The required section modulus M/f is equated to S and the resulting expression is solved for A_f as follows:

$$\frac{M}{F_b} = \frac{t_w h^2}{6} + A_f h$$

$$A_f = \frac{M}{F_b h} - \frac{t_w h}{6}$$

In practice this expression for the required flange area is frequently simplified to the more conservative value to follow, in which F_b is assumed to equal $0.60F_y$.

$$A_f \approx \frac{M}{F_b h}$$

From the estimated flange area a trial size is selected that meets the $b_f/2t_f \leq 95/\sqrt{F_y}$ requirement of Table B5.1 of the ASD Specification. It is then necessary to compute the maximum bending stress to see if the girder proportions selected are satisfactory. The computed stress must not be greater than the allowable stress F_b' for plate girders, the determination of which is described in the next few paragraphs.

Allowable Bending Stress

The largest allowable bending stress that we can expect in a plate girder is $0.60F_y$. There is usually no reduction in allowable stress for such members due to unbraced lengths of the compression flanges, because of the large sizes of plate girder cross sections. Girder flanges, however, are not continuously connected to the webs as required by Section F1.1 of the ASD Specification if F_b' is to equal $0.66F_y$. In addition it may be necessary to reduce F_b' a little further as described herein. The allowable bending stress in a plate girder is given by the following expression:

$$F_b' \leq F_b R_{PG} R_e \qquad \text{(ASD Equation G2-1)}$$

In this expression F_b is the allowable bending stress (usually $0.60F_y$) computed for the girder as though it were a rolled shape without a continuous connection between web and flanges, R_{PG} is the *plate girder reduction factor,* and R_e is the hybrid *girder reduction factor.* Each of these factors is equal to or less than 1.0.

As a plate girder bends vertically, its web on the compression side will start to deflect laterally. As a result the web will not carry as much stress as our linear flexure formula will seem to indicate, and the compression flange will carry a little more. This situation is shown in Fig. 17-10.

The bending strength of the flange is much greater than the bending strength of the web, but the transfer of some of the stress from the web does cause the flange strength to be reduced a little bit.[1] This is accounted for with the plate girder reduction factor R_{PG}.

$$R_{PG} = 1 - 0.0005 \frac{A_w}{A_f}\left(\frac{h}{t} - \frac{760}{\sqrt{F_b}}\right) \leq 1.0$$

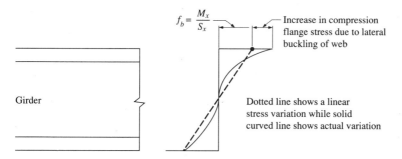

$$f_b = \frac{M_x}{S_x}$$

Increase in compression flange stress due to lateral buckling of web

Girder

Dotted line shows a linear stress variation while solid curved line shows actual variation

Figure 17-10 Flexure stress variation in a plate girder.

---

[1] "Commentary on the Specification for Structural Steel Buildings—Allowable Stress Design and Plastic Design," *ASD Manual,* p. 5-149.

The web of a hybrid plate girder yields before the maximum strength is reached in the flanges. As a result, the web contributes a smaller percentage of strength to a hybrid girder than it does to a homogeneous girder. Thus it is considered necessary to reduce the allowable bending stress in hybrid girders. This is accomplished with the hybrid girder factor R_e. It is given a value of 1.00 for nonhybrid girders.

$$R_e = \frac{12 + \left(\dfrac{A_w}{A_f}\right)(3\alpha - \alpha^3)}{12 + 2\dfrac{A_w}{A_f}}$$

In this expression $\alpha = 0.6F_{yw}/F_b \leq 1.0$. As shown in the formula for R_e, the reduction in the allowable stress is dependent upon the ratio of the web area to the flange area and also upon the ratio of F_{yw} to F_{yf}.

Tension Field Action

For this discussion the reader should refer to the Pratt truss of Fig. 17-11. The shear applied to the truss is resisted by tension in the diagonals, and at the same time the vertical members act as compression struts.

When a column approaches its buckling stage, it is on the verge of collapse, but this is not the case for a plate girder with intermediate stiffeners. After loads of such magnitude as to cause compression buckling are applied to a plate girder with intermediate stiffeners, additional loads will cause it to behave in much the same manner as does a Pratt truss under working loads. The steel in the web can resist tension as illustrated in Fig. 17-12, while the vertical stiffeners act as compression struts. This means that after compression buckling begins (which may cause a little initial waviness in the web), a new system will come into play before actual failure occurs. The designer may or may not elect to take advantage of this behavior, which is called *tension field action*.

The stiffeners keep the flanges from coming together, and the flanges keep the stiffeners from coming together. The intermediate stiffeners, which before

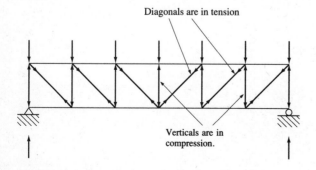

Figure 17-11 Pratt truss.

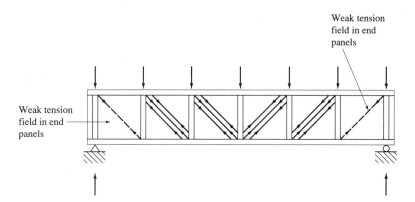

Figure 17-12 Tension field action.

initial buckling were assumed to resist no load, will after buckling resist compressive loads (or will serve as the compression verticals of a truss) due to diagonal tension. The result is that a plate-girder web can probably resist loads equal to two or three times those present at initial buckling before complete collapse occurs.

The horizontal components of diagonal tension in this "truss" will be resisted by the flange adjacent to the panel in question. *As the end panels of plate girders do not have such adjacent flanges on one side they are considered to be ineffective so far as tension field action goes.*

Until the web buckles initially, deflections are relatively small, but after initial buckling the girder's stiffness decreases considerably, and deflections may increase to several times the values estimated by the usual deflection theory.

The estimated total or ultimate shear that a *panel* (a part of the girder between a pair of stiffeners) can withstand equals the shear initially causing web buckling plus the shear which can be resisted by tension field action. The amount of tension field action is dependent on the proportions of the panels.

Designs based on tension field action may result in better economy and will provide more realistic ideas of actual strengths. One allowable shear stress is permitted by the ASD Specification if tension field action is ignored, and a larger one is permitted if tension field action is considered. The two allowable stresses are presented in the next section.

17-7 STIFFENERS

As previously described, it is usually necessary to stiffen the high thin webs of plate girders to keep them from buckling. The older riveted or bolted girders used angles connected to the webs and flanges, while for modern welded girders plates are welded to the webs and sometimes the flanges.

Stiffeners are divided into two groups: *Bearing stiffeners* transfer heavy reactions and concentrated loads to the full depth of the webs, without putting all the load on the flange connections. *Intermediate* or *nonbearing stiffeners* are placed at various intervals along the web to prevent buckling due to diagonal compression. (An intermediate stiffener placed on one side of a girder web is almost always more economical than a pair of stiffeners one on each side of a web).

Intermediate Stiffeners

Intermediate stiffeners are also referred to as nonbearing stiffeners, stability stiffeners, or transverse intermediate stiffeners. Their use is required by ASD Specification F5 if the ratio h/t_w is greater than 260 *and* the maximum web shear stress is greater than the value permitted by the equation

$$F_v = \frac{F_y}{2.89}(C_v) \le 0.40F_y \qquad \text{(ASD Equation F4-2)}$$

where

$$C_v = \frac{45{,}000k_v}{F_y\left(\dfrac{h}{t_w}\right)^2} \text{ when } C_v \text{ is less than 0.8 and}$$

$$= \frac{190}{\dfrac{h}{t_w}}\sqrt{\frac{k_v}{F_y}} \text{ when } C_v \text{ is more than 0.8}$$

$$k_v = 4.00 + \frac{5.34}{\left(\dfrac{a}{h}\right)^2} \text{ when } \frac{a}{h} \text{ is less than 1.0 and}$$

$$= 5.34 + \frac{4.00}{\left(\dfrac{a}{h}\right)^2} \text{ when } \frac{a}{h} \text{ is more than 1.0}$$

Before an allowable shear stress can be obtained for a particular girder, it is necessary to try a stiffener spacing. If the computed shear stress is less than the value given by these equations, no more intermediate stiffeners are required.

If intermediate stiffeners are properly spaced and proportioned, the designer may elect to consider tension field action; then the allowable shear stress may be increased to the following value:

$$F_v = \frac{F_y}{2.89}\left[C_v + \frac{1 - C_v}{1.15\sqrt{1 + \left(\dfrac{a}{h}\right)^2}} \right] \le 0.40F_y \qquad \text{(ASD Equation G3-1)}$$

Tension field action may not be considered for end panels in nonhybrid girders or for any panels in hybrid girders or web-tapered girders. It is thought that the end panel (which is the part of the girder up to the first interior stiffener) needs to anchor the horizontal component of the tension field action in the interior panels.

The ASD Specification imposes some additional arbitrary limits on panel aspect ratios (a/h) for plate girders even where shear stresses are small. These limitations facilitate the handling of girders during fabrication, shipping, and erection. Specification F5 states that, when intermediate stiffeners are required, the panel aspect ratio a/h may not exceed the value $260/\left(\dfrac{h}{t_w}\right)^2$ or three times the girder depth, whichever is smaller.

For the design of intermediate stiffeners the ASD Specification (G4) requires a minimum moment of inertia I_{st}, which is to be computed with reference to an axis in the plane of the web. It is applicable to either a pair of stiffeners or to a single stiffener.

$$I_{st} \geq \left(\frac{h}{50}\right)^4 \qquad \text{(ASD Equation G4-1)}$$

The ASD further stipulates that the total area of the stiffener in square inches may not be less than the following:

$$A_{st} = \frac{1 - C_v}{2}\left[\frac{a}{h} - \frac{(a/h)^2}{\sqrt{1 + \left(\dfrac{a}{h}\right)^2}}\right]YDht \qquad \text{(ASD Equation G4-2)}$$

In this expression Y is the ratio of F_{yw} to F_y of the stiffener. The term D is included because one-sided stiffeners are subject to moment as well as axial load, and as a result it is necessary to increase their cross-sectional areas to take into account their less efficient operation.

Fortunately values for this rather complicated expression for A_{st} are provided in Tables 2-36 and 2-50 of Part 2 of the Manual. The values are expressed as percentages of the web area.

Section G4 gives an expression to estimate the shear in pounds per inch which must be transferred between the intermediate stiffeners and web during shear transfer due to tension field action. This expression follows:

$$f_{vs} = h\sqrt{\left(\frac{F_{yw}}{340}\right)^3} \qquad \text{(ASD Equation G4-3)}$$

If the actual web shear ($f_v = V/ht_w$) is less than the allowable shear at the point (Equation G3-1), the shear to be transferred can be reduced in direct proportion.

Bearing Stiffeners

Bearing stiffeners are needed at the unframed ends of plate girders and also at interior concentrated loads if so required by Sections K1.2 through K1.6 of the ASD Specification. These stiffeners serve simultaneously as intermediate

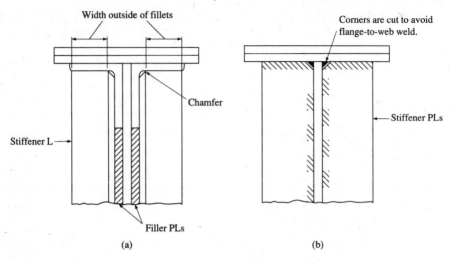

Figure 17-13 (a) Angle bearing stiffeners. (b) Welded girder stiffeners.

stiffeners. Typical bearing stiffeners are shown in Fig. 17-13. Part (a) illustrates the older angle stiffeners, which were riveted or bolted to the web and flanges, while part (b) illustrates the modern plate stiffeners with their welded connections.

Bearing stiffeners must fit tightly against the flanges being loaded and should extend out toward the edges of the flange plates. If the load normal to

Highway bypass bridge, Stroudsburg, Pa. (Courtesy of Bethlehem Steel Company.)

the girder flange is tensile, the stiffeners must be welded to the loaded flange. If the load is compressive, it is necessary to obtain a snug fit, that is, a good bearing between the flange and the stiffeners. To accomplish this goal the stiffeners may be welded to the flange, or the outstanding legs of the stiffeners may be milled.

A bearing stiffener is a special type of column which is difficult to accurately analyze because it must support the load in conjunction with the web. The amount of support provided by these two elements is difficult to estimate. Some steel specifications permit only the part of the stiffeners outside of the fillets of the flange angles or the parts outside of the flange-to-web welds (see Fig. 17-13) to be considered in supporting the bearing loads. The ASD Specification (K1.8), however, stipulates that the load or reaction may not exceed the allowable strength of a column having a cross section composed of the two stiffeners and a portion of the web of length $12t_w$ at girder ends and $25t_w$ at interior concentrated loads. It further states that the effective length of these "bearing stiffener columns" is to be $0.75h$.

17-8 FLEXURE-SHEAR INTERACTION

When a plate girder whose web depends on tension field action is subjected to relatively large shear and bending moment at the same location, the girder cannot provide full capacity in either shear or in moment. In that case an empirical interaction equation is used to check the adequacy of the girder.

Should tension field action not be considered in the design, no allowable-stress reduction due to the interaction of the two stresses is required by the ASD Specification. Should tension field action be considered in the design, however, the web must be so proportioned that the maximum tensile stress due to moment in the plane of the web does not exceed either $0.60F_y$ or

$$\left(0.825 - 0.375\frac{f_v}{F_v}\right)F_y \qquad \text{(ASD Equation G5-1)}$$

In this equation f_v is the average shear stress in the web equal to the total shear divided by the web area $\left(\dfrac{V}{ht_w}\right)$, and F_v is the allowable shear stress given by ASD Equation G3-1, which takes into account tensile field action.

The ASD Commentary (G5) states that plate girder webs can be proportioned on the basis of bending stress alone if the shearing stress does not exceed 60 percent of its permissible value; or it can be proportioned on the basis of shear alone if the bending stress does not exceed 75 percent of its allowable value. Otherwise ASD Equation G5-1 must be applied.

The Commentary also states that tension field action can only be used for webs of homogeneous girders with yield stresses larger than 65 ksi if the concurrent bending stress is not larger than 75 percent of its allowable value. This limitation is given because the webs of such girders develop more waviness when loaded to full capacity than do webs of lower-strength girders.

Arc-welded plate girders, Louisiana State Fair Grounds, Shreveport, La. (Courtesy of Lincoln Electric Company.)

17-9 EXAMPLE PLATE GIRDER DESIGN

Several plate girder design examples are presented in Part 2 of the ASD Manual. Example 17-3 was deliberately made quite similar to the first of these ASD examples, as the author feels this will enable the reader to easily follow the other ASD examples. The design presented here is for a welded girder supporting two concentrated column loads applied at its one-third points as well as a uniform load for its entire span.

The flange plates are selected for maximum moment at the girder centerline. Theoretically their sizes may be reduced towards the end of the girder where the moments are smaller, the smaller plates being butt welded to the larger plates. Space is not taken, however, to show such calculations. If the plate sizes are changed, the smaller plates (as well as the larger ones) must meet the $95/\sqrt{F_y}$ requirement for their width-thickness ratios as given in Table B5.1 of the ASD Specification.

For the reader's convenience, rather detailed references are made to the various ASD tables and specifications used in the solution.

■ Example 17-3

Design a welded plate girder for a 54-ft simple span to support a uniform load of 2 k/ft in addition to its own weight and 80-k concentrated column loads applied at the one-third points of the span. Assume the girder will have lateral support provided for its compression flange only at the reactions and the column loads. Use A36 steel and make use of tension field action as needed.

Solution.
Shear and moment diagrams:
   Assume girder weight = 200 lb/ft. Fig. 17-14 diagrams the shear and moment for the girder.
Preliminary web design:
   Assume girder depth = ($\frac{1}{10}$)(54) = 5.4 ft = 64.8 in. (say 65 in.)
   Assume web depth = 65 − 3 = 62 in.
   By ASD Specification G2 maximum h/t_w ratio for web so there is no reduction in allowable flange bending stress

$$\frac{h}{t_w} = \frac{760}{\sqrt{F_b}} = \frac{760}{\sqrt{22}} = 162$$

$$t_w = \frac{62}{162} = 0.383 \text{ in.}$$

   ASD Specification G1 states that, when no transverse stiffeners are provided or when they are provided but their spacing is $>1\frac{1}{2}h$, then

$$\frac{h}{t_w} \le \frac{14,000}{\sqrt{F_{yf}(F_{yf} + 16.5)}} = \frac{14,000}{\sqrt{36(36 + 16.5)}} = 322$$

$$t_w = \frac{62}{322} = 0.193 \text{ in.}$$

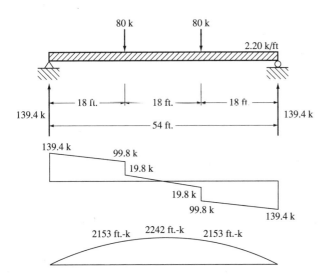

Figure 17-14

Assume minimum $t_w = \frac{5}{16}$ for corrosion.

<u>Try $\frac{5}{16} \times 62$-in. web ($A_w = 19.38$ in.^{2})</u>

$$\frac{h}{t_w} = \frac{62}{\frac{5}{16}} = 198.4$$

Preliminary flange design:

Apply approximate formula for required flange area.

$$A_f \approx \frac{M}{F_b h} = \frac{(12)(2242)}{(22)(62)} = 19.72 \text{ in.}^2 \qquad \text{(say } 1\tfrac{1}{4} \times 16\text{)}$$

$$\text{Total girder weight} = \frac{(2)(1.25 \times 16) + (\frac{5}{16})(62)}{144}(490)$$

$$= 202 \text{ lb/ft} > 200 \text{ lb} \qquad\qquad\qquad \text{OK}$$

Check for adequacy against local buckling. From Table B5.1 for limiting width-thickness ratios for compression elements,

$$k_c = \frac{4.05}{\left(\dfrac{h}{t}\right)^{0.46}} \qquad \text{since } \frac{h}{t} > 70$$

$$k_c = \frac{4.05}{\left(\dfrac{62}{\frac{5}{16}}\right)^{0.46}} = 0.355$$

Then use maximum width-thickness ratio from Table B5.1 for noncompact sections.

$$= \frac{95}{\sqrt{\dfrac{F_y}{k_c}}} = \frac{95}{\sqrt{\dfrac{36}{0.355}}} = 9.43$$

$$\frac{b_f}{2t_f} = \frac{16}{(2)(1.25)} = 6.4 < 9.43 \qquad\qquad\qquad \text{OK}$$

<u>Try flange PL$1\tfrac{1}{4} \times 16$</u>

Section properties and stress checking:

Dimensions shown in Fig. 17-15

$$I = (\tfrac{1}{12})(\tfrac{5}{16})(62)^3 + (2)(1.25 \times 16)(31.625)^2 = 46{,}212 \text{ in.}^4$$

$$S = \frac{46{,}212}{32.25} = 1433 \text{ in.}^4$$

$$f_b \text{ at centerline of girder} = \frac{(12)(2242)}{1433} = 18.77 \text{ ksi}$$

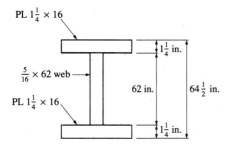

Figure 17-15

Since the web depth-thickness ratio $62/\frac{5}{16} = 198.4 > 760/\sqrt{F_b}$ from chapter G of ASD Specification $= 760/\sqrt{22} = 162$, we have to check to see if this stress is satisfactory in the compression flange as computed by ASD Equation (G2-1),

$$F_b' = F_b R_{PG} R_e$$

The compression flange is unbraced laterally for 18 ft, and it is necessary to compute an allowable compression flexural stress as required in the ASD Specification (F1.3).

Compute I_y and r_T for flange $+ \frac{1}{6}$ web.

$$I_y = (\tfrac{1}{12})(1.25)(16)^3 = 426.7 \text{ in.}^4$$
$$A_f + \tfrac{1}{6} \text{ web} = (1.25)(16) + (\tfrac{1}{6})(\tfrac{5}{16})(62) = 23.23 \text{ in.}^2$$
$$r_T = \sqrt{\frac{426.7}{23.23}} = 4.29 \text{ in.}$$

Note that $C_b = 1.0$, since the maximum moment of 2242 ft-k out in the unbraced length is greater than the moment (2153 ft-k) at either end.

$$\frac{L}{r_T} = \frac{(12)(18)}{4.29} = 50.35 < \sqrt{\frac{102 \times 10^3 \times C_b}{36}} = 53\sqrt{1.0} = 53$$

∴ F_b from the standpoint of lateral buckling is $0.6F_y = 22$ ksi (remembering that the web is noncompact, and F_b cannot be $> 0.6F_y$).

Reduced allowable bending stress in compression flange as per ASD Equation G2-1

$$R_{PG} = 1 - 0.0005 \frac{A_w}{A_f}\left(\frac{h}{t} - \frac{760}{\sqrt{F_b}}\right)$$

$$= 1 - (0.0005)\left(\frac{\frac{5}{16} \times 62}{1.25 \times 16}\right)\left(\frac{62}{\frac{5}{16}} - \frac{760}{\sqrt{22}}\right)$$

$$= 0.982$$

$R_e = 1.0$ for nonhybrid girders

$F_b' = F_b R_{PG} R_e = (22)(0.982)(1.0) = 21.6$ ksi > 18.77 ksi   OK

Repeat preceding process in the end panel.

$$f_b = \frac{(12)(2153)}{1433} = 18.03 \text{ ksi}$$

$$C_b = 1.75 + 1.05\frac{M_1}{M_2} + 0.3\left(\frac{M_1}{M_2}\right)^2 \text{ where } M_1 = 0$$

$$= 1.75 + 1.05\left(\frac{0}{2153}\right) + (0.3)\left(\frac{0}{2153}\right)^2 = 1.75$$

$$\frac{L}{r_T} = \frac{(12)(18)}{4.29} = 50.35 < 53\sqrt{1.75} = 70.11$$

$\therefore$ Allowable F_b from lateral buckling standpoint $= 0.6F_y = 22$ ksi. F_b' is the same as for center section.

$$F_b' = 21.6 \text{ ksi} > 18.03 \text{ ksi} \qquad\qquad \text{OK}$$

Use $\frac{5}{16} \times 62$ web and one PL$1\frac{1}{4} \times 16$ each flange

Bearing stiffeners at column loads:

Assume point bearing and $\frac{1}{4}$-in. web to flange weld leg sizes. Check web yielding using ASD Equation K1-2 for interior loads.

$$\frac{R}{t_w(N + 5k)} \le 0.66F_y = 24 \text{ ksi}$$

k from Fig. 17-16 $= 1\frac{1}{4} + \frac{1}{4} = 1.50$ in.

$$\frac{80}{\frac{5}{16}[0 + (5 \times 1.50)]} = 34.13 \text{ ksi} > 24 \text{ ksi}$$

$\therefore$ Bearing stiffeners are required. A similar procedure will show they are also needed at the end reactions.

If the local web yielding requirement had been satisfied, it would also be necessary for the web crippling criteria of ASD Specification K1-4 to be satisfied before we could say that bearing stiffeners are not needed.

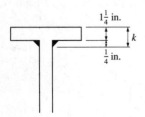

$1\frac{1}{4}$ in.

k

$\frac{1}{4}$ in.

Figure 17-16

The bearing stiffeners will be selected after a check is made to see if intermediate stiffeners are needed.

Design of intermediate stiffeners:

Shear stress in the 18-ft end panel

$$f_v = \frac{139.4}{62 \times \frac{5}{16}} = 7.19 \text{ ksi}$$

$$\frac{a}{h} = \frac{(12)(18)}{62} = 3.48$$

$$\frac{h}{t_w} = \frac{62}{\frac{5}{16}} = 198.4 > \frac{380}{\sqrt{36}} = 63.33 \text{ (ASD Section F4)}$$

$\therefore$ Determine F_v by ASD Equation F4-2.

Since $\dfrac{a}{h} > 1.0$ the following expressions are used (Table 1-36 of Part 2 of the Manual may be used also)

$$k_v = 5.34 + \frac{4.00}{\left(\dfrac{a}{h}\right)^2} = 5.34 + \frac{4.00}{(3.48)^2} = 5.67$$

$$C_v = \frac{45,000 \, k_v}{F_y\left(\dfrac{h}{t_w}\right)^2} = \frac{(45,000)(5.67)}{(36)(198.4)^2} = 0.180$$

$$F_v = \frac{F_y}{2.89}(C_v) = \frac{36}{2.89}(0.180) = 2.242 \text{ ksi} < 7.19 \text{ ksi}$$

$\therefore$ Use intermediate stiffeners

To find the position of first stiffener in end panel, go into Table 1-36 with tension field action not included, and set $F_v = 7.19$ ksi from last step. With $\dfrac{h}{t_w} = 198.4$ we find by interpolation

$$\frac{a}{h} = 0.616$$

$$a = (0.616)(62) = 38.2 \text{ in.} \qquad \text{(say 3 ft)}$$

First stiffener 3 ft from end

Check need for additional intermediate stiffeners in first panel.

$$V = 139.4 - (3)(2.20) = 132.8 \text{ k}$$

$$f_v = \frac{132.8}{(62)(\frac{5}{16})} = 6.85 \text{ ksi}$$

Distance from first intermediate stiffener to column load = 18 − 3 = 15 ft = 180 in.

$$\frac{a}{h} = \frac{180}{62} = 2.90$$

Using ASD Equation F4-2 again

$$k_v = 5.34 + \frac{4.00}{(2.90)^2} = 5.82$$

$$C_v = \frac{(45,000)(5.82)}{(36)(198.4)^2} = 0.185$$

$$F_v = \left(\frac{36}{2.89}\right)(0.185) = 2.30 \text{ ksi} < 6.85 \text{ ksi}$$

∴ More intermediate stiffeners needed. Assume the remaining stiffeners are spaced at $\frac{15}{2}$ = 7.5 ft = 90 in., and check to see if that spacing is satisfactory.

$$\frac{a}{h} = \frac{90}{62} = 1.45$$

Checking this $\frac{a}{h}$ with ASD Specification F5 and ASD Equation F5-1,

$$\text{Max } \frac{a}{h} = \left(\frac{260}{198.4}\right)^2 = 1.72 > 1.45 \qquad \text{OK}$$

Entering Table 2-36 (which includes tension field action) with $\frac{a}{h}$ = 1.45 and $\frac{h}{t_w}$ = 198.4, we find by interpolation or by using ASD Equation G3-1

$$F_v = 7.60 \text{ ksi} > 6.85 \text{ ksi} \qquad \text{OK}$$

Are intermediate stiffeners needed in center panel?

$$\frac{h}{t_w} = 198.4$$

$$\frac{a}{h} = \frac{(12)(18)}{62} = 3.48$$

But can't be >3.0 (ASD Specification F5).
<u>Try 1 stiffener at midpoint</u>

$$\frac{a}{h} = \frac{(12)(9)}{62} = 1.74$$

$$f_v = \frac{19.8}{(62)(\frac{5}{16})} = 1.02 \text{ ksi}$$

From Equation F4-2 again,

$$k_v = 5.34 + \frac{4.00}{(1.74)^2} = 6.66$$

$$C_v = \frac{(45,000)(6.66)}{(36)(198.4)^2} = 0.211$$

$$F_v = \left(\frac{36}{2.89}\right)(0.211) = 2.63 \text{ ksi} > 1.02 \text{ ksi} \qquad \text{OK}$$

Use one stiffener at centerline of middle panel

For combined shear and tension, check interaction at column load in tension field panel.

$$f_v = \frac{99.8}{(62)(\frac{5}{16})} = 5.15 \text{ ksi}$$

Apply ASD Equation G5-1.

$$F_b = \left(0.825 - 0.375\frac{5.15}{7.60}\right)36 = 20.55 \text{ ksi} > f_b \qquad \text{OK}$$

Design of stiffeners:
Intermediate stiffeners: use single plate stiffeners. Referring to ASD Specification G-4,

$$A_{st} = (\%)(A_w)(D)\left(\frac{f_v}{F_v}\right)$$

$$\frac{h}{t_w} = 198.4$$

$$\frac{a}{h} = \frac{90}{62} = 1.45$$

From Table 2-36 determine the gross area needed for stiffener as a percent of the web area. For single plate stiffeners D will equal 2.4 as given in a footnote to the table.

$$\% = 9.82$$

$$A_{st} = (0.0982)\left(62 \times \frac{5}{16}\right)(2.4)\left(\frac{6.85}{7.60}\right) = 4.12 \text{ in.}^2$$

Try one stiffener $\frac{5}{8} \times 7\frac{1}{2}$ ($A = 4.69 \text{ in.}^2 > 4.28 \text{ in.}^2$)

Width-thickness ratio $= \dfrac{7.50}{0.625} = 12 < \dfrac{95}{\sqrt{36}} = 15.83$ from ASD Table B5.1

Moment of inertia: use ASD Equation G4-1.

$$\text{Min } I_{st} = \left(\frac{h}{50}\right)^4$$

$$= \left(\frac{62}{50}\right)^4 = 2.36 \text{ in.}^4$$

Actual I_{st} using a depth from the centerline of the web to the outside of the stiffener

$$I_{st} = (\tfrac{1}{3})(0.625)(7.66)^3 = 93.6 \text{ in.}^4 > 2.36 \text{ in.}^4 \qquad \text{OK}$$

Minimum length required by ASD (G4) $= 62 - \tfrac{1}{4}$ weld: no less than $4t_w$ nor more than $6t_w$.

$$\text{Min length} = 62 - \tfrac{1}{4} - (6)(\tfrac{5}{16}) = 59.88 \text{ in.}$$
$$\underline{\text{Use } \tfrac{5}{8} \times 7\tfrac{1}{2} \times 5 \text{ ft 0 in. stiffeners}}$$

Bearing stiffeners: for end reaction, as required in Section K1.8 of ASD Specification,

$$\underline{\text{Try two } \tfrac{1}{2} \times 6 \text{ in. bars as shown in Fig. 17-17}}$$

Compute compressive stresses for stiffeners.

$$I = (\tfrac{1}{12})(\tfrac{1}{2})(12.31)^3 = 77.73 \text{ in.}^4$$

$$A_{\text{eff}} = (2)(6)(\tfrac{1}{2}) + (3.75)(\tfrac{5}{16}) = 7.17 \text{ in.}^2$$

$$r = \sqrt{\frac{77.73}{7.17}} = 3.29 \text{ in.}$$

$$KL = (0.75)(62) = 46.5 \text{ in.}$$

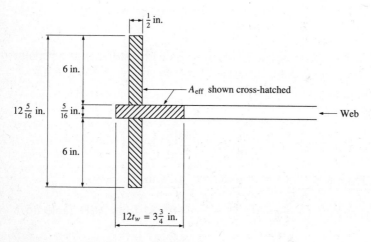

Figure 17-17 End bearing stiffener.

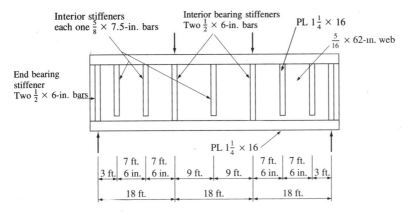

Interior stiffeners
each one $\frac{5}{8} \times$ 7.5-in. bars

Interior bearing stiffeners
Two $\frac{1}{2} \times$ 6-in. bars

PL $1\frac{1}{4} \times 16$

$\frac{5}{16} \times$ 62-in. web

End bearing
stiffener
Two $\frac{1}{2} \times$ 6-in. bars

PL $1\frac{1}{4} \times 16$

3 ft. | 7 ft. 6 in. | 7 ft. 6 in. | 9 ft. | 9 ft. | 7 ft. 6 in. | 7 ft. 6 in. | 3 ft.

18 ft. | 18 ft. | 18 ft.

Figure 17-18 Final design.

$$\frac{KL}{r} = \frac{46.5}{3.29} = 14.13$$

$F_a = 20.94$ ksi from Table C-36 in Part 3 of the Manual

$$f_a = \frac{\text{reaction}}{A_{\text{eff}}} = \frac{139.4}{7.17} = 19.44 \text{ ksi} < 20.94 \text{ ksi} \qquad \text{OK}$$

Use two end bearing stiffeners $\frac{1}{2} \times 6 \times 5$ ft $1\frac{3}{4}$ in.
with close bearing on the flange receiving the reaction.

Interior bearing stiffeners at column loads: use same bearing stiffeners as for end reactions. Loads are smaller, and there is more web area (length = $25t_w$, says ASD Section K1.8), which can be used as part of "column" area.

Design sketch of girder, Fig. 17-18 ∎

17-10 FLANGE-TO-WEB WELDS

The flange plate to web welds must be able to resist the horizontal shear flow VQ/I at any point. The designer may use intermittent welds or continuous welds. Though it might seem that intermittent welds would be used most of the time, it is rather economical for the fabricator to make use of continuous welds with his or her automatic welding equipment. Whether continuous or intermittent, the welds are designed as they were in Chapter 13.

When intermittent welds are used for plate girders, their maximum spacings are limited by the ASD Specification to the same values as are required for built-up tension and compression members (ASD Sections D2 and E4).

PROBLEMS

17-1 An A36 W16 × 45 with $\frac{7}{8}$-in.-thick cover plates is to be used to support the loads shown. Assume the beam has full lateral bracing for its compression flange and determine plate width. (*Ans.* PL$\frac{7}{8}$ × 14)

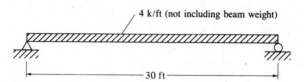

4 k/ft (not including beam weight)

|← 30 ft →|

17-2 The architects specify that a beam no greater than 16.00 in. deep be designed to support a uniform load of 8 k/ft, not including the beam weight for a 24-ft simple span. Use A572 steel with $F_y = 50$ ksi and full lateral bracing for the compression flange. Use a W14 × 90 with cover plates.

17-3 Design a 48-in.-deep welded built-up wide-flange section with no intermediate stiffeners for a 50-ft simple span to support a 3 k/ft uniform load (not including the beam weight). The section is to be framed between columns and is to have full lateral bracing for its compression flange. Use A36 steel. (*Ans.* $\frac{7}{16}$ × 46 web, PL$\frac{1}{2}$ × 16 each flange)

17-4 Repeat Prob. 17-3 with a span of 60 ft and A572 steel ($F_y = 50$ ksi).

17-5 Design a welded plate girder with A36 steel for a 60-ft span to support a uniform load of 7 k/ft (not including the girder weight). The compression flange will be laterally braced for its entire length. The girder is to be simply supported at its ends and is not framed into the columns. Do not design stiffeners. Assume girder depth $\frac{1}{10}$ span. (*Ans.* $\frac{5}{16}$ × 68 web, PL1$\frac{1}{2}$ × 18 each flange)

17-6 Repeat Prob. 17-5 using A572 steel ($F_y = 50$ ksi).

17-7 The cross section for a welded plate girder is to be selected for the span and loads shown. The design is to be made with A36 steel, assuming that lateral bracing is provided for the compression flange at the ends and concentrated load points only. Do not design stiffeners. Assume girder depth $\frac{1}{10}$ span. (*Ans.* $\frac{5}{16}$ × 56 web, PL1$\frac{1}{2}$ × 16 each flange)

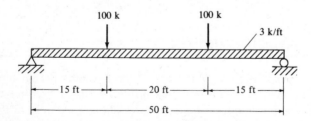

100 k   100 k   3 k/ft

|← 15 ft →|← 20 ft →|← 15 ft →|
|← 50 ft →|

17-8 Repeat Prob. 17-7 using A572 steel ($F_y = 50$ ksi).

17-9 Design end bearing stiffeners and intermediate stiffeners for the plate girder selected in Prob. 17-7. Also check flexure-shear interaction at points 8, 15, and 20 ft from the girder ends. (*Ans.* Intermediate stiffeners each end at 30, 50, and 50 in.; flexure-shear interaction OK)

18

Plastic Analysis

18-1 INTRODUCTION

All designs performed in the previous chapters of this book were based on the elastic theory. The maximum load that a structure could support was assumed to equal the load that first caused a stress somewhere in the structure to equal the yield point of the material. Engineering structures have been designed for many decades by this method with satisfactory results. The design profession, however, has long been aware that ductile materials do not fail until a great deal of yielding occurs after the yield stress is reached. Chapters 18 and 19 are devoted to a consideration of the behavior of ductile steel structures in the plastic range.

When the stress at one point in a ductile steel structure reaches the yield point, that part of the structure will yield locally, permitting some readjustment of the stresses. Should the load be increased, the stress at the point in question will remain approximately constant, thereby requiring the less stressed parts of the structure to support the load increase. It is true that statically determinate structures can resist little load in excess of the amount that causes the yield stress to first develop at some point. For statically indeterminate structures, however, the load increase can be quite large; and these structures are said to have the happy facility of spreading out overloads. due to the steel's ductility.

As early as 1914 a Hungarian, Dr. Gabor Kazinczy, recognized that the ductility of steel permitted a redistribution of stresses in an overloaded statically indeterminate structure.[1] In the United States Prof. J. A. Van den Broek introduced his plastic theory which he called limit design.[2]

In the plastic theory, rather than basing designs on the allowable-stress method, the problem is handled by considering the greatest load that can be carried by the structure acting as a unit. The resulting designs are quite interesting to the structural engineer, as they offer several advantages.

1. When the plastic-design procedure is used, there can be considerable savings in steel (perhaps as high as 10 or 15 percent for some structures) as compared to a similar design made by the allowable-stress procedure. It is true, however, that the necessary continuity connections required in plastic design may reduce the actual money savings somewhat.

2. Plastic design permits the designer to make a more accurate estimate of the maximum load that a structure can support, enabling him or her to have a better idea of the actual safety factor of the structure. The elastic method works very well for computing the stresses and strains for loads in the elastic range, but gives a very poor estimate of the actual collapse strength of a structure.

3. For many structures, plastic analysis is easier to apply than is elastic analysis.

4. Structures are often subjected to large stresses which are difficult to predict, such as those caused by settlement, erection, and so on. Plastic design provides for such situations by permitting plastic deformation.

Despite these several important advantages, plastic design has not caught on with the design profession. The plastic theory, however, has decidedly affected allowable-stress steel specifications. The 0.9 rule used for the design of continuous beams is one example. Today a related method called load and resistance factor design (LRFD) is being used by an appreciable number of designers and appears to be the steel design method of the immediate future.

The material presented in this chapter and the next provides a better understanding of some of the provisions of the ASD Specification and serves as a useful background for the LRFD method discussed in Chapters 20–23.

Despite the progress made in the field of plastic design, the method still has some disadvantages with which the designer should be completely familiar.

1. Plastic design is of little value for high-strength brittle steels. (The method is just as applicable to high-strength steels as it is to lower-

---

[1] L. S. Beedle, *Plastic Design of Steel Frames* (New York: Wiley, 1958), p. 3.

[2] J. A. Van den Broek, "Theory of Limit Design," *Proceedings ASCE* (February 1939), vol. 65, pp. 193–216.

strength structural-grade steels as long as the steels have the required ductility.)

2. Plastic design is not satisfactory for situations where fatigue stresses are a problem.
3. Columns designed by the plastic theory provide little savings.
4. Although for many statically indeterminate structures plastic analysis is simpler than elastic analysis, it should be realized that unstable plastic structures are more difficult to detect than are unstable elastic structures.

18-2 THEORY OF PLASTIC ANALYSIS

The basic theory of plastic design is that there is a major change in the distribution of stresses after the stresses at certain points in a structure reach the yield point. Those parts of the structure which have been stressed to the yield point cannot resist additional stresses. They instead will yield enough to permit the extra load or stresses to be transferred to other parts of the structure where the stresses are below the yield stress and thus in the elastic range and able to resist increased stress. Plasticity can be said to serve the purpose of equalizing stresses in cases of overload.

For this discussion the stress-strain diagram is assumed to have the idealized shape shown in Fig. 18-1. The yield point and the proportional limit are assumed to occur at the same point for this steel, and the stress-strain diagram is assumed to be a perfectly straight horizontal line in the plastic range. Beyond the plastic range there is a range of strain hardening. This latter range could theoretically permit steel members to withstand additional stress, but from a practical standpoint the strains occurring are so large that they cannot be considered.

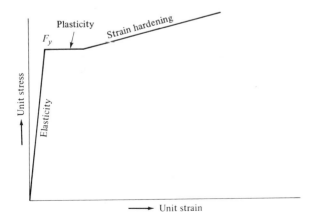

Figure 18-1 Stress-strain diagram.

18-3 THE PLASTIC HINGE

As the bending moment is increased at a particular section of a beam, there will be a linear variation of stress until the yield stress is reached in the outermost fibers. An illustration of stress variation for a rectangular beam in the elastic range is shown in Fig. 18-2. The *yield moment* of a cross section is defined as the moment that will just produce the yield stress in the outermost fiber of the section.

If the moment is increased beyond the yield moment, the outermost fibers that had previously been stressed to their yield point will continue to have the same stress but will yield, and the duty of providing the necessary additional resisting moment will fall on the fibers nearer to the neutral axis. This process will continue, with more and more parts of the beam cross section stressed to the yield point as shown by the stress diagrams of Fig. 18-2(c) and (d), until finally a fully plastic distribution is approached as shown in part (e). When the stress distribution has reached this stage, a *plastic hinge* is said to have formed, because no additional moment can be resisted at the section. Any additional moment applied at the section will cause the beam to rotate with little increase in stress.

The *plastic moment* is the moment that will produce full plasticity in a member cross section and create a plastic hinge. The ratio of the plastic moment M_p to the yield moment M_y is the *shape factor*. The shape factor equals 1.50 for rectangular sections and varies from about 1.10 to 1.20 for standard rolled-beam sections.

The development of a plastic hinge in a simple beam is shown in Fig. 18-3. The load shown is applied to the beam and increased in magnitude until the yield moment is reached and the outermost fiber is stressed to the yield point. The magnitude of the load is further increased, with the result that the outer fibers begin to yield. The yielding spreads to the other fibers away from the section of maximum moment as indicated in the figure. The length in which this yielding occurs away from the section in question is dependent on the loading conditions and the member cross section. For a concentrated load applied at the centerline of a simple beam with a rectangular cross section, yielding in the extreme fibers at the time the plastic hinge is formed will extend

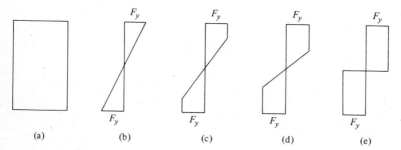

Figure 18-2 Stress variation in rectangular beam.

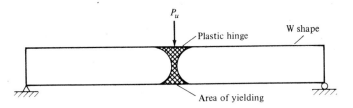

Figure 18-3

for one-third of the span. For a W shape in similar circumstances, yielding will extend for approximately one-eighth of the span. During this same period the interior fibers at the section of maximum moment yield gradually until nearly all of them have yielded and a plastic hinge is formed as shown in Fig. 18-3.

Although the effect of a plastic hinge may extend for some distance along the beam, it is assumed to be concentrated at one section for analysis purposes. For the calculation of deflections and for the design of bracing, the length over which yielding extends is quite important.

When steel frames are loaded to failure, the points where rotation is concentrated (plastic hinges) become quite visible to the observer before collapse occurs.

18-4 THE PLASTIC MODULUS

The yield moment M_y equals the yield stress times the elastic modulus. The elastic modulus equals I/c or $bd^2/6$ for a rectangular section, and the yield moment equals $F_y bd^2/6$. This same value can be obtained by considering the resisting internal couple shown in Fig. 18-4.

The resisting moment equals T or C times the lever arm between them, as follows:

$$M_y = \left(\frac{F_y bd}{4}\right)\left(\frac{2}{3}d\right) = \frac{F_y bd^2}{6}$$

The elastic section modulus can again be seen to equal $bd^2/6$ for a rectangular beam. The resisting moment of a rectangular section at full plasticity can

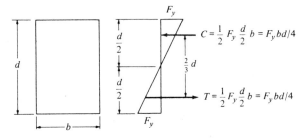

Figure 18-4 Internal couple at first yield.

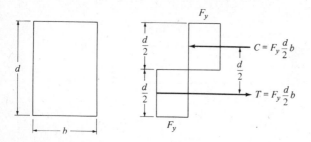

Figure 18-5 Internal couple at full plastic condition.

be determined in a similar manner (see Fig. 18-5).

$$M_p = \left(F_y \frac{d}{2} b\right)\left(\frac{d}{2}\right) = \frac{F_y b d^2}{4}$$

The plastic moment is said to equal the yield stress times the plastic modulus. From the foregoing expression for a rectangular section, the plastic modulus Z can be seen to equal $bd^2/4$. The shape factor, which equals M_p/M_y, $F_y Z/F_y S$, or Z/S is $(bd^2/4)/(bd^2/6) = 1.50$ for a rectangular section.

A study of the plastic modulus determined here shows that it equals the statical moment of the tension and compression areas about the neutral axis. Unless the section is symmetrical, the neutral axis for the plastic condition will not be in the same location as for the elastic condition. The total internal compression must equal the total internal tension. As all fibers are considered to have the same stress (F_y) in the plastic condition, the areas above and below the neutral axis must be equal. This situation does not hold for unsymmetrical sections in the elastic condition. Example 18-1 illustrates the calculations necessary to determine the shape factor for a tee beam and the ultimate uniform load w_u that the beam can support.

■ Example 18-1

Determine M_y, M_p, and Z for the steel tee beam shown in Fig. 18-6. Also calculate the shape factor and the ultimate uniform load (w_u) which can be placed on the beam for a 12-ft simple span. $F_y = 36$ ksi.

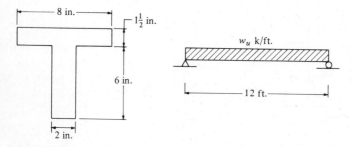

Figure 18-6

Solution.
Elastic calculations:

$$A = (8)(1\tfrac{1}{2}) + (6)(2) = 24 \text{ in.}^2$$

$$\bar{y} = \frac{(12)(0.75) + (12)(4.5)}{24} = 2.625 \text{ in. from top flange}$$

$$I = (\tfrac{1}{3})(2)(1.125^3 + 4.875^3) + (\tfrac{1}{12})(8)(1\tfrac{1}{2})^3 + (12)(1.875)^2$$
$$= 122.4 \text{ in.}^4$$

$$S = \frac{I}{C} = \frac{122.4}{4.875} = 25.1 \text{ in.}^3$$

$$M_y = F_y S = \frac{(36)(25.1)}{12} = \underline{\underline{75.3 \text{ ft-k}}}$$

Plastic calculations:
Neutral axis is at base of flange.

$$Z = (12)(0.75) + (12)(3) = \underline{\underline{45 \text{ in.}^3}}$$

$$M_p = F_y Z = \frac{(36)(45)}{12} = \underline{\underline{135 \text{ ft-k}}}$$

$$\text{Shape factor} = \frac{M_p}{M_y} \quad \text{or} \quad \frac{Z}{S} = \frac{45}{25.1} = \underline{\underline{1.79}}$$

$$M_p = \frac{w_u l^2}{8}$$

$$w_u = \frac{(8)(135)}{(12)^2} = \underline{\underline{7.5 \text{ k/ft}}}$$

The values of the plastic moduli for the standard steel beam sections are tabulated in the ASD Manual in the "Plastic Design Selection Table." These values will be used frequently in the pages to follow.

18-5 FACTORS OF SAFETY AND LOAD FACTORS

The factor of safety used in designing a particular structure should probably be selected only after a study is made of the uncertainties present in the design (a subject discussed in Section 2.8). Perhaps the usual practice, however, is to use a value that seems reasonable from the standpoint of past experience as expressed in the specifications being used. The usual safety factor considered to be present in elastic design is the one obtained by dividing the yield point of the steel by its working stress. For compact laterally supported beams of A36 steel, this safety factor is $\frac{36}{24} = 1.50$ using the ASD Specification.

If the safety factor is multiplied by the shape factor, the result is referred to as the *load factor*. For a typical W section the load factor equals the safety factor (1.50) times the average shape factor of W sections of about 1.12, or 1.68 (say 1.70). In plastic design the working loads are multiplied by the load factor to give the estimated ultimate loads, and members are proportioned with these loads on the basis of plastic or collapse strengths. This procedure is illustrated in the design examples of Chapter 19. The minimum values for load factors permitted by the ASD for design purposes are presented in Section 19-2.

18-6 THE COLLAPSE MECHANISM

A statically determinate beam will fail if one plastic hinge develops. Fig. 18-7 illustrates this fact. The simple beam of constant cross section is loaded with a concentrated load at midspan. Should the load be increased until a plastic hinge is developed at the point of maximum moment (underneath the load in this case), an unstable structure will have been created as shown in Fig. 18-7(b). Any further increase in load will cause collapse.

The plastic theory is of little advantage for statically determinate beams and frames, but it may be of decided advantage for statically indeterminate beams and frames. For a statically indeterminate structure to fail, it is necessary for more than one plastic hinge to form. The number of plastic hinges required for failure of statically indeterminate structures varies from structure to structure, but may never be less than two. The fixed-end beam of Fig. 18-8 cannot fail unless the three plastic hinges shown in the figure develop.

Although a plastic hinge may have formed in a statically indeterminate structure, the load can still be increased without causing failure if the geometry of the structure permits. The plastic hinge will act like a real hinge insofar as increased loading is concerned. As the load is increased, there is a redistribution of moment because the plastic hinge can resist no more moment. As more plastic hinges are formed in the structure, there will eventually be a sufficient number of them to cause collapse. Actually some additional load can

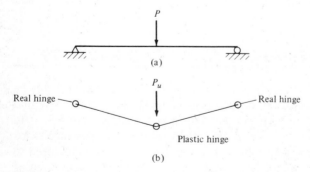

(a)

(b)

Real hinge

Real hinge

Plastic hinge

Figure 18-7

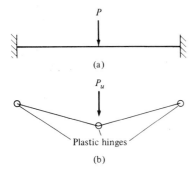

(a)

(b)

Plastic hinges

Figure 18-8

be carried after this time before collapse occurs as the stresses go into the strain hardening range, but the deflections that would occur are too large to be permissible.

The propped beam of Fig. 18-9 is an example of a structure that will fail after two plastic hinges develop. Three hinges are required for collapse, but there is a real hinge on the right end. In this beam the largest elastic moment caused by the design concentrated load is at the fixed end. As the magnitude of the load is increased, a plastic hinge will form at that point.

The load may be further increased until the moment at some other point (here it will be at the concentrated load) reaches the plastic moment. Additional load will cause the beam to collapse. The arrangement of plastic hinges and perhaps real hinges which permit collapse in a structure is called the *mechanism*. Parts (b) of Figs. 18-7, 18-8, and 18-9 show mechanisms for various beams.

After observing the large number of fixed-end and propped beams used for illustrations in this text, the student may form the mistaken idea that he or she will frequently encounter such beams in engineering practice. These types of beams, truthfully, are difficult to find in actual structures but are very convenient to use in illustrative examples. They are particularly convenient for introducing plastic analysis before continuous beams and frames are considered.

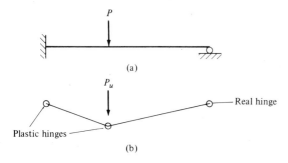

(a)

(b)

P_u

Real hinge

Plastic hinges

Figure 18-9

18-7 THE VIRTUAL-WORK METHOD

One very satisfactory method used for the plastic analysis of structures is the *virtual-work method*. The structure in question is assumed to be loaded to its ultimate capacity M_p and is then assumed to deflect through a small additional displacement. The work performed by the external loads during this displacement is equated to the internal work absorbed by the hinges. For this discussion the *small-angle theory* is used. By this theory the sine of a small angle equals the tangent of that angle and also equals the same angle expressed in radians. In the pages to follow the author uses these values interchangeably because the small displacements considered here produce extremely small rotations or angles.

As a first illustration Fig. 18-10 shows a uniformly loaded fixed-ended beam and its collapse mechanism. Owing to symmetry, the rotations at the end plastic hinges are equal, and they are represented by θ in the figure; thus the rotation at the middle plastic hinge will be 2θ.

The work performed by the total external load $(w_u L)$ is equal to $w_u L$ times the average deflection of the mechanism. The average deflection equals one-half the deflection at the center plastic hinge $(\frac{1}{2} \times \theta \times L/2)$. The external work is equated to the internal work absorbed by the hinges or to the sum of M_p at each plastic hinge times the angle through which it works. The resulting expression can be solved for M_p and w_u as follows:

$$M_p(\theta + 2\theta + \theta) = w_u L \left(\frac{1}{2} \times \theta \times \frac{L}{2} \right)$$

$$M_p = \frac{w_u L^2}{16}$$

$$w_u = \frac{16 M_p}{L^2}$$

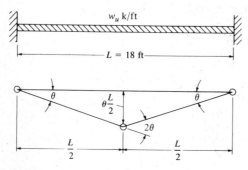

Figure 18-10

For the 18-ft span used in Fig. 18-10 these values become

$$M_p = \frac{(w_u)(18)^2}{16} = 20.25w_u$$

$$w_u = \frac{M_p}{20.25}$$

Plastic analysis can be handled in a similar manner for the propped beam of Fig. 18-11. There the collapse mechanism is shown, and the end rotations (which are equal to each other) are assumed to equal θ.

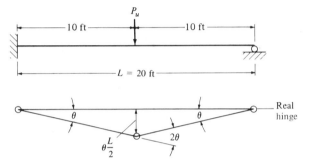

Figure 18-11

The work performed by the external load P_u as it moves through the distance $\theta L/2$ is equated to the internal work performed by the plastic moments at the hinges, with no moment at the real hinge on the right end of the beam.

$$M_p(\theta + 2\theta) = P_u\left(\theta\frac{L}{2}\right)$$

$$M_p = \frac{P_u L}{6} \quad \text{or } 3.33 \, P_u \text{ for the 20-ft beam shown}$$

$$P_u = \frac{6M_p}{L} \quad \text{or } 0.3 \, M_p \text{ for the 20-ft beam shown}$$

The fixed-end beam of Fig. 18-12 together with its collapse mechanism and assumed angle rotations is next considered. From this figure the values of M_p and P_u can be determined by virtual work as follows:

$$M_p(2\theta + 3\theta + \theta) = P_u\left(2\theta \times \frac{L}{3}\right)$$

$$M_p = \frac{P_u L}{9} \quad \text{or } 3.33 \, P_u \text{ for this beam}$$

$$P_u = \frac{9M_p}{L} \quad \text{or } 0.3 \, M_p \text{ for this beam}$$

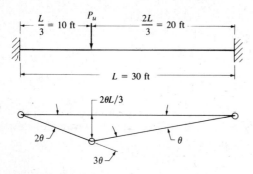

Figure 18-12

A person beginning the study of plastic analysis needs to learn to think of all the possible ways in which a particular structure might collapse. Such a habit is of the greatest importance when one begins to analyze more complex structures. In this light the plastic analysis of the propped beam of Fig. 18-13

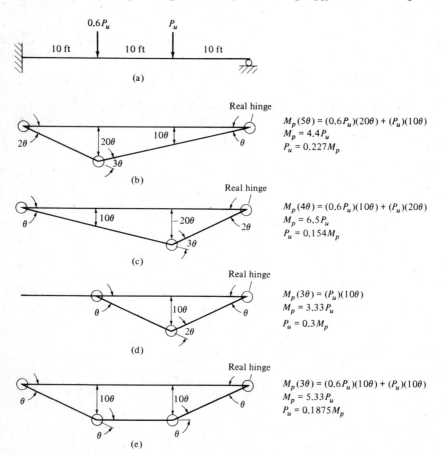

$M_p(5\theta) = (0.6P_u)(20\theta) + (P_u)(10\theta)$
$M_p = 4.4P_u$
$P_u = 0.227M_p$

$M_p(4\theta) = (0.6P_u)(10\theta) + (P_u)(20\theta)$
$M_p = 6.5P_u$
$P_u = 0.154M_p$

$M_p(3\theta) = (P_u)(10\theta)$
$M_p = 3.33P_u$
$P_u = 0.3M_p$

$M_p(3\theta) = (0.6P_u)(10\theta) + (P_u)(10\theta)$
$M_p = 5.33P_u$
$P_u = 0.1875M_p$

Figure 18-13

is considered by the virtual-work method. The beam with its two concentrated loads is shown with four possible collapse mechanisms and the necessary calculations. It is true that the mechanisms of Fig. 18-13(b), (d), and (e) do not control, but such a fact is not obvious to the average student until he or she makes the virtual-work calculations for each case. Actually the mechanism of part (e) is based on the assumption that the plastic moment is reached at both the concentrated loads simultaneously (a situation that might very well occur).

The value for which the collapse load is the smallest in terms of M_p is the correct value (or the value where M_p is the greatest in terms of P_u). For this beam the second plastic hinge forms at the P_u concentrated load, and P_u equals $0.154\, M_p$.

18-8 LOCATION OF PLASTIC HINGE FOR UNIFORM LOADINGS

There was no difficulty in locating the plastic hinge for the uniformly loaded fixed-end beam, but for other beams with uniform loads, such as propped or continuous beams, the problem may be rather difficult. For this discussion the uniformly loaded propped beam of Fig. 18-14(a) is considered.

The elastic moment diagram for this beam is shown in Fig. 18-14(b). As the uniform load is increased in magnitude, a plastic hinge will first form at the fixed end. At this time the beam will, in effect, be a "simple" beam with a

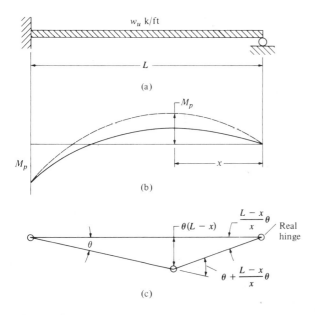

Figure 18-14

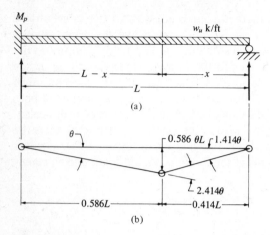

Figure 18-15

plastic hinge on one end and a real hinge on the other. Subsequent increases in the load will cause the moment to change as represented by the dotted line. This process will continue until the moment at some other point (a distance x from the right support in the figure) reaches M_p and creates another plastic hinge.

The virtual-work expression for the collapse mechanism of this beam shown in Fig. 18-14(c) is written

$$M_p\left(\theta + \theta + \frac{L-x}{x}\theta\right) = (w_u L)(\theta)(L-x)\left(\frac{1}{2}\right)$$

Solving this equation for M_p, taking $dM_p/dx = 0$, the value of x can be calculated to equal $0.414L$. This value is also applicable to uniformly loaded end spans of continuous beams with simple end supports.

The beam and its collapse mechanism are redrawn in Fig. 18-15, and the following expression for the plastic moment uses the virtual-work procedure:

$$M_p(\theta + 2.414\theta) = (w_u L)(0.586\theta L)(\tfrac{1}{2})$$
$$M_p = 0.0858 \, w_u L^2$$

PROBLEMS

18-1 to 18-10   Find the values of S and Z and the shape factor about the x axis for the sections shown.

18-1 (*Ans.* 265.4 in.^{3}, 309 in.^{3}, 1.16)

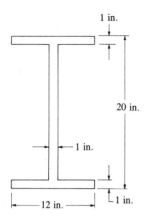

1 in.

20 in.

1 in.

1 in.

12 in.

18-2

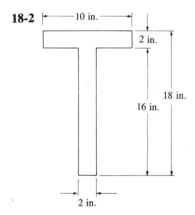

10 in.

2 in.

18 in.

16 in.

2 in.

18-3 (*Ans.* 268.7 in.³, 332 in.³, 1.24)

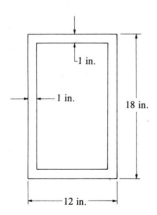

1 in.

1 in.

18 in.

12 in.

18-4

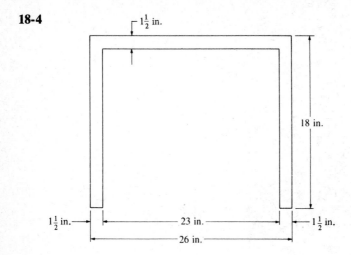

18-5 (*Ans.* 132.6 in.³, 240 in.³, 1.81)

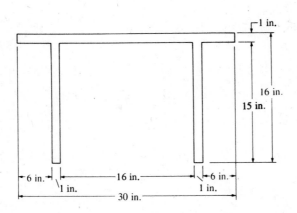

18-6

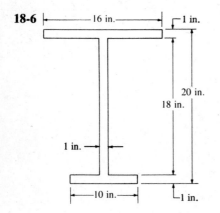

18-7 (*Ans.* 268.1 in.³, 311.8 in.³, 1.16)   **18-8**

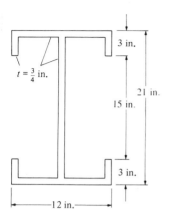

18-9 (*Ans.* 124 in.³, 168.8 in.³, 1.36)   **18-10**

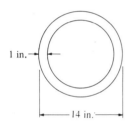

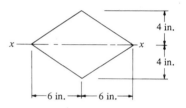

18-11 to 18-15   Determine the values of S and Z and the shape factor about the x axes for the situations described.

18-11 A W30 × 99. (*Ans.* 264.6 in.³, 307.1 in.³, 1.16)

18-12 A W27 × 114 with one PL $\frac{3}{4}$ × 16 in.³, on each flange.

18-13 Two L8 × 4 × 1s long legs vertical and back-to-back. (*Ans.* 28.0 in.³, 49.1 in.³, 1.75)

18-14 Two MC 12 × 40s back-to-back.

18-15 Four L8 × 8 × 1s arranged as shown. (*Ans.* 86.5 in.³, 142.0 in.³, 1.64)

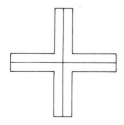

18-16 Rework Prob. 18-2 considering the *y* axis.

18-17 Rework Prob. 18-12 considering the *y* axis. (*Ans.* 83.8 in.³, 147.3 in.³, 1.76)

18-18 Rework Prob. 18-14 considering the *y* axis.

18-19 to 18-29   Select beam sections for the situations shown, using the elastic theory, A36 steel and ASD Specification, and assuming full lateral support. Calculate the factors of safety for the resulting beam sections according to the elastic and plastic theories. The loads shown in each figure are assumed to include the estimated beam weight.

18-19 (*Ans.* Without 0.9 rule: W18 × 50, 1.50, 2.27; with 0.9 rule: W21 × 44, 1.38, 2.15)

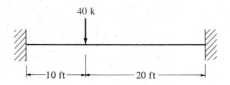

18-20

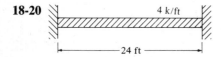

18-21 (*Ans.* Without 0.9 rule: W21 × 62, 1.50, 2.27; with 0.9 rule: W24 × 55, 1.35, 2.12)

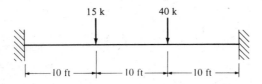

18-22

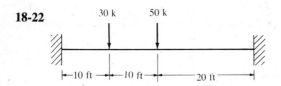

18-23 (*Ans.* Without 0.9 rule: W30 × 99, 1.63, 2.60; with 0.9 rule: W30 × 90, 1.48, 2.36)

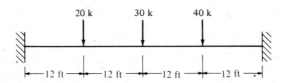

18-24

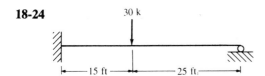

18-25 (*Ans.* Without 0.9 rule: W27 × 84, 1.55, 2.74; with 0.9 rule: W24 × 84, 1.43, 2.52)

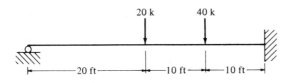

18-26

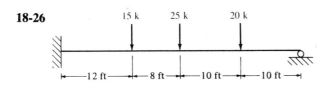

18-27 (*Ans.* Without 0.9 rule: W21 × 68, 1.49, 2.48; with 0.9 rule: W21 × 62, 1.35, 2.23)

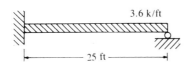

18-28

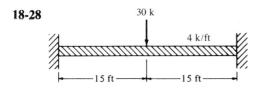

18-29 (*Ans.* Without 0.9 rule: W21 × 50, 1.55, 2.20; with 0.9 rule: W18 × 50, 1.45, 2.02)

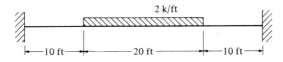

19

Plastic Analysis and Design

19-1 INTRODUCTION TO PLASTIC DESIGN

The shape factor multiplied by the safety factor was defined as the *load factor* in Chapter 18 and was shown to have an average value of approximately 1.70 for beams. To design a beam by the plastic method, the estimated working load is multiplied by the load factor; the plastic moment is calculated, and a member is selected which furnishes the required plastic modulus ($Z = M_p/F_y$). Example 19-1 presents a comparison of elastic and plastic designs for a uniformly loaded fixed-end beam.

■ Example 19-1
Design the beam shown in Fig. 19-1 by the plastic and elastic methods. The uniform load is 3 k/ft and includes the estimated beam weight. Use A36 steel and the ASD Specification, and assume full lateral support.

Solution.
Plastic analysis:

$$M_p(\theta + 2\theta + \theta) = (w_u L)\left(\frac{L}{2}\right)\left(\theta\frac{L}{2}\right)$$

$$M_p = \frac{w_u L^2}{16}$$

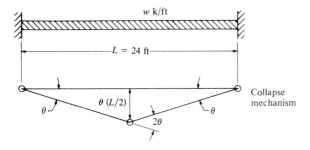

Figure 19-1

Plastic design:

$$w_u = (1.70)(3) = 5.1 \text{ k/ft}$$

$$M_p = \frac{(5.1)(24)^2}{16} = 183.6 \text{ ft-k}$$

$$Z_{reqd} = \frac{(12)(183.6)}{36} = 61.2 \text{ in.}^3$$

Use W18 × 35

Elastic design:

$$-M = \frac{wL^2}{12} = -\frac{(3)(24)^2}{12} = -144 \text{ ft-k}$$

$$+M = \frac{wL^2}{24} = \frac{(3)(24)^2}{24} = +72 \text{ ft-k}$$

$-M \text{ for design} = (0.9)(-144) = -129.6 \text{ ft-k}$

$+M \text{ for design} = 72 + (0.10)\left(\dfrac{144 + 144}{2}\right) = +86.4 \text{ ft-k}$ $\left.\begin{array}{c}\\[1.5em]\end{array}\right\}$ (ASD Section F1.1)

$$S_{reqd} = \frac{(12)(129.6)}{24} = 64.8 \text{ in.}^3$$

Use W18 × 40. Weight saving by plastic design = $\frac{5}{40}$ = 12.5% ∎

Plastic design is permissible for statically determinate structures but has little if any economic advantage. To illustrate this fact, the design of a simply supported beam by the two methods is presented in Example 19-2. The shape factor advantage (about 12 percent) has already been substantially used in the 10 percent higher allowable bending stress permitted by the ASD for compact laterally supported sections ($0.66F_y$ instead of $0.60F_y$). As only one plastic hinge is needed to cause a statically determinate structure to be unstable, there is no redistribution of load after the hinge forms.

■ **Example 19-2**

The beam shown in Fig. 19-2 is assumed to support a uniform load of 4 k/ft (including the estimated beam weight). Design the beam by the plastic and elastic methods using A36 steel and the ASD Specification, and assuming full lateral support.

Solution.
Plastic analysis:

$$M_p(2\theta) = (w_u L)\left(\frac{1}{2}\theta\frac{L}{2}\right)$$

$$M_p = \frac{W_u L^2}{8}$$

Plastic design:

$$w_u = (1.70)(4) = 6.8 \text{ k/ft}$$

$$M_p = \frac{(6.8)(20)^2}{8} = 340 \text{ ft-k}$$

$$Z_{\text{reqd}} = \frac{(12)(340)}{36} = 113.3 \text{ in.}^3$$

$$\underline{\underline{\text{Use W24} \times 55}}$$

Elastic design:

$$M = \frac{(4)(20)^2}{8} = 200 \text{ ft-k}$$

$$S_{\text{reqd}} = \frac{(12)(200)}{24} = 100 \text{ in.}^3$$

$$\underline{\underline{\text{Use W24} \times 55}}$$ ■

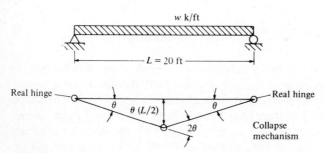

Figure 19-2

If all of the plastic hinges needed to produce the collapse mechanism for a statically indeterminate structure form at the same time, plastic design will usually yield the same section as will elastic design. For such situations there is no load redistribution after the first hinge forms. This same situation was shown to occur in statically determinate structures where the formation of one plastic hinge was sufficient to cause collapse (see Example 19-2).

Three plastic hinges are required to form a mechanism for a fixed-end beam, but the elastic moment diagram in Fig. 19-3 shows that, if such a beam is loaded with a concentrated load at its centerline, the hinges will all theoretically form at the same time. Example 19-3 illustrates the design of this beam. The student does not have to worry about overlooking a situation where there is no redistribution or where all of the plastic hinges form simultaneously. The virtual-work procedure will yield the correct expression for M_p regardless of the order in which the hinges are formed.

■ Example 19-3

Design the beam shown in Fig. 19-3 by the plastic and elastic methods using A36 steel, the ASD Specification, and full lateral support. Assume the load given takes into account the estimated beam weight.

Solution.

Plastic analysis:

$$M_p(\theta + 2\theta + \theta) = P_u(15\theta)$$
$$M_p = 3.75P_u$$

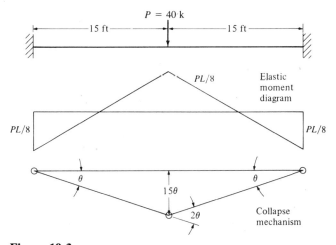

Figure 19-3

Plastic design:

$$P_u = (1.70)(40) = 68 \text{ k}$$
$$M_p = (3.75)(68) = 255 \text{ ft-k}$$

$$Z_{reqd} = \frac{(12)(255)}{36} = 85 \text{ in.}^3$$

Use W21 × 44

Elastic design (no advantage in 0.9 rule in ASD Section F1.1):

$$M = \frac{(40)(30)}{8} = 150 \text{ ft-k}$$

$$S_{reqd} = \frac{(12)(150)}{24} = 75 \text{ in.}^3$$

Use W21 × 44 ∎

Example 19-4 presents the design of a fixed-end beam with a concentrated load at a point other than midspan. For this beam there is appreciable load redistribution after the first and second hinges form.

■ Example 19-4

Design the beam shown in Fig. 19-4 by the plastic and elastic methods using A36 steel and the ASD Specification, and assuming full lateral support. Assume the load given takes into account the estimated beam weight.

Solution.
Plastic analysis:

$$M_p(1.5\theta + 2.5\theta + \theta) = (P_u)(18\theta)$$
$$M_p = 3.6P_u$$

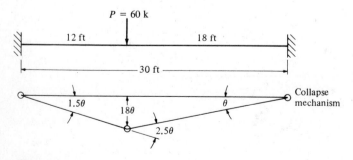

Figure 19-4

Plastic design:

$$P_u = (1.70)(60) = 102 \text{ k}$$
$$M_p = (3.6)(102) = 367.2 \text{ ft-k}$$
$$Z_{reqd} = \frac{(12)(367.2)}{36} = 122.4 \text{ in.}^3$$

Use W24 × 55

Elastic design:
    The elastic moment diagram is shown in Fig. 19-5.

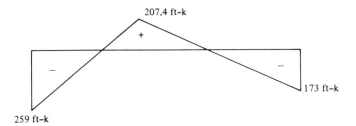

Figure 19-5

$$-M \text{ for design} = (0.9)(-259) = -233.1 \text{ ft-k}$$
$$+M \text{ for design} = +207.4 + (0.10)\left(\frac{259 + 173}{2}\right) = +229 \text{ ft-k}$$

$$S_{reqd} = \frac{(12)(233.1)}{24} = 116.6 \text{ in.}^3$$

Use W21 × 62 or W24 × 62. Weight saving by plastic design = $\frac{7}{62}$ = 11.3%

∎

19-2 ASD REQUIREMENTS FOR PLASTIC DESIGN

Chapters A to M of the ASD Specification present requirements for allowable-stress design, while Chapter N provides plastic-design requirements. Several of the more important items included are listed here.

1. Plastic design is permissible for simple and continuous beams and for braced and unbraced planar rigid frames. In addition, portions of structures rigidly constructed so as to be continuous over at least one interior support can be designed plastically.

2. The plastic design procedure can be used for steels having specified yield points up to 65 ksi or 448 MPa (A572, grade 65).

3. The load factors used in design may not be less than 1.7 for the given dead and live loads or less than 1.3 for those loads acting in conjunction with a load factor of 1.3 for any wind or seismic forces. (The 1.3 value is obtained by making use of the one-third increase in allowable stresses for wind and seismic loadings provided in ASD Section A5.2. This is the same thing as multiplying the load by 0.75 or in this case the load factor.)

4. The webs of beams and girders or columns that are not stiffened in some manner (as with diagonal stiffeners or doubler plates) shall be designed so that the ultimate shear V_u in kips does not exceed $0.55F_y t_w$, where t_w is the web thickness.

5. The members of a plastically designed structure must be prevented from local and lateral buckling until the plastic hinges develop. To prevent such buckling the ASD provides maximum width-thickness ratios for the flanges of S, W, and similar built-up shapes as well as maximum permissible depth-thickness ratios for the webs of such members.

The minimum width-thickness ratios $(b_f/2t_f)$ are given in Section N7 of the ASD Specification and equal 8.5 for $F_y = 36$ ksi, 8.0 for $F_y = 42$ ksi, and so on. In addition the width-thickness ratios of the flange plates in box sections and cover plates are not permitted to be greater than $190/\sqrt{F_y}$.

The depth-thickness ratio of the webs of members subject to plastic bending may not exceed the values given by the equations to follow, in which P_y equals the member area times F_y. If $P/P_y \leq 0.27$,

$$\frac{d}{t} = \frac{412}{\sqrt{F_y}}\left(1 - 1.4\frac{P}{P_y}\right) \qquad \text{(ASD Equation N7-1)}$$

If $P/P_y > 0.27$,

$$\frac{d}{t} = \frac{257}{\sqrt{F_y}} \qquad \text{(ASD Equation N7-2)}$$

6. To prevent lateral and torsional displacements at the plastic hinge locations associated with the collapse mechanism, the members must be braced in those locations and at distances from the hinges not exceeding the value L_{cr} given by the following equations. If $+1.0 > M/M_p > -0.5$,

$$\frac{L_{cr}}{r_y} = \frac{1375}{F_y} + 25 \qquad \text{(ASD Equation N9-1)}$$

If $-0.5 \geq M/M_p > -1.0$,

$$\frac{L_{cr}}{r_y} = \frac{1375}{F_y} \qquad \text{(ASD Equation N9-2)}$$

where

r_y = the r of the member about its weak axis, inches

M = smaller moment at the ends of the unbraced length foot-kips.

M/M_p = end moment ratio which is positive when the member is bent in reverse curvature and negative when bent in single curvature

These requirements do not have to be followed in the vicinity of the last hinge to form, but the lateral bracing requirements for elastic design must be met.

7. Web stiffeners are frequently required at plastic hinges to prevent hinges forming in a member at its point of connection to another member. For such situations it may be necessary to use web stiffeners as required by Section N6 of the ASD Specification.

8. If the edges of structural steel members in the vicinity of the plastic hinges are sheared, the edges should be ground, chipped, or planed smooth. Furthermore, holes in similar locations must be drilled or subpunched and reamed to full size.

9. There are several other ASD requirements for plastic design which refer in detail to columns, haunched members, connections, and so on.

Example 19-5 illustrates the design of a uniformly loaded propped beam using the ASD Specification. The W section selected in this problem is a compact shape, but the ratios required for such shapes are checked to illustrate their application.

■ Example 19-5
Design a section for the span and loading shown in Fig. 19-6, using A36 steel and the ASD Specification. Lateral support is assumed to be provided at 4-ft intervals. Also draw the shear and moment diagrams when the load w_u is applied to the section selected.

Solution.
Plastic analysis:

$$M_p(\theta + 2.414\theta) = (w_u)(24)(14.06\theta)(\tfrac{1}{2})$$
$$= 49.4w_u$$

Plastic design:

$$w_u = (1.70)(3) = 5.1 \text{ k/ft}$$

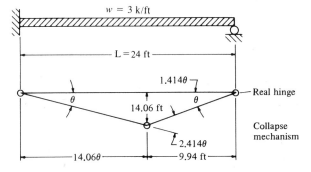

Figure 19-6

$$M_p = (49.4)(5.1) = 252 \text{ ft-k}$$

$$Z_{\text{reqd}} = \frac{(12)(252)}{36} = 84 \text{ in.}^3$$

Try W21 × 44 ($A = 13.0$ in.^{2}, $d = 20.66$ in., $b_f = 6.500$ in., $t_f = 0.450$ in., $t_w = 0.350$ in., $d/A_f = 7.06$, $r_y = 1.26$ in.)

Checking ASD requirements:

Shear and moment diagrams just before collapse shown in Fig. 19-7.

Shear

$$0.55 F_y t d = (0.55)(36.0)(0.350)(20.66) = 143.2 \text{ k} > 73.1 \text{ k} \qquad \text{OK}$$

Local buckling

$$\frac{b_f}{2t_f} = \frac{6.500}{(2)(0.450)} = 7.2 < 8.5 \qquad \text{OK}$$

$$\frac{P}{P_y} = \frac{0}{P_y} \text{ since no axial load}$$

∴ Use ASD Equation N7-1 as follows:

$$\frac{d}{t} = \frac{20.66}{0.350} = 59.0 < \frac{412}{\sqrt{36}}\left(1 - 1.4\frac{0}{P_y}\right) = 68.7 \qquad \text{OK}$$

Lateral buckling: plastic hinge at support

$$1.0 > \frac{M}{M_p} = -\frac{34}{286} = -0.119 > -0.5$$

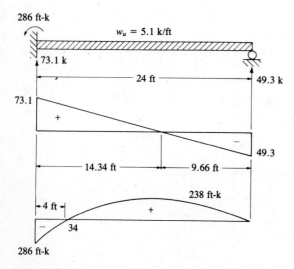

286 ft-k

$w_u = 5.1$ k/ft

73.1 k

24 ft

49.3 k

73.1

+

−

49.3

14.34 ft

9.66 ft

238 ft-k

+

4 ft

−

34

286 ft-k

Figure 19-7

∴ Use ASD Equation N9-1 as follows:

$$\frac{L_{cr}}{r_y} = \frac{4 \times 12}{1.26} = 38.1 < \frac{1375}{36} + 25 = 63.2 \qquad \text{OK}$$

Interior plastic hinge is the last one to form (∴ see ASD Section N9).

$$\left.\begin{array}{l} L_c = \dfrac{76.0 \times 6.50}{\sqrt{36}} = 82.3 \\[4mm] L_c = \dfrac{20,000}{7.60 \times 36} = 78.7 \end{array}\right\} \quad \text{Both} > 4 \times 12 = 48 \text{ in.} \qquad \text{OK}$$

$$\underline{\underline{\text{Use W21} \times 44}} \qquad\qquad \blacksquare$$

19-3 CONTINUOUS BEAMS

Continuous beams are very common in engineering structures. Their continuity causes analysis to be rather complicated in the elastic theory, and even though one of the complex "exact" methods is used for analysis, the resulting stress distribution is not nearly so accurate as is usually assumed.

   Plastic analysis is applicable to continuous structures as it is to one-span structures. The resulting values definitely give a more realistic picture of the limiting strength of a structure than can be obtained by elastic analysis. Continuous statically indeterminate beams can be handled by the virtual-work procedure as were the single-span statically indeterminate beams. As an introduction to continuous beams, Examples 19-6 and 19-7 illustrate two of the more elementary cases. These designs should be checked for shear, local buckling, and so on, as was Example 19-5, but space is not taken to show the calculations.

■ Example 19-6

Select a steel shape for the two-span beam shown in Fig. 19-8, using A36 steel and the ASD Specification, and assuming full lateral support.

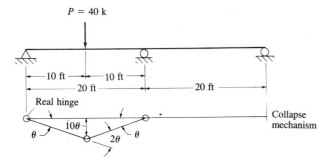

Figure 19-8

Solution.
Plastic analysis:

$$M_p(2\theta + \theta) = (P_u)(10\theta)$$
$$= 3.33P_u$$

Plastic design:

$$P_u = (1.70)(40) = 68 \text{ k}$$
$$M_p = (3.33)(68) = 226.7 \text{ ft-k}$$
$$Z_{\text{reqd}} = \frac{(12)(226.7)}{36} = 75.6 \text{ in.}^3$$

Use W18 × 40 ∎

■ Example 19-7

Select a section for the beam shown in Fig. 19-9. Use A36 steel and the ASD Specification, and assume full lateral support.

Solution.
Plastic analysis:

$$M_p(2.414\theta + \theta) = (w_u)(24)(14.06\theta)(\tfrac{1}{2})$$
$$= 49.4w_u$$

Plastic design:

$$w_u = (1.7)(4) = 6.8 \text{ k/ft}$$
$$M_p = (49.4)(6.8) = 336 \text{ ft-k}$$
$$Z_{\text{reqd}} = \frac{(12)(336)}{36} = 112 \text{ in.}^3$$

Use W18 × 55 or W24 × 55 ∎

Additional spans have little effect on the amount of work involved in the plastic design procedure. The same cannot be said for elastic design. Example

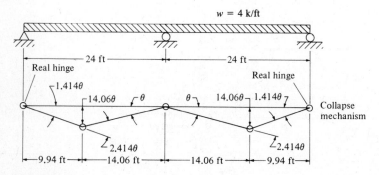

Figure 19-9

19-8 illustrates the design of a three-span beam that has a concentrated load on each span. Students from their knowledge of elastic analysis can see that plastic hinges will initially form at the first interior supports and then at the centerlines of the end spans, at which time each end span will have a collapse mechanism.

■ Example 19-8

Select a steel section for the beam and loading shown in Fig. 19-10, using A36 steel and the ASD Specification, and assuming full lateral support.

Solution.
Plastic analysis:

$$M_p(2\theta + \theta) = P_u(15\theta)$$
$$= 5P_u$$

Plastic design:

$$P_u = (1.70)(30) = 51 \text{ k}$$
$$M_p = (5)(51) = 255 \text{ ft-k}$$

$$Z_{reqd} = \frac{(12)(255)}{36} = 85 \text{ in.}^3$$

$$\underline{\underline{\text{Use W21} \times 44}}$$ ■

A uniformly loaded two-span beam with equal spans was designed in Example 19-7. A uniformly loaded two-span beam with unequal spans is designed in Example 19-9. If the same-size section is used for both spans, the longer span will obviously collapse first, and the shorter span will be decidedly overdesigned. Nevertheless this procedure is used in this example.

■ Example 19-9

Select a section for the beam shown in Fig. 19-11, using A36 steel and the ASD Specification, and assuming satisfactory lateral support.

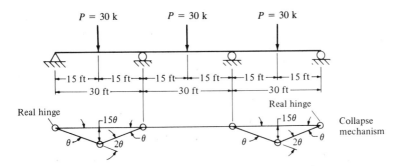

Figure 19-10

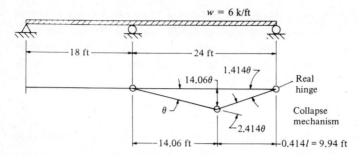

Figure 19-11

Solution.
Plastic analysis:

$$M_p(\theta + 2.414\theta) = (24w_u)(14.06\theta)(\tfrac{1}{2})$$
$$= 49.4w_u$$

Plastic design:

$$w_u = (1.70)(6) = 10.2 \text{ k/ft}$$
$$M_p = (49.4)(10.2) = 504 \text{ ft-k}$$

$$Z_{reqd} = \frac{(12)(504)}{36} = 168 \text{ in.}^3$$

Use W24 × 68 ∎

It may be more economical to select a section for the shorter span with its smaller moment and run it through both spans. Cover plates can be added in the longer span where the moment is greater than the plastic resisting moment of the section selected. This procedure is illustrated in Example 19-10 for the beam designed in Example 19-9. In the example a W21 × 44 with a plastic moment resistance of 286.2 ft-k is selected for the shorter span. The moment diagram for this beam (Fig. 19-13) shows that cover plates are needed for 15.64 ft in the right-hand span where the moment exceeds 286.2 ft-k.

The derivation of an expression for determining cover plate sizes is quite similar to the derivation presented in Section 17-2. For this discussion Z is the required plastic modulus for the entire moment, Z_s the plastic modulus for the section selected for the shorter span, d the depth of the section selected for the shorter span, t_p the thickness of one cover plate, and A_p the area of one cover plate. An expression for the required area A_p can be written as

$$Z = Z_s + 2A_p\left(\frac{d}{2} + \frac{t_p}{2}\right)$$

$$A_p = \frac{Z}{d + t_p} - \frac{Z_s}{d + t_p}$$

The cover plates should probably be extended a little distance beyond their theoretical points of cutoff. Some designers extend the plates 6 to 12 in. on each end, while others extend them until half of the strength of the plates is developed by the welds connecting them to the beam.

From a standpoint of economy some consideration should be given to using the W24 × 68 (selected for the 24-ft span in Example 19-9) for both spans. Its weight for the two spans totals 2856 lb, while the weight of the W21 × 44 with the cover plates in the longer span (selected in Example 19-10) weighs 2426 lb. The use of cover plates saved 430 lb of steel, but their connection costs may very well have exceeded the cost of 430 lb of structural steel.

■ Example 19-10

Select a section for the 18-ft span of the beam of Example 19-9 (see Fig. 19-12) and design the necessary cover plates for the 24-ft span.

Solution.

Plastic analysis for 18-ft span:

$$M_p(2.414\theta + \theta) = (18w_u)(10.55\theta)(\tfrac{1}{2})$$
$$= 27.8w_u$$

Plastic design for 18-ft span:

$$w_u = (1.70)(6) = 10.2 \text{ k/ft}$$
$$M_p = (27.8)(10.2) = 283.6 \text{ ft-k}$$

$$Z_{reqd} = \frac{(12)(283.6)}{36} = 94.5 \text{ in.}^3$$

$$\underline{\text{Use W21} \times 44 \ (Z = 95.4 \text{ in.}^3, \ d = 20.66 \text{ in.})}$$

$$\text{Actual } M_p = \frac{(36)(95.4)}{12} = 286.2 \text{ ft-k}$$

Drawing of moment diagram is shown in Fig. 19-13.

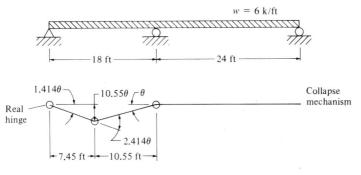

Figure 19-12

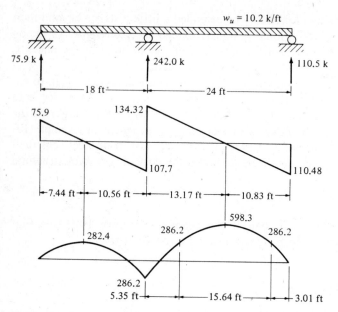

Figure 19-13

Design of cover plates for long span:

$$Z_{reqd} = \frac{(12)(598.3)}{36} = 199.4 \text{ in}^3$$

Assuming $t_p = \frac{1}{2}$ in.

$$A_p = \frac{199.4}{20.66 + 0.50} - \frac{95.4}{20.66 + 0.50} = 4.91 \text{ in.}^2$$

Try $\frac{1}{2} \times$ 10-in. cover plates ($A_p = 5.0$ in.²)

$$Z = 95.4 + (2)(5.0)(10.58) = 201.2 \text{ in.}^3 > 199.4 \text{ in.}^3 \qquad \text{OK}$$

Use $\frac{1}{2} \times$ 10-in. cover plates 17.0 ft long ■

19-4 PLASTIC ANALYSIS OF FRAMES

In this section three small frames are analyzed using the plastic method. In each case it is assumed that the columns and beams are the same sizes. Should they be different sizes, it will be necessary to take that into account in the analysis (that is, the M_p values will be different in the various members).

The pin-supported frame of Fig. 19-14(a) is statically indeterminate to the first degree. The development of one plastic hinge will cause it to become statically determinate, while the forming of a second hinge can create a mechanism. There are, however, several possible mechanisms that might feasibly occur in the frame. A possible beam mechanism is shown in Fig. 19-14(c), a

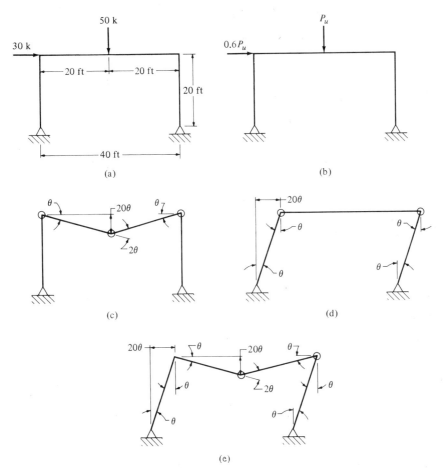

Figure 19-14 Possible mechanisms of a frame.

sidesway mechanism is shown in part (d), and a possible combination beam and sidesway mechanism in part (e). The critical condition is the one that will cause the smallest value of P_u.

Example 19-11 presents the plastic analysis of the frame of Fig. 19-14. The distances through which the loads move in the virtual-work procedure should be carefully studied. *The solution of this problem shows that superposition does not apply to plastic analysis, a point of major significance.* The situation shown in Fig. 19-14(c) added to the situation of part (d) does not equal that of part (e). In other words, the effects of different loads cannot be determined separately and added together as they can in elastic analysis. The three-fourths ASD rule, representing a one-third increase in allowable stresses when lateral loads are involved, is used in this problem.[1]

---

[1] The ASD Specification reflects the $\frac{3}{4}$ reduction by permitting the load factor of 1.7 for frames to be reduced to 1.3 when lateral loads are involved. Using these load factors in design, it is unnecessary to use the $\frac{3}{4}$ value as done in Examples 19-11 through 19-13.

■ Example 19-11

Determine an expression for the plastic moment in the frame of Fig. 19-14.

Solution.
Case (c) of figure:

$$M_p(\theta + 2\theta + \theta) = P_u(20\theta)$$
$$M_p = 5P_u$$

Case (d) of figure:

$$M_p(\theta + \theta) = (\tfrac{3}{4})(0.6P_u)(20\theta)$$
$$M_p = 4.5P_u$$

Case (e) of figure:

$$M_p(2\theta + 2\theta) = (\tfrac{3}{4})(0.6P_u)(20\theta) + (\tfrac{3}{4})(P_u)(20\theta)$$
$$M_p = 6P_u \quad \leftarrow$$

Critical value is the smallest value of P_u (or the largest value of M_p in terms of P_u).

$$M_p = 6P_u$$
$$\underline{P_u = 0.167M_p}$$ ■

Example 19-12 is similar to Example 19-11 except that there are even more possible collapse mechanisms. Fig. 19-16(a), (b), and (c) show possible beam collapse mechanisms, while part (d) shows a sidesway mechanism. Fig. 19-16(e) and (f) show possible combination sidesway and beam mechanisms. Study carefully the angles used for parts (e) and (f).

■ Example 19-12

Determine the critical value of P_u for the frame shown in Fig. 19-15.

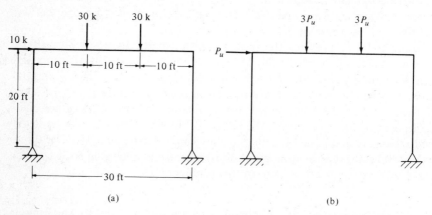

(a)   (b)

Figure 19-15

Solution. Possible collapse mechanisms are shown in Fig. 19-16.

$$\text{Critical value of } P_u = (1/16.25)M_p$$ ∎

Example 19-13 illustrates the analysis of a frame with fixed supports. The fact that the frame has two more redundants than the hinged frames of the last two examples in no way makes its plastic analysis more difficult.

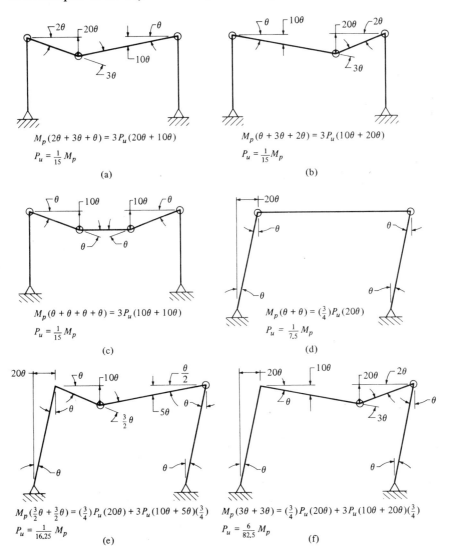

$$M_p(2\theta + 3\theta + \theta) = 3P_u(20\theta + 10\theta)$$
$$P_u = \tfrac{1}{15}M_p$$
(a)

$$M_p(\theta + 3\theta + 2\theta) = 3P_u(10\theta + 20\theta)$$
$$P_u = \tfrac{1}{15}M_p$$
(b)

$$M_p(\theta + \theta + \theta + \theta) = 3P_u(10\theta + 10\theta)$$
$$P_u = \tfrac{1}{15}M_p$$
(c)

$$M_p(\theta + \theta) = (\tfrac{3}{4})P_u(20\theta)$$
$$P_u = \tfrac{1}{7.5}M_p$$
(d)

$$M_p(\tfrac{3}{2}\theta + \tfrac{3}{2}\theta) = (\tfrac{3}{4})P_u(20\theta) + 3P_u(10\theta + 5\theta)(\tfrac{3}{4})$$
$$P_u = \tfrac{1}{16.25}M_p$$
(e)

$$M_p(3\theta + 3\theta) = (\tfrac{3}{4})P_u(20\theta) + 3P_u(10\theta + 20\theta)(\tfrac{3}{4})$$
$$P_u = \tfrac{6}{82.5}M_p$$
(f)

Figure 19-16

■ **Example 19-13**
Determine the critical value of P_u for the frame shown in Fig. 19-17.

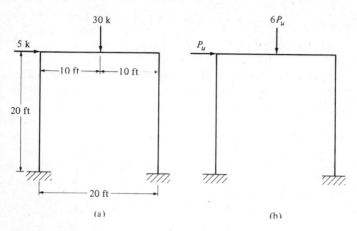

Figure 19-17

Solution. Possible collapse mechanisms are shown in Fig. 19-18.

$$\text{Critical value of } P_u = (1/15)M_p \qquad \blacksquare$$

For many frames the locations of the plastic hinges are not obvious. The usual procedure for such cases is to assume a mechanism and compute the value of M_p. Should the correct hinge location have been assumed, the value of M_p will be correct. If the wrong hinge location was assumed, the value of M_p will be too small.

Several other mechanisms can be assumed and M_p quickly computed for each. The largest value of M_p computed will be the one used. Although the mechanism finally used may not be exactly the right one, it will be satisfactory for all practical purposes. (It might be noted that the moment diagram can be drawn as a check for any one of the assumed conditions. If the moment at no point in the frame exceeds the calculated value of M_p, the correct mechanism has been used.)

For a frame supporting concentrated and uniform loads, the expected positions of the plastic hinges will be at the concentrated loads. When a structure supports only uniform loads, there will be a large (actually unlimited) number of possible hinge locations. After a few mechanism locations have been assumed and M_p determined for each, it is possible to narrow down the "critical hinge" location to within a foot or two.

Shear stresses are often high within the boundaries of the rigid connection of two or more members whose webs lie in a common plane. Such a situation occurs at the rigid frame knee in Fig. 19-19(a). Since most of the moment in a beam is resisted by the flanges, an internal resisting couple is assumed to be located at the centers of the flanges with a lever arm of $0.95d_b$ as shown in part (b). The magnitude of each of these forces V can be determined as $V = M_p/0.95d_b$.

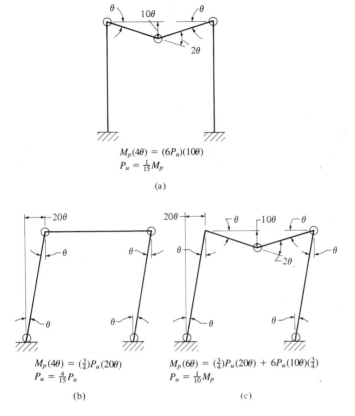

$$M_p(4\theta) = (6P_u)(10\theta)$$
$$P_u = \tfrac{1}{15}M_p$$

(a)

$$M_p(4\theta) = (\tfrac{3}{4})P_u(20\theta)$$
$$P_u = \tfrac{4}{15}P_u$$

(b)

$$M_p(6\theta) = (\tfrac{3}{4})P_u(20\theta) + 6P_u(10\theta)(\tfrac{3}{4})$$
$$P_u = \tfrac{1}{10}M_p$$

(c)

Figure 19-18

This force is applied perpendicularly to the column as a shearing force. The column must have a shearing strength equal to V. If not, it will be necessary to use a larger column, diagonal stiffeners as shown in Fig. 19-19(c), or doubler plates in contact with the web over the connection area.

The ultimate shearing strength of the column web ($0.55\ F_y t d_c$) must be equal to V or stiffening is required. From this information the required web thickness can be determined. In the expression developed, A_{bc} equals the planar area of the connection web ($d_b d_c$). If M_p is assumed to be in foot-kips, the following expression can be developed:

$$0.55 F_y t d_c = \frac{M_p(12)}{0.95 d_b}$$

$$t = \frac{23 M_p}{A_{bc} F_y}$$

Should the shearing strength of the column web be insufficient, the diagonal stiffeners shown in Fig. 19-19(c) can be used. These stiffeners act like the

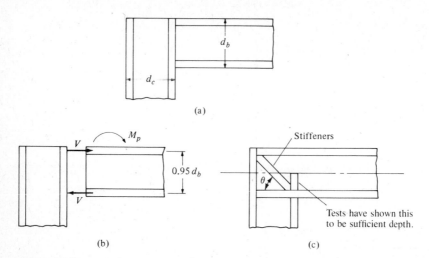

Figure 19-19 Rigid frame knee.

agonals of a truss to carry the excess shear. The flange force V must be resisted by the web and the stiffener. An expression for the required area of the stiffener A_{st} is developed as follows:

$$\frac{M_p}{0.95d_b} = 0.55F_y td_c + F_y A_{st} \cos \theta$$

$$A_{st} = \frac{1}{\cos \theta}\left(\frac{M_p}{0.95d_b F_y} - 0.55td_c\right)$$

A similar discussion can be made for the elastic design of rigid knee connections and the equations presented in Section E6 of the ASD Commentary.

PROBLEMS

Use A36 steel and assume full lateral support for all of the beams and frames shown. The working loads given in each case include estimates of the structure weight.

Probs. 19-1 to 19-8 Select sections by the elastic and plastic methods for each of the beams shown.

19-1 (*Ans.* W24 × 76; with 0.9 rule: W24 × 68, W24 × 55)

19-2

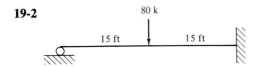

80 k

15 ft   15 ft

19-3 (*Ans.* W21 × 68; with 0.9 rule: W21 × 62, W24 × 55)

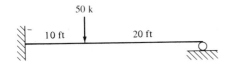

50 k

10 ft   20 ft

19-4

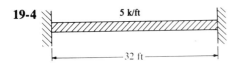

5 k/ft

32 ft

19-5 (*Ans.* W30 × 108; with 0.9 rule: W30 × 99, W24 × 84)

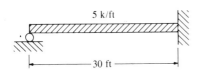

5 k/ft

30 ft

19-6

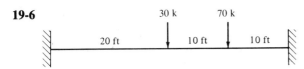

30 k   70 k

20 ft   10 ft   10 ft

19-7 (*Ans.* W30 × 116; with 0.9 rule: W30 × 108, W30 × 90)

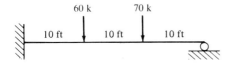

60 k   70 k

10 ft   10 ft   10 ft

19-8

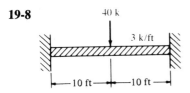

40 k

3 k/ft

10 ft   10 ft

Probs. 19-9 to 19-15   Select by the plastic method a single rolled section that will be satisfactory for the entire structure for each of the beams shown.

19-9 (*Ans.* W24 × 55)

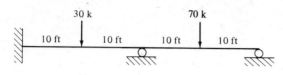

19-10

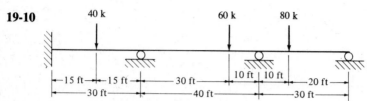

19-11 (*Ans.* W24 × 55)

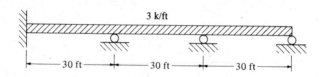

19-12

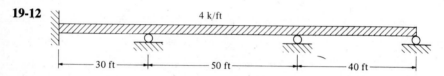

19-13 (*Ans.* W21 × 68)

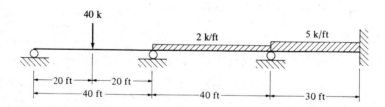

19-14

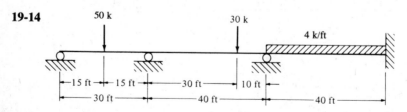

19-15 (*Ans.* W18 × 50)

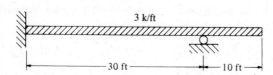

Probs. 19-16 to 19-18   Select a single rolled section for the beams shown, and draw shear and moment diagrams for the sections selected, using the collapse loads used in design. The service loads shown include the estimated beam weights. Use plastic method only.

19-16

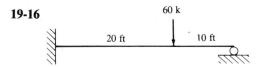

19-17 (*Ans.* W21 × 68)

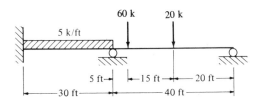

19-18

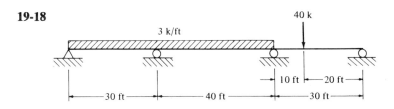

Probs. 19-19 to 19-27   Using plastic design, select a single rolled section for each of the beams shown which will be satisfactory for the entire structure. Then select a section for the span that has the smallest moment, and design cover plates as necessary for the other parts of the beam. Compare the weight differences between the two procedures. Use plastic method only.

19-19 (*Ans.* W18 × 50, W18 × 35 with PL$\frac{1}{2}$ × 5 × 1 ft 0 in. span 2)

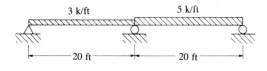

19-20

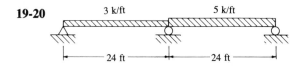

19-21 (*Ans.* W21 × 44, W16 × 40 with PL$\frac{1}{2}$ × 2 × 7 ft 0 in. span 1; first solution better)

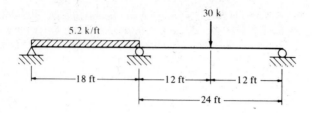

19-22

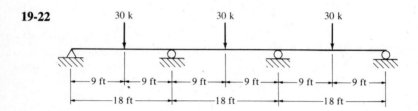

19-23 (*Ans.* W18 × 40, W18 × 35 with $\frac{1}{2}$ × 2 × 9 ft 0 in. PLs spans 1 and 3)

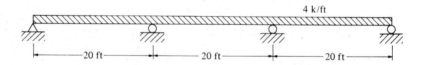

19-24
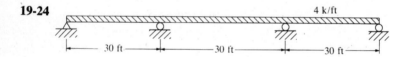

19-25 (*Ans.* W18 × 35, W14 × 22 with PL$\frac{1}{2}$ × 3 × 10 ft 0 in. in span 1 and PL$\frac{5}{8}$ × 4 × 14 ft 0 in. in span 3)

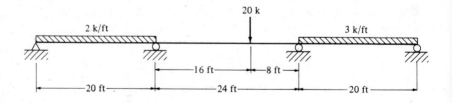

19-26
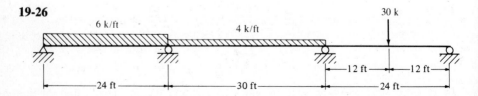

19-27 (*Ans.* W16 × 31, W12 × 26 with PL$\frac{1}{2}$ × 3 × 10 ft 0 in. in span 3)

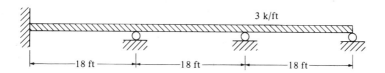

Probs. 19-28 to 19-33   Select sections using the plastic method for the frames shown, considering moment only. Assume beam and column sizes are the same for each frame.

19-28   **19-29** (*Ans.* W18 × 35)

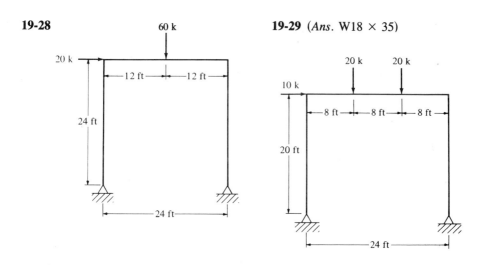

19-30

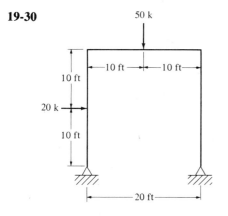

19-31 (*Ans.* W14 × 22)

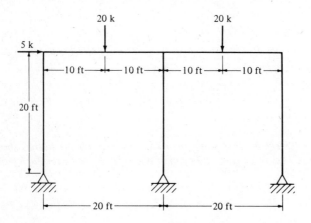

19-32

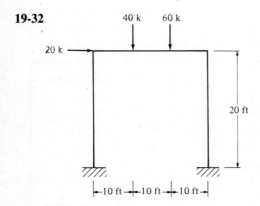

19-33 (*Ans.* W18 × 35)

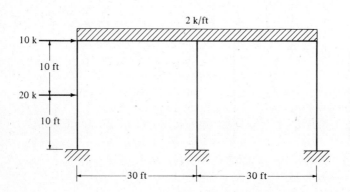

20

Tension Members—LRFD

20-1 GENERAL

Chapters 20–23 make constant reference to the *Manual of Steel Construction Load and Resistance Factor Design* (copyright 1986), published by the American Institute of Steel Construction. This manual is referred to as the LRFD Manual in the pages that follow. Reference is made here to the first edition of the manual, which is based on the September 1, 1986, "Load and Resistance Factor Design Specification for Structural Steel Buildings."

It is assumed that the reader has already studied the initial chapters of this book in which the ASD method was presented for tension members, columns, beams, and connections. Thus the author has attempted as much as possible to omit topics that are handled exactly as they were earlier. He has in mind such items as net areas, effective net areas, properties of built-up sections, and so on.

20-2 LOAD AND RESISTANCE FACTOR DESIGN

Load and resistance factor design includes many of the features commonly associated with ultimate strength design, plastic design, limit design, or collapse design. It is based on a limit states philosophy. The term *limit state* describes a condition at which a structure or some part of that structure ceases to perform its intended function. There are actually two categories of limit states: strength and serviceability.

Strength limit states are based on the safety or load-carrying capacity of structures and include plastic strengths, buckling, fracture, fatigue, overturning, and so on.

Serviceability limit states refer to the performance of structures under normal service loads and are concerned with the uses and/or occupancy of structures, including such items as excessive deflections, slipping, vibrations, and cracking.

Not only must the structure be capable of supporting the design or ultimate loads, but it must also be able to support the service or working loads in such a manner as to meet the requirements of the users or occupants of the structure. For instance, a high-rise building must be designed so that lateral deflections or drift are not excessive during ordinary storms; that is, the occupants must not become uncomfortable (frightened or seasick). As to the strength limit state, the frame will be designed so that it will safely resist the ultimate load occurring during that once in 50-year storm, although there may be some minor damage to the building and the occupants may suffer some discomfort.

The LRFD Specification concentrates on very specific requirements relating to the strength limit state and allows the designer some freedom of judgment in the serviceability area. This doesn't mean that the serviceability limit state is not significant, but by far the most important consideration (as in all structural specifications) is the life and property of the public. As a result public safety is not left to the judgment of the individual designer.

In LRFD the working or service loads (Q_i) are multiplied by certain load or safety factors (λ_i) almost always larger than 1.0 and the resulting "factored loads" used for designing the structure. The magnitudes of load factors vary depending on the type and combination of the loads and are discussed in detail in Section 20-3.

The structure is proportioned to have a design ultimate strength sufficient to support the factored loads. This strength is considered to equal the nominal or theoretical strength of the member (R_n) multiplied by a resistance or overcapacity factor (ϕ), which is normally less than 1.0, with which the designer attempts to account for the uncertainties in material strengths, dimensions, and construction. Furthermore these factors have been adjusted somewhat to ensure more uniform reliability in design as described in Section 20-5. Values of ϕ factors for different situations are discussed in detail in Section 20-4.

For a particular member the preceding information can be summarized as follows: (sum of products of load effects and load factors) ≤ (resistance factor) (nominal resistance),

$$\Sigma \lambda_i Q_i \leq \phi R_n$$

The left-hand side of this expression refers to the load effects on the structure, while the right-hand side refers to the resistance or capacity of the member.

In Canada the LRFD Method has been used since 1974 and, in fact, is the only method used there since 1978. Much progress has been made in European countries toward the goal of formulaing LRFD procedures in their national codes.[1] In the U.S. the use of LRFD has gradually increased since the AISC published its specifications in 1986.

At Washington University in St. Louis, a research project was conducted on LRFD from 1969 through 1976 under the direction of T. V. Galambos and M. K. Ravindra. At the conclusion of the project, a document was published entitled "Proposed Criteria for Load and Resistance Factor Design of Steel Building Structures."[2]

20-3 LOAD FACTORS

The purpose of load factors is to increase the loads to account for the uncertainities involved in estimating the magnitudes of dead or live loads. For instance, How close in percentage could you the reader estimate the worst wind or snow loads that will ever be applied to the building which you are now occupying?

The value of the load factor used for dead loads is smaller than the one used for live loads because designers can estimate so much more accurately the magnitudes of dead loads than they can the magnitudes of live loads. In this regard the student will notice that loads which remain in place for long periods will be less variable in magnitude, while those which are applied for brief periods, such as wind or seismic loads, will have larger variations. It is hoped that the LRFD design procedure will make the designer more conscious of load variations than he or she is with the allowable-stress design method.

The LRFD Specification presents load factors and load combinations that were selected to be used with the recommended minimum loads given in the ASCE Specification 7-88 (formerly the American National Standards Institute or ANSI Specification A58.1).[3]

The usual load combinations to be considered with the LRFD are given in LRFD Specification A4.1 by formulas A4-1 and A4-2. In these expressions the abbreviations used are D for dead loads, L for live loads, L_r for roof live loads, S for snow loads, and R for initial rainwater or ice, not including a ponding contribution. The letter U represents the ultimate load.

$$U = 1.4D \qquad \text{(LRFD Formula A4-1)}$$

$$U = 1.2D + 1.6L + 0.5(L_r \text{ or } S \text{ or } R) \qquad \text{(LRFD Formula A4-2)}$$

---

[1] M. K. Ravindra and T. V. Galambos, "Load and Resistance Factor Design for Steel," *Journal of Structural Division,* ASCE, 104, ST9 (September 1978), pp. 1337–1353.

[2] Research Report 45, C.E. Dept., Washington University, St. Louis, May 1976.

[3] *American Society of Civil Engineers Minimum Design Loads for Buildings and Other Structures,* ASCE 7–88 (New York: ASCE), p. 3.

Impact loads will be included only in the second of these combinations. Should wind (W) or earthquake (E) forces be involved, the following combinations need to be considered:

$$U = 1.2D + 1.6(L_r \text{ or } S \text{ or } R) + (0.5L \text{ or } 0.8W) \quad \text{(LRFD Formula A4-3)}$$
$$U = 1.2D + 1.3W + 0.5L + 0.5(L_r \text{ or } S \text{ or } R) \quad \text{(LRFD Formula A4-4)}$$
$$U = 1.2D + 1.5E + (0.5L \text{ or } 0.2S) \quad \text{(LRFD Formula A4-5)}$$

It is necessary to consider impact loading only in Formula A4-5. There is a change in the value of the load factor for L for Formulas A4-3, A4-4, and A4-5 for garages, public assembly areas, and all areas where the live load exceeds 100 psf. For such cases 1.0 is to be used.

To account for possibilities of uplift, another load combination is given in the LRFD Specification. This condition is included to cover cases where tension forces develop owing to overturning moments. It will govern only for tall buildings where high lateral loads are present. In this combination the dead loads are reduced by 10 percent to take into account situations where they may have been overestimated.

$$U = 0.9D - (1.3W \text{ or } 1.5E) \quad \text{(LRFD Formula A4-6)}$$

The magnitudes of the loads (D, L, L_r, etc.) should be obtained from the governing building code or from ASCE 7-88 (*Minimum Design Loads for Buildings and Other Structures.*) Whenever applicable the live loads used for design should be the reduced values specified for large floor areas, multistory buildings, and so on.

A brief illustration of the calculation of the critical loading for a building column using the LRFD load combinations is presented in Example 20-1. The largest value obtained is to be used for the design.

■ Example 20-1

The various axial loads for a building column have been computed according to the applicable building code with the following results: dead load = 200 k, load from roof = 50 k (live load or snow or rainwater), live load from floors (has been reduced as applicable for large floor area and multistory columns) = 250 k, wind = 80 k, and earthquake = 60 k. Determine the critical design load using the six LRFD load combinations.

Solution.

A4-1 $U = (1.4)(200) = 280 \text{ k}$
A4-2 $U = (1.2)(200) + (1.6)(250) + (0.5)(50) = 665 \text{ k} \leftarrow$
A4-3(a) $U = (1.2)(200) + (1.6)(50) + (0.5)(250) = 445 \text{ k}$
A4-3(b) $U = (1.2)(200) + (1.6)(50) + (0.8)(80) = 384 \text{ k}$
A4-4 $U = (1.2)(200) + (1.3)(80) + (0.5)(250) + (0.5)(50) = 494 \text{ k}$
A4-5 $U = (1.2)(200) + (1.5)(60) + (0.5)(250) = 455 \text{ k}$

A4-6(a) $U = (0.9)(200) - (1.3)(80) = 76$ k
A4-6(b) $U = (0.9)(200) - (1.5)(60) = 90$ k

The critical factored load combination or design strength required for this column is 665 k as determined by LRFD Formula A4-2. It will be noted that the results of Formula A4-6 do not indicate an uplift problem.   ■

20-4 RESISTANCE FACTORS

To accurately estimate the ultimate strength of a structure, it is considered necessary to take into account the uncertainties in material strengths, dimensions, and workmanship. With a resistance factor the designer attempts to show that the strength of a member cannot be computed exactly because of imperfections in analysis theory (remember the imperfect assumptions you made to analyze trusses and other members), variations in material properties, and imperfect dimensions of members.

This is done by multiplying the theoretical ultimate strength (called the *nominal strength* here) of each member by a resistance or overcapacity factor ϕ, which is almost always less than 1.0. These values are 0.85 for columns, 0.75 or 0.90 for tension members, 0.90 for bending or shear in beams, and so on.

Typical resistance factors from the LRFD Specification are shown in Table 20-1. The magnitudes of the resistance factors given in the LRFD Specification are based on research recommendations from investigators at Washington University in St. Louis.

TABLE 20-1 TYPICAL RESISTANCE FACTORS

RESISTANCE OR ϕ FACTORS	SITUATIONS
1.00	Bearing on the projected areas of pins, web yielding under concentrated loads, slip-resistant bolt shear values
0.90	Beams in bending and shear, fillet welds with stress parallel to weld axis, groove welds base metal
0.85	Columns, web crippling, edge distance and bearing capacity at holes
0.80	Shear on effective area of full-penetration groove welds, tension normal to effective area of partial-penetration groove welds
0.75	Bolts in tension, plug, or slot welds, fracture in the net section of tension members
0.65	Bearing on bolts (other than A307)
0.60	Bearing A307 bolts, bearing on concrete foundations

20-5 RELIABILITY AND THE LRFD SPECIFICATION

The word *reliability* as used in Chapters 20–23 refers to the estimated percentage of times that the strength of a structure will equal or exceed the maximum loading applied to that structure during its estimated life (say 50 years).

In this section the author briefly describes how

1. LRFD investigators developed a procedure for estimating the reliability of given designs.
2. They set what to them were desirable reliability percentages for different situations.
3. They were able to adjust resistance or ϕ factors, so that steel designers are able to obtain the reliability percentages established in 2 above.

Before this discussion proceeds, a few comments are presented concerning the word *failure* as it is used in this discussion of reliability. Let us assume that a designer states that his or her designs are 99.7 percent reliable (and this is the approximate value obtained with most LRFD designs). This means that, if this person were to design 1000 different structures, 3 of them would probably be overloaded at some time during their estimated 50-year lives and would fail. The reader probably and quite reasonably thinks this an unacceptably high rate of failure.

To the author 99.7 percent reliability doesn't mean that 3 of 1000 structures are going to fall flat on the ground. To him it means that those structures at some time will be loaded into the plastic range and perhaps the strain-hardening range. As a result deformations may be quite large during the overloading and some slight damage may occur. It is not anticipated that any of these structures will completely collapse. (The reader who is unfamiliar with statistics might say he or she wants 100 percent reliability in design, but this is statistically an impossible goal, as we will see in the paragraphs to follow.)

For this discussion it is assumed that we make a study of the reliability of a large number of steel structures designed at various times and with different past editions of the ASD Specification. To do this we will compute the resistance or strength R of each structure as well as the maximum loading Q expected during the life of the structure. The structure will be deemed safe if $R \geq Q$.

The actual values of R and Q are random variables, and it is therefore impossible to say with 100 percent certainty that R is always equal to or larger than Q for a particular structure. No matter how carefully a structure is designed and constructed, there will always be some little chance that Q will be greater than R or that the strength limit state will be exceeded. The goal of the preparers of the LRFD Specification was to keep this chance to a very small and consistent percentage.

Thus the magnitudes of both resistances and loads are uncertain. If we were to plot a curve of R/Q values for a large number of structures, the result would be a typical bell-shaped probability curve with mean values R_m and Q_m

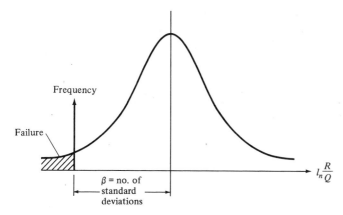

Figure 20-1

and a standard deviation. If at any location $R < Q$, we will have exceeded the strength limit state.

Such a curve is plotted logarithmically and shown in Fig. 20-1. It is to be remembered that the logarithm of 1.0 is 0, and thus if $\ln R/Q < 0$ the strength limit state has been exceeded. Such a situation is represented by the shaded area on the diagram. The smaller the shaded area, the more reliable is the structure. Another way of expressing this fact is to say that the larger the number of standard deviations from the mean values to the shaded area, the greater the reliability. In the figure the number of standard deviations is represented by β and is called the *reliability index*.

Even though the probable values of R and Q are not known very well, a formula has been developed with which values of β can be reasonably calculated.[4,5,6]

$$\beta = \frac{\ln(R_m/Q_m)}{\sqrt{V^2_R + V^2_Q}}$$

In this expression R_m and Q_m are respectively the mean resistance and load effects while V_R and Q_R are respectively the corresponding coefficients of variation.

As a result of the preceding work, it is now possible to design a particular element in accordance with a certain edition of the ASD Specification and

[4] C. A. Cornell, "A Probability-Based Structural Code," *Journal, American Concrete Institute* 66, no. 12 (December 1969).

[5] B. Ellingwood, T. V. Galambos, J. G. MacGregor, and C. A. Cornell, "Development of a Probability-Based Load Criterion for American National Standard A58," National Bureau of Standards Special Publication 577 (June 1980).

[6] M. K. Ravindra and T. V. Galambos, "Load and Resistance Factor Design for Steel," *Journal of Structural Division,* ASCE, 104, ST9, (September 1978), pp. 1337–1353.

with the appropriate statistical information compute the value of β for the design. This process is called *calibration*.

The percentage of structures for which the design strengths equal or exceed the worst anticipated loading will vary as we examine designs made in accordance with the requirements of different editions of the ASD Specification. Furthermore, our calculations will show that this reliability will vary for the designs of different types of members (such as columns and beams) made with the same edition of the ASD Specification.

Based upon the calculations of reliability described here, the investigators decided to use consistent β values in this new specification. These were the values they selected:

1. $\beta = 3.00$ for member subject to gravity loading.
2. $\beta = 4.50$ for connections. (This value reflects the usual practice that connections should be stronger than the members they connect.)
3. $\beta = 2.5$ for members subject to gravity plus wind loadings. (This value reflects an old practice that safety factors do not have to be as large for cases where lateral loads, which are of shorter duration, are involved.)
4. $\beta = 1.75$ for members subject to gravity loads plus earthquake loads.

Then the values of ϕ for the various parts of the specification were adjusted so the β values shown here would be obtained in design. In effect this causes most LRFD designs to be almost identical with those obtained by the allowable-stress method when the live-load–dead-load ratio is 3.

20-6 ADVANTAGES OF LRFD

The average person looking at this material will ask, Will LRFD save money compared with allowable-stress design (ASD)? The answer is that it often will, particularly if the live loads are small compared with the dead loads.

It should be noted, however, that the AISC has introduced LRFD not for the specific purpose of obtaining immediate economic advantages but because it helps provide a more uniform reliability for all steel structures whatever the loads. It is also written in a form that facilitates the introduction of the advances in knowledge that will occur through the years in structural steel design.

Despite this stated objective of the AISC, the author would like to continue for a little while with the topic of possible steel weight savings with LRFD. In ASD the same safety factor was used for a dead loads and live loads, whereas in LRFD a much smaller safety or load factor is used for dead loads (because they generally can be determined so much more accurately than can live loads). As a result, a weight of steel comparison for ASD and LRFD designs will necessarily depend on the ratio of the live loads to dead loads.

For the usual building the live-load–dead-load ratio varies from approximately 0.25 to as high as 4.0 or even a little higher for some very light structures. Low-rise steel buildings and preengineered buildings generally will fall in the upper range of these ratios. In ASD we use the same safety factors for dead and live loads regardless of their ratio to each other. Thus with ASD heavier members will result, and the factor of safety will increase as the live-load–dead-load ratio decreases.

It can be shown that for the lower range of L to D values, that is, less than 3, there can be steel weight savings with LRFD perhaps as much as one-sixth for tension members and columns and perhaps as much as one-tenth for beams. On the other hand if we have very high L and D ratios, there is almost no increase in steel weight with LRFD compared with ASD.[7]

20-7 DESIGN STRENGTH OF TENSION MEMBERS

A ductile steel member without holes subject to a tensile load can resist without fracture a load larger than its gross cross-sectional area times its yield stress because of strain hardening. However, a tension member loaded until strain hardening is reached will lengthen a great deal before fracture, a fact that will in all probability take away the member's usefulness and may even cause failure of the structural system of which the member is a part.

If, on the other hand, we have a tension member with bolt holes, it can fail by fracture at the net section through the holes. This failure load may very well be smaller than the load required to yield the gross section away from the holes. It is to be realized that the portion of the member where we have a reduced cross-sectional area due to the presence of holes normally is very short compared with the total length of the member. Though the strain-hardening situation is quickly reached at the net section portion of the member, yielding there may not really be a limit state of significance because the overall change in length of the member due to yielding in this small part of the member length may be negligible.

As a result of the preceding information, the LRFD Specification (D1) states that the design strength of a tension member $\phi_t P_n$ is to be a smaller of the values obtained by substituting into the following two expressions:

For the limit state of yielding in the gross section (which is intended to prevent excessive elongation of the member)

$$P_u = \phi_t F_y A_g \qquad \text{with } \phi_t = 0.9$$

For fracture in the net section where bolt or rivet holes are present

$$P_u = \phi_t F_u A_e \qquad \text{with } \phi_t = 0.75$$

F_u is the specified minimum tensile stress and A_e is the effective net area that can be assumed to resist tension at the section through the holes. The calculation of A_e values was discussed in Chapter 3.

The design strengths presented here are not applicable to threaded steel rods or to members with pin holes (as in eyebars).

---

[7] J. A. Edinger, "Introduction to the Proposed AISC Load and Resistance Factor Design Specification," *Engineering Journal*, AISC, 21, no. 1 (1st quarter, 1984), pp. 62–65.

It is not likely that stress fluctuations will be a problem in the average building frame, because the changes in load in such structures usually occur only occasionally and produce relatively minor stress variations. Full design wind or earthquake loads occur so infrequently that they are not considered in fatigue design. Should there be frequent variations or even reversals in stress, however, the matter of fatigue must be considered. This topic was discussed in Section 4.5 for the ASD method. The LRFD design procedure, which is presented in Appendix K of the LRFD Specification, is identical.

Example 20-2 illustrates the calculations necessary for determining the tensile design strength of a W section.

■ Example 20-2

Determine the design strength of a W10 × 45 with two lines of $\frac{3}{4}$-in.-diameter bolts in each flange, using A36 steel and the LRFD Specification. There are assumed to be at least three bolts in each line, and the bolts are not staggered with respect to each other.

Solution.   Using a W10 × 45 ($A_g = 13.3$ in.^{2}, $d = 10.10$ in., $b_f = 8.020$ in., $t_f = 0.620$ in.)

$$P_u = \phi_t F_y A_g = (0.90)(36)(13.3) = 430.9 \text{ k} \leftarrow$$
$$A_n = 13.3 - (4)(\tfrac{7}{8})(0.620) = 11.13 \text{ in.}^2$$
$$U = 0.90 \quad \text{since } b_f > \tfrac{2}{3}d$$
$$A_e = UA_n = (0.90)(11.13) = 10.02 \text{ in.}^2$$
$$P_u = \phi_t F_u A_e = (0.75)(58)(10.02) = 435.9 \text{ k}$$
$$\underline{\underline{\text{Design strength } P_u = 430.9 \text{ k}}}$$ ■

When splice or gusset plates are used as statically loaded tensile connecting elements, their strength shall be determined from the following expression (LRFD Specification J5.2): For yielding of welded, bolted, or riveted connection elements

$$\phi_t R_n = \phi_t A_g F_y \quad \text{with } \phi_t = 0.9$$

For fracture of bolted or riveted connection elements

$$\phi_t R_n = \phi_t A_n F_u \quad \text{with } \phi_t = 0.75 \text{ and } A_n \le 0.85 A_g$$

The A_n to be used in the second of these expressions may not exceed 85 percent of A_g. In Example 20-3 the strength of a pair of tensile connecting plates is computed.

■ Example 20-3

The A36 tension member of Example 20-2 is assumed to be connected at its ends with two $\frac{3}{8}$ × 12-in. plates as shown in Fig. 20-2. If two lines of $\frac{3}{4}$-in. bolts are used in each plate, determine the maximum tensile force which the plates can transfer.

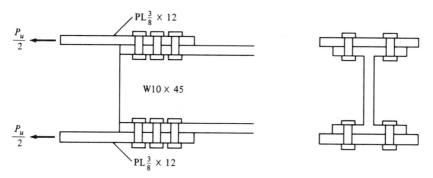

Figure 20-2

Solution.

$P_u = \phi_t F_y A_g = (0.9)(36)(2 \times \frac{3}{8} \times 12) = 291.6$ k

A_n of 2 plates $= (\frac{3}{8} \times 12 - \frac{7}{8} \times 2 \times \frac{3}{8})2 = 7.69$ in.^{2}

$0.85 A_g = (0.85)(2 \times \frac{3}{8} \times 12) = 7.65$ in.$^2 = A_n$

$P_u = \phi_t F_u A_n = (0.75)(58)(7.65) = 332.7$ k

$$P_u = 291.6 \text{ k}$$

■

20-8 DESIGN OF TENSION MEMBERS

The determination of the design strength of various tension members was pre-sented in the last section. In this section the selection of members to support given tensile loads is described.

For tension members other than rods the LRFD Specification B7 recom-mends maximum slenderness ratios of 300. Such members may on occasion be subjected to compressive forces caused by wind or earthquake. This specifi-cation states that these compressive forces may not exceed 50 percent of the design compressive strengths of the members. (It will later be learned that, for slenderness ratios greater than 200, design compressive stresses will be very small, in fact less than 5.33 ksi for A36 steel.) Furthermore, members in this category do not have to meet other compression member requirements such as the preferable maximum KL/r value of 200.

The *recommended* maximum slenderness ratio of 300 is not applicable to tension rods. Maximum L/r values for rods are left to the designer's judg-ment. If a maximum value of 300 was specified for them, they would seldom be used because of their extremely small radii of gyration.

Remember that out-of-straightness within reasonable limits does not ap-preciably affect the strength of tension members. Tensile loads tend to reduce out-of-straightness, whereas compression loads tend to amplify it. As a result the LRFD Specification is rather liberal in its requirements for tension mem-bers, including those subject to short-term wind and earthquake compression forces.

The design strength P_u of tension members has been specified as being the lesser of $\phi_t F_y A_g$ or $\phi_t F_u A_e$. For convenience in selecting member sections, these expressions may be written in a little different fashion as described herein.

To satisfy the first of these expressions the minimum gross area must be at least

$$\min A_g = \frac{P_u}{\phi_t F_y} \tag{1}$$

To satisfy the second expression the minimum value of A_e must be at least

$$\min A_e = \frac{P_u}{\phi_t F_u}$$

And since $A_e = UA_n$ the minimum value of A_n is

$$\min A_n = \frac{\min A_e}{U} = \frac{P_u}{\phi_t F_u U}$$

Then the minimum A_g for the second expression must at least equal the minimum value of A_n plus the estimated hole areas.

$$\min A_g = \frac{P_u}{\phi_t F_u U} + \text{estimated hole areas} \tag{2}$$

The designer can substitute into Equations (1) and (2), taking the larger value of A_g so obtained for an initial size estimate. It is, however, well to notice that the maximum preferable slenderness ratio L/r is 300. From this value it is easy to compute the least permissible value of r for a particular design, that is, the value of r for which the slenderness ratio will be exactly 300. It is undesirable to consider a section whose least r is less than this value because its slenderness ratio would exceed the preferable maximum value of 300.

$$\min r = \frac{L}{300} \tag{3}$$

For the two examples to follow the author has substituted into the two ordinary gravity load expressions:

$$P_u = 1.4 D$$
$$P_u = 1.2 D + 1.6L$$

As the first of these expressions will not control unless the dead load is more than eight times as large as the live load, the first expression is omitted for the remaining problems in this text (unless $D > 8L$).

In Example 20-4 a W section is selected for a given set of tensile loads. For this first application of the tension design formulas, the author has narrowed the problem down to one series of W shapes so the reader can concentrate on the application of the formulas and not become lost in considering W8s, W10s, W14s, and so on. Exactly the same procedure can be used for trying these other series as is used here for the W12.

■ Example 20-4

Select a 30-ft-long W12 section of A36 steel to support a tensile service dead load $P_D = 140$ k and a tensile service live load $P_L = 80$ k. As shown in Fig. 20-3 the member is to have two lines of bolts in each flange for $\frac{7}{8}$-in. bolts (at least three in a line).

Solution.   Considering the two ordinary load conditions:

$$P_u = 1.4\, P_D = (1.4)(140) = 196 \text{ k}$$
$$P_u = 1.2\, P_D + 1.6\, P_L = (1.2)(140) + (1.6)(80) = 296 \text{ k} \leftarrow$$

Computing the minimum A_g required:

$$\text{Min } A_g = \frac{P_u}{\phi_t F_y} = \frac{296}{(0.90)(36)} = 9.14 \text{ in.}^2$$

$$\text{Min } A_g = \frac{P_u}{\phi_t F_u U} + \text{estimated hole areas}$$

Assume $U = 0.90$ and assume flange thickness is about 0.520 in. after looking at W12 sections in LRFD Manual which have areas of about 9.14 in.²

$$\text{Min } A_g = \frac{296}{(0.75)(58)(0.90)} + (4)(1.00)(0.520) = 9.64 \text{ in.}^2 \leftarrow$$

$$\text{Preferable min } r = \frac{L}{300} = \frac{(12)(30)}{300} = 1.2 \text{ in.}$$

Try W12 × 35 ($A_g = 10.3$ in.², $d = 12.50$ in., $b_f = 6.56$ in., $t_f = 0.520$ in., $r_y = 1.54$ in.)
Checking:

$$P_u = \phi_t F_y A_g = (0.90)(36)(10.3) = 333.7 \text{ k} > 296 \text{ k} \qquad\qquad \text{OK}$$

$$P_u = \phi_t F_u U A_n \text{ with } U = 0.85 \text{ since } \frac{b_f}{d} < \frac{2}{3},$$

$$\text{and } A_n = 10.3 - (4)(1.00)(0.520) = 8.22 \text{ in.}^2$$
$$P_u - (0.75)(58)(0.85)(8.22) = 303.9 \text{ k} > 296 \text{ k} \qquad\qquad \text{OK}$$

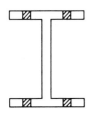

Figure 20-3

$$\frac{L}{r} = \frac{(12)(30)}{1.54} = 234 < 300 \qquad\qquad \text{OK}$$

$$\text{Use W12} \times 35 \qquad\qquad \blacksquare$$

20-9 BLOCK SHEAR

The design strength of a tension member is not always controlled by $\phi_t F_y A_g$ or $\phi_t F_u A_e$ or by the strength of the bolts or welds with which the member is connected. It may instead be controlled by its *block shear* strength as described in this section.

The failure of a member may occur along a path involving tension on one plane and shear on a perpendicular plane as shown in Fig. 20-4, where several possible block shear failures are shown.

It is unlikely that fracture will occur on both planes at the same time. It seems logical to assume that the load will cause the yield strength to be reached on one plane while the other plane is stressed past yielding and on to fracture. Thus it does not seem realistic to add the fracture strengths on both planes together to determine the block shear strength of a particular member.

The member shown in Fig. 20-5(a) has a large shear area and a small tensile area, and the primary resistance to a block shear failure is shearing and

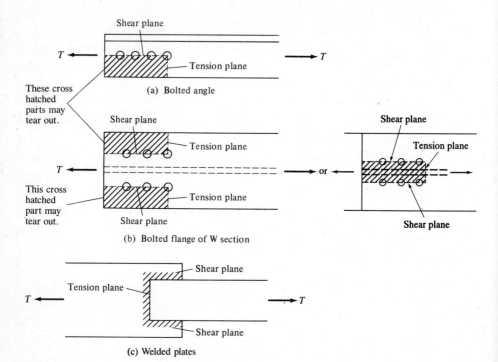

Figure 20-4 Block shear failures.

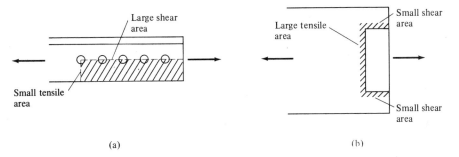

Figure 20-5

not tensile. The LRFD Specification states that it is logical to assume that, when a shear fracture occurs on this large shear-resisting area, the small tensile area has yielded.

In Fig. 20-5(b) a member is shown which, so far as block shear goes, has a large tensile area and a small shearing area. The LRFD says that for this case the primary resisting force against a block shear failure will be tensile and not shearing. Thus a block shear failure cannot occur until the tensile area fractures. At that time it seems logical to assume that the shear area has yielded.

Based on the preceding discussion the LRFD Specification (J5.2c) states that the block shear design strength of a particular member is to be determined by (a) computing the tensile fracture strength on the net section in one direction and adding to that value the shear yield strength on the gross area on the perpendicular segment and (b) computing the shear fracture strength on the gross area subject to tension and adding it to the tensile yield strength on the net area subject to shear on the perpendicular segment. The larger value determined in (a) and (b) is the block shear strength.

Test results show that this procedure gives good results. Furthermore it is consistent with the calculations previously used for tension members where gross areas are used for one limit state of yielding ($\phi_t F_y A_g$) and net area for the fracture limit state ($\phi_t F_u A_e$). The LRFD Commentary (J4) states that the block shear strength of a member P_{bs} is to be taken as the larger of the values given at the end of this paragraph. For tension fracture and shear yielding

$$P_{bs} = \phi \text{ (tensile fracture strength + shear yielding)}$$
$$= \phi(F_u A_{nt} + 0.6\, F_y A_{vg}) \qquad \text{(LRFD Formula C-J4-1)}$$

and for shear fracture and tension yielding

$$P_{bs} = \phi \text{ (shear fracture strength + tension yielding)}$$
$$= \phi(F_y A_{tg} + 0.6\, F_u A_{ns}) \qquad \text{(LRFD Formula C-J4-2)}$$

where

$$\phi = 0.75$$
$$A_{vg} = \text{gross area subjected to shear}$$
$$A_{tg} = \text{gross area subjected to tension}$$

A_{ns} = net area subjected to shear

A_{nt} = net area subjected to tension

Example 20-5 illustrates the determination of the block shear strength for an angle.

■ Example 20-5

The A36 tension member shown in Fig. 20-6 is connected with three $\frac{3}{4}$-in. bolts. Determine the block shearing strength of the member and its tensile strength.

Solution.
Tension fracture + shear yielding:

$$P_{bs} = \phi(F_u A_{nt} + 0.6\,F_y A_{vg})$$
$$= 0.75[(58)(\tfrac{1}{2})(2.50 - \tfrac{1}{2} \times \tfrac{7}{8}) + (0.6)(36)(10)(\tfrac{1}{2})]$$
$$= 125.9 \text{ k}$$

Shear fracture + tension yielding:
Note that there are $2\frac{1}{2}$ holes in the net area shear plane shown in Fig. 20-6.

$$P_{bs} = \phi(F_y A_{tg} + 0.6\,F_u A_{ns})$$
$$= 0.75[(36)(2.5)(\tfrac{1}{2}) + (0.6)(58)(10.0 - 2.5 \times \tfrac{7}{8})(\tfrac{1}{2})]$$
$$= 135.7 \text{ k} \leftarrow$$

Tensile strength of angle:

$$P_u = \phi_t F_y A_g = (0.9)(36)(4.75) = 153.9 \text{ k} \leftarrow$$
$$A_n = 4.75 - (1)(\tfrac{7}{8})(\tfrac{1}{2}) = 4.31 \text{ in.}^2$$

$$U = 1 - \frac{0.987}{8} = 0.88 \qquad (0.85 \text{ given in LRFD})$$

$$P_u = \phi_t F_u A_e = (0.75)(58)(0.88 \times 4.31) = 165 \text{ k}$$

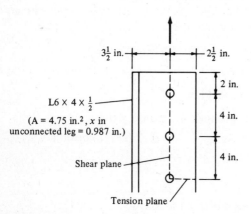

$3\frac{1}{2}$ in.

$2\frac{1}{2}$ in.

2 in.

L6 × 4 × $\frac{1}{2}$

(A = 4.75 in.^{2}, x in
unconnected leg = 0.987 in.)

4 in.

Shear plane

4 in.

Tension plane

Figure 20-6

P_u for member = larger of P_{bs} values (135.7 k) or smaller of tensile strength values (153.9 k).

$$\underline{\underline{P_u = 135.7 \text{ k}}}$$ ■

PROBLEMS

20-1 to 20-5   Determine the tensile design strength of each of the sections, using A36 steel and neglecting block shear. Assume that there are two lines of bolts in each flange with at least three bolts in a line.

20-1 W18 × 40, $\frac{7}{8}$-in. bolts. (*Ans.* 358.7 k)

20-2 W10 × 49, $\frac{3}{4}$-in. bolts.

20-3 W21 × 101, 1-in. bolts. (*Ans.* 965.5 k)

20-4 WT18 × 115, $\frac{3}{4}$-in. bolts.

20-5 WT10.5 × 34, $\frac{7}{8}$-in. bolts. (*Ans.* 319.3 k)

20-6 Repeat Prob. 20-3 with $\frac{3}{4}$-in. bolts, F_y = 50 ksi, and F_u = 70 ksi.

20-7 Repeat Prob. 20-4 with 1-in. bolts and only two bolts in each line. (*Ans.* 1010.1 k)

20-8 A single-angle tension member (7 × 4 × $\frac{3}{4}$) has two gage lines in its long leg and one in its short leg for $\frac{3}{4}$-in. bolts arranged as shown. Determine the tensile design strength P_u of this member if A36 steel is used and if block shear is neglected.

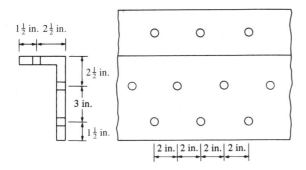

20-9 Determine the tensile design strength P_u of the pair of 6 × 6 × $\frac{1}{2}$ angles made from A242 steel shown. Standard gages are to be used as determined from the LRFD Manual for the $\frac{3}{4}$-in. bolts. Neglect block shear. There are at least three bolts in each line. (*Ans.* 496.6 k)

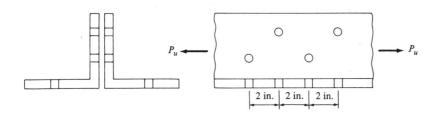

20-10 Repeat Prob. 20-9 with $F_y = 50$ ksi, $F_u = 70$ ksi, and $\frac{7}{8}$-in. bolts.

20-11 to 20-15 Select sections as indicated. Assume there are two lines of bolts in each flange (at least three in a line), and use A36 steel.

PROBLEM	SECTION	LENGTH (ft)	P_D (k)	P_L (k)	BOLT DIAMETER (in.)	
20-11	W14	22	200	250	$\frac{3}{4}$	(*Ans.* W14 × 68)
20-12	W12	20	240	100	$\frac{3}{4}$	
20-13	W12	17	120	150	$\frac{7}{8}$	(*Ans.* W12 × 45)
20-14	W10	25	100	170	1	
20-15	W8	23	150	200	$\frac{3}{4}$	(*Ans.* W8 × 58)

20-16 Repeat Prob. 20-11 using A441 steel.

20-17 Select the lightest American Standard channel that will safely support the service tensile loads $P_D = 80$ k and $P_L = 120$ k. The member is 15 ft long and is assumed to have one line of holes for 1-in. bolts in each flange. Use A36 steel and assume there are at least three holes in each line. (*Ans.* C15 × 33.9)

20-18 A welded tension member is to support a design load $P_u = 620$ k and is to consist of two channels placed 12 in. back-to-back with the flanges turned out. Select the lightest standard channels available using A36 steel. Assume $U = 0.75$. The member is to be 30 ft long.

20-19 A36 steel is to be used in selecting a single-angle member to resist service tensile loads of $P_D = 90$ k and $P_L = 120$ k. The member is to be 18 ft long and is assumed to be connected with one line of four $\frac{7}{8}$-in. bolts in one leg. (*Ans.* L8 × 8 × $\frac{5}{8}$)

20-20 Repeat Prob. 20-19 using a pair of equal-leg angles and A36 steel. Assume angle legs are touching and assume one hole for a $\frac{7}{8}$-in. bolt is to be taken out of each angle. Also assume $U = 0.85$. Consider only the pairs of angles listed in the double-angle tables of Part 1 of the Manual.

20-21 Design member L_2L_3 of the truss shown. It is to consist of a pair of angles with a $\frac{3}{8}$-in. gusset plate between the angles at each joint. Use A36 steel and assume one line of three $\frac{3}{4}$-in. bolts in each angle leg. Consider only the angles shown in the double-angle tables of the Manual. For each load $P_D = 20$ k and $P_r = 14$ k (roof load). (*Ans.* 2L6 × 6 × $\frac{3}{8}$s)

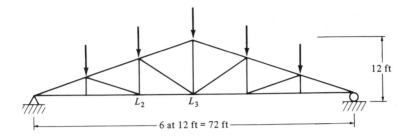

6 at 12 ft = 72 ft

20-22 Select a single-angle tension member to resist the service loads $P_D = 60$ k and $P_L = 60$ k. The member is to be 15 ft long and is to be connected with one line of four $\frac{7}{8}$-in. bolts. Assume $F_y = 40$ ksi and $F_u = 60$ ksi.

20-23 Repeat Prob. 20-18, assuming that one line of of $\frac{7}{8}$-in. bolts will be used in each flange with at least three bolts in each line. Also design tie plates. Assume distance or gage from back of channel to center of line of bolts is 2 in. U is to be determined from LRFD Specification B3. (*Ans.* 2C15 × 33.9s, $\frac{3}{16}$ × 6 × 1 ft 0 in. tie plates spaced 15 ft 0 in. on centers)

20-24 A W10 × 45 is connected at its ends with the plates shown. Determine the block shear strength of the member if A36 steel is used and if it is connected with six $\frac{3}{4}$-in. bolts in each flange as shown. Compare the result with the tensile design strength of the member. Do not check plate strength.

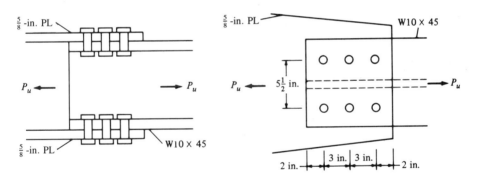

20-25 The 7 × 4 × $\frac{3}{8}$-in. angle shown is connected with three $\frac{7}{8}$-in. bolts. If the angle consists of A36 steel, determine its block shear strength. Compare the results with the design tensile strength of the member. (*Ans.* 103.8 k)

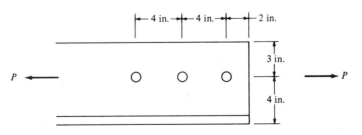

20-26 Determine the design strength of the A36 W10 × 60 section shown. Include block shear. Bolts are $\frac{7}{8}$-in.

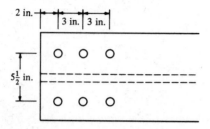

20-27 Compute the design strength of the 6 × 6 × $\frac{1}{2}$-in. angle shown if it consists of A36 steel. Consider block shear as well as the tensile strength of the angle. Assume vertical leg is unconnected for determining U. (*Ans.* 186.3 k)

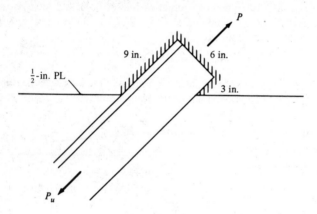

21

Compression Members—LRFD

21-1 LONG, SHORT, AND INTERMEDIATE COLUMNS

A column subject to an axial compression load will shorten in the direction of the load. If the load is increased until the column buckles, the shortening will stop, and the column will suddenly bend or deform laterally and may at the same time twist in a direction perpendicular to its longitudinal axis.

The strength of a column and the manner in which it fails are greatly dependent on its effective length. A very short stocky steel column may be loaded until the steel yields and perhaps on into the strain-hardening range. As a result it can support about the same load in compression that it can in tension.

As the effective length of a column increases, its buckling stress will decrease. If the effective length exceeds a certain value, the buckling stress will be less than the proportional limit of the steel. Columns in this range are said to fail *elastically*.

As shown in Section 5-5 very long steel columns will fail at loads which are proportional to the bending rigidity of the column (EI) and independent of the steel. For instance, a long column constructed with a 36-ksi yield stress steel will fail at just about the same load as one constructed with a 100-ksi yield stress steel.

Columns are sometimes classed as being long, short, or intermediate. A brief discussion of each of these classifications is presented in the paragraphs to follow.

Long Columns

The Euler formula predicts very well the strength of long columns where the axial buckling stress remains below the proportional limit. Such columns will buckle *elastically*.

Short Columns

For very short columns the failure stress will equal the yield stress, and no buckling will occur. (For a column to fall into this class it would have to be so short as to have no practical application. Thus no further reference is made to them here.)

Intermediate Columns

For intermediate columns some of the fibers will reach the yield stress and some will not. The members will fail by both yielding and buckling, and their behavior is said to be *inelastic*. (For the Euler formula to be applicable for such columns it would have to be modified according to the reduced modulus concept or the tangent modulus concept to account for the presence of residual stresses.)

In Section 21-2 formulas are presented with which the LRFD Specification estimates the strength of columns in these different ranges.

21-2 COLUMN FORMULAS

The LRFD Specification provides one formula (the Euler equation) for long columns with inelastic buckling and a parabolic empirical equation for short and intermediate columns. With these equations, a critical or buckling stress F_{cr} is determined for a compression member. Once this stress is computed for a particular compression member, it is multiplied by the cross-sectional area of the member to obtain the member's nominal strength. The design strength of the member can then be determined as follows:

$$P_u = \phi_c P_n = \phi_c F_{cr} A_g \qquad \text{with } \phi_c = 0.85$$

One LRFD formula for F_{cr} is for inelastic buckling, and the other is for elastic buckling. In both equations λ_c in easily remembered form is $\sqrt{F_y/F_e}$, where F_e is the Euler stress $\pi^2 E/(KL/r)^2$. Substituting this value for F_e we get the form of λ_c given in the LRFD Manual.

$$\lambda_c = \frac{KL}{r\pi} \sqrt{\frac{F_y}{E}} \qquad \text{(LRFD Formula E2-4)}$$

Both equations for F_{cr} include the estimated effects of residual stresses and initial out-of-straightness of the members. The inelastic formula that follows is an empirical or test-result formula:

$$F_{cr} = (0.658^{\lambda_c^2})F_y \text{ for } \lambda_c \leq 1.5 \qquad \text{(LRFD Formula E2-2)}$$

The other equation is for elastic or Euler buckling and is the familiar Euler equation multiplied by 0.877 to estimate the effect of out-of-straightness.

$$F_{cr} = \left(\frac{0.877}{\lambda_c^2}\right)F_y \text{ for } \lambda_c > 1.5 \qquad \text{(LRFD Formula E2-3)}$$

These equations are represented graphically in Fig. 21-1.

After looking at these equations the reader might think that their use would be very tedious and time-consuming with a pocket calculator. Such calculations, however, rarely have to be made because the LRFD Manual provides computed $\phi_c F_{cr}$ values for steels with $F_y = 36$ ksi and 50 ksi for KL/r values from 1 to 200 and has shown the results in Table 3-36 and 3-50 of Part 6 of the LRFD Manual. Furthermore the LRFD Specification has another table in Part 6 (Table 4) from which the user may obtain values for steels with any F_y values. To use this table we calculate λ_c for the steel in question, look in Table 4 for $\phi_c F_{cr}/F_y$, and solve the expression for $\phi_c F_{cr} = (F_y)$ (value read in table).

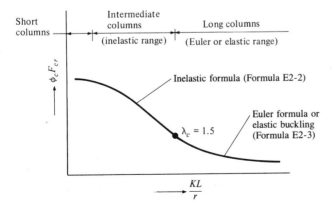

Figure 21-1 Graphical representation of LRFD column equations.

21-3 MAXIMUM SLENDERNESS RATIOS

In Part 6, Section B7, the LRFD Specification states that compression members *preferably* should be designed with KL/r ratios not exceeding 200. The reader might note from LRFD Tables 3-36 and 3-50 that design stresses $\phi_c F_{cr}$ for KL/r values of 200 are both 5.33 ksi. Should slenderness ratios larger than these be used, the $\phi_c F_{cr}$ values will be very small and will require the user to substitute into the column formulas provided in Section 21-2.

21-4 EXAMPLES

In this section the design strengths of three columns are considered. In Example 21-1(a) the author determines the strength of a W section. The value of K is determined as described in Section 5-9, the effective slenderness ratio is computed, and the design stress of the member $\phi_c F_{cr}$ is selected from the appropriate LRFD table and multiplied by the cross-sectional area of the column.

It will be noted that the LRFD in its Part 2 has further simplified the calculations required by computing the column design strength $\phi_c F_{cr} A_g$ for each of the steel shapes normally used as columns for F_y values of 36 and 50 ksi for most of the commonly used effective lengths or KL values given in feet. The use of these tables is illustrated in Example 21-1(b).

■ Example 21-1
(a) Using the column design stress values shown in Table 3-36, Part 6, of the LRFD Manual, determine the design strength ($P_u = \phi_c P_n$) of the A36 axially loaded column shown in Fig. 21-2,
(b) Repeat the problem using the column tables of Part 2 of the LRFD Manual.

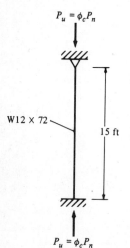

$P_u = \phi_c P_n$

W12 × 72

15 ft

$P_u = \phi_c P_n$

Figure 21-2

Solution.

(a) Using a W12 × 72 ($A = 21.1$ in.², $r_y = 3.04$ in.):

$$K = 0.80 \text{ from Table C-C2.1, Part 6, LRFD Manual}$$

$$\frac{KL}{r} = \frac{(0.80)(12 \times 15)}{3.04} = 47.37$$

$$\phi_c F_{cr} = 27.19 \text{ ksi from Table 3-36, Part 6, LRFD Manual}$$

$$P_u = \phi_c P_n = \phi_c F_{cr} A_g = (27.19)(21.1) = 573.7 \text{ k}$$

(b) Entering column tables Part 2 of Manual with $K_y L_y$ in feet

$$K_y L_y = (0.80)(15) = 12 \text{ ft}$$

$$P_u = \phi_c P_n = 574 \text{ k}$$ ∎

In Example 21-2 the author illustrates the computations necessary to determine the design strength of a built-up column section. Several special requirements for built-up column sections are described in Section 21-8 of this chapter.

■ Example 21-2

Determine the design strength $\phi_c P_n$ of the axially loaded column shown in Fig. 21-3 if $KL = 19$ ft and A36 steel is used.

Solution.

$$A = (20)(\tfrac{1}{2}) + (2)(12.6) = 35.2 \text{ in.}^2$$

$$\bar{y} \text{ from top} = \frac{(10)(0.25) + (2)(12.6)(9.50)}{35.2} = 6.87 \text{ in.}$$

$$I_x = (2)(554) + (25.2)(2.63)^2 + (\tfrac{1}{12})(20)(\tfrac{1}{2})^3 + (10)(6.62)^2 = 1721 \text{ in.}^4$$

$$I_y = (2)(14.4) + (12.6)(6.877)^2(2) + (\tfrac{1}{12})(\tfrac{1}{2})(20)^3 = 1554 \text{ in.}^4$$

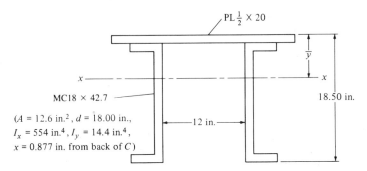

Figure 21-3

$$\text{Least } r = \sqrt{\frac{1554}{35.2}} = 6.64 \text{ in.}$$

$$\frac{KL}{r} = \frac{(12)(19)}{6.64} = 34.34$$

$$\phi_c F_{cr} = 28.76 \text{ ksi}$$

$$\phi_c P_n = (28.76)(35.2) = \underline{\underline{1012 \text{ k}}}$$ ∎

Example 21-3 illustrates the calculations necessary to determine the design strength of a column with different unbraced lengths in the x and y directions.

■ Example 21-3
(a) Using Table 3-36 of Part 6 of the LRFD Manual, determine the design strength $\phi_c P_n$ of the A36 axially loaded W14 × 90 shown in Fig. 21-4. Because of its considerable length this column is braced perpendicular to its weak or y axis at the points shown in the figure. These connections are assumed to permit rotation of the member but to prevent translation or sidesway.
(b) Repeat part (a) using the column tables of Part 2 of the LRFD Manual.

Solution.
(a) Using a W14 × 90 ($A = 26.5$ in.^{2}, $r_x = 6.14$ in., $r_y = 3.70$ in.)
Determining effective lengths:

$$K_x L_x = (0.80)(32) = 25.6 \text{ ft}$$
$$K_y L_y = (1.0)(10) = 10 \text{ ft} \leftarrow$$
$$K_y L_y = (0.80)(12) = 9.6 \text{ ft}$$

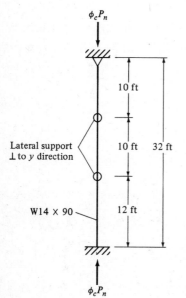

Figure 21-4

Computing slenderness ratios:

$$\left(\frac{KL}{r}\right)_x = \frac{(12)(25.6)}{6.14} = 50.03 \leftarrow$$

$$\left(\frac{KL}{r}\right)_y = \frac{(12)(10)}{3.70} = 32.43$$

$$\phi_c F_{cr} = 26.82 \text{ ksi}$$

$$\phi_c P_n = (26.82)(26.5) = \underline{710.7 \text{ k}}$$

(b) Noting from part (a) solution that there are two different KL values:

$$K_x L_x = 25.6 \text{ ft}$$

$$K_y L_y = 10 \text{ ft}$$

Controlling $K_y L_y$ for use in tables is either 10 ft or $K_x L_x / r_x / r_y$:

$\dfrac{r_x}{r_y}$ for W14 × 90 from column tables = 1.66

$$\frac{K_x L_x}{r_x / r_y} = \frac{25.6}{1.66} = 15.42 \text{ ft} \leftarrow$$

From column tables with $K_y L_y = 15.42$ ft we find by interpolation

$$\phi_c Pn = \underline{711 \text{ k}} \qquad \blacksquare$$

21-5 DESIGN OF AXIALLY LOADED COLUMNS

In the pages that follow the designs of several axially loaded columns are presented. Included are the selections of single W shapes and W shapes with cover plates. Also included are the designs of sections whose unbraced lengths in the x and y directions are different.

In Example 21-4 a column with $KL = 10$ ft is selected using the LRFD formulas. An effective slenderness ratio of 50 is assumed, the design stress for that value is determined from Table 3-36 of Part 6 of the LRFD Manual, and the resulting stress is divided into the factored column load to obtain an estimated column area. After a trial section is selected with approximately that area, its actual slenderness ratio and design strength are determined. The first estimated size in Example 21-4, though quite close, is a little too small, and the next larger section in that series of shapes is tried and found to be satisfactory.

■ Example 21-4

Using A36 steel, select the lightest W14 available for the service loads $P_D = 100$ k and $P_L = 160$ k. $KL = 10$ ft.

Solution.

$$P_u = (1.2)(100) + (1.6)(160) = 376 \text{ k}$$

Assume $KL/r = 50$:

$$\phi_c F_{cr} \text{ from Table 3-36, Part 6} = 26.83 \text{ ksi}$$

$$A_{reqd} = \frac{376}{26.83} = 14.01 \text{ in.}^2$$

Try W14 × 48 ($A = 14.1$ in.^{2}, $r_y = 1.91$ in.)

$$\frac{KL}{r} = \frac{(12)(10)}{1.91} = 62.83$$

$$\phi_c F_{cr} = 24.86 \text{ ksi}$$

$$\phi_c P_n = (24.86)(14.1) = 350 \text{ k} < 376 \text{ k} \qquad\qquad \text{NG}$$

Try W14 × 53 ($A = 15.6$ in.^{2}, $r_y = 1.92$ in.)

$$\frac{KL}{r} = \frac{(12)(10)}{1.92} = 62.5$$

$$\phi_c F_{cr} = 24.91 \text{ ksi}$$

$$\phi_c P_n = (24.91)(15.6) = 388.6 \text{ k} > 376 \text{ k} \qquad\qquad \text{OK}$$

$$\underline{\text{Use W14} \times 53} \qquad\qquad \blacksquare$$

21-6 LRFD DESIGN TABLES

For Example 21-5, Part 2 of the LRFD Manual is used to select various column sections from tables without the necessity of using a trial-and-error process. These tables provide axial design strengths ($\phi_c P_n$) for various practical effective lengths of the steel sections commonly used as columns (Ws, Ms, Ss, pipes, tubes, pairs of angles, and structural tees). The values are given with respect to the least radius of gyration for steels with $F_y = 36$ ksi and 50 ksi, *with the exception of square and rectangular tubes which are available only in 46-ksi steel.*

The resulting tables are very simple to use. The designer takes the KL value for the weaker direction in feet, enters the table in question from the left-hand side, and moves horizontally across the table. Under each section is listed the design strength $\phi_c P_n$ for that KL and the steel yield stress. As an illustration it is assumed that we have a factored design load $P_u = \phi_c P_n = 800$ k, $K_y L_y = 12$ ft, and we want to select the lightest available W14 section using A36 steel. We enter the tables with $KL = 12$ ft in the left column and read from left to right for A36 steel the numbers 6260, 5700, 5170 k, and so on until several pages later where the consecutive values 823 and 749 k are found. The 749 k value is not sufficient, and we go back to the 823 k, which falls under the W14 × 99. A similar procedure can be followed for the other available shapes.

Example 21-5 illustrates the selection of W sections as well as pipes and tubes. It is possible to support a given column load with a standard pipe column; or with an extra strong pipe column ($\times$ strong) which has a smaller diameter but thicker walls, thus is heavier and more expensive; or with a double extra strong pipe column ($\times\times$ strong) which has an even smaller diameter and even thicker walls and heavier weight.

■ Example 21-5

Using the column tables of Part 2 of the LRFD Manual,
(a) Select the lightest W section available for the loads, steel and KL of Example 21-4.
(b) Select the lightest standard, extra strong, and double extra strong pipe columns for the situation of part (a) of this problem.
(c) Select the lightest square and rectangular tubes satisfactory for the situation of part (a), except use an F_y of 46 ksi.

Solution.
(a) Enter tables with $K_y L_y = 10$ ft and $P_u = \phi_c P_n = 376$ k:

$$W14 \times 53 \ (\phi_c P_n = 389 \text{ k})$$
$$W12 \times 53 \ (\phi_c P_n = 422 \text{ k})$$
$$W10 \times 49 \ (\phi_c P_n = 392 \text{ k}) \leftarrow$$
$$W8 \times 58 \ (\phi_c P_n = 441 \text{ k})$$
$$\underline{\underline{\text{Use W10} \times 49}}$$

(b) Pipe columns:

$$\underline{12 \text{ pipe std}} \ (\phi_c P_n = 429 \text{ k}) \ \text{wt} = 49.56 \text{ lb/ft}$$
$$\underline{\text{Pipe 10}\times \text{ strong}} \ (\phi_c P_n = 465 \text{ k}) \ \text{wt} = 54.74 \text{ lb/ft}$$
$$\underline{\text{Pipe 6}\times\times \text{ strong}} \ (\phi_c P_n = 399 \text{ k}) \ \text{wt} = 53.16 \text{ lb/ft}$$

(c) Square and rectangular tubing ($F_y = 46$ ksi):

$$\underline{8 \times 8 \times \tfrac{3}{8}} \ (\phi_c P_n = 392 \text{ k}) \ \text{wt} = 37.60 \text{ lb/ft}$$
$$\underline{8 \times 6 \times \tfrac{1}{2}} \ (\phi_c P_n = 404 \text{ k}) \ \text{wt} = 42.05 \text{ lb/ft} \qquad ■$$

The selection of a column with different $K_x L_x$ and $K_y L_y$ values is presented in Example 21-6. It is solved first by trial and error and then directly with the LRFD tables. For the latter solution we enter the column tables with $K_y L_y$ and select a shape. Then we divide our $K_x L_x$ by the shape's r_x/r_y value to determine an effective $K_y L_y$. If this value is larger than our actual $K_y L_y$, we use it to reenter the tables and repeat the selection process.

■ Example 21-6

Select the lightest satisfactory W12 for the following conditions: A36 steel, $P_u = 670$ k, $K_x L_x = 26$ ft, and $K_y L_y = 13$ ft.
(a) By trial and error
(b) Using LRFD tables

Solution.

(a) Using trial and error
Assume $KL/r = 50$:

$$\phi_c F_{cr} = 26.83 \text{ ksi}$$

$$A_{reqd} = \frac{670}{26.83} = 24.97 \text{ in.}^2$$

Try W12 × 87 ($A = 25.6$ in.^{2}, $r_x = 5.38$ in., $r_y = 3.07$ in.

$$\left(\frac{KL}{r}\right)_x = \frac{(12)(26)}{5.38} = 57.99 \leftarrow$$

$$\left(\frac{KL}{r}\right)_y = \frac{(12)(13)}{3.07} = 50.81$$

$$\phi_c F_{cr} = 25.63 \text{ ksi}$$

$$\phi_c P_n = (25.63)(25.6) = 656 \text{ k} < 670 \text{ k} \qquad\qquad \text{NG}$$

A subsequent check of the next W12 section (96 lb) shows it will work.

<u>Use W12 × 96</u>

(b) Using LRFD tables
Enter tables with $K_y L_y = 13$ ft:
Try W12 × 87 ($r_x/r_y = 1.75$)

$$(K_y L_y)\left(\frac{r_x}{r_y}\right) = (13)(1.75) = 22.75 < K_x L_x$$

$\therefore K_x L_x$ controls.
Reenter tables with new $K_y L_y$:

$$K_y L_y = \frac{K_x L_x}{r_x/r_y} = \frac{26}{1.75} = 14.86$$

<u>Use W12 × 96</u> ∎

21-7 BUILT-UP MEMBERS WITH COMPONENTS IN CONTACT WITH EACH OTHER

Should a column consist of two equal-size plates as shown in Fig. 21-5 and should those plates not be connected, each plate will act as a separate column and each will resist approximately half of the total column load. In other words, the total moment of inertia of the column will equal two times the moment of inertia of one plate. The two "columns" will act the same and have equal deformations as shown in Fig. 21-5(b).

Should the two plates be connected sufficiently to prevent slippage on each other as shown in Fig. 21-6, they will act as a unit. Their moment of inertia may be computed for the whole built-up section as shown in the figure and will be four times as large as it was for the column of Fig. 21-5, where slipping between the plates was possible. The reader should also notice that the plates of the column of Fig. 21-6 will deform different amounts as the column bends laterally.

Should the plates be connected in a few places, it would appear that the strength of the resulting column would be somewhere in between the two cases just described.

Reference to Fig 21-5(b) shows that the greatest displacement between the two plates occurs at their ends and the least displacement occurs at middepth. As a result slip-critical connectors placed at column ends have the greatest strengthening effect while those placed at middepth have the least effect. (Slip-critical connections were discussed in Chapter 11.)

Should the plates be fastened together at their ends with slip-critical connectors, those ends will deform together, and the column will take the shape shown in Fig. 21-7.

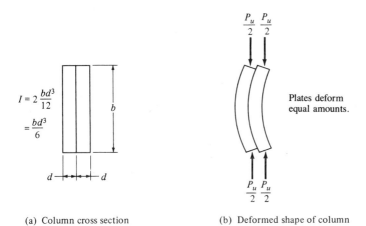

(a) Column cross section   (b) Deformed shape of column

Figure 21-5 Column consisting of two plates not connected to each other.

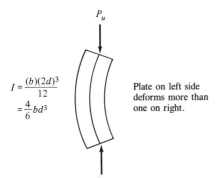

Figure 21-6 Column consisting of two plates fully connected to each other.

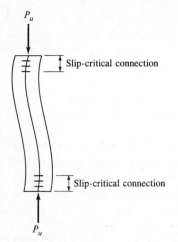

P_u

Slip-critical connection

Slip-critical connection

P_u

Figure 21-7 Column consisting of two plates fastened with slip-resistant connectors.

If the column were to bend in the S shape shown, its K factor would theoretically equal 0.5, and its KL/r value would be the same as the one for the continuously connected column of Fig. 21-6.[1]

$$\frac{KL}{r} \text{ for the column of Fig. 21-6} = \frac{(1)(L)}{\sqrt{\frac{4}{6}bd^3/2bd}} = 1.732L$$

$$\frac{KL}{r} \text{ for the end-fastened column of Fig. 21-7} = \frac{(0.5)(L)}{\sqrt{\frac{1}{6}bd^3/2bd}} = 1.732L$$

Thus the design stresses are equal for the two cases, and the columns theoretically would carry the same loads. This is true for the particular case described here but is not applicable for the common case where the parts of Fig. 21-7 begin to separate.

21-8 CONNECTION REQUIREMENTS FOR BUILT-UP COLUMNS WHOSE COMPONENTS ARE IN CONTACT WITH EACH OTHER

Several requirements concerning built-up columns are presented in LRFD Specification E4. When such columns consist of different components which are in contact with each other and which are bearing on base plates or milled surfaces, they must be connected at their ends with bolts or welds. If welds are used, the weld lengths must at least equal the maximum width of the member. If bolts or rivets are used, they may not be spaced longitudinally more than four diameters on center, and the connection must extend for a distance at least equal to $1\frac{1}{2}$ times the maximum width of the member.

---

[1] J. A. Yura, "Elements for Teaching Load and Resistance Factor Design" (Chicago: AISC, August 1987), pp. 17–19.

The LRFD Specification also requires the use of welded, bolted, or riveted connections between the end ones described in the last paragraph. These must be sufficient to provide for the transfer of calculated stresses. If it is desired to have a close fit over the entire faying surfaces between the components, it may be necessary to place the connectors even closer than is required for shear transfer.

When the component of a built-up column consists of an outside plate, the LRFD Specification provides specific maximum spacings for fastening. If intermittent welds are used along the edges of the components or if bolts or rivets are provided along all gage lines at each section, their maximum spacing may not be greater than $127/\sqrt{F_y}$ times the thickness of the thinner outside plate or 12 in. Should these fasteners be staggered, however, they may not be spaced farther apart on each gage line than $190/\sqrt{F_y}$ times the thickness of the thinner part or 18 in.

Other spacing values are given in LRFD Specification E4 for built-up members made from unpainted weathering steel.

For the discussion to follow, a represents the distance between connectors and r_i is the minimum radius of gyration of an individual component of the column. According to test results if the ratio a/r_i is ≤ 50, all the parts of a built-up column will act together as a unit. If the ratio is greater than 50, the LRFD Specification requires that a modified or larger column slenderness ratio $(KL/r)_m$ be used to determine the design stress in place of the value used if the whole cross section is fully effective $(KL/r)_0$. The modified value is

$$\left(\frac{KL}{r}\right)_m = \sqrt{\left(\frac{KL}{r}\right)_0^2 + \left(\frac{a}{r_i} - 50\right)^2} \qquad \text{(LRFD Formula E4-2)}$$

This equation is only applicable to the buckling axis where connectors are required to resist the shear.

It is to be clearly remembered that the design strength of a built-up column will be reduced if the spacing of the connectors is such that one of the components of the column can buckle before the whole column buckles. Such a situation will be prevented if the ratio a/r_i is kept $\leq$ the governing ratio, that is, the lesser value of $(KL/r)_x$ or $(KL/r)_y$ for the whole member.

For built-up columns whose components are connected at their ends only with snug-tight bolts (described in Chapter 11), the LRFD Specification states that the following slenderness ratio is to be used:

$$\left(\frac{KL}{r}\right)_m = \sqrt{\left(\frac{KL}{r}\right)_0^2 + \left(\frac{a}{r_i}\right)^2} \qquad \text{(LRFD Formula E4-1)}$$

The column tables for double angles given in Part 2 of the LRFD Manual were developed on the basis of the modified slenderness ratio and also upon flexural-torsional buckling as described in Section 21.9 and in the Appendix.

Example 21-7 illustrates the design of a column consisting of a W section with cover plates welded to its flanges as shown in Fig. 21-8. It is assumed that the plates are connected to the W section at its ends and at intermediate points so the parts act together and $a/r_i \leq 50$. As this type of section is not shown in the column tables of the LRFD Manual, it is necessary to use a trial-and-error procedure. An effective slenderness ratio is assumed; $\phi_c F_{cr}$ for that KL/r is determined and divided into the column design load to estimate the total area required. The area of the W section is subtracted from the estimated total area to obtain the estimated cover plate area. A cover plate size is selected to provide the estimated area, and then $\phi_c P_n$ is calculated for the whole section, after which it may be necessary to revise the cover plate size and try again.

■ Example 21-7

It is desired to design a column for $P_u = 1740$ k, using A36 steel and $KL = 14$ ft. A W12 × 120 (for which $\phi_c P_n = 928$ k from the Part 2 tables of the LRFD Manual) is on hand. Design cover plates to be welded to the W section as shown in Fig. 21-8 to enable the column to support the required load. Assume $a/r_i \leq 50$.

Solution.   Assume $KL/r = 40$:

$$\phi_c F_{cr} = 28.13 \text{ ksi}$$

$$A_{\text{reqd}} = \frac{1740}{28.13} = 61.86 \text{ in.}^2$$

$$-A \text{ of W12} \times 120 = \underline{-35.30}$$
$$A \text{ of 2 cover plates} = 26.56 \text{ in.}^2 \text{ or } 13.28 \text{ in.}^2 \text{ each}$$

Try 1 PL $\frac{3}{4}$ × 18 each flange²

$$A = 35.30 + (2)(\tfrac{3}{4})(18) = 62.3 \text{ in.}^2$$
$$I_x = 1070 + (2)(\tfrac{1}{12})(\tfrac{3}{4})(18)(\tfrac{3}{4})^3 + (2)(\tfrac{3}{4})(18)(6.935)^2$$
$$= 2369 \text{ in.}^4$$
$$I_y = 345 + (2)(\tfrac{1}{12})(\tfrac{3}{4})(18)^3 = 1074 \text{ in.}^4$$
$$r_y = \sqrt{\frac{1074}{62.3}} = 4.15 \text{ in.}$$

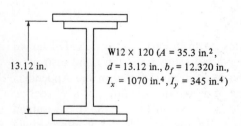

13.12 in.

W12 × 120 (A = 35.3 in.²,
d = 13.12 in., b_f = 12.320 in.,
I_x = 1070 in.⁴, I_y = 345 in.⁴)

Figure 21-8

$$\frac{KL}{r} = \frac{(12)(14)}{4.15} = 40.48$$

$$\phi_c F_{cr} = 28.07 \text{ ksi}$$

$$\phi_c P_n = (28.07)(62.3) = 1749 \text{ k} > 1740 \text{ k} \qquad \text{OK}$$

Use W12 × 120 with 1 cover plate $\frac{3}{4}$ × 18 each flange[2]   ■

21-9 FLEXURAL-TORSIONAL BUCKLING OF COMPRESSION MEMBERS

Axially loaded compression members can theoretically fail in three different fashions: by flexural buckling, by torsional buckling, or by flexural-torsional buckling.

Flexural buckling (also called Euler buckling) is the situation considered up to this point in our column discussions where we have computed slenderness ratios for the principal column axes and determined $\phi_c F_{cr}$ for the highest ratios so obtained. Doubly symmetrical column members (such as W sections) are subject only to flexural buckling and torsional buckling.

As torsional buckling can be very complex, it is very desirable to prevent its occurrence. This may be done by careful arrangements of the members and by providing bracing to prevent lateral movement and twisting. If sufficient end supports and intermediate lateral bracing are provided, flexural buckling will always control. The values given in the LRFD column tables for W, M, S, tube, and pipe sections are based on flexural buckling.

Open sections such as Ws, Ms, and channels have little torsional strength, but box beams have a great deal. Thus if a torsional situation is encountered, it may be well to use box sections or to make box sections out of W sections by adding welded side plates (⊔⊓). Another way in which torsional problems can be reduced is to shorten the lengths of members that are subject to torsion.

For a singly symmetrical section such as a tee or double angle, Euler buckling may occur about the x or y axes. For equal-leg single angles Euler buckling may occur about the z axis. For all these sections flexural-torsional buckling is definitely a possibility and may control. (It will always control for unequal-leg single-angle columns.) The values given in the LRFD column load tables for double-angle and structural tee sections were computed for buckling about the weaker of the x or y axes and for flexural-torsional buckling.

Usually symmetrical members such as W sections are used as columns. Torsion will not occur in such sections if the lines of action of the lateral loads pass through their shear centers. The *shear center* is that point in the cross section of a member through which the resultant of the transverse loads must pass

---

[2] Many other plate sizes could be selected.

so that no torsion will occur. The calculations necessary to locate shear centers were presented in Chapter 9. The shear centers of the commonly used doubly symmetrical sections occur at their centroids. This is not necessarily the case for other sections such as channels and angles. Shear center locations for several types of sections are shown in Fig. 21-9. Also shown in the figure are the coordinates x_0 and y_0 for the shear center of each section with respect to its centroid. These values are needed to solve the flexural-torsional formulas, presented in the appendix.

Even though loads pass through shear centers, torsional buckling may still occur. If you load any section through its shear center, no torsion will occur—but you still compute torsional buckling strength for these members; that is, buckling load does not depend on the nature of the axial or transverse loading, rather it depends on the cross-section properties, column length, and support conditions.

The average designer does not consider the torsional buckling of symmetrical shapes or the flexural-torsional buckling of unsymmetrical shapes. The usual feeling is that these conditions don't control the critical column loads or at least don't affect them very much. Should we have unsymmetrical columns or even symmetrical columns made up of thin plates, however, we will find that torsional buckling or flexural-torsional buckling may significantly reduce column capacities.

In Appendix E of the LRFD Specification a long list of formulas is presented for computing the flexural-torsional strength of column sections. The values given for column design strengths ($\phi_c P_n$ values) for double angles and tees in Part 2 of the LRFD Manual make use of these formulas. The LRFD does not provide tables for single-angle columns. It is stated in the manual that

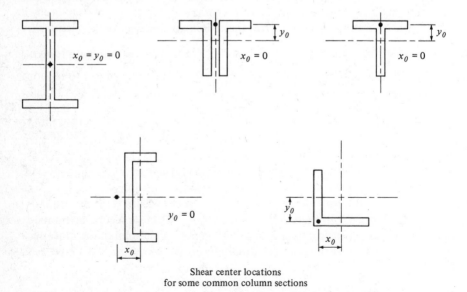

Shear center locations
for some common column sections

Figure 21-9 Shear center locations for some common column sections.

this is because of the difficulty of loading such members concentrically. This is a rather puzzling statement because, even if double angles or tees are loaded concentrically, flexural-torsional buckling may occur. Why then does the LRFD Manual not provide similar loading tables for single angles? The LRFD feels that in practice actual eccentricities for single-angle members are rather large, and they feel that neglect of these eccentricities may lead to the use of some much underdesigned members. This discussion boils down to the fact that the designer using single-angle struts is going to have to wade through the flexural-torsional formulas and make some allowances for eccentricities of loading. A sample problem of this type is presented on pages 2-49 and 2-50 of the LRFD Manual. In addition an example appears in the Appendix of this text (Example A-3).

PROBLEMS

21-1 Using A36 steel and the LRFD Manual, determine the design strength $\phi_c P_n$ for each of the compression members.
(a) W12 × 87 with KL = 9 ft (*Ans.* 734 k)
(b) W14 × 120 with KL = 17 ft (*Ans.* 924 k)
(c) S12 × 50 with KL = 10 ft (*Ans.* 220.2 k)

21-2 Using the LRFD Specification, determine the design strength $\phi_c P_n$ for each of the columns.
(a) W10 × 45 with fixed ends, L = 16 ft 6 in., A36 steel
(b) W14 × 82 with pinned ends, L = 20 ft 0 in., A36 steel
(c) W12 × 79 with one end fixed and the other pinned, L = 24 ft 6 in., F_y = 50 ksi
(d) W14 × 132 with fixed ends, L = 22 ft 0 in., F_y = 50 ksi
(e) Pipe 10 Std. with pinned ends, L = 20 ft 0 in., A36 steel
(f) Two L8 × 8 × 1s separated $\frac{3}{8}$ in. (for gusset PLs at ends), pinned ends, L = 24 ft 6 in., A36 steel

21-3 A W12 × 106 with a $\frac{3}{4}$ × 12-in. cover plate bolted to each flange is to be used for a column with KL = 18 ft. Find its design strength $\phi_c P_n$ if A36 steel is used. (*Ans.* $\phi_c P_n$ = 1191.6 k)

21-4 to 21-7   Determine the design strength $\phi_c P_n$ for the axially loaded compression members shown, which consist of A36 steel.

21-4

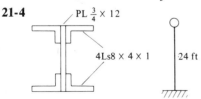

PL $\frac{3}{4}$ × 12

4Ls8 × 4 × 1   24 ft

21-5 (*Ans.* 920.3 k)

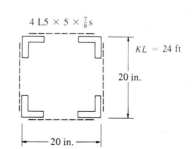

4 L5 × 5 × $\frac{7}{8}$ s

KL = 24 ft

20 in.

20 in.

21-6

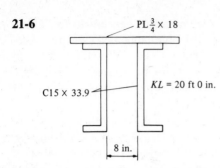

PL$\frac{3}{4}$ × 18

C15 × 33.9

KL = 20 ft 0 in.

8 in.

21-7 (*Ans.* 1424.1 k)

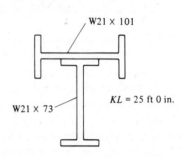

W21 × 101

W21 × 73

KL = 25 ft 0 in.

21-8 A 27-ft axially loaded W10 × 68 column is laterally supported perpendicular to its y axis at middepth. Determine $\phi_c P_n$ for this column if A36 steel is used and if effective length factors = 1.0 are assumed in each case.

21-9 to 21-11 Use a trial-and-error procedure in which a KL/r value is estimated, a value of $\phi_c F_{cr}$ determined from the appropriate LRFD table (3-36 or 3-50 in Part 6 of the Manual), the estimated column area determined, a trial section selected, its P_u computed, and then another size tried if necessary, and so on.

21-9 Select the lightest available W12 section to support the axial compression loads $P_D = 100$ k and $P_L = 160$ k if $KL = 15$ ft and A36 steel is used. (*Ans.* W12 × 58)

21-10 Select the lightest available W14 section to support the axial loads $P_D = 180$ k and $P_L = 320$ k if $KL = 16$ ft and A36 steel is used.

21-11 Repeat Prob 21-10 with $F_y = 50$ ksi. (*Ans.* W14 × 90)

21-12 to 21-24 Take advantage of all available column tables in the LRFD Manual, particularly those in Part 2.

21-12 Repeat Prob. 21-9.

21-13 Repeat Prob. 21-10. (*Ans.* W14 × 99)

21-14 Repeat Prob. 21-11.

21-15 Several building columns are to be designed using A36 steel and the LRFD Specification. Select the lightest available W sections for these columns.
(a) $P_u = 600$ k, $L = 12$ ft, pinned-end supports (*Ans.* W12 × 79)
(b) $P_u = 400$ k, $L = 14$ ft, fixed-end supports (*Ans.* W10 × 49)
(c) $P_u = 660$ k, $L = 16$ ft 6 in., fixed at bottom, pinned at top (*Ans.* W12 × 87)
(d) $P_u = 1750$ k, $L = 15$ ft, pinned-end supports (*Ans.* W12 × 230)

21-16 Select the lightest W section available in A36 steel to serve as a pinned-end column to support the following axial loads: $P_D = 200$ k, $P_L = 100$k and P_W due to wind = 220 k. Assume KL is 14 ft.

21-17 A W section is to be selected to support an axial compressive load $P_u = 1600$ k. The member, which is to be 24 ft long and is to be pinned top and bottom, has lateral support (pinned) supplied in the weak direction at middepth. Select the lightest W12 and W14 sections using A36 steel. (*Ans.* W12 × 210, W14 × 211)

21-18 Repeat Prob. 21-17 with $F_y = 50$ ksi.

21-19 Repeat Prob. 21-17 if pinned lateral support is provided in the weak direction at the one-third points and the total column length is changed to 33 ft. (*Ans.* W14 × 233; W12 × 230)

21-20 A 27-ft column is laterally supported in the weak direction at its middepth. Select the lightest W section that can adequately support the axial gravity loads P_D = 100 k and P_L = 300 k using A36 steel. Assume all K's are 1.0.

21-21 Repeat Prob. 21-20 with F_y = 50 ksi. (*Ans.* W12 × 65)

21-22 A W14 section of A36 steel is to be selected to support the axial compressive loads P_D = 150 k and P_L = 320 k. The member which is to be 30 ft long is to be fixed top and bottom and is to have lateral support at its one-third points perpendicular to the y axis (pinned).

21-23 Using A36 steel (except F_y = 46 ksi for square and rectangular tubing), select the lightest available rolled sections (W, M, S, HP, square, rectangular, or round tubing) that are adequate for the following situations:
   (a) P_u = 420 k, L = 14 ft, pinned ends (*Ans.* W10 × 60, tube 10 × 10 × $\frac{3}{8}$, etc.)
   (b) P_u = 375 k, L = 15 ft, fixed ends (*Ans.* W10 × 49, tube 8 × 8 × $\frac{3}{8}$, etc.)
   (c) P_u = 680 k, L = 20 ft, one end pinned and the other fixed (*Ans.* W14 × 90, tube 14 × 14 × $\frac{3}{8}$, etc.)

21-24 Assuming axial loads only, select W sections for an interior column of the frame shown. Use A36 steel and the LRFD Specification. Each column section can be used for one or two stories before it is spliced, whichever seems advisable. Miscellaneous data: concrete weighs 150 lb/ft³. *LL* on roof = 30 psf. Roofing = 6 psf. *LL* on interior floors = 80 psf. Partition load on interior floors = 15 psf. All points pinned. Frames 30 ft on center.

21-24

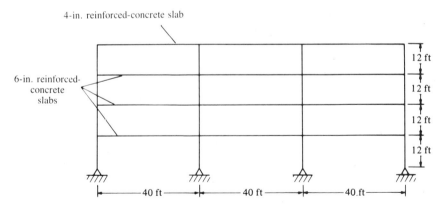

4-in. reinforced-concrete slab

6-in. reinforced-concrete slabs

12 ft

12 ft

12 ft

12 ft

40 ft   40 ft   40 ft

22

Design of Beams—LRFD

22-1 INTRODUCTION

In this chapter the buckling moments of a series of compact ductile steel beams with different lateral bracing situations are considered. First the beams will be assumed to have continuous lateral bracing for their compression flanges. Next the beams will be assumed to be braced laterally at short intervals. Then the beams will be assumed to be braced at larger and larger intervals.

In Fig. 22-1 a typical curve showing the nominal resisting or buckling moments of one of these beams with varying unbraced lengths is shown. An examination of this figure will show that the beams have three distinct ranges or zones of buckling depending on their lateral bracing situation. If we have continuous or closely spaced lateral bracing, the beams will buckle plastically and fall into what is classified as zone 1 buckling. As the distance between lateral bracing is increased further, the beams will begin to fail inelastically at smaller moments and fall into zone 2. Finally, with even larger unbraced lengths, the beams will fail elastically and fall into zone 3. The first few sections of this chapter are concerned with these three types of buckling, while shear, deflection, and continuous beams are discussed in the last part of the chapter.

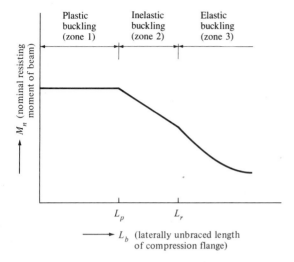

Figure 22-1 Nominal moment as a function of unbraced length of compression flange.

Plastic Buckling (Zone 1)

If we were to take a compact beam whose compression flange is continuously braced laterally, we would find that we could load it until its full plastic moment M_p is reached; further loading then produces a redistribution of moments as was described in Chapter 18. In other words, the moments in these beams can reach M_p and then develop a rotation capacity sufficient for moment redistribution.

If we now take one of these compact beams and provide closely spaced intermittent lateral bracing for its compression flanges, we will find that we can still load it until the plastic moment plus moment redistribution is achieved if the spacing between the bracing does not exceed a certain value called L_p herein. (The value of L_p is dependent on the dimensions of the beam cross section and on its yield stress.)

Inelastic Buckling (Zone 2)

If we now further increase the spacing between points of lateral bracing, the section may be loaded until some but not all of the compression fibers are stressed to F_y. The section will have insufficient rotation capacity to permit full moment redistribution and thus will not permit plastic analysis. In other words, in this zone we can bend the member until the yield stress is reached in some but not all of its compression elements before buckling occurs. This is referred to as inelastic buckling.

As we increase the unbraced length, we will find that the moment the section resists will decrease until finally it will buckle before the yield stress is anywhere reached. The maximum unbraced length at which we can still reach F_y at one point is the end of the inelastic range. It is shown as L_r in Fig. 22-1; its value is dependent upon the properties of the beam cross section as well as on the yield and residual stresses of the beam. At this point, as soon as we have a moment which theoretically causes the yield stress to be reached anywhere (actually it is less than F_y because of residual stresses), the section will buckle.

150 Federal Street, Boston, Mass. (Courtesy of Owen Steel Company, Inc.)

Elastic Buckling (Zone 3)

If the unbraced length is greater than L_r, the section will buckle elastically before the yield stress is reached anywhere. As the unbraced length is further increased, the buckling moment becomes smaller and smaller. As the moment is increased in such a beam, the beam will deflect more and more transversely until a critical moment value M_{cr} is reached. At this time the beam cross section will twist, and the compression flange will move laterally. The moment M_{cr} is provided by the torsional resistance and the warping resistance of the beam, as will be discussed in Section 22-6.

22-2 PLASTIC BUCKLING, ZONE 1

In this section and the next, two beam formulas for plastic buckling (zone 1) are presented, while in Sections 22-4 through 22-6 formulas are presented for inelastic buckling (zone 2) and elastic buckling (zone 3).

If the unbraced length L_b of the compression flange of a compact I- or C-shaped section including hybrid members does not exceed L_p (if elastic analysis is being used) or L_{pd} (if plastic analysis is being used), then the member's bending strength about its major axis may be determined as follows:

$$M_n = M_p = F_y Z$$
$$M_u = \phi_b M_n = \phi_b F_y Z \qquad \text{with } \phi_b = 0.9$$

For elastic analysis L_b may not exceed the value L_p if M_n is to equal $F_y Z$.

$$L_p = \frac{300 r_y}{\sqrt{F_{yf}}} \qquad \text{(LRFD Formula F1-4)}$$

For plastic analysis L_b (which is defined as the laterally unbraced length of the compression flange at plastic hinge locations associated with failure mechanisms) may not exceed the value L_{pd} if M_n is to equal $F_y Z$.

$$L_{pd} = \frac{3600 + 2200(M_1/M_p)}{F_y} r_y \qquad \text{(LRFD Formula F1-1)}$$

In this expression M_1 is the smaller moment at the end of the unbraced length of the beam, and the ratio M_1/M_p is positive when the moments cause the member to be bent in double curvature (⌢⌣) and negative if they bend it in single curvature (⌣⌣). Only steels with F_y values (F_y is the specified minimum yield stress of the compression flange) of 65 ksi or less may be considered. Higher-strength steels may not be ductile.

There is no limit of the unbraced length for circular or square cross sections or for I-shaped beams bent about their minor axes. (If I-shaped sections are bent about their minor or y axes they will not buckle before the full plastic moment M_p about the y axis is developed.) Section F of the LRFD Specification provides other values for L_p and L_{pd} for solid rectangular bars and symmetric box beams.

For these sections to be compact the width-thickness ratios of the flanges and webs of I- and C-shaped sections are limited to the maximum values to follow which are taken from Table B5.1 of the LRFD Specification.

$$\lambda_p = \frac{b_f}{2t_f} \leq \frac{65}{\sqrt{F_y}} \quad \text{for flanges}$$

$$\lambda_p = \frac{h_c}{t_w} \leq \frac{640}{\sqrt{F_y}} \quad \text{for webs}$$

In this expression h_c is the distance from the web toe of the fillet in the top of the web to the web toe of the fillet in the bottom of the web (that is, twice the distance from the neutral axis to the inside face of the compression flange less the fillet or corner radius).

22-3 DESIGN OF BEAMS, ZONE 1

Included in the items that need to be considered in beam design are moments, shears, deflections, crippling, lateral bracing for the compression flanges, and fatigue. Beams will probably be selected that provide sufficient design moment capacities ($\phi_b M_n$) and then checked to see if any of the other items are critical. The factored moment will be computed, and a section having that much design moment capacity will be initially selected from the LRFD Manual.

From the table entitled "Load Factor Design Selection Table for Shapes Used as Beams" in Section 3 of the LRFD Manual, steel shapes having sufficient plastic moduli to resist certain moments can quickly be selected. The sections are arranged in various groups having certain ranges of plastic moduli. The bold-type section at the top of each group is the lightest section in that group, and the others are arranged in the order of their plastic moduli.

The examples to follow illustrate the design of compact steel beams whose compression flanges have full lateral support or bracing, thus permitting plastic analysis. For the selection of such sections the designer may enter the tables either with the required plastic modulus or with the factored design moment (if $F_y = 36$ or 50 ksi).

■ Example 22-1

Select a beam section for the span and loading shown in Fig. 22-2, assuming full lateral support is provided for the compression flange by the floor slab above (that is, $L_b = 0$). Use A36 steel with $F_y = 36$ ksi.

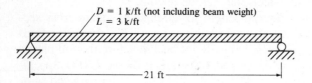

Figure 22-2

Solution.   Assume beam weight = 55 lb/ft:

$$w_u = (1.2)(1.055) + (1.6)(3) = 6.07 \text{ k/ft}$$

$$M_u = \frac{(6.07)(21)^2}{8} = 334.6 \text{ ft-k}$$

$$Z_{x\,reqd} = \frac{M_u}{\phi_b F_y} = \frac{(12)(334.6)}{(0.9)(36)} = 123.9 \text{ in.}^3$$

<u>Use W24 × 55</u>   ■

■ Example 22-2

The 5-in. reinforced-concrete slab shown in Fig. 22-3 is to be supported with steel W sections 8 ft 0 in. on centers. The beams, which will span 20 ft, are assumed to be simply supported. If the concrete slab is designed to support a live load of 100 psf, determine the lightest steel section required to support the slab. It is assumed that the compression flange of the beam will be fully supported laterally by the concrete slab. The concrete weighs 150 lb/ft³. Use A36 steel.

Solution.   Dead loads:

$$\text{Slab} = (\tfrac{5}{12})(150)(8) = 500 \text{ lb/ft}$$
$$\text{Estimated beam wt} = \underline{26 \text{ lb/ft}}$$
$$\text{Total} = 526 \text{ lb/ft}$$

$$w_u = (1.2)(526) + (1.6)(8 \times 100) = 1911 \text{ lb/ft} = 1.911 \text{ k/ft}$$

$$M_u = \frac{(1.911)(20)^2}{8} = 95.55 \text{ ft-k}$$

$$Z_{x\,reqd} = \frac{(12)(95.55)}{(0.9)(36)} = 35.4 \text{ in.}^3$$

<u>Use W12 × 26</u>   ■

The LRFD Specification (B1) does not require the subtraction of holes in either flange of a beam, provided the hole area in any one flange does not ex-

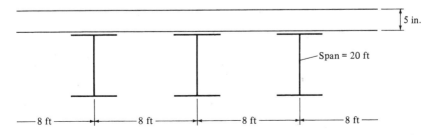

Figure 22-3

ceed 15 percent of the gross area of that flange, and then the deduction is only for the area in excess of 15 percent. Furthermore the LRFD does not make a distinction between holes in the compression and tension flanges. Although the 15 percent value is permitted by the LRFD, some specifications (notably the bridge ones) and a good many designers have not adopted the idea and follow the more conservative practice of deducting all holes.

The usual practice is to subtract the same area of holes from both flanges whether they are present or not. For a section with two holes in the tension flange only, the properties of the section would be computed based on the subtraction of two holes from the tension flange and two holes from the compression flange. (If bolts are placed only in the compression flange, they are neglected.) Example 22-3 illustrates this method. Again the author has made a few preliminary calculations in estimating the member size.

■ Example 22-3

Select the lightest section available for the beam of Fig. 22-4, assuming that it will be necessary to punch holes for two 1-in. bolts in the tension flange at the point of maximum moment. Use A36 steel and assume full lateral bracing for the compression flange.

Solution.   Assume beam weight = 84 lb/ft:

$$w_u = (1.2)(1.284) + (1.6)(2.3) = 5.22 \text{ k/ft}$$

$$M_u = \frac{(5.22)(30)^2}{8} = 587.3 \text{ ft-k}$$

$$Z_{x \, reqd} = \frac{(12)(587.3)}{(0.9)(36)} = 217.5 \text{ in.}^3$$

It appears from the table that we should use a W24 × 84, but to account for the holes the author estimated that a little deeper section would be required and thus assumed a W27 × 84 for the following calculations.

Try A W27 × 84 (see Fig. 22-5 for dimensions).

Assuming two holes in each flange, Z of holes about the neutral axis is calculated.

Area of 2 holes in each flange = $(2)(1\frac{1}{8})(0.640) = 1.44 \text{ in.}^2$

−15 percent of flange area as per Specification B1 of LRFD

Area to be subtracted = $1.44 - (0.15)(9.96)(0.640) = 0.484 \text{ in.}^2$

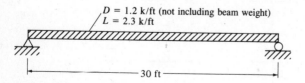

D = 1.2 k/ft (not including beam weight)
L = 2.3 k/ft

30 ft

Figure 22-4

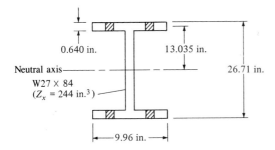

Figure 22-5

Calculation of Z_{net} by taking hole area from both flanges:

$$Z_{net} = 244 - (0.484)(13.035)(2) = 231.4 \text{ in.}^3 > 217.5 \text{ in.}^3 \quad \text{OK}$$

$$\underline{\underline{\text{Use W27} \times 84}} \quad \blacksquare$$

22-4 INTRODUCTION TO INELASTIC BUCKLING, ZONE 2

When intermittent lateral bracing is supplied for the compression flange of a beam section such that the member can be bent until the yield strain is reached in some but not all of its compression elements before lateral buckling occurs, we have inelastic buckling. In other words, the bracing is insufficient to permit the member to reach a full plastic strain distribution before buckling occurs.

Because of the presence of residual stresses (discussed in Section 5-2), yielding will begin in a section at applied stresses equal to $F_{yw} - F_r$, where F_{yw} is the yield stress of the web, and F_r equals the compressive residual stress assumed equal to 10 ksi for rolled shapes and 16.5 ksi for welded shapes. It should be noted that the definition of plastic moment $F_y Z$ in zone 1 is not affected by residual stresses because the sum of the compressive residual stresses equals the sum of the tensile residual stresses in the section and the net effect is theoretically zero.

If the unbraced length L_b of a compact I- or C-shaped section is larger than L_p, the beam will fail inelastically unless L_b is greater than a distance L_r (to be discussed) beyond which the beam will fail elastically before F_y is anywhere reached (thus falling into zone 3).

Bending Coefficients   In the formulas for inelastic and elastic buckling presented in these next few sections, we will use a term C_b. This is a moment coefficient that is included in the formulas to account for the effect of different moment gradients on lateral-torsional buckling. In other words, lateral buckling may be appreciably affected by the end restraint and loading conditions of the member.

As an illustration, the reader can see that the moment in the unbraced beam of Fig. 22-6(a) causes a worse compression flange situation than does the moment in the unbraced beam of part (b). For one reason, the upper flange of the beam in part (a) is in compression for its entire length, while in (b) the length of the "column," that is, the length of the upper flange which is in compression, is much less (thus a much shorter "column").

For the simply supported beam of part (a) of the figure, C_b is taken as 1.0, while for the beam of part (b) it is taken as larger than 1.0. The basic moment capacity equations for zones 2 and 3 were developed for laterally unbraced beams subject to single curvature with $C_b = 1.0$. Frequently, beams are not bent in single curvature, with the result that they can resist more moment. To handle this situation the LRFD Specification provides moment or C_b coefficients larger than 1.0 which are to be multiplied by the computed M_n values. The results are higher moment capacities. The designer who conservatively says, "I'll always use $C_b = 1.0$," is missing out on the possibility of significant savings in steel weight for some situations. *When using C_b values, the designer should clearly understand that the moment capacity obtained by multiplying M_n by C_b may not be larger than the plastic M_n of zone 1 which is equal to F_yZ.* This situation is illustrated in Fig. 22-7.

The value of C_b is determined from the expression to follow, in which M_1 is the smaller and M_2 the larger of the bending moments at the ends of the unbraced length taken about the strong axis of the member. Should the moment at any point within the unbraced length be larger than the end moments, C_b shall be taken as 1.0. The ratio M_1/M_2 is considered positive if M_1 and M_2 have the same sign (reverse curvature bending) and negative if they have opposite signs (single curvature bending).

$$C_b = 1.75 + 1.05\left(\frac{M_1}{M_2}\right) + 0.3\left(\frac{M_1}{M_2}\right)^2 \le 2.3$$

C_b is equal to 1.0 for unbraced cantilever beams and for beams that have a moment over an appreciable part of their unbraced span equal to or larger than

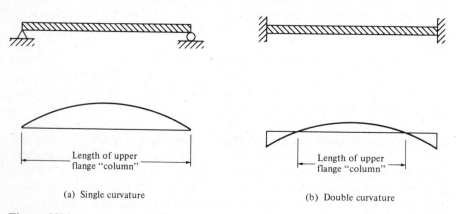

(a) Single curvature   (b) Double curvature

Figure 22-6

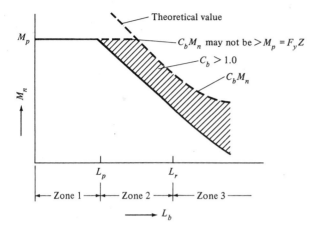

Figure 22-7

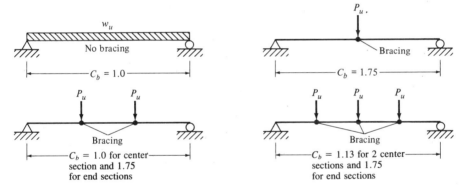

Figure 22-8

the larger of the segment's end moments. Some typical values of C_b are shown in Fig. 22-8 for different beam and loading conditions. Table 7 in Part 6 of the LRFD Manual presents calculated values of C_b for different M_1/M_2 ratios.

22-5 MOMENT CAPACITIES, ZONE 2

As the unbraced length of the compression flange of a beam is increased beyond L_p, the moment capacity of the section will become smaller and smaller. Finally at an unbraced length L_r the section will buckle elastically as soon as the yield stress is reached at any point. Owing to the rolling operation, however, there is a residual stress in the section equal to F_r. Thus the elastically computed stress caused by bending can only reach $F_{yw} - F_r$. Assuming $C_b =$

1.0, the permissible moment capacity for a compact I- or C-shaped section bent about its x axis may be determined as follows if $L_b = L_r$:

$$M_u = \phi_b M_r = \phi_b S_x (F_{yw} - F_r)$$

L_r is a function of several of the section's properties such as its cross-sectional area, modulus of elasticity, yield stress, and warping and torsional properties. The very complex formulas needed for its computation are given in the LRFD Specification (F2), and space is not taken to show them here. Fortunately numerical values have been determined for sections normally used as beams and are given in the "Load Factor Design Selection Table."

Going backward from an unbraced length of L_r toward an unbraced length L_p, we can see that buckling does not occur when the yield stress is first reached. We are in the inelastic range (zone 2) where there is some penetration of the yield stress into the section from the extreme fibers. For these cases when the unbraced length falls between L_p and L_r, the moment capacity will fall approximately on a straight line between $M_u = \phi_b F_y Z$ and L_p and $\phi_b S_x (F_{yw} - F_r)$ at L_r. For intermediate values of the unbraced length the moment capacity may be determined by straight line proportions or by substituting into the expression at the end of this paragraph. If C_b is larger than 1.0, the section will resist additional moment but not more than $\phi_b F_y Z = \phi_b M_p$

$$\phi_b M_n = C_b [\phi_b M_p - BF(L_b - L_p)] \le \phi_b M_p$$

BF is a factor given in the "Load Factor Design Selection Table" for each section which enables us to do the interpolation or proportioning with a simple formula.

Example 22-4 illustrates the determination of the moment capacities of a section with L_b between L_p and L_r, while Example 22-5 demonstrates the design of a beam in the same range.

■ Example 22-4

Determine the moment capacity of a W24 × 62 with $F_y = 36$ ksi and again with $F_y = 50$ ksi if $L_b = 8.0$ ft and $C_b = 1.0$.

Solution.   For $F_y = 36$ ksi:
From the "Load Factor Design Selection Table" for a W24 × 62

$$L_p = 5.8 \text{ ft}$$
$$L_r = 17.2 \text{ ft}$$
$$\phi_b M_r = 255 \text{ ft-k}$$
$$\phi_b M_p = 413 \text{ ft-k}$$
$$BF = 13.8 \text{ k}$$

Since $L_b > L_p < L_r$ the section is in zone 2 for inelastic buckling and $\phi_b M_n$ can be determined as follows:

$$\phi_b M_n = C_b [\phi_b M_p - BF(L_b - L_p)]$$
$$= 1.0[413 - (13.8)(8.0 - 5.8)] = \underline{382.6 \text{ ft-k}}$$

Or directly by proportions:

$$\phi_b M_n = 255 + \left(\frac{17.2 - 8.0}{17.2 - 5.8}\right)(413 - 255) = 382.5 \text{ ft-k}$$

For $F_y = 50$ ksi:

From the "Load Factor Design Selection Table"

$$L_p = 4.9 \text{ ft}$$
$$L_r = 13.3 \text{ ft}$$
$$\phi_b M_r = 393 \text{ ft-k}$$
$$\phi_b M_p = 574 \text{ ft-k}$$
$$BF = 21.4 \text{ k}$$

Since $L_b > L_p < L_r$,

$$\phi_b M_n = 1.0[574 - (21.4)(8.0 - 4.9)] = \underline{507.7 \text{ ft-k}} \qquad \blacksquare$$

■ **Example 22-5**

Select the lightest available section for a factored moment of 290 ft-k if $L_b = 10.0$ ft. Use A36 steel and assume $C_b = 1.0$.

Solution. Enter the "Load Factor Design Selection Table" and notice that $\phi_b M_p$ for a W21 × 50 is 297 ft-k but L_p is 5.4 ft $< L_b$ of 10.0 ft. Also $\phi_b M_r = 184$ ft-k and $BF = 10.5$ k.

$$\phi_b M_n = 1.0[297 - (10.5)(10.0 - 5.4)]$$
$$= 248.7 \text{ ft-k} < 290 \text{ ft-k} \qquad\qquad \text{NG}$$

Moving up in the "Load Factor Design Selection Table," try a W24 × 55 (with $L_p = 5.6$ ft, $L_r = 16.6$ ft, $\phi_b M_r = 222$ ft-k, $\phi_b M_p = 362$ ft-k, and $BF = 12.7$ k).

$$\phi_b M_n = 1.0[362 - (12.7)(10.0 - 5.6)]$$
$$= 306.1 \text{ ft-k} > 290 \text{ ft-k} \qquad\qquad \text{OK}$$
$$\text{Use W24 × 55}$$

Note: A much easier solution is presented in the next section. ■

22-6 ELASTIC BUCKLING, ZONE 3

If the unbraced length of the compression flange of a beam section is greater than L_r, the section will buckle elastically before the yield stress is reached anywhere in the section. In Section F1.4 of the LRFD Specification the classic equation for determining this flexural-torsional buckling moment called M_{cr} is presented.

$$M_{cr} = C_b \frac{\pi}{L_b} \sqrt{EI_y GJ + \left(\frac{\pi E}{L_b}\right)^2 I_y C_w} \qquad \text{(LRFD Formula F1-13)}$$

In this equation G is the shear modulus of elasticity of the steel = 11,200 ksi, J is a torsional constant (in.^{4}), while C_w is the warping constant (in.^{6}). The values of J and C_w are provided for rolled sections in the tables entitled "Torsion Properties" in Part 1 of the LRFD Manual.

This expression is applicable to compact doubly symmetric I-shaped members, channel sections loaded in the plane of their webs, and I-shaped singly symmetric sections with their compression flanges larger than their tension ones $\left(\mathbf{I}\right)$. Expressions are also given in Section F1.4 and F1.5 of the LRFD Specification for M_{cr} in the elastic range for other sections such as solid rectangular bars, symmetric box sections, tees, and double angles.

Example 22-6 illustrates the computation of $\phi_b M_{cr}$ for an elastic buckling situation.

■ Example 22-6

Compute $M_u = \phi_b M_{cr}$ for a W18 × 97 consisting of A36 steel if the unbraced length L_b is 44 ft. Assume $C_b = 1.0$.

Solution.   Note L_b = 44 ft > L_r = 38.1 ft from "Load Factor Design Selection Table" (Part 3, LRFD Manual).

The following values for the W18 × 97 are also obtained from the LRFD Manual: $I_y = 201$ in.^{4}, $J = 5.86$ in.^{4}, and $C_w = 15,800$ in.^{6}

$$\phi_b M_n = \phi M_{cr} = M_u =$$

$$(0.9)(1.0)\left(\frac{\pi}{12 \times 44}\right) \sqrt{(29 \times 10^3)(201)(11,200)(5.86) + \left(\frac{\pi \, 29 \times 10^3}{44 \times 12}\right)^2 (201)(15,800)}$$

$$= 3698.9 \text{ in.-k} = \underline{\underline{308.2 \text{ ft-k}}}$$

This value may be checked with the LRFD Manual charts described in the next few paragraphs. There we will be able to read $\phi M_n' = M_u = 308$ ft-k. ■

The LRFD Specification (F1.4) also presents the elastic buckling equation in an alternate form as follows:

$$M_{cr} = \frac{C_b S_x X_1 \sqrt{2}}{L_b/r_y} \sqrt{1 + \frac{X_1^2 X_2}{2(L_b/r_y)^2}} \qquad \text{(LRFD Formula F1-13)}$$

where

$$X_1 = \frac{\pi}{S_x} \sqrt{\frac{EGJA}{2}} \qquad \text{(LRFD Formula F1-8)}$$

$$X_2 = 4\frac{C_w}{I_y}\left(\frac{S_x}{GJ}\right)^2 \qquad \text{(LRFD Formula F1-9)}$$

Values of X_1 and X_2 are shown for W shapes in the "Properties for W Shapes" section of Part 1 of the LRFD Manual.

Fortunately the values of $\phi_b M_{cr} = \phi_b M_n$ for the sections normally used as beams are computed for different unbraced lengths and the results plotted as curves in Part 3 of the LRFD Manual. The values not only cover unbraced lengths in the elastic range but also the inelastic range, enabling us very easily to handle the problems presented in the last section as well as the ones in this section which fall in zone 3. The moments are plotted for F_y values of 36 ksi and 50 ksi and for $C_b = 1.0$.

The curve for a typical W section is shown in Fig. 22-9. For each of the shapes L_p is indicated with a solid circle (●) while L_r is shown with a hollow circle (○).

The charts were developed without regard to such things as shear, deflection, and so on, items that may occasionally control the design as described later in this chapter. The curves extend to unbraced lengths equal to 30 times the section depths. As such they cover almost all of the unbraced lengths encountered in practice. If C_b is greater than 1.0, the values given will be magnified somewhat as illustrated in Fig. 22-7.

To select a member it is necessary only to enter the chart with the unbraced length L_b and the factored design moment M_u. For an illustration let's assume $F_y = 36$ ksi and that we wish to select a beam with $L_b = 20.0$ ft for a moment $M_u = 590$ ft-k. We enter charts in Part 3 entitled "Beam Design Moments" and go through the pages where $F_y = 36$ ksi until we find in the left-hand column $\phi_b M_n = 590$ ft-k. We proceed up from the bottom of the chart for an unbraced length $= 20.0$ ft until we intersect a horizontal line from the 590 ft-k. Any section to the right and above this intersection point will have a greater unbraced length and a greater moment capacity.

Moving up and to the right we first encounter a W21 × 101. In this area of the charts this section is shown with a dashed line. This section will provide the necessary moment capacity, but the dashed line indicates that it is in an uneconomical range. If we proceed farther upward and to the right, the first solid line that we encounter will represent the lightest available section. In this case it is a W30 × 99. Another illustration of the use of these charts is presented in Example 22-7.

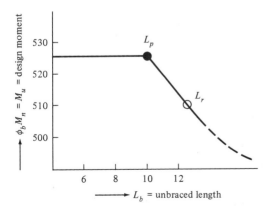

Figure 22-9 Beam design moments for a W section with different unbraced lengths.

■ Example 22-7

Using A36 steel, select the lightest available section for the beam of Fig. 22-10 which has lateral bracing provided for its compression flange only at its ends.

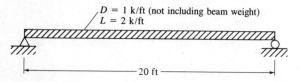

Figure 22-10

Solution.   Assume beam weight = 60 lb/ft:

$$w_u = (1.2)(1.060) + (1.6)(2) = 4.472 \text{ k/ft}$$

$$M_u = \frac{(4.472)(20)^2}{8} = 223.6 \text{ ft-k}$$

Noting $C_b = 1.0$, enter the "Beam Design Moments" charts with $L_b = 20$ ft and $M_u = 223.6$ ft-k.

Use W18 × 60   ■

For the example that follows, C_b is greater than 1.0. For such a situation the reader should look back to Fig. 22-7. There he or she will see that the design moment strength of a section can go to $\phi_b C_b M_n$ when $C_b > 1.0$ but may under no circumstances exceed $\phi_b M_p = \phi_b F_y Z$.

To handle such a problem we calculate an effective moment as follows (the numbers being taken from Example 22-8):

$$M_{eff} = \frac{M_u}{C_b} = \frac{868.7}{1.75} = 496.4 \text{ ft-k}$$

Then we enter the charts with our unbraced length of 17 ft and with $M_{eff} = 496.4$ ft-k and there select a W27 × 84. We must, however, check to see that our M_u does not exceed $\phi_b F_y Z$ for the section. In this case it does, and we must keep going until we find the lightest section which has a $\phi_b M_n \geq 496.4$ ft-k at $L_b = 17$ ft and a $\phi_b F_y Z \geq 868.7$ ft-k.

■ Example 22-8

Using A36 steel, select the lightest available section for the situation shown in Fig. 22-11. Bracing is provided only at the ends and centerline of the member, and thus $C_b = 1.75$.

Solution.   Assume beam weight = 108 lb/ft:

$$w_u = (1.2)(0.108) = 0.1296 \text{ k/ft}$$

$$M_u = \frac{(100)(34)}{4} + \frac{(0.1296)(34)^2}{8} = 868.7 \text{ ft-k}$$

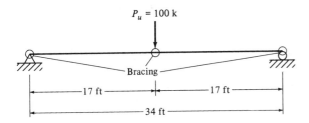

Figure 22-11

Entering the "Beam Design Moments" charts with $M_{u\,eff} = 868.7/1.75 = 496.4$ ft-k and $L_b = 17$ ft, we find a W27 × 84 will do.

But $\phi_b M_n = 868.7$ ft-k may not exceed $\phi_b F_y Z$ of the section. It does for a W24 × 84, as can be seen in the Load Factor Design Selection Table, and so we select a larger section from that table.

<u>Use W30 × 108</u> ∎

22-7 NONCOMPACT SECTIONS

There is one W section listed in the "Load Factor Design Selection Table" that is noncompact when F_y is 36 ksi. It is the W6 × 15, and the LRFD Manual indicates that it is noncompact with an asterisk. There are six noncompact sections in the same table when $F_y = 50$ ksi. These sections which are listed with a dagger (†) are the W14 × 99, W14 × 90, W12 × 65, W10 × 12, W6 × 15, and W8 × 10.

The LRFD Specification Appendix F1.7 provides formulas for the reduced values L_p' and $\phi_b M_n'$ for these sections. These values are tabulated in the "Load Factor Design Selection Table" and are shown as L_p and $\phi_b M_n$.

The designer will have little trouble with noncompact sections when F_y is 36 or 50 ksi. He or she should be very careful for cases where F_y values are greater than 50 ksi, however, and will have to use the complicated formulas of the Appendix.

22-8 DESIGN OF CONTINUOUS BEAMS

Section A5 of the LRFD Specification permits the design of beams analyzed either on the basis of elastic analysis with factored loads or by plastic analysis with the same ultimate loads. Plastic analysis is permitted only for sections with yield stresses no greater than 65 ksi.

Both theory and tests show clearly that continuous ductile steel members meeting the requirements for compact sections with sufficient lateral bracing supplied for their compression flanges have the desirable ability of being able to redistribute moments caused by overloads. If plastic analysis is used, this advantage is automatically included in the analysis.

If elastic analysis is used, the LRFD handles the redistribution by a rule of thumb that approximates the real plastic behavior. The LRFD Specification (A5) states that for continuous compact sections the design *may* be made on the basis of nine-tenths of the maximum negative moments caused by gravity loads, which are maximum at points of support, if the positive moments are increased by one-tenth of the average negative moments at the adjacent supports. *The 0.9 factor is applicable only to gravity loads and not to lateral loads such as those caused by wind and earthquake.* (The factor can also be applied to columns that have axial stresses less than $0.15F_y$.) This moment reduction does not apply to members consisting of A514 steel, to hybrid girders, or to moments produced by loading on cantilevers.

Example 22-9 illustrates the design of a three-span continuous beam analyzed by plastic analysis and elastic analysis.

■ Example 22-9

The beam shown in Fig. 22-12 is assumed to consist of A36 steel. (a) Select the lightest W section available using plastic analysis and assuming full lateral support is provided for its compression flanges. (b) Design the beam using elastic analysis with the factored loads and the 0.9 rule and assuming full lateral support is provided for its compression flanges. (c) Design the beam using elastic analysis with the factored loads and the 0.9 rule and assuming full lateral support is provided for the upper flange *but* for the lower flange only at support points.

Solution.

(a) Plastic analysis and design (see Fig. 22-13):

$$\text{Span 1}\begin{cases} M_u 4\theta = (30w_u)(\tfrac{1}{2})(15\theta) + (P_u)(15\theta) \\ M_u = 56.25w_u + 3.75P_u \\ M_u = (56.25)(60) + (3.75)(50) \\ M_u = 525 \text{ ft-k} \end{cases}$$

$$\text{Span 2}\begin{cases} M_u(4\theta) = (40w_u)(\tfrac{1}{2})(20\theta) \\ M_u = 100w_u = (100)(6.0) \\ M_u = 600 \text{ ft-k} \quad \longleftarrow \end{cases}$$

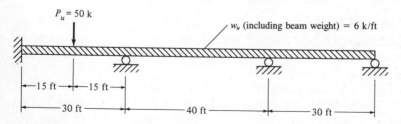

Figure 22-12

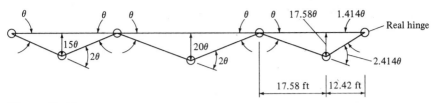

Figure 22-13

$$\text{Span 3}\begin{cases} M_u(3.414\theta) = (30w_n)(\tfrac{1}{2})(17.58\theta) \\ \qquad M_u = 77.24w_u = (77.24)(6.0) \\ \qquad\qquad = 463.4 \text{ ft-k} \end{cases}$$

$$Z_{reqd} = \frac{M_u}{\phi_b F_y} = \frac{(12)(600)}{(0.9)(36)} = 222.2 \text{ in.}^3$$

Use W24 × 84

(b) Elastic analysis and design—full lateral bracing both flanges (see Fig. 22-14):

Maximum negative moment = $(0.9)(759.3) = -683.4$ ft-k ←

$$\text{Maximum positive moment} = 443.2 + \frac{1}{10}\left(\frac{754.3 + 759.3}{2}\right) = +518.9 \text{ ft-k}$$

$$Z_{reqd} = \frac{(12)(683.4)}{(0.9)(36)} = 253.1 \text{ in.}^3$$

Use W24 × 94

(c) Elastic analysis and design—full lateral support supplied for top flange but only at support points for bottom flange:

Referring to the moment diagram for the part (b) solution (Fig. 22-14), we see that the greatest unbraced length in the negative moment region is 8.44 ft and the reduced M_u is 683.4 ft-k.

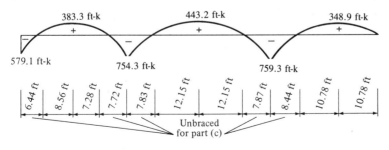

Figure 22-14

From the "Load Factor Design Selection Table" with $M_u = 683.4$ ft-k, we try a W24 × 94. Its $L_p = 8.3$ ft < 8.44 ft.

Use W27 × 94

NG ■

22-9 SHEAR

Fig. 22-15(a) shows the variation in shear stresses across the cross section of an I-shaped member, while (b) shows the shear stress variation in a member with a rectangular cross section. It can be seen in (a) that the shear in I-shaped sections is primarily resisted by the web.

If the load is increased on an I-shaped section until the bending yield stress is reached in the flange, the flange will be unable to resist shear stress, and it will be carried in the web. If the moment is further increased, the bending yield stress will penetrate farther down into the web, and the area of web that can resist shear will be further reduced. Rather than assuming the nominal shear stress is resisted by part of the web, the LRFD Specification assumes that a reduced shear stress is resisted by the entire web area. This web area A_w is equal to the overall depth of the member d times the web thickness t_w.

In the shear strength expressions to follow, F_{yw} is the specified minimum yield stress of the web, while k is a web plate buckling coefficient defined in LRFD Specification F2, and ϕ_v is equal to 0.9. Different expressions are given

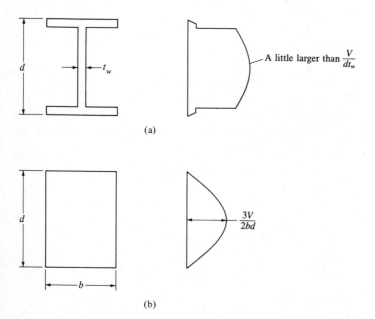

Figure 22-15 Variations in shear stresses for W and rectangular sections.

for different h/t_w ratios where shear failures would be plastic, inelastic, or elastic.

1. Web yielding. All W and channel shapes in the LRFD Manual fall into this classification.

    If $\dfrac{h}{t_w} \le 187\sqrt{\dfrac{k}{F_{yw}}} = 70$ for A36 steel,

    $\phi_v V_n = \phi_v 0.6 F_{yw} A_w = 19.4\, dt_w$ in kips for A36 steel

2. Inelastic buckling of web

    If $187\sqrt{\dfrac{k}{F_{yw}}} < \dfrac{h}{t_w} \le 234\sqrt{\dfrac{k}{F_{yw}}} = 87$ for A36 steel,

    $\phi_v V_n = \phi_v 0.6 F_{yw} A_w \dfrac{187\sqrt{k/F_{yw}}}{h/t_w}$

3. Elastic buckling of web

    If $\dfrac{h}{t_w} > 234\sqrt{\dfrac{k}{F_{yw}}} = 87$ for A36 steel,

    $\phi_v V_n = \phi_v A_w \dfrac{26{,}400\,k}{(h/t_w)^2} = \dfrac{118{,}800}{(h/t_w)^2} dt_w$ in kips for A36 steel

In Example 22-10 a beam's shear strength is checked.

■ **Example 22-10**
A W24 × 55 ($d = 23.57$ in., $t_w = 0.395$ in.) consisting of A36 steel is used for the beam and loading of Fig. 22-16. Check its adequacy in shear.

Solution.

$$V_u = (10)(7.3) = 73 \text{ k}$$
$$\phi_v V_n = 19.4 dt_w = (19.4)(23.57)(0.395)$$
$$= 180.6 \text{ k} > 73 \text{ k} \qquad\qquad \text{OK}$$

Note. This same value can be taken from the tables entitled "Beams W Shapes" in Part 3 of the LRFD Manual.   ■

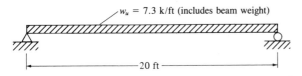

$w_u = 7.3$ k/ft (includes beam weight)

—20 ft—

Figure 22-16

22-10 DEFLECTIONS

A very common deflection expression is the one for the centerline deflection of uniformly loaded simple beams:

$$\Delta = \frac{5wL^4}{384EI}$$

To use deflection expressions such as this, the reader must be very careful to use consistent units. Example 22-11 illustrates the application of the preceding expression. The author has changed all units to pounds and inches. Thus the uniform load given in the problem as so many kips per foot is changed to so many pounds per inch.

■ Example 22-11

In Example 22-1 a W24 × 55 (I_x = 1350 in.) was selected for a 21-ft simple span to support a service dead load of 1 k/ft and a service live load is 3 k/ft. Is the centerline deflection of this selection satisfactory for the service live load if the maximum permissible value is $\frac{1}{360}$ of the span?

Solution.

$$\Delta_{centerline} = \frac{5wL^4}{384EI} = \frac{(5)(3000/12)(12 \times 21)^4}{(384)(29 \times 10^6)(1350)} = 0.335 \text{ in.}$$

Checking:

$$\text{maximum permissible } \Delta = \left(\frac{1}{360}\right)(12 \times 21) = 0.70 \text{ in.} > \underline{\underline{0.335 \text{ in.}}} \quad \text{OK} ■$$

In Part 3 (page 3-24) of the LRFD Manual the following simple formula for determining maximum beam deflections for W, M, HP, S, C, and MC sections for several different loading conditions is presented:

$$\Delta = \frac{ML^2}{C_1 I_x}$$

In this expression M is the service load moment in ft-k, C_1 is a constant whose value can be determined from Fig. 22-17, L is the span length (ft), and I_x is the moment of inertia (in.^{4}).

If we want to use this LRFD expression for the beam of Example 22-11, C_1 from Fig. 22-17(a) would be 161, the centerline bending moment is $wL^2/8 = (3)(21)^2/8 = 165.375$ ft-k, and the centerline deflection for the service live load would be

$$\Delta_{centerline} = \frac{ML^2}{C_1 I_x} = \frac{(165.375)(21)^2}{(161)(1350)} = 0.336 \text{ in.}$$

Some specifications handle the deflection problem by requiring certain minimum depth-span ratios. For example, the AASHTO suggests the depth-

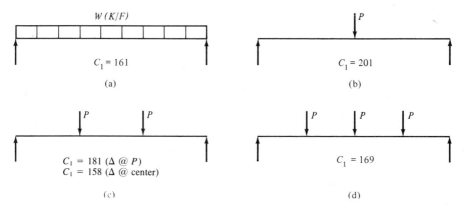

Figure 22-17 Loading constants. From American Institute of Steel Construction, *Manual of Steel Construction Load and Resistance Factor Design* (Chicago: AISC, 1986), fig. 3.3, p. 3-24. Reprinted with the permission of AISC.

span ratio be limited to a minimum value of 1/25. A shallower section is permitted, but it should have sufficient stiffness to prevent a deflection greater than would have occurred if the 1/25 ratio had been used. The LRFD Manual in Table 3-1 presents recommended span-depth ratios for simply supported uniformly loaded beams with different service dead-load to live-load (or to total load) ratios, which if followed will effectively limit deflections.

Deflections may very well control the sizes of beams for longer spans or for short ones where deflection limitations are severe. To assist the designer in selecting sections where deflections may control, the LRFD Manual includes in Part 3 a table entitled "Moment of Inertia Selection Table" in which the I_x values are given in numerically descending order for all of the sections normally used as beams. In this table the sections are arranged in groups, with the lightest section in each group printed in heavy type. Example 22-12 presents the design of a beam where deflection controls the design.

■ Example 22-12
Select the lightest available section with $F_y = 36$ ksi to support a service dead load of 1.2 k/ft and a service live load of 3 k/ft for a 30-ft simple span. The section is to have full lateral bracing for its compression flange, and the maximum total service load deflection is not to exceed 1/1500 the span length.

Solution. After some scratchwork assume beam weight = 194 lb/ft.

$$w_u = (1.2)(1.394) + (1.6)(3) = 6.473 \text{ k/ft}$$

$$M_u = \frac{(6.473)(30)^2}{8} = 728.2 \text{ ft-k}$$

$$Z_{reqd} = \frac{(12)(728.2)}{(0.9)(36)} = 269.7 \text{ in.}^3$$

Try W27 × 94 (I_x = 3270 in.^{4})

$$\text{Maximum permissible } \Delta = \left(\frac{1}{1500}\right)(12 \times 30) = 0.24 \text{ in.}$$

$$\text{Actual } \Delta_{\text{centerline}} = \frac{ML^2}{C_1 I_x} = \frac{\dfrac{(4.394)(30)^2}{8}(30)^2}{(161)(3270)}$$

$$= 0.845 \text{ in.} > 0.24 \text{ in.} \qquad\qquad \text{NG}$$

Find maximum I_x required to limit deflection to 0.24 in.:

$$I_x = \left(\frac{0.845}{0.24}\right)(3270) = 11{,}513 \text{ in.}^4$$

From "Moment of Inertia Selection Table,"

<u>Use W36 × 194</u>   ■

22-11 WEBS AND FLANGES WITH CONCENTRATED LOADS

When steel members have concentrated loads applied that are perpendicular to one flange and symmetric to the web, their flanges and webs must have sufficient flange and web design strength in the areas of flange bending, web yielding, web crippling, and sidesway web buckling. Should a member have concentrated loads applied to both flanges, it must have sufficient web design strength in the areas of web yielding, web crippling, and column web buckling. In this section formulas for determining strengths in these areas are presented.

Local Flange Bending

The nominal tensile load that may be applied through a plate welded to the flange of a W section is to be determined by the expression to follow, in which F_{yf} is the specified minimum yield stress of the flange (ksi) and t_f is the flange thickness (in.).

$$R_n = 6.25t_f^2 F_{yf}$$
$$\phi = 0.90 \qquad\qquad \text{(LRFD Formula K1-1)}$$

It is not necessary to check this formula if the length of loading across the beam flange is less than 0.15 times the flange width b_f.

Local Web Yielding

The nominal strength of the web of a beam at the web toe of the fillet when a concentrated load or reaction is applied is to be determined by one of the following two expressions, in which k is the distance from the outer edge of the flange to the web toe of the fillet, N is the bearing length (in.) of the force, F_{yw} is the specified minimum yield stress (ksi) of the web, and t_w is the thickness of the web.

If the force is a concentrated load or reaction that causes tension or compression and is applied at a distance greater than the member depth from the end of the member,

$$R_n = (5k + N)F_{yw}t_w$$

$$\phi = 1.0 \qquad \text{(LRFD Formula K1-2)}$$

If the force is a concentrated load or reaction applied at or near the end of the member,

$$R_n = (2.5k + N)F_{yw}t_w$$

$$\phi = 1.0 \qquad \text{(LRFD Formula K1-3)}$$

Reference to Fig. 22-18 clearly shows where these expressions were obtained. The nominal strength R_n equals the length over which the force is assumed to be spread when it reaches the web toe of the fillet times the web thickness times the yield stress of the web.

Web Crippling

Should concentrated loads be applied to a member with an unstiffened web, the nominal web crippling strength of the web is to be determined by the appropriate equation of the two that follow (in which d is the overall depth of the member). If web stiffeners are provided and extend for at least half of the web depth, web crippling does not need to be checked.

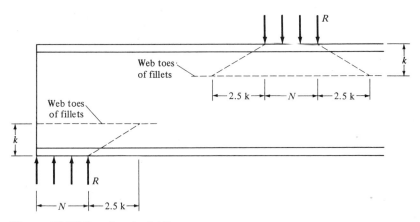

Figure 22-18 Local web yielding.

If the concentrated load is applied at a distance not less than $d/2$ from the end of the member,

$$R_n = 135t_w^2\left[1 + 3\left(\frac{N}{d}\right)\left(\frac{t_w}{t_f}\right)^{1.5}\right]\sqrt{\frac{F_{yw}t_f}{t_w}}$$

$$\phi = 0.75 \qquad\qquad\qquad \text{(LRFD Formula K1-4)}$$

If the concentrated load is applied at a distance less than $d/2$ from the end of the member,

$$R_n = 68t_w^2\left[1 + 3\left(\frac{N}{d}\right)\left(\frac{t_w}{t_f}\right)^{1.5}\right]\sqrt{\frac{F_{yw}t_f}{t_w}}$$

$$\phi = 0.75 \qquad\qquad\qquad \text{(LRFD Formula K1-5)}$$

Sidesway Web Buckling

Should compressive loads be applied to laterally braced compression flanges, the web will be put in compression, and the tension flange may buckle as shown in Fig. 22-19.

It has been found that sidesway web buckling will not occur if the flanges are restrained against rotation with $(d_c/t_w)/(\ell/b_f) > 2.3$ or if $(d_c/t_w)/(\ell/b_f) >$ 1.7 when flange rotation is not restrained. In these expressions d_c is the web depth between the web toes of the fillets, that is, $d - 2k$, and ℓ is the largest laterally unbraced length along either flange at the point of the load.

It is also possible to prevent sidesway web buckling with properly designed lateral bracing or stiffeners at the load point. The LRFD Commentary (K1.5) suggests that local bracing for *both* flanges be designed for 1 percent of the magnitude of the concentrated load applied at the point. If stiffeners are used, they must extend from the point of load for at least one-half of the member depth and should be designed to carry the full load. Flange rotation must be prevented if the stiffeners are to be effective.

If members are not restrained against relative movement by stiffeners or lateral bracing and are subject to concentrated compressive loads, their strength may be determined as follows:

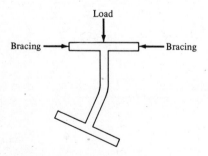

Figure 22-19 Sidesway web buckling.

When the loaded flange is braced against rotation and $(d_c/t_w)/(\ell/b_f)$ is less than 2.3,

$$R_n = \frac{12,000t_w^3}{h}\left[1 + 0.4\left(\frac{d_c/t_w}{\ell/b_f}\right)^3\right]$$

$\phi = 0.85$ (LRFD Formula K1-6)

When the loaded flange is not restrained against rotation and $(d_c/t_w)/(\ell/b_f)$ is less than 1.7,

$$R_n = \frac{12,000t_w^3}{h}\left[0.4\left(\frac{d_c/t_w}{\ell/b_f}\right)^3\right]$$

$\phi = 0.85$ (LRFD Formula K1-7)

■ Example 22-13

Select a W section for moment with A36 steel for the beam shown in Fig. 22-20. Lateral bracing will be provided at the ends and concentrated loads. Determine the minimum bearing length required for web yielding and web crippling at the reactions. Service loads are shown in the figure.

Solution.   Assume beam weight = 44 lb/ft:

$$M_u = \frac{(1.2)(1.044)(15)^2}{8} + (1.6)(25)(5) = 235.2 \text{ ft-k}$$

$$Z_{\text{reqd}} = \frac{(12)(235.2)}{(0.9)(36)} = 87.1 \text{ in.}^3$$

Girders for all-welded Connecticut expressway bridge. (Courtesy of Lincoln Electric Company.)

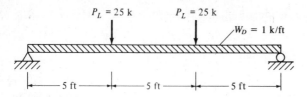

Figure 22-20

Use W21 × 44 (L_p = 5.3 ft > L_b of 5.0 ft, d = 20.66 in., t_w = 0.350 in., t_f = 0.450 in., k = $1\frac{3}{16}$ in.)
Required bearing length to prevent web yielding at support:

$$R_n = (1.2)(1.044)\left(\frac{15}{2}\right) + (1.6)(25) = 49.4 \text{ k}$$

$$R_n' = \frac{R_u}{\phi} = \frac{49.4}{1.0} = 49.4 \text{ k}$$

$$R_n = (2.5k + N)F_{yw}t_w$$
$$49.4 = (2.5 \times 1\tfrac{3}{16} + N)(36)(0.350)$$
$$N = 0.95 \text{ in.}$$

Required bearing length to prevent web crippling at support:

$$R_n = \frac{R_u}{\phi} = \frac{49.4}{0.75} = 65.9 \text{ k}$$

$$65.9 = (68)(0.350)^2\left[1 + 3\left(\frac{N}{20.66}\right)\left(\frac{0.350}{0.450}\right)^{1.5}\right]\sqrt{\frac{(36)(0.450)}{0.350}}$$

$$N = 1.63 \text{ in.}$$

<u>Use 4-in. minimum bearing length at support</u>   ■

In Part 3 of the LRFD Manual sets of tables entitled "Beams W Shapes," "Beams S Shapes," and so on, are presented with which the calculations for this beam and other beams in this chapter can be greatly simplified. The definition of the terms used in these tables is given on pages 3–25 through 3–29 in the LRFD Manual together with several useful sample problems.

For instance, for the sections normally used as beams and with yield stresses of 36 and 50 ksi, the values $\phi_b M_p$, $\phi_v V_n$, L_p, and L_u are provided along with the quantities ϕR_1, ϕR_2, $\phi_r R_3$, and $\phi_r R_4$. The latter values are useful in making the calculations necessary to check web yielding and web crippling. If we have a W21 × 44 (t_w = 0.350 in., k = $1\frac{3}{16}$ in; f_{yw} = 36 ksi) bearing on a length N of 3.5 in., we can compute ϕR_n for web yielding as follows:

$$\phi R_n = \phi(2.5k + N)F_{yw}t_w$$
$$= (1.0)(2.5 \times 1\tfrac{3}{16} + 3.5)(36)(0.350) = 81.5 \text{ k}$$

Or we can use the table values for ϕR_1 and ϕR_2.

$$\phi R_n = \phi R_1 + \phi R_2 N = 37.4 + (12.6)(3.5) = 81.5 \text{ k}$$

22-12 UNSYMMETRICAL BENDING

If a load is not perpendicular to one of the principal axes, it may be broken into components which are perpendicular to those axes and the moments about each axis, M_{ux} and M_{uy}, determined as shown in Fig. 22-21.

To check the adequacy of members bent about both axes simultaneously, the LRFD Specification provides an equation in Section H1. The equation to follow is for combined bending and axial tension or compression if $P_u/\phi P_n < 0.2$:

$$\frac{P_u}{2\phi P_n} + \left(\frac{M_{ux}}{\phi_b M_{nx}} + \frac{M_{uy}}{\phi_b M_{ny}}\right) \leq 1.0 \qquad \text{(LRFD Formula H1-1b)}$$

Since for the problem discussed here P_u is equal to zero, the formula reduces to

$$\frac{M_{ux}}{\phi_b M_{nx}} + \frac{M_{uy}}{\phi_b M_{ny}} \leq 1.0$$

Example 22-14 illustrates the design of beams subjected to unsymmetical bending. To illustrate the trial-and-error nature of the problem the author did not do quite as much scratchwork as in some of his earlier examples. The first design problems of this type which the student attempts may well take several trials. Consideration needs to be given to the question of lateral support for the compression flange. Should the lateral support be of questionable nature, the engineer should reduce the design moment resistance by means of one of the expressions previously given for that purpose.

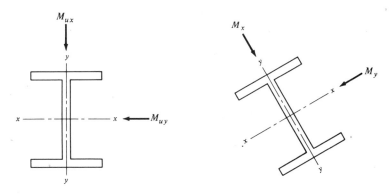

Figure 22-21

■ Example 22-14

A certain beam in its upright position is estimated to have a vertical bending moment $M_{ux} = 160$ ft-k and a lateral bending moment $M_{uy} = 35$ ft-k. These moments include the effect of the estimated beam weight. The loads are assumed to pass through the centroid of the section. Select a W24 shape of A36 steel that can resist these moments, assuming full lateral support for the compression flange.

Solution.   Try W24 × 62 ($\phi_b M_p = \phi_b M_{nx} = 413$ ft-k, $Z_y = 15.7$ in.³)

$$\phi_b M_{ny} = \frac{(0.9)(36)(15.7)}{12} = 42.39 \text{ ft-k}$$

$$\frac{M_{ux}}{\phi_b M_{nx}} + \frac{M_{uy}}{\phi_b M_{ny}} = \frac{160}{413} + \frac{35}{42.39} = 1.21 > 1.00 \qquad \text{NG}$$

Try W24 × 68 ($\phi_b M_p = \phi_b M_{nx} = 478$ ft-k, $Z_y = 24.5$ in.³)

$$\phi_b M_{ny} = \frac{(0.9)(36)(24.5)}{12} = 66.15 \text{ ft-k}$$

$$\frac{160}{478} + \frac{35}{66.15} = 0.864 < 1.00 \qquad \text{OK}$$

$$\underline{\text{Use W24} \times 68}$$ ■

It will be noted in the solution for Example 22-14 that, though the procedure used will yield a section that will adequately support the moments given, the selection of the absolutely lightest section listed in the LRFD Manual could be quite lengthy because of the two variables Z_x and Z_y which affect the size. If we have a large M_{ux} and a small M_{uy}, the most economical section will probably be quite deep and rather narrow, whereas if we have a large M_{uy} in proportion to M_{ux}, the most economical section may be rather wide and shallow.

PROBLEMS

22-1 to 22-8   Select the most economical sections using A36 steel unless otherwise specified and assuming full lateral bracing for the compression flanges. Working or service loads are given for each case, and beam weights are not included.

22-1 (*Ans.* W27 × 94)

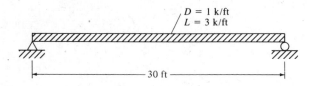

$D = 1$ k/ft
$L = 3$ k/ft

30 ft

22-2

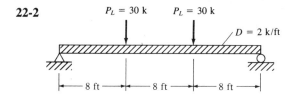

$P_L = 30$ k $P_L = 30$ k $D = 2$ k/ft

8 ft — 8 ft — 8 ft

22-3 Repeat Prob. 22-2 using $F_y = 50$ ksi. (*Ans.* W24 × 62)

22-4

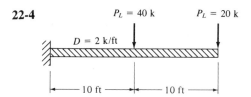

$P_L = 40$ k $P_L = 20$ k $D = 2$ k/ft

10 ft — 10 ft

22-5 (*Ans.* W36 × 150)

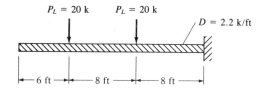

$P_L = 20$ k $P_L = 20$ k $D = 2.2$ k/ft

6 ft — 8 ft — 8 ft

22-6

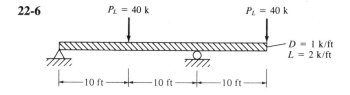

$P_L = 40$ k $P_L = 40$ k $D = 1$ k/ft $L = 2$ k/ft

10 ft — 10 ft — 10 ft

22-7 (*Ans.* W27 × 84)

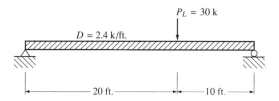

$P_L = 30$ k $D = 2.4$ k/ft.

20 ft. — 10 ft.

22-8 The illustration shows the arrangement of beams and girders which is used to support a 6-in. reinforced-concrete floor for a small industrial building. Design the beams and girders, assuming they are simply supported. Assume full lateral support and a live load of 120 psf. Concrete weight is 150 lb/ft³.

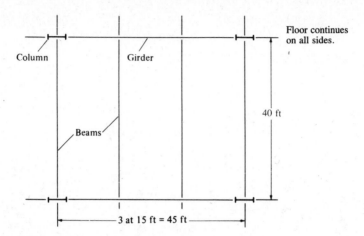

22-9 A beam consists of a W16 × 40 with a $\frac{1}{2}$ × 12-in. cover plate welded to each flange. Determine the design uniform load w_u the member can support in addition to its own weight for a 30-ft simple span. Use A36 steel. (*Ans.* w_u = 4.03 k/ft)

22-10 Repeat Prob. 22-9 with F_y = 50 ksi.

22-11 Select a section for a 30-ft simple span to support a uniform service dead load of 2 k/ft and a live uniform service load of 3 k/ft if two holes for 1-in. bolts are assumed in each flange at the section of maximum moment. Follow LRFD Specification B1 as to the 15 percent rule. (*Ans.* W30 × 108)

22-12 Rework Prob. 22-1 with D = 2k/ft assuming two holes for $1\frac{1}{4}$-in. bolts pass through each flange at the point of maximum moment. Use the LRFD 15 percent rule.

22-13 The section shown has two 1-in. bolts passing through each flange. Find the design or factored load w_u which the section can support in addition to its own weight for a 30-ft simple span if it consists of a steel with F_y = 50 ksi. Ignore the 15 percent rule of LRFD Specification B1. (*Ans.* 12.17 k/ft)

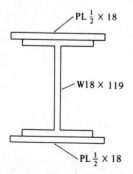

22-14 to 22-27   For these problems different values of L_b are given. Dead loads do not include beam weights.

22-14 Determine the design moment strength $\phi_b M_n$ of a W21 × 57 for simple spans of 5, 12, and 25 ft if lateral bracing for the compression flanges is provided at the ends only. Use A36 steel.

22-15 Using A36 steel, select the lightest satisfactory section for the situation shown, if bracing is provided at the ends only. (*Ans.* W33 × 118)

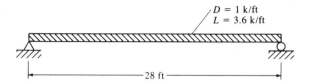

$$D = 1 \text{ k/ft}$$
$$L = 3.6 \text{ k/ft}$$

28 ft

22-16 Select the lightest available section for the situation shown if F_y = 50 ksi. Lateral bracing is provided at the ends only.

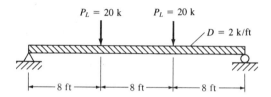

P_L = 20 k P_L = 20 k

D = 2 k/ft

8 ft   8 ft   8 ft

22-17 Repeat Prob. 22-16 with lateral bracing supplied at each of the concentrated loads as well as at the ends of the span. (*Ans.* W24 × 55)

22-18 A W36 × 150 consisting of A36 steel is used for a simple span of 30 ft and has lateral bracing provided at its ends only. If the only dead load present is the beam weight, what is the largest service concentrated live load which can be placed at the beam centerline?

22-19 Repeat Prob. 22-18 with lateral bracing supplied at beam ends and at the concentrated load. (*Ans.* P_L = 129.1 k)

22-20 If F_y is 36 ksi, select the lightest section for the situation shown if lateral bracing is supplied at the fixed end only.

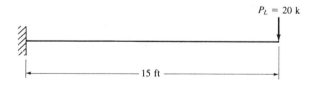

P_L = 20 k

15 ft

22-21 Using a steel with F_y = 50 ksi, what uniformly distributed service live load can a simply supported W33 × 241 carry for a span of 40 ft when (a) the compression flange is braced laterally for its full length and (b) lateral bracing is supplied at the ends only? (*Ans.* [a] 10.82 k/ft [b] 6.57 k/ft)

22-22 What uniform service live load can be placed on a W14 × 109 (F_y = 50 ksi) which has lateral bracing provided at its ends only? The beam is simply supported and has a span of 30 ft. The uniform service dead load is 1 k/ft plus the beam weight.

22-23 A W33 × 130 is used to support the loads shown. Using A36 steel and neglecting the dead weight of the beam, find if the beam is overloaded.

(a) If lateral support is provided for the entire span (*Ans.* $\phi_b M_p = 1260$ ft-k, OK)

(b) If lateral support is provided for the beam ends only (*Ans.* $\phi_b M_p = 755$ ft-k OK)

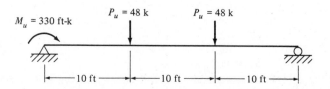

22-24 Redesign the beam of Prob. 22-23 with lateral support provided at the ends and at the concentrated loads.

22-25 The W30 × 99 shown has lateral support supplied at its centerline as well at its ends. If F_y is 50 ksi, determine the maximum permissible service live load P_L. Neglect beam weight. (*Ans.* $P_L = 73.1$ k)

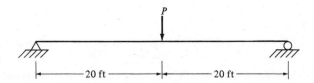

22-26 Repeat Prob. 22-25 using A36 steel if a uniform service dead load of 0.8 k/ft not including the beam weight is placed on the beam for its full length and if lateral bracing is provided at the ends and centerline of the beam.

22-27 Using A36 steel, select the lightest available section for the situation shown if bracing is provided only at the ends and centerline. (*Ans.* W33 × 118)

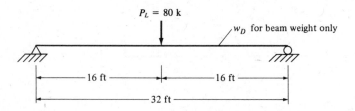

22-28 A W33 × 221 is required for a certain span but, because of a strike in the steel mills, cannot be obtained on time; however, an extra W33 × 118 of sufficient length is available along with a good supply of plates. Select $\frac{7}{8}$-in.-thick cover plates to be welded to the flanges of the W33 × 118 to provide a satisfactory substitute for the original beam. Use A36 steel and assume full lateral bracing is to be provided for the compression flange.

22-29 A W30 × 173 has been specified for use on a certain job. By mistake a W30 × 124 was shipped to the field. This beam must be erected today. Assuming that 1-in.-thick plates are obtainable immediately, select cover plates to be welded to the flanges to obtain the necessary section. Use A36 steel and assume full lateral bracing is supplied for the compression flange. (*Ans.* W30 × 124, 1 PL1 × 8 each flange)

22-30 Repat Prob. 22-28, assuming the original beam was to have been of a steel with $F_y = 50$ ksi and the substitute material is available only in A36 steel. Assume also that only $1\frac{1}{2}$-in. plates are available.

22-31 Design a beam of A36 steel with a depth no greater than 12.00 in. to support a uniform load of $w_u = 12.5$ k/ft (includes beam weight) for a 20-ft simple span. The member is to have full lateral bracing for its compression flange. (*One ans.* W10 × 39 with 1 PL1 × 18 each flange)

22-32 to 22-36   Considering moment only and assuming full lateral support for the compression flanges, select the lightest A36 sections available. The loads shown include the effect of the beam weights. Use elastic analysis, factored loads, and 0.9 rule.

22-32

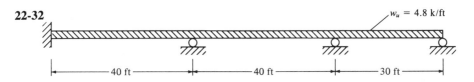

22-33 (*Ans.* W24 × 94)

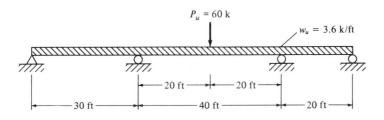

22-34

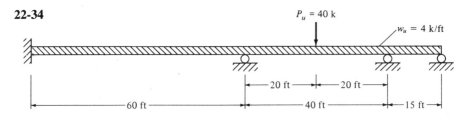

22-35 (*Ans.* W24 × 94)

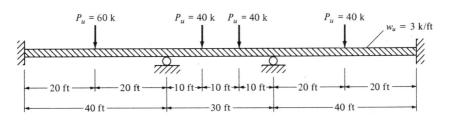

22-36

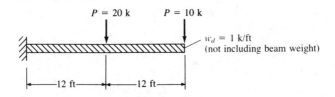

22-37 Repeat Prob. 22-32 using plastic analysis. (*Ans.* W24 × 76 or W24 × 68, which is slightly overstressed)

22-38 Repeat Prob. 22-34 using plastic analysis.

22-39 Repeat Prob. 22-36 using plastic analysis. (*Ans.* W24 × 55)

22-40 Select the lightest available W section (A36 steel) for the working live loads and span shown. Lateral support is provided only at the 12-ft points, and a maximum deflection (under working loads) equal to 1/1500 of the 24-ft span is permitted. Neglect beam weight. Consider moment, shear, and deflections.

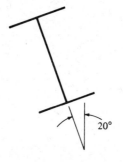

22-41 Select the lightest available W30 section consisting of A36 steel to resist a gravity moment $M_{ux} = 400$ ft-k and a lateral bending moment of $M_{uy} = 100$ ft-k. The section is assumed to have full lateral support. (*Ans.* W30 × 132)

22-42 The 20-ft simple beam shown has full lateral support for its compression flange and consists of A36 steel. The beam supports a gravity service dead load of 1.2 k/ft (includes beam weight) and a gravity live load of 3.4 k/ft. The loads are assumed to act through the center of gravity of the section. Select the lightest available W36 section.

22-43 A simply supported W24 × 146 consisting of A36 steel is supporting a 300-k service live load as shown. If the length of bearing at the left support is 8 in. and at the concentrated load is 12 in., check the beam for shear, web yielding, and web crippling. (*Ans.* All items except web crippling at concentrated load NG)

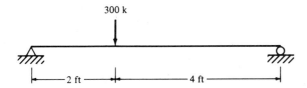

300 k

2 ft ‖ 4 ft

22-44 A 32-ft beam with full lateral support for its compression flange is supporting a moving service concentrated load of 32 k. Using A36 steel, select the lightest section available for moment. Then check to see if the section is satisfactory in shear and compute the minimum length of bearing required at the supports from the standpoint of web yielding and web crippling.

22-45 A 40-ft simple beam which supports a service concentrated load $P_L = 30$ k at midspan is laterally unbraced except at its ends and centerline. If the maximum permissible centerline deflection under service loads equals $1/1000$ of the span, select the most economical W section of A36 steel considering moment, shear, and deflection. Neglect beam weight. (*Ans.* W33 × 118)

22-46 Design a beam for a 24-ft simple span to support the working uniform loads $w_D = 1.2$ k/ft (includes beam weight) and $w_L = 2.1$ k/ft. The maximum permissible deflection under working loads is $1/1200$ of the span. Use A36 steel and consider moment, shear, and deflection. The beam is to be braced laterally for its full length.

22-47 Select the lightest available W section of A36 steel for the span and service loads shown. The beam will have full lateral support for its compression flange. Its maximum service load centerline deflection may not exceed $1/1500$ of the span under working loads. Consider moment, shear, and deflection only. (*Ans.* W36 × 135)

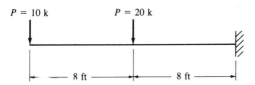

$P = 10$ k $P = 20$ k

8 ft ‖ 8 ft

22-48 Select the lightest section available if $F_y = 50$ ksi for the span and working loads shown, if the section is to be fully braced laterally and have a maximum service load deflection of $1/800$ of its span length. Neglect beam weight in all calculations.

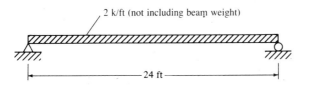

2 k/ft (not including beam weight)

24 ft

22-49 If the maximum permissible service load deflection for the fully braced beam shown is 1/1200 of the span, select the lightest available section using A36 steel. Consider moment, shear, and deflections. Working loads are shown. (*Ans.* W36 × 150)

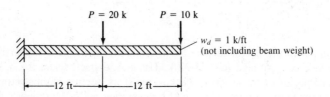

23

Connections—LRFD

23-1 INTRODUCTION

In this chapter a brief discussion of the analysis and design of high-strength bolts, rivets, and welds is presented using the LRFD Specification. For a general discussion of these connections (history, installation, etc.) the reader is referred to Chapters 11–13.

23-2 HIGH-STRENGTH BOLTS IN BEARING-TYPE CONNECTIONS—LOADS PASSING THROUGH CENTER OF GRAVITY OF CONNECTIONS

Shearing Strength

In bearing-type connections it is assumed that the loads to be transferred are larger than the frictional resistance caused by tightening the bolts with the result that the members slip a little on each other, putting the bolts in shear and bearing. The design strength of a bolt in single shear equals ϕ times the nominal shearing strength of the bolt in kips per square inch times its cross-sectional area. The LRFD ϕ values for shear are 0.65 for high-strength bolts and rivets and 0.60 for A307 common bolts.

The nominal shearing strengths of bolts and rivets are given in Table J3.2 of the LRFD Specification. For A325 bolts the values are 54 ksi if threads are not excluded from shear planes and 72 ksi if threads are excluded. (The corresponding values are 67.5 and 90 ksi for A490 bolts.) Should a bolt be in double shear, its shearing strength is considered to be twice its single shear value.

Bearing Strength

The design strength of a bolt in bearing equals ϕ times the nominal bearing strength of the connected part in kips per square inch times the diameter of the bolt times the thickness of the member that bears against the bolt. (For countersunk bolts and rivets one-half of the depth of the countersink should be deducted, says LRFD Specification J3.2.) When the distance L in the direction of the force from the center of a regular or oversized hole (or from the center of the end of a slotted hole) to the edge of a connected part is not less than $1\frac{1}{2}$ times the bolt diameter d and the distance center to center of the holes is not less than $3d$ and two or more bolts are used in the direction of the line of force, the bearing strength is

$\phi R_n = \phi 2.4 dt F_u$ for standard or short-slotted holes $\phi = 0.75$

$\phi R_n = \phi 2.0 dt F_u$ for long-slotted holes perpendicular to the load $\phi = 0.75$

Should deformations around the hole not be a design consideration, the two preceding expressions may be replaced with

$$\phi R_n = \phi 3.0 dt F_u \qquad \text{with } \phi = 0.75$$

Though the design bearing strength of a bolt with a small end distance is reduced, *the strengths of the other bolts in the connection are not reduced.* The value of ϕR_n for a single bolt or for two or more bolts in the line *each with an end distance less than $1\frac{1}{2}d$* is to be determined with the following expression:

$$\phi R_n = \phi L t F_u \qquad \text{with } \phi = 0.75$$

Tests of bolted joints have shown that neither the bolts nor the metal in contact with the bolts actually fails in bearing. However, these tests have shown that the efficiency of the connected parts in tension and compression is affected by the magnitude of the bearing stress. Therefore, the nominal bearing strengths given by the LRFD Specification are values above which it is judged the strength of the connected parts is impaired. In other words, these apparently very high design bearing stresses are not really bearing stresses at all but rather indexes of the efficiencies of the connected parts. If bearing stresses larger than the values given are permitted, the holes seem to elongate more than about $\frac{1}{4}$ in. and impair the strength of the connections.

Minimum Connection Strength

The LRFD Specification (J1.5) states that, except for lacing, sag rods, and girts, connections must be provided with design strengths sufficient to support factored loads of at least 10 kips.

Example 23-1 illustrates the calculations involved in determining the strength of a bearing-type connection. Using a similar procedure the number of bolts required for a certain loading condition is calculated in Example 23-2.

The values given for the strength of bolts in this chapter, whether bearing-type or slip-critical, can be obtained from the tables entitled "Shear Design Load in Kips" and "Bearing Design Load in Kips" in Part 5 of the LRFD Manual.

■ Example 23-1

Determine the design strength P_u of the bearing-type connection shown in Fig. 23-1. The steel is A36, the bolts are $\frac{7}{8}$-in. A325, the holes are standard sizes, the threads are excluded from the shear plane, edge distances are $>1\frac{1}{2}d$, and the distance center to center of the holes is $>3d$.

Solution. Design strength of plates:

$$A_g = (\tfrac{1}{2})(12) = 6.0 \text{ in.}^2$$
$$A_n = 6.00 - (2)(1.0)(\tfrac{1}{2}) = 5.0 \text{ in.}^2 = A_e$$
$$P_u = \phi_t F_y A_g = (0.9)(36)(6.0) = 194.4 \text{ k}$$
$$P_u = \phi_t F_u A_e = (0.75)(58)(5.0) = 217.5 \text{ k}$$

Bolts in single shear and bearing on $\frac{1}{2}$ in.:

$$P_u = \phi(0.6)(72)(4) = (0.65)(0.6)(72)(4) = 112.3 \text{ k} \leftarrow$$
$$P_u = \phi 2.4 dt F_u = (0.75)(2.4)(\tfrac{7}{8})(\tfrac{1}{2})(58)(4) = 182.7 \text{ k}$$
$$\text{Design } P_u = 112.3\text{k}$$

■

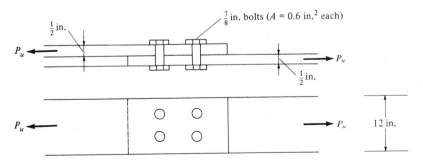

Figure 23-1

■ Example 23-2

How many $\frac{3}{4}$-in. A325 bolts in standard-size holes with threads excluded from the shear plane are required for the bearing-type connection shown in Fig. 23-2? Use A36 steel and assume edge distances to be $>1\frac{1}{2}d$ and the distance center to center of holes to be $>3d$.

Solution. Bolts in double shear and bearing on $\frac{3}{4}$ in.:

Design shear strength per bolt = $(0.65)(2 \times 0.44)(72) = 41.2$ k $\leftarrow$

Design bearing strength per bolt = $(0.75)(2.4)(\frac{3}{4})(\frac{3}{4})(58) = 58.7$ k

$$\text{Number of bolts required} = \frac{190}{41.2} = 4+$$

<u>Use 5 or 6 bolts (depending on arrangement)</u> ■

Where cover plates are bolted to the flanges of W sections, the bolts must carry the longitudinal shear on the plane between the plates and the flanges. With reference to the cover-plated beam of Fig. 23-3 the unit longitudinal

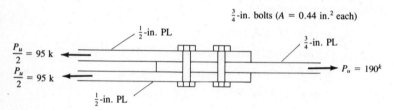

Figure 23-2

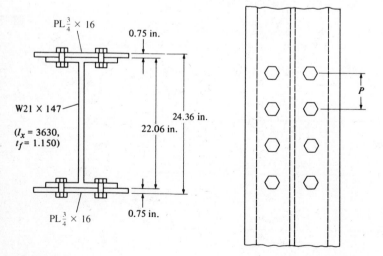

Figure 23-3

shearing stress to be resisted between a cover plate and the W flange can be determined with the expression $f_v = VQ/Ib$. The total shear force across the flange for a 1-in. length of the beam equals $(b) (1.0) (VQ/Ib) = VQ/I$.

The LRFD Specification (E4) provides a maximum permissible spacing for bolts used in the outside plates of built-up members. It equals the thinner outside plate thickness times $127/\sqrt{F_y}$ or 12 in., whichever is smaller.

The spacing of pairs of bolts in Fig. 23-3 can be determined by dividing the design strength of two bolts by the shear force to be taken per inch at a particular section. The theoretical spacings will vary as the external shear varies along the span. Example 23-3 illustrates the calculations involved in determining bolt spacing for a cover-plated beam.

■ Example 23-3

At a certain section in the cover-plated beam of Fig. 23-3 the external factored shear V_u is 275 k. Determine the spacing required for $\frac{7}{8}$-in. A325 bolts used in a bearing-type connection. Assume the $1\frac{1}{2}d$ and $3d$ edge and center-to-center distance requirements are met and that the bolt threads are excluded from the shear plane. The steel is A36.

Solution.

$$I_x = 3630 + (2)(16 \times \tfrac{3}{4})(11.405)^2 = 6752 \text{ in.}^4$$

Factored shear to be taken per inch

$$= \frac{V_u Q}{I}$$

$$= \frac{(275)(16 \times \tfrac{3}{4} \times 11.405)}{6752} = 5.574 \text{ k/in.}$$

Bolts in single shear and bearing on $\frac{3}{4}$ in.:

Design shear strength for 2 bolts $= (2)(0.65)(0.6)(72) = 56.16 \text{ k} \leftarrow$
Design bearing strength for 2 bolts $= (2)(0.75)(2.4)(\tfrac{7}{8})(\tfrac{3}{4})(58) = 137 \text{ k}$

$$p = \frac{56.16}{5.574} = 10.07 \text{ in.} \qquad \text{(say 10 in. on center)}$$

$$\text{Max } p = \frac{3}{4} \frac{127}{\sqrt{36}} = 15.88 \text{ in. or 12 in.}$$

Place bolts 10 in. on centers

23-3 HIGH-STRENGTH BOLTS IN SLIP-CRITICAL CONNECTIONS—LOADS PASSING THROUGH CENTER OF GRAVITY OF CONNECTIONS

The situations where slip-critical connections are desirable were described in Section 11-5. They are particularly useful for fatigue-type situations where members are subjected to constantly fluctuating loads. *Slip-critical connections must be checked both at service loads and at factored load conditions. (a) The design slipping resistances must be equal to or greater than the computed service load slipping forces. (b) The design strengths as bearing-type connections must be equal to or greater than the factored forces.*

If bolts are tightened to their required tensions for slip-critical connections (see Table 11-1), there is very little chance of their bearing against the plates which they are connecting. Tests show that there is very little chance of slip occurring unless there is a calculated shear of at least 50 percent of the total bolt tension. This means as we have said all along that slip-critical bolts are not stressed in shear; however, the LRFD Specification provides allowable shear strengths (they are really permissible friction values on the faying surfaces) so that the designer can handle the connections in just about the same manner he or she uses for bearing-type connections. The LRFD Specification assumes the bolts are in "shear" and no bearing, and the nominal shear strengths for high-strength bolts are given in Table 23-1. $\phi = 1.0$ except for long-slotted holes with load parallel to slot, where it is 0.85.

The preceding discussion concerning slip-critical joints does not present the whole story, because during erection the joints may be assembled with bolts, and as the members are erected their weights will often push the bolts against the side of the holes before they are tightened and put the bolts in some bearing and shear.

TABLE 23-1 NOMINAL SLIP-CRITICAL SHEAR STRENGTH, KSI, OF HIGH-STRENGTH BOLTS

TYPE OF BOLT	NOMINAL SHEAR STRENGTH		
	STANDARD-SIZE HOLES	OVERSIZED AND SHORT-SLOTTED HOLES	LONG-SLOTTED HOLES[a]
A325	17	15	12
A490	21	18	15

Source: American Institute of Steel Construction, *Manual of Steel Construction Load and Resistance Factor Design* (Chicago: AISC, 1986), Table J3.4, p. 6-69. Reprinted with the permission of AISC.

Note: Class A (slip coefficient 0.33). Clean mill scale and blast-cleaned surfaces with class A coatings. For design strengths with other coatings see RCSC "Load and Resistance Factor Design Specification for Structural Joints Using ASTM A325 or A490 Bolts."

[a] Tabulated values are for the case of load application transverse to the slot. When the load is parallel to the slot, multiply tabulated values by 0.85.

Example 23-4 presents the design of a slip-critical connection for a lap joint. First the number of bolts required for the service load limit state of no slippage is determined. Then the number of bolts required for the factored load limit state is computed, assuming the slip resistance is overcome and the bolts subjected to bearing and shear.

■ Example 23-4

It is desired to design a slip-critical connection for the plates shown in Fig. 23-4 to resist the axial service loads $P_D = 30$ k and $P_L = 50$ k using 1-in. A325 high-strength bolts with threads excluded from the shear plane and with standard-size holes, $L > 1\frac{1}{2}d$, and distance center to center of bolts $> 3d$.

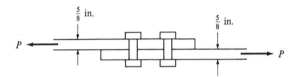

Figure 23-4

Solution. Slip-critical design (service loads):

Bolts in "single shear" and no bearing

"ss" strength of one bolt $= (\phi)(0.785)(17) = (1.0)(0.785)(17) = 13.35$ k

$$\text{Number of bolts required} = \frac{80}{13.35} = 5.99 \quad \text{(say 6)}$$

Design as bearing-type connection (factored loads):

$$P_u = (1.2)(30) + (1.6)(50) = 116 \text{ k}$$

Bolts in single shear and bearing on $\frac{5}{8}$ in.

ss strength of one bolt $= (0.65)(0.785)(72) = 36.8$ k $\leftarrow$

Bearing strength of one bolt $= (0.75)(2.4)(1.0)(\frac{5}{8})(58) = 65.25$ k

$$\text{Number of bolts required} = \frac{116}{36.8} = 3^+$$

Use 6 bolts ■

■ Example 23-5

The connection shown in Fig. 23-5 is made with $\frac{7}{8}$-in. A325 bearing-type bolts in standard-size holes with the threads excluded from the shear planes. The beam and gusset plates consist of A36 steel. Check the following items: (a) the tensile strengths of the W section and the gusset plates, (b) the strength of the bolts in single shear and bearing, and (c) the block shear strength of the W section crosshatched areas shown in Fig. 23-5(b).

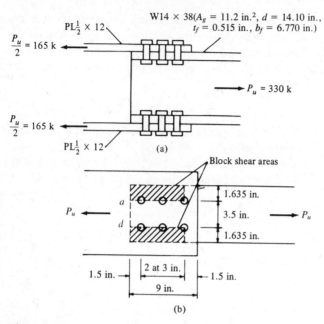

Figure 23-5

Solution.
(a) Tensile design strength of W section:

$$P_u = \phi_t F_y A_g = (0.9)(36)(11.2) = \underline{362.9 \text{ k}} > 330 \text{ k} \qquad \text{OK}$$
$$A_n = 11.2 - (4)(1)(0.515) = 9.14 \text{ in.}^2$$
$$U = 0.85 \quad \text{since } b_f < \tfrac{2}{3}d$$
$$A_e = (0.85)(9.14) = 7.77 \text{ in.}^2$$
$$P_u = \phi_t F_u A_e = (0.75)(58)(7.77) = \underline{338 \text{ k}} > 330 \text{ k} \qquad \text{OK}$$

Tensile design strength of gusset plates:

$$P_u = \phi_t F_y A_g = (0.9)(36)(\tfrac{1}{2} \times 12)(2) = \underline{388.8 \text{ k}} > 330 \text{ k} \qquad \text{OK}$$
$$A_n \text{ of 2 plates} = [(\tfrac{1}{2})(12) - (2)(1)(\tfrac{1}{2})]2 = 10 \text{ in.}^2 \leftarrow$$
$$0.85 A_g = (0.85)(\tfrac{1}{2})(12)(2) = 10.2 \text{ in.}^2$$
$$P_u = \phi_t F_u A_n = (0.75)(58)(10) = \underline{435 \text{ k}} > 330 \text{ k} \qquad \text{OK}$$

(b) Bolts in single shear and bearing on $\tfrac{1}{2}$ in.:

ss design strength of bolts $= (0.65)(0.6)(72)(12) = \underline{337 \text{ k}} > 330 \text{ k} \qquad$ OK

Bearing design strength of bolts

$$= (0.75)(2.4)(\tfrac{7}{8})(\tfrac{1}{2})(58)(12) = \underline{548.1 \text{ k}} > 330 \text{ k} \qquad \text{OK}$$

(c) Block shearing strength for the W section:
Tension fracture and shear yielding

$$P_{bs} = \phi[F_u A_{nt} + 0.6F_y A_{vg}]$$
$$= 0.75[(58)(\tfrac{1}{2})(1.635 - \tfrac{1}{2} \times 1) + (0.6)(36)(7\tfrac{1}{2})(\tfrac{1}{2})]4$$
$$= \underline{\underline{341.8 \text{ k}}} > 330 \text{ k} \qquad\qquad\qquad \text{OK}$$

Shear fracture and tension yielding

$$P_{bs} = \phi(F_y A_{tg} + 0.6F_u A_{ns})$$

Noting there are $2\tfrac{1}{2}$ holes on the net area shear plane shown in Fig. 23-5(b).

$$P_{bs} = 0.75[(36)(1.635)(\tfrac{1}{2}) + (0.6)(58)(7.5 - 2.5 \times 1)(\tfrac{1}{2})]4$$
$$= \underline{\underline{349 \text{ k}}} > 330 \text{ k} \qquad\qquad\qquad \text{OK}$$

Note: A subsequent calculation for block shear along section *abcd* shows it to be satisfactory also. ∎

The calculations for riveted connections and for connections made with A307 common bolts are made almost exactly like those for high-strength bolts in bearing-type connections. The only differences are that the shear strength values for these connections are much smaller and the shear ϕ factor for A307 bolts is 0.60 instead of the 0.65 used for rivets and high-strength bolts. The LRFD Specification does not permit the design of slip-critical joints using rivets or common bolts.

23-4 BOLTS SUBJECTED TO ECCENTRIC SHEAR

Examples 23-6 and 23-7 provide illustrations of the use of the ultimate strength tables of the LRFD Manual for both analysis and design for bolts subjected to eccentric shear.

∎ Example 23-6
The bearing-type $\tfrac{7}{8}$-in. A325 bolts of the connection of Fig. 23-6 each have a design shear strength $R_u = (0.65)(0.60)(72) = 28.1$ k. Determine the value of P_u using the ultimate strength tables entitled "Eccentric Loads on Fastener Groups" in Part 5 of the LRFD Manual.

Solution.

$$P_u = C \times \phi r$$
$$C = 2.10 \text{ from Table XI}$$
$$P_u = (2.10)(28.1) = \underline{\underline{59 \text{ k}}} \qquad\qquad \text{OK} \quad ∎$$

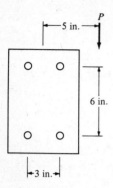

Figure 23-6

■ Example 23-7

Determine the number of $\frac{7}{8}$-in. A325 bolts in standard-size holes required for the connection shown in Fig. 23-7. Use A36 steel and assume the connection is to be a bearing type with threads excluded from the shear plane. Further assume that the bolts are in single shear and bearing on $\frac{1}{2}$ in. Use the ultimate strength analysis method presented in the LRFD Manual.

Solution.   Using Table XI page 5-66 of LRFD Manual:

$$x_0 = 5.5 \text{ in.}$$

Bolts in single shear and bearing on $\frac{1}{2}$ in.:

$$\phi r_v = \text{shear design strength per fastener}$$
$$= (0.65)(0.6)(72) = 28.1 \text{ k} \leftarrow$$
$$\phi r_v = \text{bearing design strength per fastener}$$
$$= (0.75)(2.4)(\tfrac{7}{8})(\tfrac{1}{2})(58) = 45.7 \text{ k}$$

$$C = \frac{P_u}{\phi r_v} = \frac{100}{28.1} = 3.56$$

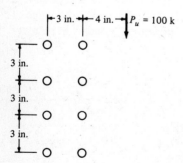

Figure 23-7

By interpolation in the table we need 4 bolts in each line.

Note: If the situation faced by the designer does not fit the ultimate strength tables given in the LRFD Manual, it is recommended that he or she use the conservative elastic procedure described in Chapter 12 to handle the problem whether it is analysis or design. ■

23-5 BOLTS SUBJECTED TO SHEAR AND TENSION

Equations for bearing-type bolts subjected to combined shear and tension are presented in Table 23-2. In these expressions f_v and f_t are respectively the computed shear and tension stresses in the bolts due to the factored loads. The maximum values given in the table (as the 68 ksi for A325 bolts with threads not excluded from the shear plane) equal ϕ (which is 0.75) times the nominal strength of the bolts if they are subjected to external tensile loads only.

Example 23-8 illustrates the calculations involved in checking a high-strength bearing-type bolted connection for shear and tension.

■ Example 23-8

The tension member shown in Fig. 23-8 is connected to the column shown with eight $\frac{7}{8}$-in. A325 high-strength bolts in a bearing-type connection with the threads excluded from the shear plane and standard-size holes. Is this a sufficient number of bolts to resist the applied load according to the LRFD Specification?

TABLE 23-2 TENSION STRESS LIMIT (F_t), KSI, FOR FASTENERS IN BEARING-TYPE CONNECTIONS

DESCRIPTION OF FASTENERS	THREADS INCLUDED IN THE SHEAR PLANE	THREADS EXCLUDED FROM THE SHEAR PLANE
A307 bolts	$39 - 1.8f_v \leq 30$	
A325 bolts	$85 - 1.8f_v \leq 68$	$85 - 1.4f_v \leq 68$
A490 bolts	$106 - 1.8f_v \leq 84$	$106 - 1.4f_v \leq 84$
Threaded parts A449 bolts over $1\frac{1}{2}$-in. diameter	$0.73F_u - 1.8f_v \leq 0.56F_u$	$0.73F_u - 1.4f_v \leq 0.56F_u$
A502, grade 1, rivets	$44 - 1.3f_v \leq 34$	
A502, grade 2, rivets	$59 - 1.3f_v \leq 45$	

Source: American Institute of Steel Construction, *Manual of Steel Construction Load and Resistance Factor Design* (Chicago: AISC, 1986), p. 6-68, Table J3.3.

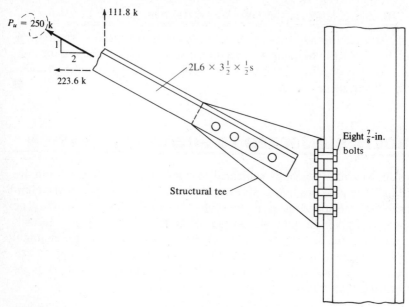

Figure 23-8 Combined shear and tension connection.

Solution.

$$\text{Shearing stress } f_v = \frac{111.8}{(8)(0.6)} = 23.29 \text{ ksi}$$

$$\text{Tension stress } f_t = \frac{223.6}{(8)(0.6)} = 46.58 \text{ ksi}$$

$$\text{Limiting tension stress } F_t = 85 - (1.4)(23.29) = 52.39 \text{ ksi}$$

$$> 46.58 \text{ ksi} \qquad \text{OK}$$

Connection is satisfactory ∎

When an axial tension force is applied to a slip-critical connection, the clamping force will be reduced and the design shear strength must be decreased in some proportion to the loss in clamping or prestress. This is accomplished in the LRFD Specification (J3.5) by requiring that the nominal slip-critical shear strengths given in LRFD Table J3.4 be multiplied by the reduction factor $(1 - T/T_b)$. T is the service tensile force applied to one bolt, and T_b is the minimum pre-tension load for one bolt in a slip-critical connection as given in Table 23-1 (LRFD Table J3.1). For such a situation ϕ is equal to 1.0 unless we have long-slotted holes with the load applied in the direction of the slot. In the latter case ϕ is to be taken as 0.85, says the LRFD Specification.

For bolts in slip-critical connections with standard holes the nominal shear strengths at service load conditions are to be determined as follows:

$$\text{For A325 bolts} \le \left(1 - \frac{T}{T_b}\right)(17)$$

$$\text{For A490 bolts} \le \left(1 - \frac{T}{T_b}\right)(21)$$

■ Example 23-9

Rework Example 23-8 assuming that a slip-critical connection is to be used. The service loads are $P_D = 75$ k and $P_L = 90$ k.

Solution.

$$\text{Service load} \quad P = 75 + 90 = 165 \text{ k}$$

$$P_v = \left(\frac{1}{\sqrt{5}}\right)(165) = 73.8 \text{ k}$$

$$P_h = \left(\frac{2}{\sqrt{5}}\right)(165) = 147.6 \text{ k}$$

$$f_v = \frac{73.8}{(8)(0.6)} = 15.38 \text{ ksi}$$

$$f_h = \frac{147.6}{(8)(0.6)} = 30.75 \text{ ksi}$$

$$\text{Limiting shear strength} = \left(1 - \frac{T}{T_b}\right)(17)\,\text{where}\,T = \frac{2}{\sqrt{5}}(75 + 90) = 147.6\,\text{k}$$

$$= \left(1 - \frac{147.6/8}{39}\right)(17) = 8.95 \text{ ksi} < 15.38 \text{ ksi} \quad \text{NG}$$

Connection is unsatisfactory if slip-critical ■

The LRFD tensile design strength of bolts, rivets, and threaded parts is given by the expression to follow, which is independent of any initial tightening force:

$$P_u = \phi F_t A_g$$

When fasteners are loaded in tension, there is usually some bending due to the deformation of the connected parts. As a result the value of ϕ in this expression is a rather small 0.75. Table J3.2 of the LRFD Specification gives values of F_t for the different kinds of connectors, with the values of rivets and threaded parts being quite conservative.

In this expression A_g is the nominal body area of a rivet or the unthreaded portion of a bolt or its threaded part not including upset rods.

If an upset rod is used, the nominal tensile strength of the threaded portion is set equal to $0.75 F_u$ times the cross-sectional area at its major thread di-

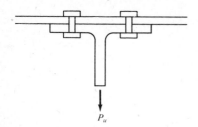

Figure 23-9 Hanger connection.

ameter. This value must be larger than F_y times the nominal body area of the rod before upsetting.

Example 23-10 illustrates the determination of the strength of a tension connection.

■ Example 23-10

Determine the design tensile strength of the bolts of the hanger connection of Fig. 23-9 if eight $\frac{7}{8}$-in. A490 high-strength bolts are used.

Solution.

$$P_u = (0.75)(8)(0.6)(112.5) = \underline{\underline{405 \text{ k}}}$$ ■

The total load applied to a tensile connection shall be the sum of the factored external loads plus any tension forces that result from prying action as discussed in Chapter 12.

23-6 ALLOWABLE STRENGTH OF RIVETED CONNECTIONS—RIVETS IN SHEAR

The nominal tension and shearing strengths for rivets and A307 bolts are given for static loads only in Table 23-3. Notice that the nominal shear strength of the A307 bolts is not affected if the bolt threads are in the shear plane.

Examples 23-11 and 23-12 illustrate the calculations necessary to determine the design strengths of existing connections or to design riveted connections. Little comment is made here concerning A307 bolts: all the calculations for these fasteners are made exactly as they are for rivets except that the shearing strengths given by the LRFD Specification are different. Only one brief example with these common bolts (23-13) is included.

■ Example 23-11

Determine the design strength P_u of the bearing-type connection shown in Fig. 23-10. A36 steel and A502, grade 1, rivets are used in the connection, and it

TABLE 23-3   LRFD NOMINAL TENSILE AND SHEARING STRENGTHS FOR
RIVETS AND A307 BOLTS

FASTENER TYPE	TENSILE STRENGTH (KSI)	SHEARING STRENGTH IN BEARING-TYPE CONNECTIONS (KSI)
A502, grade 1, hot-driven rivets	45.0, $\phi = 0.75$	36.0, $\phi = 0.65$
A502, grade 2 or 3, hot-driven rivets	60.0, $\phi = 0.75$	48.0, $\phi = 0.65$
A307 bolts	45.0, $\phi = 0.75$	27.0, $\phi = 0.60$

Source: American Institute of Steel Construction, *Manual of Steel Construction Load and Resistance Factor Design* (Chicago: AISC, 1986), Table J3.2, p. 6-67. Reprinted with the permission of AISC.

is assumed that standard-size holes are used and that edge distances and center-to-center distances are $> 1\frac{1}{2}d$ and $3d$ respectively.

Solution.   Design tensile force applied to plates:

$$A_g = (\tfrac{1}{2})(10) = 5.00 \text{ in.}^2$$
$$A_n = [(\tfrac{1}{2})(10) - (2)(\tfrac{7}{8})(\tfrac{1}{2})] = 4.125 \text{ in.}^2 = A_e$$
$$P_u = \phi_t F_y A_g = (0.90)(36)(5.00) = 162 \text{ k}$$
$$P_u = \phi_t F_u A_e = (0.75)(58)(4.125) = 179.4 \text{ k}$$

Rivets in single shear and bearing on $\frac{1}{2}$ in.:

Design shearing strength of rivets $= (0.65)(0.44)(36)(4) = 41.2 \text{ k} \leftarrow$
Design bearing strength of rivets $= (0.75)(2.4)(\tfrac{3}{4})(\tfrac{1}{2})(58)(4) = 156.6 \text{ k}$
$$\underline{\underline{P_u = 41.2 \text{ k}}}$$ ■

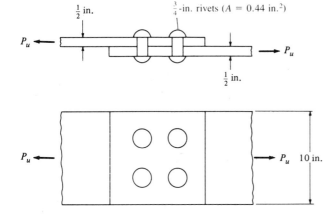

$\frac{1}{2}$ in. $\frac{3}{4}$-in. rivets ($A = 0.44$ in.^{2})

P_u P_u

$\frac{1}{2}$ in.

P_u P_u 10 in.

Figure 23-10

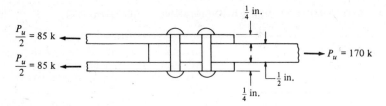

Figure 23-11

■ **Example 23-12**

How many $\frac{7}{8}$-in. A502, grade 1, rivets are required for the connection shown in Fig. 23-11, if the plates are A36 and if standard-size holes are used and the edge distances and center-to-center distances are $> 1\frac{1}{2}d$ and $3d$ respectively?

Solution. Bolts in double shear and bearing on $\frac{1}{2}$ in.:

Design shear strength of 1 rivet $= (0.75)(2 \times 0.6)(36) = 28.1 \text{ k} \leftarrow$
Design bearing strength of 1 rivet $= (0.75)(2.4)(\frac{7}{8})(\frac{1}{2})(58) = 45.7 \text{ k}$

$$\text{Number of rivets required} = \frac{170}{28.1} = 6.05$$

<u>Use six $\frac{7}{8}$-in. rivets</u>   ■

■ **Example 23-13**

Repeat Example 23-12 using $\frac{7}{8}$-in. A307 bolts.

Solution. Bolts in double shear and bearing on $\frac{1}{2}$ in.:

Design shear strength of 1 bolt $= (0.60)(2 \times 0.6)(27.0) = 19.44 \text{ k} \leftarrow$
Design bearing strength of 1 bolt $= (0.75)(2.4)(\frac{7}{8})(\frac{1}{2})(58) = 45.7 \text{ k}$

$$\text{Number of bolts required} = \frac{170}{19.44} = 8.74$$

<u>Use nine or ten $\frac{7}{8}$-in. A307 bolts</u>   ■

23-7 LRFD REQUIREMENTS FOR WELDS

When welds are made, the electrode material should have properties of the base metal. If the properties are comparable, the weld metal is referred to as the "matching" base metal. Table 4.11 of AWS D1.1 gives details concerning matching weld metals.

   Table 23-4 provides nominal strengths for various types of welds including fillet welds, plug and slot welds, and complete-penetration and partial-penetration groove welds.

The design strength of a particular weld is taken as the lower value of ϕF_W (where F_W is the nominal strength of the weld) and ϕF_{BM} (where F_{BM} is the nominal strength of the base material).

For fillet welds the nominal strength for stress on the effective area of the weld is $0.60F_{EXX}$ (where F_{EXX} is the classification strength of the weld metal) and ϕ is 0.75. If we have tension or compression parallel to the axis of the weld, the nominal strength of the base metal is F_y and ϕ is 0.90. The design shear strength of members being connected is taken as $\phi F_n A_{ns}$, where ϕ is 0.75, F_n is $0.6F_u$, and A_{ns} is the net area subject to shear.

The filler metal electrodes for shielded arc welding are listed as E60XX, E70XX, and so on. In this classification the letter E represents an electrode, while the first set of digits (e.g., 60 or 70) indicates the minimum tensile strength of the weld in kips per square inch. The remaining digits designate the position, current, and other necessary information for a certain electrode.

In addition to the nominal stresses given in Table 23-4, there are several other LRFD provisions applying to welding. Among the more important are the following:

1. The minimum length of a fillet weld may not be less than four times the nominal leg size of the weld. Should its length actually be less than this value, the weld size considered effective must be reduced to one-quarter of the weld length.

2. The maximum size of a fillet weld along edges of material less than $\frac{1}{4}$ in. thick equals the material thickness. For thicker material it may not be larger than the material thickness less $\frac{1}{16}$ in., unless the weld is specially built out to give a full throat thickness. For a plate with a thickness of $\frac{1}{4}$ in. or more it is desirable to keep the weld back at least $\frac{1}{16}$ in. from the edge so that the inspector can clearly see the edge of the plate and thus accurately determine the dimensions of the weld throat.

3. The minimum permissible sizes for fillet welds in the LRFD Specification are given in Table 23-5. They vary from $\frac{1}{8}$ in. for $\frac{1}{4}$ in. or thinner material up to $\frac{5}{16}$ in. for material over $\frac{3}{4}$ in. thick. The smallest practical weld size is about $\frac{1}{8}$ in., and the most economical size is probably about $\frac{1}{4}$ or $\frac{5}{16}$ in. The $\frac{5}{16}$-in. weld is about the largest size that can be made in one pass with the shielded arc welded process (SMAW). It is about $\frac{1}{2}$ in. with the submerged arc process (SAW).

These minimum sizes were not developed on the basis of strength considerations but rather on the fact that thick materials have a quenching or rapid cooling effect on small welds. If this happens, the result is often a loss in weld ductility. In addition the thicker material tends to restrain the weld material from shrinking as it cools, with the result that weld cracking can be a problem.

4. When practical, end returns (sometimes called boxing) should be made for fillet welds as shown in Fig. 23-12. The length of returns should not be less than twice the nominal size of the weld. When they are not used, it is considered good practice by some designers to subtract two times the weld size from the effective weld length. End returns are useful in reducing the high

TABLE 23-4 DESIGN STRENGTH OF WELDS

TYPES OF WELD AND STRESS[a]	MATERIAL	RESIS-TANCE FACTOR ϕ	NOMINAL STRENGTH F_{BM} OR F_w	REQUIRED WELD STRENGTH LEVEL[b]
COMPLETE-PENETRATION GROOVE WELD				
Tension normal to effective area	Base	0.90	F_y	"Matching" weld must be used.
Compression normal to effective area	Base	0.90	F_y	Weld metal with a strength level equal to or less than "matching" weld metal may be used.
Tension or compression parallel to axis of weld				
Shear on effective area	Base Weld electrode	0.90 0.80	$0.60F_y$ $0.60F_{EXX}$	
PARTIAL-PENETRATION GROOVE WELDS				
Compression normal to effective area	Base	0.90	F_y	Weld metal with a strength level equal to or less than "matching" weld metal may be used.
Tension or compression parallel to axis of weld[c]				
Shear parallel to axis of weld	Base[d] Weld electrode	0.75	$0.60F_{EXX}$	
Tension normal to effective area	Base Weld electrode	0.90 0.80	F_y $0.60F_{EXX}$	
FILLET WELDS				
Stress on effective area	Base[d] Weld electrode	0.75	$0.60F_{EXX}$	Weld metal with a strength level equal to or less than "matching" weld metal may be used.
Tension or compression parallel to axis of weld[d]	Base	0.90	F_y	
PLUG OR SLOT WELDS				
Shear parallel to faying surfaces (on effective area)	Base[d] Weld electrode	0.75	$0.60F_{EXX}$	Weld metal with a strength level equal to or less than "matching" weld metal may be used.

Source: American Institute of Steel Construction, *Manual of Steel Construction Load and Resistance Factor Design* (Chicago: AISC, 1986), Table J2.3, p. 6-65. Reprinted with the permission of AISC.

[a] For definition of effective area, see Sec. J2 of the LRFD Specification.

[b] For "matching" weld metal, see Table 4.1.1, AWS D1.1. Weld metal one strength level stronger than "matching" weld metal will be permitted.

[c] Fillet welds and partial-penetration groove welds joining component elements of built-up members, such as flange-to-web connections, may be designed without regard to the tensile or compressive stress in these elements parallel to the axis of the welds.

[d] The design of connected material is governed by Sec. J4 of the LRFD Specification.

TABLE 23-5   MINIMUM SIZE OF FILLET WELDS

MATERIAL THICKNESS OF THICKER PART JOINED (IN.)	MINIMUM SIZE OF FILLET WELD[a] (IN.)
To $\frac{1}{4}$ inclusive	$\frac{1}{8}$
Over $\frac{1}{4}$ to $\frac{1}{2}$	$\frac{3}{16}$
Over $\frac{1}{2}$ to $\frac{3}{4}$	$\frac{1}{4}$
Over $\frac{3}{4}$	$\frac{5}{16}$

Source: American Institute of Steel Construction, *Manual of Steel Construction Load and Resistance Factor Design* (Chicago: AISC, 1986), p. 6-62, Table J2.5.

[a] Leg dimension of fillet welds.

stress concentrations which occur at the ends of welds, particularly for connections where there is considerable vibration and eccentricity of load. The LRFD Specification, however, says that the effective length of fillet welds includes the length of the end returns used.

5. When longitudinal fillet welds are used for the connection of plates or bars, their length may not be less than the perpendicular distance between them because of shear lag, discussed in Chapter 3. Furthermore, the distance between fillet welds may not be greater than 8 in. for end connections unless the member is designed on the basis of its effective area in accordance with LRFD Specification B3.

6. For lap joints the minimum amount of lap permitted is equal to five times the thickness of the thinner part joined but may not be less than 1 in.

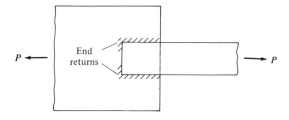

Figure 23-12

The purpose of this minimum lap is to keep the joint from rotating excessively (see Fig. 11-1[a]) when the connected parts are loaded.

23-8 DESIGN OF SIMPLE FILLET WELDS

Examples 23-14 through 23-16 illustrate the calculations necessary to determine the strength of various fillet-welded connections, while Example 23-17 presents the design of such a connection. In these and other problems weld lengths are selected no closer than to the nearest $\frac{1}{4}$ in., because closer work cannot be expected in shop or field.

■ Example 23-14

Determine the design strength of a 1-in. length of $\frac{5}{16}$-in. fillet weld using (a) SMAW and (b) SAW. Use E70 electrodes with a minimum tensile strength of 70 ksi $= F_{EXX}$.

Solution.
(a) SMAW:

Effective throat thickness of weld $= (0.707)(\frac{5}{16}) = 0.221$ in.

$$\text{Design strength} = \phi F_w = (0.75)(0.60 \times 70)(0.221)(1.0)$$
$$= 6.96 \text{ k/in.}$$

(b) SAW:

From LRFD Section J2.2a the effective throat thickness of weld $= \frac{5}{16}$ in.

$$\text{Design strength} = (0.75)(0.60 \times 70)(\tfrac{5}{16})(1.0)$$
$$= 9.84 \text{ k/in.} \qquad ■$$

Fillet welds may not be designed with a stress that is greater than the design stress on the adjacent members being connected. If the external force applied to the member (tensile or compressive) is parallel to the axis of the weld metal, the design strength may not exceed the axial design strength of the member.

Examples 23-14 and 23-15 illustrate the calculations necessary to determine the design strength of plates connected with SMAW and SAW fillet welds. In each of these examples the shearing strength per inch of the welds controls and is multiplied by the total length of the welds to give the total capacity of the connections.

■ Example 23-15

What is the design strength of the connection shown in Fig. 23-13 if A36 steel and E70 electrodes are used? The $\frac{7}{16}$-in. fillet welds shown were made by the SMAW process.

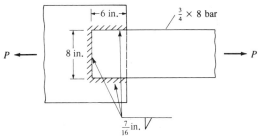

Figure 23-13

Solution.

Effective throat thickness $= (0.707)(\frac{7}{16}) = 0.309$ in.
Capacity of weld/in. $= \phi F_w = (0.75)(0.60 \times 70)(0.309)(1.0)$
$\qquad\qquad = 9.73$ k/in. $\leftarrow$
Total capacity of weld $= (9.73)(20) = 194.65$ k $\leftarrow$
Design strength of plate $= \phi F_y A_g = (0.90)(36)(\frac{3}{4} \times 10)$
$\qquad\qquad = 243$ k
$\underline{\text{Design capacity} = 194.6 \text{ k}}$ ∎

■ Example 23-16
Repeat Example 23-15 using SAW welds.

Solution.

Effective throat thickness $= (0.707)(\frac{7}{16}) + 0.11 = 0.419$ in.
Capacity of weld/in. $= \phi F_w = (0.75)(0.60 \times 70)(0.419)(1.0)$
$\qquad\qquad = 13.20$ k/in.
Total capacity of weld $= (13.20)(20) = 264$ k
Design strength of plate $= \phi F_y A_g = (0.90)(36)(\frac{3}{4} \times 8)$
$\qquad\qquad = 194.4$ k $\leftarrow$
$\underline{\text{Design capacity} = 194.4 \text{ k}}$ ∎

■ Example 23-17
Using A36 steel and E70 electrodes, design SMAW fillet welds to resist a full-capacity load on the $\frac{3}{8} \times 6$-in. member shown in Fig. 23-14.

Solution.

$P_u = \phi_t F_y A_g = (0.90)(36)(\frac{3}{8} \times 6) = 72.9$ k
Maximum weld size $= \frac{3}{8} - \frac{1}{16} = \frac{5}{16}$ in. (LRFD Section J2.2b)
Minimum weld size $= \frac{3}{16}$ in. (Table 23-5)
$\underline{\text{Use } \frac{5}{16}\text{-in. weld}}$

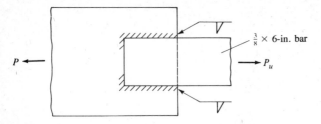

Figure 23-14

Effective throat thickness of weld $= (0.707)(\frac{5}{16}) = 0.221$ in.

Capacity of weld per in. $= \phi F_w$

$$= (0.75)(0.60 \times 70)(0.221)(1.0)$$

$$= 6.96 \text{ k/in.}$$

Length required $= \dfrac{72.9}{6.96} = 10.47$ in.

Use end returns not less than $2 \times \frac{5}{16} = \frac{10}{16}$ in. (say 1 in.)

Leaving $10.47 - 2.0 = 8.47$ in. or $4\frac{1}{2}$ in. each side.

However, must use $6 - 1 = 5$-in.-long welds

each side as required by LRFD Specification J2.2b. ∎

23-9 DESIGN OF FILLET WELDS FOR TRUSS MEMBERS

Should the members of a welded truss consist of single angles, double angles, or similar shapes and be subjected to static axial loads only, the LRFD Specification (J1.6) permits the connections to be designed by the same procedures described in the preceding section. The designers can select the weld size, calculate the total length of the weld required, and place the welds around the member ends as they see fit.

Should a welded connection be subjected to varying stresses (such as those occurring in a bridge member), it is considered desirable to place the welds so that their centroid will coincide with the centroid of the member (or the resulting torsion must be accounted for in design). If the member being connected is symmetrical, the welds will be placed symmetrically; if not, the welds will not be symmetrical.

The force in an angle, such as the one shown in Fig. 23-15, is assumed to act along its center of gravity. If the center of gravity of weld resistance is to coincide with the angle force, the weld must be asymmetrically placed, or in this figure L_1 must be longer than L_2. (When angles are connected by rivets or bolts, there is usually an appreciable amount of eccentricity, but in a welded joint eccentricity can be fairly well eliminated.) The information necessary to

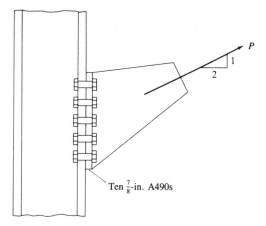

Ten $\frac{7}{8}$-in. A490s

Figure 23-15

handle this type of weld design can be easily expressed in equation form, but only the theory behind the equations is presented here.

For the angle shown in Fig. 23-15 the force acting along line L_2 (designated here as P_2) can be determined by taking moments about L_1. The member force and the weld resistance are to coincide, and the moments of the two about any point must be zero. If moments are taken about L_1, the force P_1 (which acts along line L_1) will be eliminated from the equation and P_2 can be determined. In a similar manner P_1 can be determined by taking moments along L_2 or by $\Sigma V = 0$. An angle with end and side fillet welds may be handled in a similar manner. The center of gravity and resistance of the end weld are known and can be easily included in the moment equations.

It is also to be remembered from our fatigue discussion in Chapter 4 that, should the estimated number of loading cycles during the structure's estimated life exceed 20,000, it will be necessary to study the stress range of the connection at service loads. This study may very well result in larger connections as required by Appendix K of the LRFD Specification.

23-10 SHEAR AND TORSION

Fillet welds are frequently loaded with eccentrically applied loads, with the result that the welds are subjected to either shear and torsion or to shear and bending. The tables of Part 5 of the LRFD Manual entitled "Eccentric Loads on Weld Groups" were developed to handle these types of problems. With the tables and the expression to follow, the ultimate strength P_u of a particular connection can be determined. In the formula C is a coefficient taken from the table, ϕ is 0.75, C_1 is a coefficient depending on the electrode number (it is 1.0 for E70XX), D is the weld size in sixteenths of an inch, and l is the length of the vertical weld.

$$P_u = CC_1 Dl$$

The LRFD Manual includes tables for both vertical and inclined loads (at angles from the vertical of 0°, 45°, and 75°). The user is warned not to interpolate for angles in between these values because the results may be quite un-conservative. Therefore, he or she is advised to use the value given for the next lower angle. Should the connection arrangement being considered not be covered by the tables, it is suggested that the conservative elastic procedure previously described be used. A simple illustration of these ultimate strength tables is presented in Example 23-18.

■ Example 23-18

For the bracket shown in Fig. 23-16(a), determine the fillet weld size required if E70 electrodes and the SMAW process are used. The center of gravity of the weld has been determined and is shown in part (b) of the figure. Use the LRFD tables.

Solution.

$$e = 11.11 \text{ in.}$$

$$l = 10 \text{ in.}$$

$$a = \frac{11.11}{10} = 1.111$$

$$k = \tfrac{4}{10} = 0.4$$

$C = 0.799$ from Table XXII, Part 5, LRFD Manual

$C_1 = 1.0$ from same table

$$D = \text{weld size required} = \frac{P_u}{CC_1 l}$$

$$= \frac{25}{(0.799)(1.0)(10)} = 3.13 \text{ sixteenths}$$

$$= 0.196 \text{ in. (as compared with } 0.274 \text{ in.}$$
by the elastic method)

Use $\tfrac{1}{4}$ in.

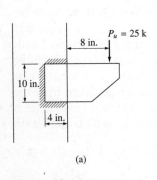

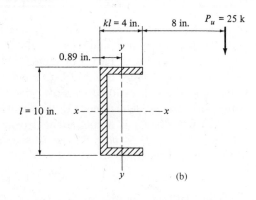

(a)   (b)

Figure 23-16

PROBLEMS

23-1 to 23-19   The following information is to be used unless otherwise indicated: (a) LRFD Specification; (b) standard-size holes; (c) threads of bolts excluded from shear plane; (d) members with clean mill scale surfaces (class A); (e) edge distances $>1\frac{1}{2}d$ and distances center to center of holes $>3d$.

23-1 The plates used in the lap joint shown are each $\frac{3}{4} \times 14$ in. and consist of A36 steel. Determine the design load P_u which can be resisted if $\frac{3}{4}$-in. A325 bolts are used in a bearing-type connection. (*Ans.* 186.1 k)

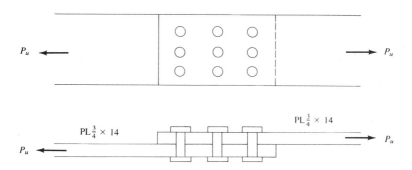

23-2 Rework Prob. 12-1 if 1-in. A490 bolts are used in a bearing-type connection.

23-3 If $\frac{7}{8}$-in. A325 bolts are used in the bearing-type connection shown, determine its design tensile capacity. Steel is A36. (*Ans.* 388.8 k)

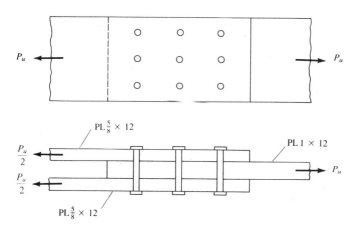

23-4 The truss member shown consists of two C12×25s (A36 steel) connected to a 1-in. gusset plate. How many $\frac{3}{4}$-in. A325 bolts are required to develop the full design tensile capacity of the member if it is used as a bearing-type connection?

23-4

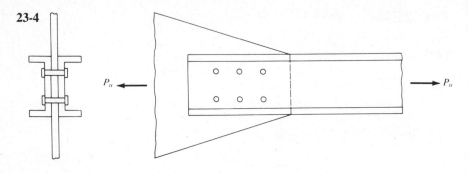

23-5 Rework Prob. 23-4 using $\frac{7}{8}$-in. A490 bolts and A572 steel ($F_y = 50$ ksi, $F_u = 65$ ksi). (*Ans.* 7.75, say 8 bolts)

23-6 For the connection shown, P_u equals 800 k. Determine the number of 1-in. A325 bolts required for a bearing-type connection. Use A36 steel.

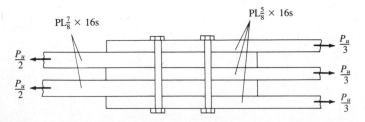

23-7 Rework Prob. 23-6 using $\frac{7}{8}$-in. A490 bolts. (*Ans.* 11.37, say 12 bolts)

23-8 Determine the number of bolts required for a slip-critical connection for the plates shown. The axial service loads are $P_D = 50$ k and $P_L = 110$ k; the bolts are $\frac{3}{4}$-in. A325 high-strength bolts, and A36 steel is to be used.

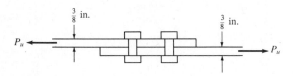

23-9 Repeat Prob. 23-8 with the threads not excluded from the shear plane. (*Ans.* 21.30, say 22 bolts)

23-10 How many $\frac{3}{4}$-in. A325 high-strength bolts are required for the slip-critical connection shown? $P_D = 100$ k and $P_L = 200$ k. Use A36 steel.

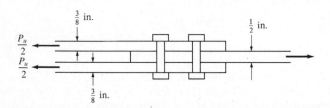

23-11 Repeat Prob. 23-6 with $P_D = 100$ k and $P_L = 180$ k and using $\frac{7}{8}$-in. A325 bolts in a slip-critical connection. Use A36 steel. (*Ans.* 13.70, say 14 or 15 bolts)

23-12 For the beam shown, what is the required spacing of $\frac{3}{4}$-in. A325 bolts in a bearing-type connection at a section where the external shear V_u is 260 k? Use A36 steel.

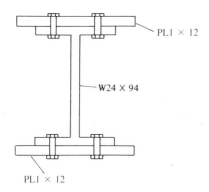

23-13 The cover-plated section shown is used to support a uniform load $w_D = 14$ k/ft (includes beam weight effect) and $w_L = 8$ k/ft for an 18-ft simple span. If 1-in. A325 bolts are used in a slip-critical connection, work out a spacing diagram for the entire span. Use A36 steel. (*Ans.* Use $6\frac{1}{2}$-in. spacing as controlled by LRFD Specification E4)

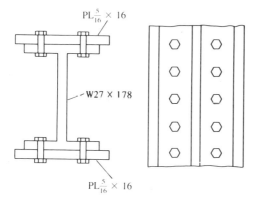

23-14 If the load shown passes through the center of gravity of the bolt group, how large can it be if bearing type bolts are used?

23-15 Is the connection of Prob. 23-14 satisfactory if the bolts are slip critical and if $P_D = 50$ k and $P_L = 70$ k? (*Ans.* Connection is satisfactory)

23-15

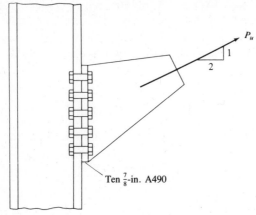

Ten $\frac{7}{8}$-in. A490

23-16 Repeat Prob. 23-15 using A325 bolts.

23-17 Determine the number of $\frac{7}{8}$-in. A325 bolts required in the angles and in the flange of the W shape shown if a bearing-type (snug-tight) connection is used. Use A36 steel. (*Ans.* 4 bolts in angles, 6 in column)

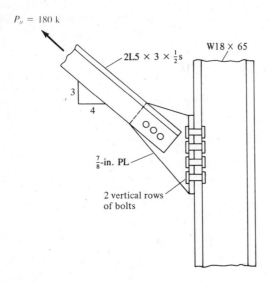

$P_u = 180$ k

2L5 × 3 × $\frac{1}{2}$s

W18 × 65

$\frac{7}{8}$-in. PL

2 vertical rows of bolts

23-18 Determine P_u of the plate shown, using the ultimate strength tables entitled "Eccentric Loads on Fastener Groups" in Part 5 of the LRFD Manual. The bolts are $\frac{7}{8}$-in. A325 and are in single shear and bearing on $\frac{1}{2}$-in. A36 steel.

23-18

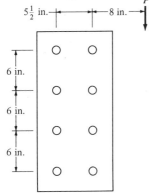

$5\frac{1}{2}$ in. — 8 in. — P

6 in.

6 in.

6 in.

23-19 Determine P_u of the plate shown, using the ultimate strength tables entitled "Eccentric Loads on Fastener Groups" in Part 5 of the LRFD Manual. The bolts are $\frac{3}{4}$-in. A325 and are in double shear and bearing on $\frac{3}{4}$-in. A36 steel. (*Ans.* 197.7 k)

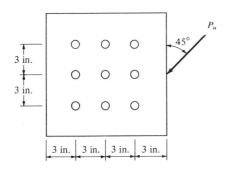

P_u

45°

3 in.

3 in.

3 in. | 3 in. | 3 in. | 3 in.

Probs. 23-20 through 23-26 use A36 steel.

23-20 Determine the design strength P_u of the connection shown if $\frac{3}{4}$-in. A502, grade 1, rivets are used.

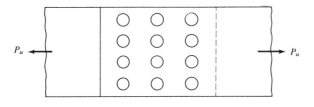

P_u P_u

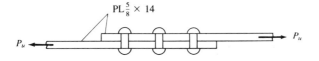

PL$\frac{5}{8}$ × 14

P_u P_u

23-21 The truss tension member shown consists of a single-angle $5 \times 3 \times \frac{5}{16}$ angle and is connected to a $\frac{1}{2}$-in. gusset plate with five $\frac{7}{8}$-in. A502, grade 1, rivets. Determine P_u. (*Ans.* 70.35 k)

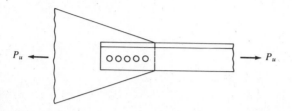

23-22 How many $\frac{3}{4}$-in. A502, grade 1, rivets are needed to carry the load shown?

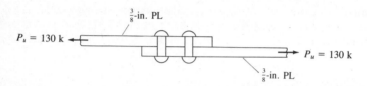

23-23 Repeat Prob. 23-22 using A307 bolts. (*Ans.* 18.16, say 18 or 20 bolts)

23-24 How many A502, grade 1, rivets with $\frac{3}{4}$-in. diameters needed to be used for the butt joint shown?

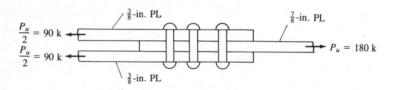

23-25 How many 1-in. A307 bolts are required for the connection shown? Factored loads are shown. (*Ans.* 15.72, say 16 bolts)

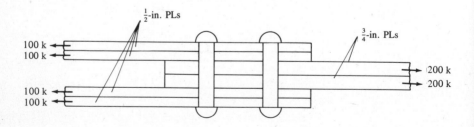

23-26 For the connection shown, $P_u = 650$ k. Determine the number of 1-in. A502, grade 2, rivets required.

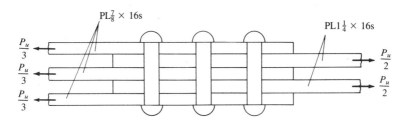

PL$\frac{7}{8}$ × 16s

PL1$\frac{1}{4}$ × 16s

$\frac{P_u}{3}$

$\frac{P_u}{3}$

$\frac{P_u}{3}$

$\frac{P_u}{2}$

$\frac{P_u}{2}$

23-27 A $\frac{1}{4}$-in. fillet weld (SMAW process) is used to connect the members shown. Determine the design load that can be applied to this connection according to the LRFD Specification if the steel is A36 and E70 electrodes are used. Include plate strength in calculations. (*Ans. $P_u = 97.2$ k*)

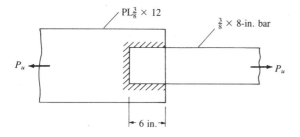

PL$\frac{3}{8}$ × 12

$\frac{3}{8}$ × 8-in. bar

P_u

P_u

← 6 in. →

23-28 Rework Prob. 23-27 using the SAW process.

23-29 Rework Prob. 23-27 using A572, grade 65 steel and E80 electrodes. (*Ans. $P_u = 127.4$ k*)

23-30 Design maximum-size fillet welds to develop the full strength of the A36 bar shown. Use E70 electrodes and the SMAW process.

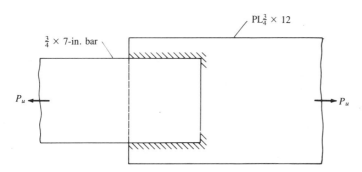

PL$\frac{3}{4}$ × 12

$\frac{3}{4}$ × 7-in. bar

P_u

P_u

23-31 Rework Prob. 23-30 using the SAW process. (*Ans. Use 8-in. welds each side, including end returns, as required by LRFD Section J2.2b*)

23-32 Rework Prob. 23-30 using side welds and a vertical weld at the end of the $\frac{3}{4}$×7 bar. Also use A572, grade 65, steel and E80 electrodes.

23-33 Rework Prob. 23-30 using side welds and welds at the end of the $\frac{3}{4}$×7 bar at the connection. (*Ans. Total $L = 11.11$ in., say 12 in.*)

23-34 Design maximum-size side SMAW fillet welds to develop the full tensile strength of an L6×4×$\frac{1}{2}$ using E70 electrodes and A36 steel. The member is connected on the 6-in. leg and is subject to alternating loads.

23-35 Rework Prob. 23-34 using side welds and a weld at the end of the angle. (*Ans.* $\frac{7}{16}$-in. weld, $2\frac{1}{2}$ in. and 8 in. on sides and 6 in. on end of angle)

23-36 One leg of an 8×8×$\frac{3}{4}$ angle is to be connected with side welds and a weld at the end of the angle to a plate behind to develop the full capacity of the angle (A36). Balance the SMAW fillet welds around the center of gravity of the angle, use maximum weld size, and assume E70 electrodes.

23-37 It is desired to design $\frac{5}{16}$-in. SMAW fillet welds necessary to connect the back of a C10×30 made from A36 steel to a $\frac{3}{8}$-in. gusset plate. End, side, and slot welds may be used to develop the full tensile capacity of the channel. Use E70 electrodes. It is assumed that due to space limitations the channel can lap over the gusset plate by a maximum of 10 in. (*Ans.* $\frac{5}{16}$-in. fillet welds and one $1\frac{1}{16}$×$2\frac{1}{2}$-in. slot weld)

23-38 Rework Prob. 23-37 using A572, grade 60, steel and E70 electrodes and $\frac{3}{8}$-in. fillet welds.

23-39 Using the elastic method, determine the maximum force per inch to be resisted by the fillet weld shown. (*Ans.* 17.66 k/in.)

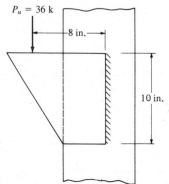

23-40 Using the elastic method, determine the maximum force per inch to be resisted by the fillet welds shown. Also determine the required weld thickness using A36 steel, E70 electrode, and SMAW welds.

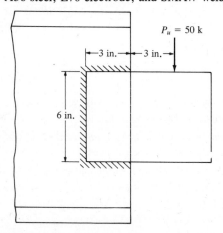

23-41 Determine the maximum eccentric load P_u that can be applied to the connection shown if $\frac{5}{16}$-in. SMAW fillet welds are used. Assume plate thickness $= \frac{1}{2}$ in., E70 electrodes, and A36 steel.   **(a)** Use elastic method.   **(b)** Use LRFD tables and ultimate strength method. (*Ans.* [a] 20.59 k [b] 29.45 k)

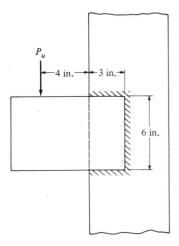

23-42 Rework Prob. 23-41 using $\frac{3}{8}$-in. fillet welds and a vertical weld 10 in. high.

23-43 Determine the SMAW fillet weld size required for the connection of Prob. 23-39 with the load increased to 40 k and the height of the weld to 12 in. Use A36 steel and E70 electrodes.   **(a)** Use elastic method.   **(b)** Use LRFD tables and ultimate strength method. (*Ans.* [a] 0.617 in., say $\frac{5}{8}$ in. [b] 0.412 in., say $\frac{7}{16}$ in.)

23-44 Determine the SMAW fillet weld size required for the connection shown. Use E70 electrodes and A36 steel. What angle thickness should be used?   **(a)** Use elastic method.   **(b)** Use LRFD tables and ultimate strength method.

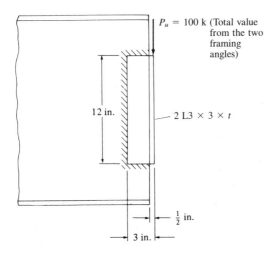

23-45 Determine the length of $\frac{1}{4}$-in. SMAW fillet welds 12 in. on center required to connect the cover plates for the section shown at a point where the external shear V_u is 170 k. Use A36 steel and E70 electrodes. (*Ans.* 5.73 in., say 6 in.)

23-45

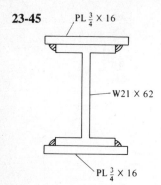

PL $\frac{3}{4}$ × 16

W21 × 62

PL $\frac{3}{4}$ × 16

23-46 The welded built-up section shown has an external shear of 750 k at a particular section. Determine the fillet weld size required to fasten the plates to the web if the SMAW process is used. Use E70 electrodes and A36 steel.

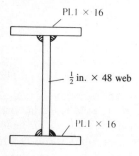

PL.1 × 16

$\frac{1}{2}$ in. × 48 web

PL.1 × 16

Appendix

A-1 SLENDER COMPRESSION ELEMENTS

Section B5 of the ASD Specification is concerned with the local buckling of compression elements. In that section elements are classified as being compact, noncompact, or slender. As compact and noncompact elements have previously been discussed, this section is concerned only with the slender ones. A brief summary of the ASD method for determining allowable stresses for such members is presented in the next few paragraphs.

Should b/t values exceed the values given in ASD Table B5.1 for noncompact elements, the elements will be slender, and allowable stresses will have to be reduced. Allowable stresses are determined with the usual ASD column and beam equations *but the yield stress F_y in those expressions is replaced with a reduced value F_{ye}*. Its value is determined from the appropriate expression of the ones to be listed here.

To obtain F_{ye}, the yield stress F_y is multiplied by a reduction factor Q that is determined by multiplying together two reduction factors Q_s and Q_a. In other words, $F_{ye} = QF_y = Q_aQ_sF_y$. Three cases are considered in the Appendix to the ASD Specification.

1. For members consisting of unstiffened elements only, Q_s is to be determined with the appropriate formulas presented in this appendix, and $Q_a = 1.0$.
2. For members consisting of stiffened elements only, $Q_s = 1.0$, and Q_a is to be determined with the formulas to be presented in this appendix.
3. For members consisting of some stiffened elements and some unstiffened ones, Q_s and Q_a are determined with the appropriate formulas to be presented in this appendix.

Values of Q_s

Expressions for computing Q_s (which is referred to as the axial stress reduction factor for slender elements) are given in Appendix B of the ASD Manual. Equations are given for the following types of members: (a) single angles, (b) angles or plates projecting from columns or other compression members and for the projecting elements of the compression flanges of beams and girders, and (c) the stems of tees. In these equations the following terms are used:

$b = $ width of unstiffened compression element, in.

$t = $ thickness of unstiffened compression element, in.

$F_y = $ specified minimum yield stress, ksi

$k_c = $ compression element restraint coefficient

$$k_c = \frac{4.05}{\left(\dfrac{h}{t}\right)^{0.46}} \quad \text{if } \frac{h}{t} > 0, \text{ otherwise } k_c = 1.0$$

Only the Q_s expressions for case (b) are presented here. These are for angles or plates projecting from columns or other compression members or from the projecting elements of the compression flanges of beams or girders.

$$\text{When } 95.0 \sqrt{\frac{F_y}{k_c}} < \frac{b}{t} < 195 \sqrt{\frac{F_y}{k_c}}$$

$$Q_s = 1.293 - 0.00309\left(\frac{b}{t}\right)\sqrt{\frac{F_y}{k_c}} \qquad \text{(ASD Equation A-B5-3)}$$

$$\text{When } \frac{b}{t} > 195 \sqrt{\frac{F_y}{k_c}}$$

$$Q_s = \frac{26{,}200\,k_c}{F_y\left(\dfrac{b}{t}\right)^2} \qquad \text{(ASD Equation A-B5-4)}$$

In Example A-1 the value of Q_s is computed for a pair of angles that is used as a compression member. The resulting value can be checked in the double-angle tables of Part 1 of the Manual. This same double-angle member is further considered in Examples A-2 and A-3 as to lateral-torsional buckling, and the Q_s obtained here is used.

■ Example A-1

The pair of L8×6×$\frac{1}{2}$s, long legs back-to-back separated by $\frac{3}{8}$ in., shown in Fig. A-1, is used as a compression member. Compute Q_s for this member, which is assumed to be made from A36 steel.

Solution.   As shown in ASD Appendix B5.2a, if

$$\frac{76}{\sqrt{F_y}} < \frac{b}{t} < \frac{155}{\sqrt{F_y}}$$

then ASD Equation A-B5-3 is used to compute Q_s for angles projecting from a column.

$$\frac{76}{\sqrt{36}} = 12.67 < \frac{8}{\frac{1}{2}} = 16 < \frac{155}{\sqrt{36}} = 25.83$$

$$Q_s = 1.340 - 0.00447\left(\frac{b}{t}\right)\sqrt{F_y}$$

$$= 1.340 - (0.00447)(16)(\sqrt{36})$$

$$= \underline{\underline{0.911}}$$

Checks with value in double angle tables.

    Note: If the 8-in. legs were not separated, b/t would equal $6/\frac{1}{2} = 12 <$ 12.67 and Q_s would equal 1.0.   ■

A-2 FLEXURAL-TORSIONAL BUCKLING OF COMPRESSION MEMBERS—ASD SPECIFICATION

Axially loaded compression members can theoretically fail in three different fashions: by flexural buckling, by torsional buckling, or by flexural-torsional buckling.

    Flexural buckling (also called Euler buckling) is the situation considered in the column chapters of this book. For such cases the slenderness ratios are

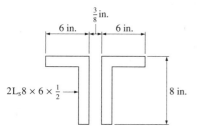

Figure A-1

computed for each of the principal column axes, and F_a is determined for the highest ratio so obtained. Doubly symmetrical columns (such as W sections) are subject only to flexural buckling and torsional buckling. Torsion will not occur in such sections if the lines of action of the loads pass through the shear centers of the sections. The locations of shear centers for several sections are shown in Fig. A-2.

As torsional buckling can be very complex, it is very desirable to prevent its occurrence. This may be done by careful arrangements of the members and by providing bracing to prevent lateral movement and twisting. If sufficient end supports and intermediate lateral bracing are provided, flexural buckling will always control. The values given in the ASD column tables for W, M, S, and pipe sections are all based on flexural buckling.

For a singly symmetrical section such as a tee or double angle, Euler buckling may occur about the x or y axis. For equal-leg single angles, Euler buckling may occur about the z axis. For each of these sections flexural-torsional buckling is definitely a possibility and may control. (It will always control for unequal-leg single-angle columns.) The values given in both the ASD and LRFD column load tables for double-angle and structural tee sections were computed for buckling about the weaker of the x and y axis and for flexural-torsional buckling.

The average designer does not consider the torsional buckling of symmetrical shapes or the flexural-torsional buckling of unsymmetrical shapes. The usual feeling is that these conditions don't control the critical column loads or at least don't affect them very much. Should we have unsymmetrical columns or even symmetrical columns made up of thin plates, however, we will find that torsional buckling or flexural-torsional buckling may significantly reduce column capacities.

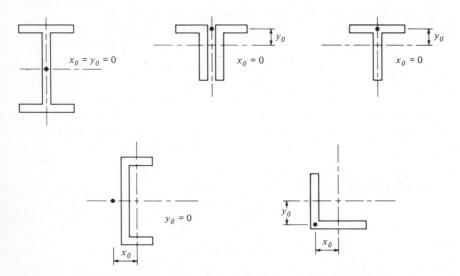

Figure A-2 Shear center locations for some common column sections.

Neither the ASD nor the LRFD Manual provides tables for single-angle compression members. It is stated in the manuals that such tables are not provided because of the difficulty of loading such members concentrically. As a result, the designer using single-angle struts is going to have to wade through the very complex flexural-torsional formulas and make some allowances for eccentricities of loading. Excellent examples of this type are presented in Part 2 of the LRFD Manual and Part 3 of the ASD Manual. A set of tables for the design of single equal leg angles is available.[1]

In Appendix E of the LRFD Specification a long list of formulas is presented for computing the flexural-torsional strength of column sections. The values given for column design strengths for double angles and tees in the LRFD Manual ($\phi_c P_n$ values) and the allowable column load values given for those same sections in the ASD Manual make use of these formulas. In this section the ASD values are considered. The steps and equations involved follow.

The Euler equation is usually written as

$$F = \frac{\pi^2 E}{\left(\dfrac{KL}{r}\right)^2}$$

It can be solved for $\dfrac{KL}{r}$:

$$\frac{KL}{r} = \pi \sqrt{\frac{E}{F}}$$

For flexural-torsional buckling this form of the equation is used, except that F is replaced with F_e, which is the critical flexural-torsional elastic buckling stress. The determination of F_e is described later in this section. Having its value, an equivalent slenderness ratio is determined:

$$\left(\frac{KL}{r}\right)_{eq} = \pi \sqrt{\frac{E}{F_e}}$$

Once this equivalent slenderness ratio is determined, it is used to determine the allowable compression stress with the usual column formulas or tables. It should be clearly noted that (as often happens for single and double angles and tees), if b/t exceeds the values given in Table B5.1 of the ASD Specification, the Q reduction factors will come into play. If Q is involved we need to calculate C_c' as follows:

$$C_c' = \sqrt{\frac{2\pi^2 E}{QF_y}}$$

---

[1] W. W. Walker, "Tables for Equal Single Angles in Compression," *Engineering Journal*, AISC, 2nd quarter, no. 2 (1991), pp 65–68.

We use that value in the applicable equation of the two equations that follow for axially loaded compression members:

If $\dfrac{Kl}{r} > C_c'$

$$F_a = \frac{Q\left[1 - \dfrac{\left(\dfrac{KL}{r}\right)^2}{2C_c'^2}\right]F_y}{\dfrac{5}{3} + \dfrac{3\left(\dfrac{KL}{r}\right)}{8C_c'} - \dfrac{\left(\dfrac{KL}{r}\right)^3}{8C_c'^3}}$$ (ASD Equation A-B5-11)

Or if $KL/r > C_c'$,

$$F_a = \frac{12\pi^2 E}{23\left(\dfrac{KL}{r}\right)^2}$$ (ASD Equation A-B5-12)

The formulas for determining F_e for different steel sections are given at the end of this section. Once again, note that these values are taken from the LRFD Specification and thus the formula numbers given are LRFD formula numbers. For doubly symmetric shapes,

$$F_e = \left[\frac{\pi^2 E C_w}{(K_z L)^2} + GJ\right]\frac{1}{I_x + I_y}$$ (LRFD Formula A-E3-5)

For singly symmetric shapes where y is the axis of symmetry,

$$F_e = \frac{F_{ey} + F_{ez}}{2H}\left[1 - \sqrt{1 - \frac{4F_{ey}F_{ez}H}{(F_{ey} + F_{ez})^2}}\right]$$ (LRFD Formula A-E3-6)

For unsymmetrical sections F_e is the lowest root of the following cubic equation:

$$(F_e - F_{ex})(F_e - F_{ey})(F_e - F_{ez}) - F_e^2(F_e - F_{ey})\left(\frac{x_0}{r_0}\right)^2$$

$$- F_e^2(F_e - F_{ex})\left(\frac{y_0}{r_0}\right)^2 = 0$$ (LRFD Formula A-3E-7)

A perhaps more convenient form of LRFD Formula A-E3-7 follows:

$$HF_e^3 + \left[\frac{1}{r_0^2}(y_0 F_{ex} + x_0^2 F_{ey}) - (F_{ex} + F_{ey} + F_{ez})\right]F_e^2$$

$$+ (F_{ex}F_{ey} + F_{ex}F_{ez} + F_{ey}F_{ez})F_e - F_{ex}F_{ey}F_{ez} = 0$$

where

K_z = effective length factor for torsional buckling

G = shear modulus, ksi

C_w = warping constant, in.^{6}

J = torsional constant, in.^{4}

$$\bar{r}_0^2 = x_0^2 + y_0^2 + \frac{I_x + I_y}{A}$$ (LRFD Formula A-E3-8)

$$H = 1 - \left(\frac{x_0^2 + y_0^2}{\bar{r}_0^2}\right)$$ (LRFD Formula A-E3-9)

$$F_{ex} = \frac{\pi^2 E}{(KL/r)_x^2}$$ (LRFD Formula A-E3-10)

$$F_{ey} = \frac{\pi^2 E}{(KL/r)_y^2}$$ (LRFD Formula A-E3-11)

$$F_{ez} = \left[\frac{\pi^2 E C_w}{(K_z L)^2} + GJ\right]\frac{1}{A\bar{r}_0^2}$$ (LRFD Formula A-E3-12)

The values of C_w, J, $\bar{r}_0$, and H are provided for many sections in the "Flexural-Torsional Properties" tables of Part 1 of the Manual.

Example A-2 illustrates the application of the appropriate equations for determining the allowable flexural-torsional buckling strength of a double-angle member where Q_s was determined.

■ Example A-2

Determine the allowable flexural buckling load and the allowable flexural-torsional buckling load for an 18-ft pinned-end column consisting of two L8×6×$\frac{1}{2}$s, long legs back-to-back with a $\frac{3}{8}$-in. gusset plate between them. The angles are made from A36 steel and are connected to each other at 6-ft intervals with two fully tensioned high-strength bolts as shown in Fig. A-3. The cross section of the member is shown in Fig. A-4. $G = 11,200$ ksi and $K = 1.0$.

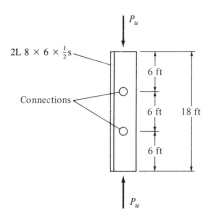

Figure A-3

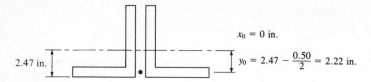

$x_0 = 0$ in.

2.47 in.

$y_0 = 2.47 - \dfrac{0.50}{2} = 2.22$ in.

Figure A-4

Solution.   Using two L8×6×$\frac{1}{2}$s, long legs back-to-back, $\frac{3}{8}$-in. gusset plate between ($A = 13.5$ in.^{2}, $r_x = 2.56$ in., r_z for one $L = 1.30$ in., and $r_y = 2.44$ in.)

Allowable flexural buckling strength:

$$\frac{K_y L_y}{r_y} = \frac{(1.0)(12 \times 18)}{2.44} = 88.52$$

F_a from Table C-36 in Part 3 of Manual $= 14.38$ ksi

Allowable $P = (14.38)(13.5) = \underline{\underline{194.1 \text{ k}}}$

Allowable flexural-torsional buckling load:
From flexural-torsional properties tables in Part 1 of Manual

$$J = 0.584 \text{ in.}^4 \text{ per L}$$
$$C_w = 2.280 \text{ in.}^6 \text{ per L}$$
$$r_0 = 4.18 \text{ in. for 2 Ls}$$
$$H = 0.718 \text{ for 2 Ls}$$

Checking table values of r_0 and H

$$\bar{r}_0^2 = x_0^2 + y_0^2 + r_x^2 + r_y^2 = 0^2 + 2.22^2 + 2.56^2 + 2.44^2$$
$$\bar{r}_0 = 4.18 \text{ in.}$$

$$H = 1 - \frac{0^2 + 2.22^2}{(4.18)^2} = 0.718$$

Computing values of F_{ey}, F_{ez}, and F_e

$$\frac{a}{r_z} = \frac{72}{1.30} = 55.38 > 50$$

$\therefore$ Use $\left(\dfrac{KL}{r}\right)_m$ rather than $\dfrac{K_y L_y}{r_y}$.

$$\left(\frac{KL}{r}\right)_y = \frac{(1.0)(12 \times 18)}{2.44} = 88.52$$

$$\left(\frac{KL}{r}\right)_m = \sqrt{(88.52)^2 + (55.38 - 50)^2} = 88.68$$

$$F_{ey} = \frac{(\pi)^2(29,000)}{(88.68)^2} = 36.40 \text{ ksi}$$

$$F_{ez} = \left[\frac{(\pi)^2(29,000)(2 \times 2.28)}{(1.0 \times 12 \times 18)^2} + 11,200 \times 2 \times 0.584 \right] \frac{1}{(13.5)(4.18)^2}$$

$$= 55.58 \text{ ksi}$$

$$F_e = \frac{36.40 + 55.58}{(2)(0.718)} \left[1 - \sqrt{1 - \frac{(4)(36.40)(55.58)(0.718)}{(36.40 + 55.58)^2}} \right]$$

$$= 28.20 \text{ ksi}$$

$$\left(\frac{KL}{r} \right)_{eff} = \pi \sqrt{\frac{\pi}{F_e}} = \pi \sqrt{\frac{29 \times 10^3}{28.20}} = 100.75$$

Then $\dfrac{b}{t} = \dfrac{8}{\frac{1}{2}} = 16 > \dfrac{76}{\sqrt{F_y}} = \dfrac{76}{\sqrt{36}} = 12.67$

∴ It is a slender element.

From double-angle tables (also Example A-1) we determine $Q_s = 0.911$. Then to find C_a from Table 3 in ASD Appendix we calculate

$$C_c' = \sqrt{\frac{2\pi^2 E}{Q_s F_y}} = 132.12$$

$$\frac{\dfrac{KL}{r}}{C_c'} = \frac{100.75}{132.12} = 0.763$$

And from the table $C_a = 0.374$. Then allowable flexural-torsional buckling stress and allowable load on column

$$F_a = C_a Q_s F_y = (0.374)(0.911)(36) = 12.27 \text{ ksi}$$
$$\text{Allowable } P = (12.27)(13.5) = \underline{\underline{165.6 \text{ k}}}$$

Checks with value in ASD column tables.   ■

A-3 FLEXURAL-TORSIONAL BUCKLING OF COMPRESSION MEMBERS—LRFD SPECIFICATION

To obtain the flexural-torsional strength of a column using the LRFD Specification, the same steps are taken to determine F_e as were described in Section A-2. Then we calculate $P_u = \phi_c P_n = \phi_c A_g F_{cr}$ with $\phi_c = 0.85$ and F_{cr} to be determined from the formulas to be given here.

$$\lambda_e = \sqrt{\frac{F_y}{F_e}} \qquad \text{(LRFD Formula A-E3-4)}$$

If $\lambda_e \sqrt{Q} \le 1.5$

$$F_{cr} = Q(0.658^{Q\lambda_e^2})F_y \qquad \text{(LRFD Formula A-E3-2)}$$

If $\lambda_e\sqrt{Q} > 1.5$

$$F_{cr} = \left(\frac{0.877}{\lambda_e^2}\right)F_y \qquad \text{(LRFD Formula A-E3-3)}$$

in which $Q = 1.0$ for elements meeting the width-thickness ratios λ_r of LRFD Section B5.1 and if not calculated as described in LRFD Appendixes E3 and B5.3.

In Example A-3 the author has used the above formulas to compute the lateral-torsional strength of the pair of angles considered in Examples A-1 and A-2.

■ Example A-3
Determine the flexural buckling strength and the flexural-torsional buckling strength for the angle of Example A-2, using the LRFD Specification.

Solution.   Using the data previously determined in Example A-2:
Flexural buckling

$$\left(\frac{KL}{r}\right)_y = \frac{(1.0)(12 \times 18)}{2.44} = 88.52$$

$$\phi_c F_{cr} = 20.26 \text{ ksi}$$
$$\phi_c P_n = (20.26)(13.5) = \underline{273.5 \text{ k}}$$

Flexural-torsional buckling

$$\lambda_e = \sqrt{\frac{F_y}{F_e}} = \sqrt{\frac{36}{28.20}} = 1.130 < 1.5$$

$$\phi_c F_{cr} = (0.85)(0.911 \times 36)(0.658)^{0.911 \times 1.130^2} = 17.14 \text{ ksi}$$

where F_y is multiplied by $Q_s = 0.911$ from LRFD Manual page 1–89 as a slender unstiffened element or as determined in Example A-1.

$$\phi_c P_n = (17.14)(13.5) = \underline{231.4 \text{ k}}$$

Does not check LRFD table value page 2-59 because bolt spacings are different.   ■

A-4 PONDING

It has been claimed that almost 50 percent of the lawsuits faced by building designers are concerned with roofing systems.[2] Ponding, a problem with many flat roofs, is one of the common subjects of such litigation. If water accumu-

---

[2] Gary Van Ryzin, "Roof Design: Avoid Ponding by Sloping to Drain," *Civil Engineering* (New York: ASCE, January 1980), pp. 77–81.

lates more rapidly on a roof than it runs off, ponding results because the increased load causes the roof to deflect into a dish shape that can hold more water, which causes greater deflections, and so on. This process continues until equilibrium is reached or until collapse occurs. Ponding can be caused by increasing deflections, clogged roof drains, settlement of footings, warped roof slabs, and so on.

The best way to prevent ponding is to use appreciable roof slopes ($\frac{1}{4}$ in. per ft or more) together with good drainage facilities. Supposedly more than two-thirds of the flat roofs in the United States have slopes less than $\frac{1}{4}$ in. per ft. The construction of roofs with slopes this large will increase building costs by only a few percent as compared to perfectly flat roofs. The supporting girders for flat roofs with long spans should definitely be cambered to reduce the possibility of ponding (as well as the sagging that is so disturbing to the people occupying a building).

Section K2 of both the ASD and LRFD Specifications states that, unless roof surfaces have sufficient slopes to areas of free drainage or sufficient individual drains to prevent water accumulation, the strength and stability of the roof systems during ponding conditions must be investigated. The very detailed work of Marino[3] forms the basis of the ponding provisions of the ASD and LRFD Specifications. Many other useful references are available.[4,5,6]

The amount of water that can be retained on a roof depends on the flexibility of the framing. The specifications state that a roof system can be considered stable and not requiring further investigation if the following expressions are satisfied:

$$C_p + 0.9C_s \leq 0.25 \qquad \text{(ASD Equation K2-1)}$$
$$I_d \geq 25(S^4)10^{-6} \qquad \text{(ASD Equation K2-2)}$$

where

$C_p = 32\,L_s L_p^4/10^7 I_p$

$C_s = 32\,SL_s^4/10^7 I_s$

L_p = length of primary members, ft

L_s = length of secondary members, ft

S = spacing of secondary members, ft

I_p = moment of inertia of primary members, in.[4]

[3] F. J. Marino, "Ponding of Two-Way Roof System," *Engineering Journal*, AISC, 3rd quarter, no. 3 (1966), pp. 93–100.

[4] L. B. Burgett, "Fast Check for Ponding," *Engineering Journal*, AISC, 10, no. 1 (1st quarter, 1973), pp. 26–28.

[5] J. Chinn, "Failure of Simply-Supported Flat Roofs by Ponding of Rain," *Engineering Journal*, AISC, no. 2, 2nd quarter (1965), pp. 38–41.

[6] J. L. Ruddy, "Ponding of Concrete Deck Floors," *Engineering Journal*, AISC, 23, no. 2 (3rd quarter, 1986), pp. 107–115.

I_s = moment of inertia of secondary members, in.[4]

I_d = moment of inertia of the steel deck (if one is used) supported on the secondary members, in.[4] per ft

Should steel roof decks be used, their I_d must at least equal the value given by Equation K2-2. If the roof decking is the secondary system (i.e., no secondary beams, joists, etc.), it should be handled with Equation K2-1.

Some other ASD and LRFD requirements in applying these expressions follow:

1. The moment of inertia I_s must be decreased by 15 percent for trusses and steel joists.
2. Steel decking is considered to be a secondary member supported directly by the primary members.
3. The total bending stress caused by dead loads plus gravity live loads plus ponding may not exceed $0.80F_y$ for either primary or secondary members.
4. Stresses caused by wind or seismic forces do not have to be considered in the ponding calculations.

Should moments of inertia be needed for open-web joists, they can be computed from the member cross sections or perhaps more easily backfigured from the resisting moments and allowable stresses given in the joist tables. (Since $M_R = FI/c$, we can compute $I = M_R c/F$.)

In effect these equations reflect *stress indexes* or percentage stress increases. For instance, here we are considering the percentage increase in stress in the steel members caused by ponding. If the stress in a member increases from $0.60\,F_y$ to $0.80\,F_y$, we say that the stress index U is

$$U = \frac{0.80F_y - 0.60F_y}{0.60F_y} = 0.33$$

The terms C_p and C_s are respectively the approximate stiffnesses of the primary and secondary support systems. ASD Equation K2-1 ($C_p + 0.9C_s \leq 0.25$), which gives us an approximate stress index during ponding, is limited to a maximum value of 0.25. Should we substitute into this equation and obtain a value no greater than 0.25, ponding will supposedly not be a problem. Should the index be larger than 0.25, however, it will be necessary to conduct a further investigation. One method of doing this is presented in the ASD and LRFD Commentaries and will be described later in this section.

Example A-4 presents the application of ASD Equation K2-1 to a roofing system.

■ Example A-4

Check the roof system shown in Fig. A-5 for ponding, using the ASD and LRFD Specifications.

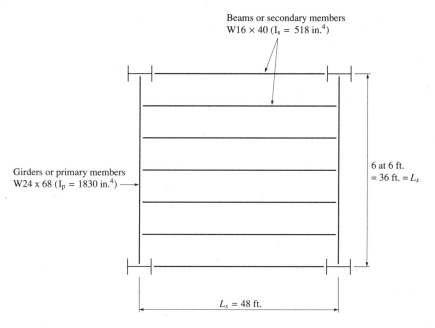

Figure A-5

Solution.

$$C_p = \frac{32\ L_s L_p^{\,4}}{10^7 I_p} = \frac{(32)(48)(36)^4}{(10^7)(1830)} = 0.141$$

$$C_s = \frac{32\ S L_s^{\,4}}{10^7 I_s} = \frac{(32)(6)(48)^4}{(10^7)(518)} = 0.197$$

$$C_p + 0.9C_s = 0.141 + (0.9)(0.197) = 0.318 > 0.25$$

Indicates insufficient strength and stability, and thus a more precise method of checking should be used.   ∎

The curves in the ASD Commentary (C-K2) provide a design aid for use when we need to compute a more accurate flat-roof framing stiffness than is given by the specification provision that $C_p + 0.9C_s \leq 0.25$.

For this application the term f_o represents the computed bending stress due to the supported loads not including the ponding effect. The following stress indexes are computed for the primary and secondary members:

$$U_p = \left(\frac{0.80F_y - f_o}{f_o}\right)_p$$

$$U_s = \left(\frac{0.80F_y - f_o}{f_o}\right)_p$$

We enter Fig. C-K2.1 in Section K of the ASD Commentary with our computed U_p and move horizontally to the calculated C_s value of the secondary members. Then we go vertically to the abscissa scale and read the flexibility constant. If this value is more than our calculated C_p value computed for the primary members, the stiffness is sufficient. The same process is followed by Fig. C-K2.2, where we enter with our computed U_s and C_p values and pick from the abscissa the flexibility constant, which should be no less than our C_s value. This procedure is illustrated in Example A-5.

■ Example A-5

Recheck the roof system considered in Example A-4 using the ASD Commentary curves. Assume that as ponding begins, f_o is 75 percent of an allowable stress of $0.60 F_y$ for A36 steel.

Solution.   Checking girders:

$$f_o = (0.75)(0.60)(36) = 16.2 \text{ ksi}$$

$$U_p = \frac{0.80F_y - f_o}{f_o} = \frac{(0.80)(36) - 16.2}{16.2} = 0.778$$

For $U_p = 0.778$ and $C_s = 0.197$, we read in Fig. C-K2.1 $C_p = 0.28 >$ our calculated C_p of 0.141.

$$\therefore \text{Girders   OK}$$

Checking beams:

$$f_o = 16.2 \text{ ksi}$$
$$U_s = 0.778$$

For $U_s = 0.778$ and $C_p = 0.141$, we read in Fig. C-K2.2 $C_s = 0.26 >$ our calculated C_s of 0.197.

$$\therefore \text{Beams   OK} \qquad ■$$

A-5 MOMENT-RESISTING COLUMN BASES

Frequently column bases are designed to resist bending moments as well as axial loads. An axial load causes compression between a base plate and the supporting footing while a moment increases the compression on one side and decreases it on the other side. For average moments the forces may be transferred to the footing through flexure of the base plate, but when they are very large, stiffened or booted connections may be used. For a small moment the entire contact area between the plate and the supporting footing will remain in compression. This will be the case if the resultant load falls within the middle third of the plate length in the direction of bending.

Fig. A-6(a) and (b) shows base plates suitable for resisting relatively small moments. For these cases the moments are sufficiently small to permit their

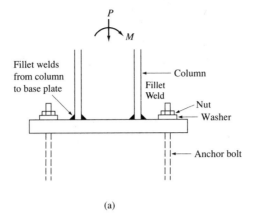

(a)

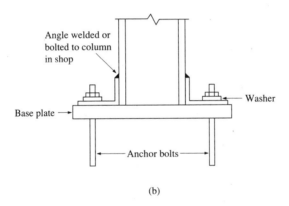

(b)

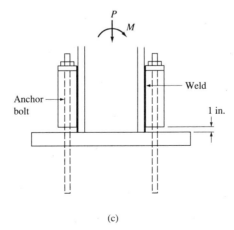

(c)

Figure A-6 Moment resisting column bases

transfer to the footings by bending of the base plates. The anchor bolts may or may not have calculable stresses, but they are nevertheless considered necessary for good construction practice. They are definitely needed to hold the columns firmly in place and upright during the initial steel erection process. Temporary guy cables are of course necessary during erection. The anchor bolts should be substantial and capable of resisting unforeseen erection forces. Sometimes these small plates are attached to the columns in the shop and sometimes they are shipped loose to the job and carefully set to the correct elevations in the field.

Should the eccentricity ($e = M/P$) be sufficiently large so that the resultant falls outside the middle third of the plate, there will be an uplift on the other side of the column, putting the anchor bolts on that side in tension.

The moment will be transferred from the column into the footing by means of the anchor bolts, embedded a sufficient distance into the footing to develop the forces. This embedment should be calculated as required by reinforced-concrete design methods.[7] The booted connection shown in Fig. A-6(c) is assumed to be welded to the column. The boots are generally made of angles or channels and are not as a rule connected directly to the base plate. Rather, the moment is transmitted from the column to the foundation by means of the anchor bolts. When booted connections are used, the base plates are normally shipped loose to the job and carefully set to the correct elevations in the field.

The capacity of these connections to resist rotation is dependent on the lengths of the anchor bolts, which are available to deform elastically. This capacity can be increased somewhat by pre-tensioning the anchor bolts. (This is similar to our prestress discussion for high-strength bolts in Section 12-3.) Actually prestressing is not very dependable and is not usually done because of the long-term creep in the concrete.

When a moment-resisting or rigid connection between a column and its footing is used, it is absolutely necessary for the supporting soil or rock beneath the footing to be appreciably noncompressible, or the column base will rotate as shown in Fig. A-7. If this happens, the rigid connection between the column and the footing is useless. For the purpose of this section the subsoil is assumed to be capable of resisting the moment applied to it without appreciable rotation.

As a first numerical example, a column base plate is designed for an axial load and a relatively small bending moment such that the resultant load falls between the column flanges. Assumptions are made for the width and length of the plate, after which the pressures underneath the plate are calculated and compared to the permissible value. If the pressures are unsatisfactory, the dimensions are changed and the pressures recalculated, until the values are satisfactory. The moment in the plate is calculated and the plate thickness determined. The critical section for bending is assumed to be at the center of the

---

[7] *Building Code Requirements for Reinforced Concrete (ACI 318-89) and Commentary (ACI 318R-89)* (Detroit: American Concrete Institute, 1989), pp. 188–189, 250–251.

Figure A-7

flange on the side where the compression is highest. Various designers will assume the point of maximum moment is located at some other point, as at the face of the flange or the center of the anchor bolt.

The moment is calculated for a 1-in.-wide strip of the plate and is equated to its resisting moment. The resulting expression is solved for the required thickness of the plate as follows:

$$M = \frac{F_b I}{c} = \frac{(F_b)(\frac{1}{12})(1)(t)^3}{t/2}$$

$$t = \sqrt{\frac{6M}{F_b}}$$

■ Example A-6

Design a moment-resisting base plate to support a W14 × 120 column with an axial load of 300 k and a bending moment of 115 ft-k. Use A36 steel with $F_b = 27$ ksi and a concrete footing with $f'_c = 2.5$ ksi and $F_p = (0.45)(2.5) = 1.125$ ksi.

Solution.   Using a W14 × 120 ($d = 14.48$ in., $t_w = 0.590$ in., $b_f = 14.670$ in., $t_f = 0.940$ in.)

$$e = \frac{(12)(115)}{300} = 4.60 \text{ in.}$$

∴ The resultant falls between the column flanges and within the middle third of the plate.

Try a 20 × 28-in. plate (after a few trials)

$$f = -\frac{P}{A} \pm \frac{Pec}{I} = -\frac{300}{(20)(28)} \pm \frac{(300)(4.60)(14)}{(\frac{1}{12})(20)(28)^3}$$

$$= -0.536 \pm 0.528 \begin{cases} -1.064 < 1.125 \text{ ksi} \\ -0.008 \text{ ksi} \end{cases} \qquad \text{OK}$$

Taking moments to right at center of right flange (see Fig. A-8):

$$M = (2.86)\left(\frac{7.23}{3}\right) + (3.85)\left(\frac{2}{3} \times 7.23\right) = 25.45 \text{ in.-k}$$

$$t = \sqrt{\frac{6M}{F_b}} = \sqrt{\frac{(6)(25.45)}{27}} = 2.38 \text{ in.}$$

Use PL2$\frac{1}{2}$ × 20 × 2 ft 4 in.   ■

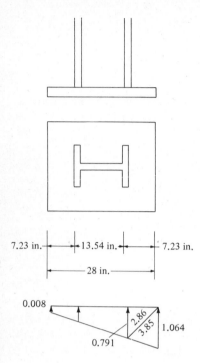

7.23 in.— —13.54 in.— —7.23 in.

—28 in.—

0.008

2.86

3.85

1.064

0.791

Figure A-8

The moment considered in Example A-7 is of such a magnitude that the resultant load falls outside the column flange. As a result, there will be uplift on one side, and the anchor bolt will have to furnish the needed tensile force to provide equilibrium.

In this design the anchor bolts are assumed to have no significant tension due to tightening. As a result, they are assumed not to affect the force system. As the moment is applied to the column, the pressure shifts toward the flange on the compression side. It is assumed that the resultant of this compression is located at the center of the flange.

■ Example A-7

Repeat Example A-6 with the same column and allowable stresses but with the moment increased from 115 ft-k to 225 ft-k. Refer to Fig. A-9.

Solution. Using a W14 × 120 ($d = 14.48$ in., $t_w = 0.590$ in., $b_f = 14.670$ in., $t_f = 0.940$ in.)

$$e = \frac{(12)(225)}{300} = 9.00 \text{ in.}$$

∴ The resultant falls outside the column flange.

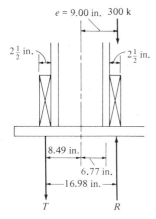

Figure A-9

Taking moments about the center of the right flange:

$$(300)(9.00 - 6.77) - 15.26T = 0$$
$$T = 43.84 \text{ k}$$

$$\text{Anchor bolt } A_{\text{reqd}} = \frac{43.84}{(0.33)(58)} = 2.29 \text{ in.}^2$$

Use two $1\frac{1}{4}$-in. diameter bolts each side

Checking pressure under the right flange:

$$R = 300 + 43.84 = 343.84 \text{ k}$$

$$f_a = \frac{343.84}{(14.67)(0.940)} = 24.93 \text{ ksi} < 27 \text{ ksi} \qquad \text{OK}$$

Approximate plate size, assuming a triangular pressure distribution:

$$A_{\text{reqd}} = \frac{343.84}{1.125/2} = 611.3 \text{ in.}^2$$

Try a 30-in.-long plate

The load is $\frac{30}{2} - 6.77 = 8.23$ in. from edge of plate, and thus the pressure triangle will be 24.69 in. long and the required plate will be

$$\frac{343.84}{\frac{1}{2} \times 1.125 \times 24.69} = 24.76 \text{ in.} \qquad \text{(say 26 in.)}$$

If the plate is made 26 in. wide, the pressure zone will have an area of $26 \times 24.69 = 641.94$ in.^{2} and the maximum pressure will be twice the average pressure or

$$\frac{343.84}{641.94} \times 2 = 1.071 \text{ ksi}$$

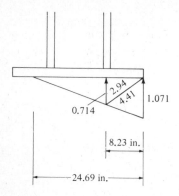

Figure A-10

Taking moments to right at center of flange (see Fig. A-10):

$$M = (2.94)\left(\frac{8.23}{3}\right) + (4.41)\left(\frac{2}{3} \times 8.23\right) = 32.26 \text{ in.-k}$$

$$t = \sqrt{\frac{(6)(32.26)}{27}} = 2.68 \text{ in.}$$

<u>Use PL$2\frac{3}{4}$ × 26 × 2 ft 6 in.</u>

Design of weld from column to base PL (see Fig. A-11)

Total length of fillet weld each flange
$$= (2)(14.67) - 0.590 = 28.75 \text{ in.}$$

$$C = T = \frac{(12)(225)}{13.54} = 199.4 \text{ k}$$

Strength of 1-in. long 1-in. fillet weld
$$= (1.0)(0.707)(0.3 \times 70)(1.0) = 14.85 \text{ k/in.}$$

Weld size required $= \dfrac{199.4}{(28.75)(14.85)} = 0.468$ in.

<u>Use $\frac{1}{2}$ in.</u>

■

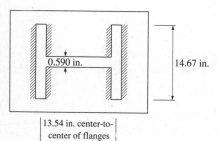

Figure A-11 Fillet welds from column to base plate

In these examples the variation of stress in the concrete supporting the columns has been assumed to vary in a triangular or straight-line fashion. It is possible to work with an assumed ultimate concrete theory where the concrete in compression under the plate is assumed to fail at a stress of $0.85f_c'$. Examples of such designs are available in several texts.[8] More detailed information concerning moment-resisting bases is available in several places.[9,10]

[8] W. McGuire, *Steel Structures* (Englewood Cliffs, N.J.: Prentice Hall, 1968), pp. 987–1004.

[9] C. G. Salmon, L. Schenker, and B. G. Johnston, "Moment-Rotation Characteristics of Column Anchorages," *Transactions ASCE*, 122 (1957), pp. 132–154.

[10] J. T. DeWolf and E. F. Sarisley, "Column Base Plates with Axial Loads and Moments," *Journal of Structural Division*, ASCE, 106, ST11 (November 1980), pp. 2167–2184.

Index